Chemistry

Foundation

for Class VIII

Atul Singhal

Professor, Institute of Science and Technology

Kansas, Missourie, USA

Universities Press

CHEMISTRY FOUNDATION FOR CLASS VIII

UNIVERSITIES PRESS (INDIA) PRIVATE LIMITED

Registered office
3-6-747/1/A & 3-6-754/1, Himayatnagar, Hyderabad 500 029, Telangana, India
info@universitiespress.com; www.universitiespress.com

Distributed by
Orient Blackswan Private Limited

Registered office
3-6-752 Himayatnagar, Hyderabad 500 029, Telangana, India

Other offices
Bengaluru, Chennai, Guwahati, Hyderabad, Kolkata,
Mumbai, New Delhi, Noida, Patna, Visakhapatnam

First published 2023

501839

ISBN: 978-93-93330-32-1

Typeset in Times New Roman PS Std 10 *by*
SRS Publishing Services, Puducherry

Printed in India by
B.B. Press, Noida 201 301

Published by
Universities Press (India) Private Limited
3-6-747/1/A & 3-6-754/1, Himayatnagar, Hyderabad 500 029, Telangana, India

Dedicated to
My Grandparents, Parents and Teachers

Preface

The Chemistry Foundation series comprises the recommended preparatory books for students aspiring to crack the prestigious examinations such as NEET, JEE (Main and Advanced), NTSE, KVPY and the Olympiads. They provide class-tested material and practice problems that help aspirants in understanding important theories, concepts and as a consequence develop problem-solving skills to attempt these examinations with confidence. The series is written in lucid language and aims to assist students in understanding the concepts even without the help of an instructor as most of the questions have been solved.

The series strictly adheres to the latest pattern of both CBSE and ICSE syllabi and the content is developed in such a way that a student will not only be able to understand all concepts but also apply them in the entrance examinations. All the chapters in this series also conform to the pattern laid out in the NCERT textbooks. The objective of this series is to provide the course material in a structured sequence that will help students prepare in an organised manner. The unique feature of this series is its huge repository of multiple choice questions that cover all important topics in every chapter. Each practice exercise is further divided into two sections—***Practice Questions:*** Analyse Your Concepts (School Exam Based) and ***Competition Window:*** Objective Type [for NEET, JEE (Main and Advanced), NTSE, KVPY and the Olympiads].

This book contains a supporting mobile App that provides chapter-wise practice tests for every chapter.

Salient Features of the Series

- Structured as per the syllabi of CBSE, ICSE and the topics required for the entrance examinations.
- Based on the latest exam patterns.
- Includes True–False, Fill in the blanks, Match the Following, Very short, Short, Long answer questions including higher order thinking questions, case study based and practical-based questions for school exams.
- Includes chapter-wise NCERT exercises and exemplar questions with solutions.
- Includes objective type questions; topic-wise, miscellaneous and advanced MCQs (single and multiple choice); comprehension passage or study case based, integer type and matrix match.

The purpose of this series is two-fold—to strengthen the foundation in chemistry and thus build the students' confidence to ace the examinations. I welcome suggestions and constructive criticism from the readers. Students can share their feedback at singhal.atul1974@gmail.com.

Atul Singhal

Acknowledgements

The contentment that accompanies the successful completion of my work would remain essentially incomplete if I fail to mention the people whose constant support has encouraged me. I am grateful to all my revered teachers, especially, Late J K Mishra, Dr D K Rastogi, Late A K Rastogi and my honourable guide, Dr S K Agarwala. Their knowledge and wisdom assisted me in no small measure in presenting this work. I express my immense gratitude to my colleagues for collaborating so nicely and constructively during the various stages of this project. I am extremely thankful to Ronit Singh Kushwaha for typing out the manuscript.

I am indebted to my father, B K Singhal, mother Usha Singhal, brothers Amit Singhal and Katar Singh, who have been my motivators at every step. Their never-ending affection has provided me with moral support and encouragement while writing this book. Last but not the least, I express my deepest gratitude to my wife Urmila, my daughters Khushi and Shanvi and my little, but witty-beyond-years son Shashwat, who always supported me during my work.

I would like to express my heartfelt gratitude to Dr Gita S Dattatri for her brilliant editing and upgradations. I would also like to show my appreciation to Kallol Das, Thomas Mathew Rajesh and Madhu Reddy for providing all the necessary help and assistance.

Dr Atul Singhal
singhal.atul1974@gmail.com

Contents

CHAPTER 1

Matter

LEARNING OBJECTIVES

After studying this unit, you will be able to understand:

- Matter and its type
- Properties of solids, liquids and gases
- Changes in states of matter
- Effect of temperature and pressure on state of matter
- Melting, freezing, vaporisation, condensation, and so on
- Types of latent heat
- Sublimation and liquefaction

Introduction

Everything in our universe is made up of material which is called *matter*. The air we breathe, the food we eat, clouds, stars, plants and animals, even a small drop of water or a particle of sand—everything is matter. When we look around, we observe that all these occupy space, that is, they have volume and mass. Modern day scientists have evolved two types of classification of matter, based on their physical properties and chemical nature. In this chapter we shall learn all about matter based on its physical properties. The chemical aspects of matter will also be taken up in subsequent chapters.

Matter

Anything that occupies space and has mass is known as *matter*. Everything around us is made up of tiny pieces or particles. The particles that form it are atoms or molecules. For example, food, water, air, clothes, table, chair, plants and trees. On the basis of their chemical properties, matter is classified as elements, compounds and mixtures.

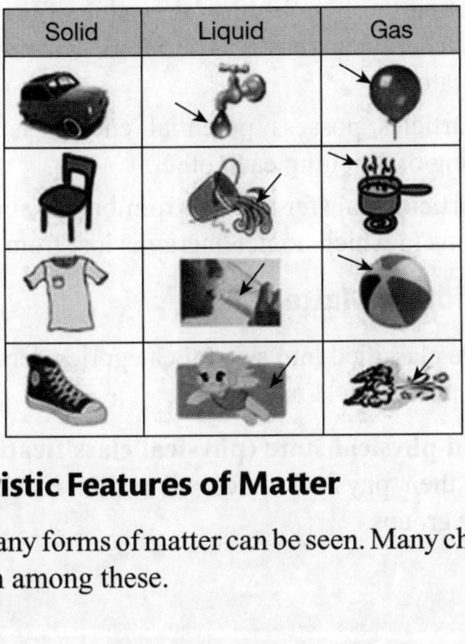

Characteristic Features of Matter

In nature, many forms of matter can be seen. Many characteristics are common among these.

(i) Mass: Any form of matter has mass. Mass is the amount of matter present in any species. For example, even though air is invisible, it has some mass.

(ii) Space: Any form of matter also occupies space. For example, a chair in a room, a book on a bookshelf, a ring in a box. All these things definitely occupy some space.

(iii) Matter cannot be destroyed or created: We can say that matter cannot be destroyed or created. During all physical and chemical changes, the total mass of the matter before and after the change remains constant (Lavoisier's *law of conservation of mass*).

Physical Nature of Matter

Every substance in nature consists of a special set of properties that allows us to recognise and distinguish it from other substances. Such properties of matter can be categorised as *physical* or *chemical*. The measurement of physical properties is possible without changing the identity and composition of the substance. Such properties are colour, hardness, odour, boiling point, melting point, density, and so on.

Chemical properties represent the way by which a substance can change or react to give other substances. Flammability is such a property and is the ability of a substance to burn in air or in the presence of oxygen (O_2).

Physical properties

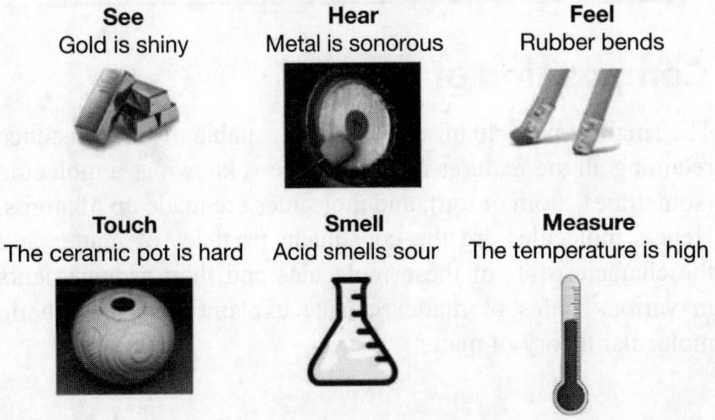

Chemical properties

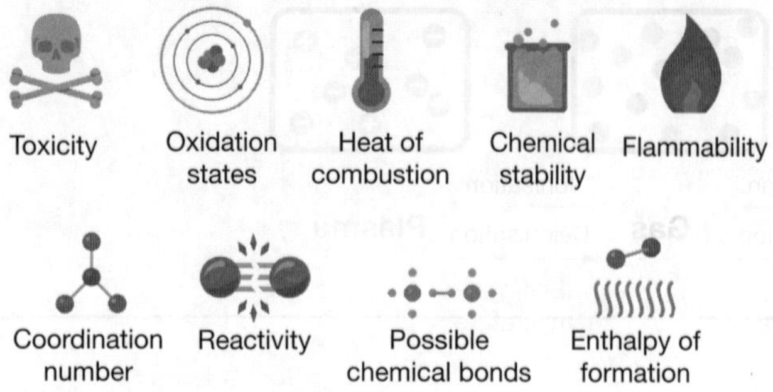

Particle Nature of Matter

As per as ancient Indian and Greek philosophers, 'matter is composed of very tiny particles that cannot be further sub-divided'. Scientifically, John Dalton was first to explain the nature and composition of matter.

Based on the results of diffusion and Brownian motion experiments, the existence of particles in matter and their motion can be proved. Here are a few properties which prove the particle nature of matter.

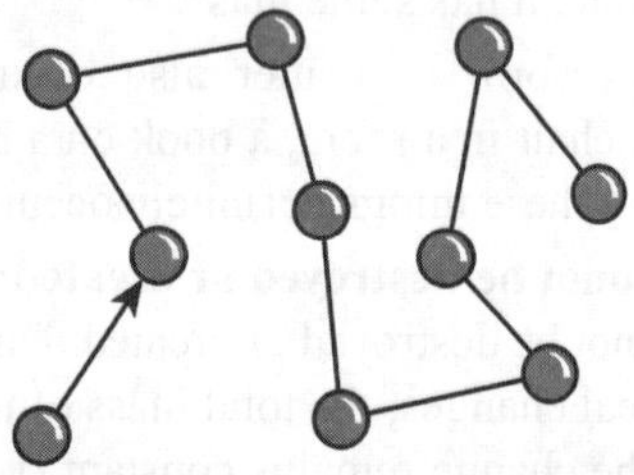

The movement of pollen grains in water is very rapid and occurs in a zig zag manner since pollen grains are hit by fast moving water molecules. As a result, pollen grain molecules can move in the net force direction on the surface of water (similar to Brownian motion).

KNOWLEDGE BOOSTER

Diffusion

Diffusion is the flow of molecules from the higher concentration to the lower concentration side. For example, perfume spray diffuses in all possible directions in a room.

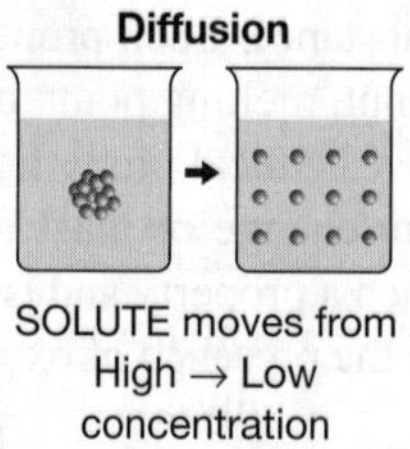

Composition of Matter

The smallest particle of matter that is capable of free existence retaining all the features of a substance is known as a molecule (sometimes, atom or ion), and molecules are made up of atoms. Hence, molecules are the constituent particles of matter and the characteristics of these molecules and their arrangements in various states of matter can be explained by the kinetic molecular theory of matter.

The light and heat that comes from the sun and stars, and electricity that can make lamps and motors run, all these are not matter but are examples of energy.

Kinetic Molecular Theory of Matter

This theory visualises that matter is made up of tiny constituents – atoms, molecules or ions – that are in random motion and have space between them.

Postulates of the Kinetic Theory of Matter

The important postulates of this theory are:

(i) Matter is made up of small particles that may be atoms, ions or molecules. The constituent particles of a given substance are identical in all respects.

(ii) There is empty space between the constituent particles (molecules, atoms or ions) which is known as intermolecular space.

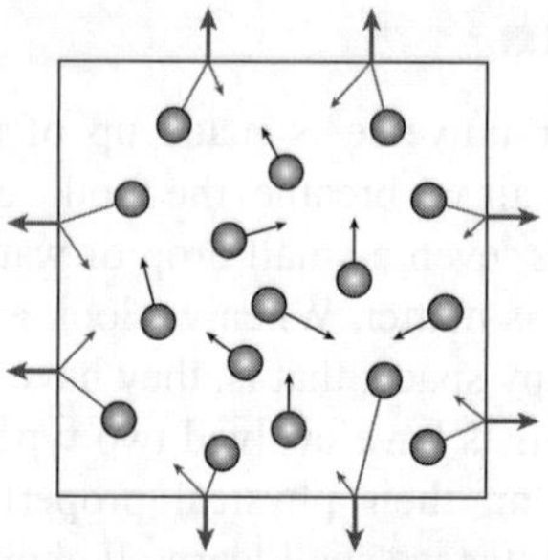

(iii) These molecules exert an attractive force upon one other called *intermolecular force of attraction*. The magnitude of this force depends upon the state of the substance (solid, liquid and gas) and also on the magnitude of intermolecular spaces.

(iv) The force of attraction that exists between similar molecules is known as *cohesive force* and that which exists between dissimilar molecules is known as *adhesive force*.

(v) These molecules are always in a state of motion and they possess kinetic energy due to their motion.

(vi) On heating these molecules, the value of their speed and kinetic energy increases. Therefore, Kinetic energy $\propto$ Temperature.

(vii) The particles possess potential energy as a result of attracting or repelling each other.

(viii) The particles transfer energy from one to another during collisions in which no net energy is lost from the system.

Classification of Matter

Matter can be classified into several categories depending upon its physical and chemical nature.

(i) Based on physical state (physical classification): On the basis of their physical states, all matter can be classified into four groups.

1. Solids 2. Liquids 3. Gases 4. Plasma

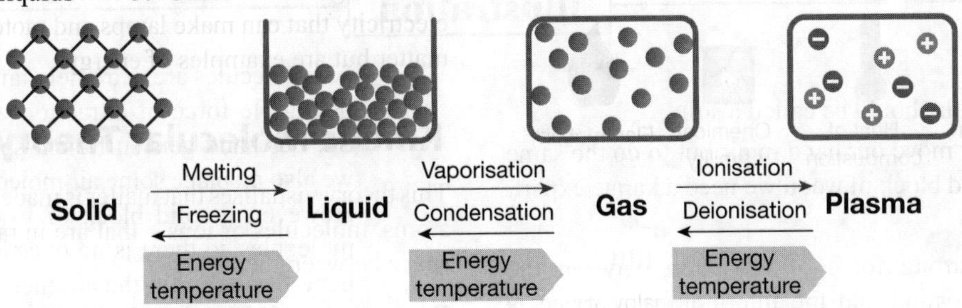

(ii) Based on purity/composition (chemical classification): On the basis of purity, matter can be classified as shown in the flowchart.

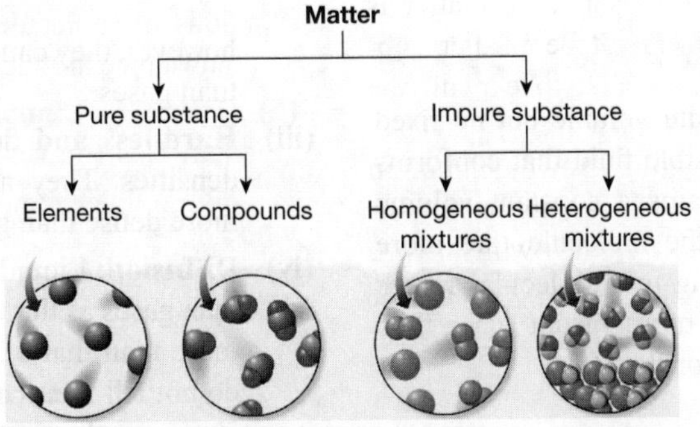

Illustrations

1. Which of the following are examples of matter?

Stool, air, love, toffies, smell, hate, peanuts, thought, heat, soft drink, smell of deodorant.

Solution: Stool, air, toffies, smell, peanuts, soft drink and smell of deodorant are classified as matter.

2. A diver is able to cut through water in a swimming pool. Which property of matter does this observation show?

Solution: The phenomenon of cutting through water by the diver shows that matter has space between its particles.

The molecules in a solid are closely packed together and contain the least amount of kinetic energy. A solid is characterised by incompressibility, structural rigidity, high mechanical strength and resistance to a force applied to the surface. In a solid, the constituent species are closely packed or held together by strong forces; so, in between the constituents, there are very small gaps and they can neither cannot move nor flow. For example, ice, wood, coal, stone, iron, steel, brick, and so on.

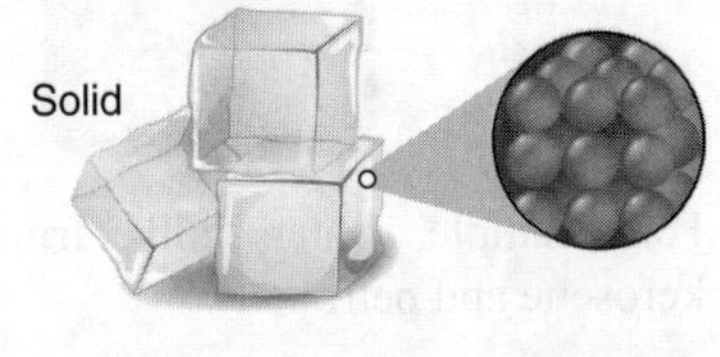

Solids

Solids are characterised by the following properties.

(i) Shape and volume: A solid has a fixed shape and a fixed volume as the constituents (atoms, ions or molecules) are arranged in a definite pattern held together by strong intermolecular forces. This regular arrangement gives them a definite geometry and volume.

(ii) Compressibility: Solids are rigid and they cannot be compressed easily as the constituents are closely packed; intermolecular distance or space in them is very low and cannot be further reduced.

(iii) Hardness and density: Solids are quite hard and they have high densities since they have close packed arrangement.

(iv) Diffusion: A solid cannot diffuse into another solid easily as very strong intermolecular forces of attraction are present in solids. For example, if two metal blocks are bound together tightly and kept undisturbed for a few years, then the particles of one metal are found to have diffused into the other metal.

(v) Melting points: Solids have a high melting point as a huge amount of energy is required to overcome the strong intermolecular forces present in their constituents.

(vi) Motion: Solids do not flow, so they cannot acquire the shape of the vessel in which they are kept. They do not show any motion as the constituents are strongly held at their respective positions by strong intermolecular forces. They can only vibrate.

Illustration

1. Give reasons:
   (a) A wooden chair should be called a solid.
   (b) We can easily move our hand in air but to do the same through a solid block of wood we need a karate expert.

Solution:

(a) There is a strong force of attraction between the molecules of wood and the intermolecular space is the least. So, a wooden chair has a definite shape and volume and it should be called a solid.

(b) Air molecules are extremely far from each other due to negligible force of attraction working between them. So, our hand gets sufficient space to move in air and we also displace some air molecules by applying force. However, a solid block of wood has closely packed molecules so there is no question of the movement of hand through it in the absence of suitable force in the proper direction.

Liquids

A liquid is the only state with a definite volume but no fixed shape. A liquid is a nearly incompressible fluid that conforms to the shape of its container but retains a constant volume independent of pressure. In liquids, the molecules are more closely held than in gases due to strong intermolecular forces (but less than in solids). Liquids have more vacant space than solids. In liquids, molecules can move or flow.

For example, water, milk, fruit juice, ink, groundnut oil, kerosene and petrol.

Properties of Liquids

Liquids are characterised by the following properties.

(i) **Shape and volume:** A liquid will have a fixed volume but cannot have a fixed shape. A liquid takes the shape of the vessel in which it is placed as the liquid molecules are not held strongly enough to have a fixed position or definite geometry. However, the intermolecular force is strong enough for them to have a definite volume.

(ii) **Compressibility:** Like solids, liquids cannot be compressed much because the vacant space is not much; however, they can be compressed more than solids but less than gases.

(iii) **Hardness and density:** Liquids have moderate to high densities. They are usually less dense than solids, but more dense than gases.

(iv) **Diffusion:** Liquids are more diffusible than solids but less than gases as the vacant spaces between their molecules is more than that of solids but less than that of gases. They do not fill their container completely. Diffusion in liquids is slower than in gases as the particles in liquids move slower as compared to the particles in gases. The rate of diffusion in liquids is much faster than that in solids because the particles in a liquid move much more freely, and have greater spaces between them as compared to particles in the solids.

For example, the spreading of blue colour of copper sulphate into water on its own, is due to the diffusion of copper sulphate particles into water.

(v) **Melting points:** Liquids have lower melting points than solids but higher than that of gases as the intermolecular forces are weaker than in solids but stronger than in gases.

(vi) **Motion:** Liquids generally flow or show motion easily as there is enough space between their molecules. However, their motion is less than in gases.

(vii) **Evaporation:** Liquid molecules can evaporate and change into vapour at room temperature as the kinetic energy of liquid molecules can overcome the intermolecular force present between molecules. The rate of evaporation is not the same for all liquids as the kinetic energies of liquid molecules differ.

Illustrations

1. Give two reasons to justify the following.
   (a) Water at room temperature is a liquid.
   (b) An iron almirah is a solid at room temperature.

Solution:

(a) Water is a liquid at room temperature because it has a tendency to flow. It takes the shape of the container in which it is filled, but its volume remains the same.

(b) An iron almirah is a solid at room temperature because its shape and volume are definite. It is hard and rigid. Its density is high.

2. Explain the following:
   (a) Liquids generally have lower density compared to solids. But you must have observed that ice floats on water.

(b) When a liquid is transferred from a small vessel to a large vessel at the same temperature, what will be the effect on vapour pressure?

Solution:

(a) Liquids have lower density than solids. Water is also a liquid so it should also have less density than ice. However, this is not so because of the cage-like structure of ice. That is, the presence of vacant spaces between water (H_2O) molecules when they link with each other in ice. The number of these spaces is comparatively less in water. Being more porous than water, ice is lighter than water and floats over the surface of water.

(b) There will be no effect on vapour pressure when a liquid is transferred from a small vessel to a large vessel at same temperature as vapour pressure does not depend on the size of the vessel.

Gases

The gaseous state is one of the four fundamental states of matter. A gas has no definite volume and no definite shape. Gases are compressible and diffusible. Their molecules have large spaces in between them and are not closely packed due to weak interactions. A pure gas may be made up of individual atoms, elemental molecules made from one type of atom or compound molecules made from a variety of atoms. A gas mixture, such as air, contains a variety of pure gases. Examples of gases are air, oxygen, hydrogen, nitrogen, helium, and so on.

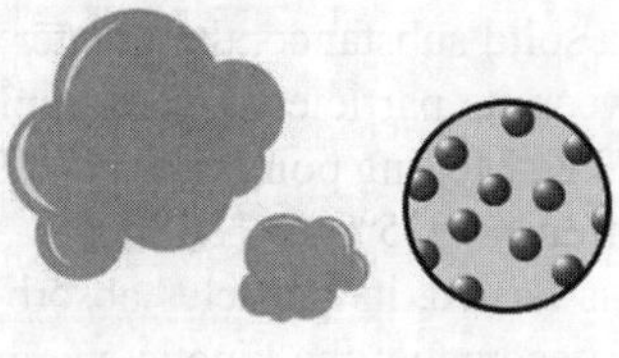

Properties of Gases

Gases are characterised by the following properties.

(i) Shape and volume: Gases have neither a fixed shape nor a fixed volume. They acquire the shape and volume of the vessel in which they are stored. This is because gaseous molecules have very weak interactions amongst each other and can thus move freely and occupy any amount of space.

(ii) Compressibility: Gases can be compressed easily by applying external pressure as the space between the gas molecules is large due to weak intermolecular forces between them. During compression, the gas molecules come closer to each other and occupy less space.

(iii) Density: Gases have very low densities and they are very light.

(iv) Diffusion: Gases can diffuse easily as the gaseous molecules are weakly held and have enough kinetic energy to diffuse. The gases can fill their container completely. For example, if LPG (Liquified Petroleum Gas) leaks, its molecules can spread all around the kitchen and it can be detected by the smell.

For example, when someone opens a bottle of perfume in one corner of a room, its smell spreads through the whole room quickly. The smell of perfume spreads due to the diffusion of perfume vapours into air.

(v) Homogeneous nature: Gases have similar composition in all parts, so they are homogeneous in nature. A gaseous mixture is always homogenous as it is contains only the gas phase.

(vi) Liquefaction: A gas can be liquified by cooling and by applying pressure.

Table 1.1 Comparison of characteristic properties of solids, liquids and gases

	Property	Solids	Liquids	Gases
1	**Shape**	Definite	Attain the shape of the container, but do not necessarily occupy all of it	Attain the shape of the container by occupying the entire space available to them
2	**Volume**	Definite	Definite	Attain the volume of the container
3	**Compressibility**	Almost nil	Very low	Very large
4	**Fluidity or rigidity**	Rigid	Fluid	Fluid
5	**Density**	High	Low	Very low
6	**Diffusion**	Generally do not diffuse	Diffuse slowly	Diffuse rapidly
7	**Free surfaces**	Any number of free surfaces	Only one free surface	No free surface

Illustrations

1. Give reasons:
   (a) A gas completely fills the vessel in which it is kept.
   (b) A gas exerts pressure on the walls of the container.

   Solution:

   (a) The forces of attraction between the molecules of gases is negligible. So, molecules of gases occupy the maximum space available to them. High kinetic energy possessed by their molecules also helps.

   (b) The motion of particles is random and occurs at very high speed in the gaseous state. Due to this random movement, the particles hit each other and also the walls of the container. The pressure exerted by the gas is due to this force exerted by the particles per unit area on the walls of the container.

2. (a) How can you separate gases from a gaseous mixture?
   (b) How can a gas be liquefied?

   Solution:

   (a) Gases can be separated from the gaseous mixture by diffusion as the rate of diffusion of each gas is different.

   (b) A gas can be liquefied by cooling or lowering of temperature and by applying pressure. This is known as liquefication of a gas.

Change of State of Matter

A substance may exist in any of the three states of matter (solid, liquid or gas) depending upon the conditions of temperature and pressure. By changing the temperature and pressure, the state of a substance can be changed. A solid on heating usually changes into a liquid which on further heating changes into gas. Similarly, a gas on cooling condenses into a liquid which on further cooling changes into a solid.

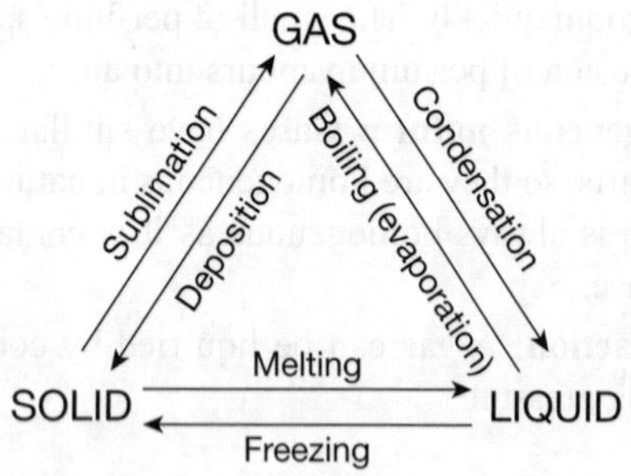

The most familiar and common example is water. It exists in all the three states—solid: ice, liquid: water and gas: water vapour. Ice is a solid and may be melted to form water (liquid) which on further heating changes into steam (gas). These changes can also be reversed on cooling.

$$\underset{\text{(Solid)}}{\text{Ice}} \underset{\text{Cool}}{\overset{\text{Heat}}{\rightleftharpoons}} \underset{\text{(Liquid)}}{\text{Water}} \underset{\text{Cool}}{\overset{\text{Heat}}{\rightleftharpoons}} \underset{\text{(Gas)}}{\text{Steam}}$$

Effect of Temperature Change

By increasing the temperature (heating), a solid can be converted into the liquid state, and the liquid can be converted into the gaseous or vapour state. By decreasing the temperature (cooling), a gas can be converted into the liquid state, and a liquid can be converted into the solid state.

Solid State to Liquid State (Melting)

The process by which the solid state of a substance is changed into its liquid state on heating is known as melting or fusion. Melting is done by providing heat to the solid substance.

Melting Point and Melting

The temperature at which a solid substance melts and changes into a liquid at atmospheric pressure is known as the *melting point* of the substance. The melting point of a solid is a measure of the force of attraction between its particles. The higher the melting point of a solid substance, the greater will be the force of attraction between its particles. For example, Melting point of ice = 0ºC or 273 K, Melting point of wax = 63ºC or 336 K and Melting point of iron = 1535ºC or 1808 K.

When a solid is heated, its particles absorb heat energy and they become more energetic; the kinetic energy of the particles increases. Due to the increase in kinetic energy, the particles start vibrating more strongly with greater speed. The energy supplied by heat overcomes the intermolecular forces of attraction between the particles. As a result, the particles leave their mean position and break away from each other. When this happens, the solid melts and a liquid is formed. This process is known as *melting*.

Liquid State to Gaseous State (Boiling or Vaporisation)

The process by which the liquid state of a substance is changed into its gaseous state on heating, is known as boiling or vaporisation.

Boiling Point

The temperature at which a liquid boils and changes rapidly into a gas at atmospheric pressure, is called the *boiling point* of the liquid. The boiling point of a liquid is a measure of the force of attraction between its particles. The higher the boiling point of a liquid, the greater will be the force of attraction between its particles. For example, Boiling point of water = 100ºC or 373 K, Boiling point of alcohol = 78ºC or 351 K and Boiling point of mercury = 357ºC or 630 K.

In a liquid, most of the particles are close together. When we supply heat energy to the liquid, the particles of the liquid start vibrating fast. Some of the particles become so energetic that they can overcome the attractive forces of the particles around them. Hence, they become free to move and escape from the liquid. When this happens, the liquid evaporates, that is, starts changing into gas. At the boiling point, the particles of a liquid have sufficient kinetic energy to overcome the forces of attraction holding them together and separate into individual particles. The liquid then boils to form a gas.

Liquid State to Vapour State (Evaporation)

The process of converting a liquid into vapour at any temperature below its boiling point is known as *evaporation*. It is a surface phenomenon as it is confined to only surface molecules. It is a slower process than boiling.

Difference Between Boiling and Evaporation

(i) Boiling can take place only at a particular temperature for a liquid while evaporation can occur at all temperatures.

(ii) Boiling is a bulk phenomenon. That is, bubble formation occurs even below the surface. Evaporation is a surface phenomenon in which bubble formation occurs only on the surface of liquid.

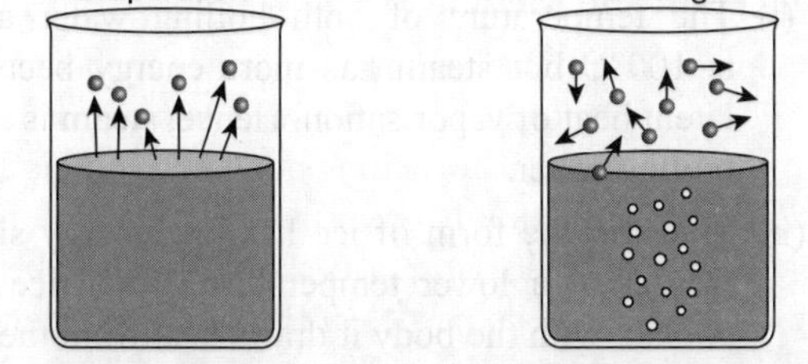

Liquid State to Solid State (Freezing)

The process of converting a liquid into a solid by cooling is known as freezing. Freezing means solidification. Freezing is the reverse of melting. So, the freezing point of a liquid is the same as the melting point of its solid form. For example, Melting point of ice = 0ºC or 273 K and Freezing point of water = 0ºC or 273 K.

Freezing Point

The temperature at which a liquid gets converted into a solid at atmospheric pressure is known as the *freezing point* of a liquid.

When a liquid substance is subjected to cooling, heat is extracted from the liquid. Therefore, the kinetic energy of the molecules decreases which results in a decrease in temperature. On reaching a certain temperature, further extraction of heat from the liquid results in a decrease in potential energy instead of decreasing the kinetic energy. As a decrease in potential energy leads to an increase in intermolecular forces of attraction due to a decrease in intermolecular space, the molecular arrangement of the liquid changes to that of a solid.

Gaseous State to Liquid State (Condensation)

The process of conversion of a gaseous substance into its liquid state is known as *condensation*. It can be carried out by cooling the gas below a particular temperature. When heat is extracted from the gas, the kinetic energy of the molecules decreases so the temperature decreases. When a sufficiently low temperature is reached, further extraction of heat from the gas does not reduce the kinetic energy. As a result, an increase in intermolecular forces of attraction brings the molecules closer; thus, at this point, the gas passes into the liquid state.

Solid State to Gaseous State (Sublimation)

The process of changing of a solid substance directly into its gaseous or vapour form on heating is known as *sublimation*, and of the vapour form into solid on cooling is known as *deposition* or *desublimation.* The solid substance which undergoes sublimation is said to *sublime* while the solid deposit of a substance which has sublimed is known as a *sublimate*. Sublimation can be represented as:

$$\text{Solid} \underset{\text{Cooling}}{\overset{\text{Heating}}{\rightleftarrows}} \text{Vapour (or gas)}$$

For example, when solid ammonium chloride is heated, it directly changes into ammonium chloride vapours. And when hot ammonium chloride vapour is cooled, it directly changes into solid ammonium chloride. The following substances also undergo sublimation—iodine, camphor, naphthalene and anthracene.

Effect of Pressure Change

The three states of matter differ in the intermolecular forces and intermolecular distances between the constituent particles (atoms, ions, molecules). Gases are said to be compressible since the space between the gaseous particles decreases on applying pressure. A substance may exist in any of the three different states of matter depending upon the conditions of temperature and pressure.

- If the melting point of a substance is above the room temperature under atmospheric pressure, it is said to be a solid.
- If the boiling point of a substance is above the room temperature under atmospheric pressure, it is classified as a liquid.

If the boiling point of the substance is below the room temperature under atmospheric pressure, it is called a gas.

Latent Heat

The word 'latent' means 'hidden'. The heat energy that has to be supplied to change the state of a substance is known as its *latent heat.* Latent heat does not raise the temperature but latent heat always has to be supplied to change the state of a substance since every substance has some forces of attraction between its particles that hold them together. Thus, if the state of a substance has to be changed, it is necessary to break these forces of attraction between its constituents by supplying heat. The latent heat does not increase the kinetic energy of the particles of the substance.

Types of Latent Heat

There are two types of latent heat.

(i) Latent heat of fusion: The heat needed to convert a solid substance into its liquid state is known as *latent heat of fusion*. In other words, the latent heat of fusion of a solid is the quantity of heat in joules required to convert 1 kilogram of the solid to liquid, without any change in temperature. For example, the latent heat of fusion of ice is 3.34×105 J/kg or 80 cal/g.

(ii) **Latent heat of vaporisation:** The heat needed to convert a liquid substance into its vapour state is known as *latent heat of vaporisation*. In other words, the latent heat of vaporisation of a liquid is the quantity of heat in joules required to convert 1 kilogram of the liquid to its vapour or gaseous state, without any change in temperature. For example, the latent heat of vaporisation of water is 22.5 × 105 J/kg or 540 cal/g.

Illustrations

1. Fill in the blanks.
   (a) Evaporation of a liquid at room temperature leads to a ______ effect.
   (b) At room temperature, the forces of attraction between the particles of a solid substances are ______ than those in the gaseous state.
   (c) The arrangement of particles is less ordered in the ______ state. However, there is no order in the state.
   (d) ______ is the change of gaseous state directly to solid state without going through the ______ state.
   (e) The phenomenon of change of a liquid into the gaseous state at any temperature below its boiling point is called ______.

Solution:

(a) Cooling
(b) Stronger
(c) Liquid, gaseous
(d) Desublimation or deposition, liquid
(e) Evaporation

2. Explain these observations.
   (i) Khushi was making tea in a kettle. Suddenly she felt intense heat from the puff of steam gushing out of the spout of the kettle. She wondered whether the temperature of the steam was higher than that of the water boiling in the kettle.
   (ii) Water as ice has a cooling effect, whereas water as steam may cause severe burns.

Solution:

(i) The temperature of both boiling water and steam is 100°C but steam has more energy because of its latent heat of vaporisation. Hence, steam is hotter than boiling water.
(ii) Water in the form of ice has low energy since water freezes at a lower temperature. When ice comes in contact with the body it draws heat from the body and produces a cooling effect. In the case of steam, the water molecules have high energy. The high energy of steam is transformed into heat and may cause severe burns.

KNOWLEDGE BOOSTER

Law of conservation of mass

Law of conservation of mass was introduced by Lavoisier (Father of chemistry) in 1774. According to this law of conservation of mass, 'matter can neither be created nor destroyed in a chemical reaction'.

For example, Lavoisier observed that on heating mercuric oxide, it produced free mercury and oxygen. The sum of masses of mercury and oxygen was found to be equal to the mass of mercuric oxide

Mercuric oxide → Mercury + Oxygen
100 g 92.6 g 7.4 g

Mass of the reactant = 100 g

Mass of the products = 92.6 + 7.4 g = 100.0 g = 100 g

Here the total mass of the products formed is equal to the total mass of the reactants undergoing reaction, hence this data is in agreement with the law of conservation of mass.

Physical and Chemical Changes

Chemistry mainly deals with the determination of the nature and properties of the non-living matter that surrounds us and the preparation of new substances. Chemists are mostly interested in the changes that these substances can undergo on subjecting them to conditions that they normally cannot encounter, such as high temperature and high pressure. For example, on boiling, water changes from the liquid to the vapour form. Plants change raw material from the air and soil into stems and leaves.

These changes can be further divided into two types—physical and chemical changes.

Physical Change

In the previous topic, we learnt about a few physical properties of matter. The properties that can be observed and specified, such as colour, hardness, rigidity, fluidity, density, melting point and boiling point are the physical properties. The interconversion of states is a physical change because these changes occur without a change in composition and there is no change in the chemical nature of the substance. A change in which no new substances are formed but the physical form of the substance changes, is known as a *physical change*. The product formed in such changes is chemically identical to the starting substance.

For example, when ice is heated, it changes into liquid water and on further heating, water changes into steam. However, water in the solid form (ice), liquid form or in the gaseous form (steam), is chemically the same substance (H_2O).

- This transformation represents a physical change and such a change can be reversed easily. For example, steam

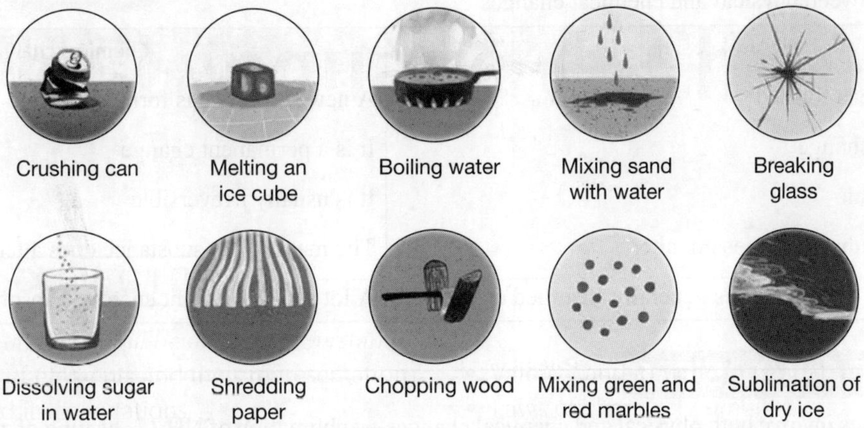

on cooling forms liquid water, which on further cooling changes into ice.

- The following processes involve physical change. Making of ice cream, sublimation of camphor, melting of wax, breaking of a glass pane, a rock rolling down a hill, bending of a glass tube by heating.

Chemical Change

A change in which one or more substances change into new substance(s) is known as a *chemical change*. It cannot be reversed easily. this process causes a change in the chemical composition which brings about a change in the chemical properties also. A chemical change is also called a *chemical reaction*.

For example,

- Both water and cooking oil are liquids but their chemical characteristics are different. They differ in odour and inflammability. Oil burns in air whereas water extinguishes fire. So, it is the chemical property of oil that makes it different from water (burning is a chemical change).
- When electricity is passed through water, it decomposes into two new substances, hydrogen and oxygen. Similarly rusting of iron and calcination of limestone are also examples of chemical changes.
- Blood clotting, burning of charcoal, digestion of food, rusting of iron, formation of biogas, burning of fuel like petrol, ripening of fruits, drying of paint and burning of magnesium ribbon in air.

A chemical change involves a change in a way that the atoms in the molecules are chemically bonded to one another. A chemical bond is the force of attraction that holds two or more atoms together. In a chemical change, old bonds are broken and new ones are formed between the atoms. In methane, one carbon has bonds with four H-atoms and in oxygen two oxygen atoms are bonded together. When the reaction occurs, C–H and O=O are broken and new bonds C=O and H–O are formed to give CO_2 and H_2O.

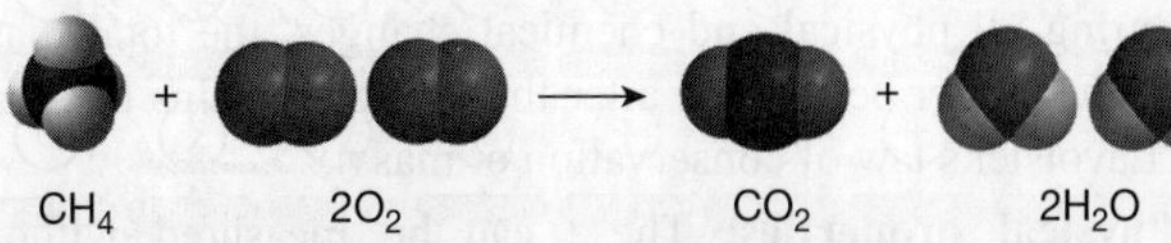

KNOWLEDGE BOOSTER

Henry Cavendish (1731–1810) produced water by reacting oxygen and hydrogen gas through a chemical change.

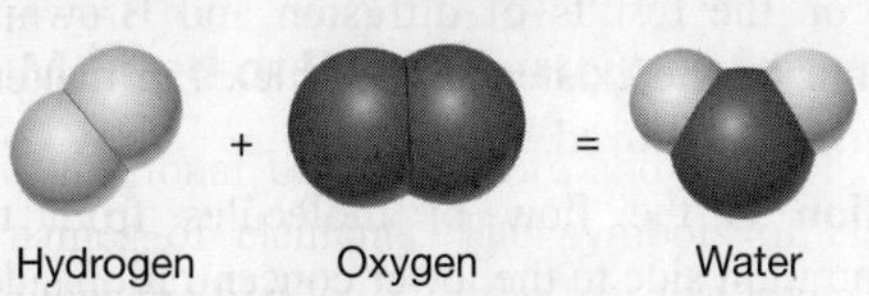

$$2H_2 + O_2 \rightarrow 2H_2O$$

Table 1.2 Differences between physical and chemical changes

	Physical change	Chemical change
1	No new substance is formed	A new substance is formed
2	It is a temporary change	It is a permanent change
3	It is easily reversible	It is usually irreversible
4	The mass of the substance does not alter	The mass of the substance does alter
5	Very little heat or light energy is generally absorbed or evolved	A lot of heat or light energy is absorbed or evolved

KNOWLEDGE BOOSTER

The following processes involve both physical and chemical changes—sublimation of NH_4Cl, heating of zinc hydroxide, heating of sodium nitrate and burning of a candle.

Illustrations

1. Classify the following as physical or chemical changes.

(a) Dissolving salt in water (b) burning of magnesium ribbon in air (c) crystallisation of copper sulfate (d) conversion of milk into curd (e) burning of coal (f) electrolysis of sodium chloride solution by passing current (g) rusting of iron nails

Solution: (a) Physical change (b) Chemical change (c) Physical change (d) Chemical change (e) Chemical change (f) Chemical change (g) Chemical change

2. Classify each of the following as a physical or a chemical change. Give reasons.

(a) Drying of a shirt in the sun.

(b) Rising of hot air over a radiator.

(c) Burning of kerosene in a lantern.

(d) Change in the colour of black tea on adding lemon juice to it.

(e) Churning of milk cream to get butter.

Solution:

(a), (b), (e): Physical changes because there is no change in chemical composition.

(c), (d): Chemical changes because new substances are formed.

CHAPTER AT A GLANCE

- Everything in our universe is made up of material which has been named **matter**.
- Anything which occupies space and has mass is known as **matter**.
- during all physical and chemical changes, the total mass of the matter before and after the change remains constant (Lavoisier's law of conservation of mass).
- **Physical properties:** These can be measured without changing the identity and composition of the substance. Such properties are colour, hardness, odour, boiling point, melting point, density.
- **Chemical properties:** It is the way by which a substance changes or reacts to give other substances.
- Based on the results of diffusion and Brownian motion experiments, the existence of particles in matter and their motion can be proved.
- **Diffusion** is the flow of molecules from the higher concentration side to the lower concentration side.
- The smallest particle of matter that is capable of free existence retaining all the features is known as a **molecule**.
- On the basis of the physical state, all the matter can be classified into four groups—solids, liquids, gases and plasma.
- Chemical classification of matter shows the existence of pure substance (elements and compounds), and impure substances (homogenous and heterogenous mixtures).
- **Solids:** The constituents are closely packed, they have definite shape and volume, are quite rigid, are hard with high density, high BP, MP.
- **Liquids:** They have a definite volume but no definite shape, less density and hardness than solids, diffuse more than solids and have moderate BP and MP.
- **Gases:** A gas has no definite volume and no definite shape. Gases have high compressibility, low density, high diffusion rate and homogenous nature.
- **State conversion:** State conversion is possible as follows:

$$\underset{\text{(Solid)}}{\text{Ice}} \underset{\text{Cool}}{\overset{\text{Heat}}{\rightleftharpoons}} \underset{\text{(Liquid)}}{\text{Water}} \underset{\text{Cool}}{\overset{\text{Heat}}{\rightleftharpoons}} \underset{\text{(gas)}}{\text{Steam}}$$

- The temperature at which a solid substance melts and changes into a liquid at atmospheric pressure is known as the **melting point** of the substance and this process is known as **melting**.

- The temperature at which a liquid boils and changes rapidly into a gas at atmospheric pressure is called the **boiling point** of the liquid and this process is called **vaporisation**.
- The process of converting a liquid into vapour at any temperature below its boiling point is known as **evaporation**.
- The temperature at which a liquid gets converted into a solid at atmospheric pressure is known as the **freezing point** of a liquid.
- The process of conversion of a gaseous substance into its liquid state is known as **condensation**.
- The process of changing of a solid substance directly into its gaseous or vapour form on heating is known as **sublimation**. The process of changing of a vapour directly into its solid form on cooling is known as **desublimation** or **deposition**.
- The heat energy which has to be supplied to change the state of a substance is known as its **latent heat**.
- The heat needed to convert a solid substance into its liquid state is known as **latent heat of fusion**.
- The heat needed to convert a liquid substance into its vapour state is known as **latent heat of vaporisation**.
- A change in which no new substances are formed but the physical form of the substance changes is known as a **physical change**.
- A change in which one or more substances change into new substance(s) is known as a **chemical change**.

PRACTICE QUESTIONS

Analyse Your Concepts (School Exam Based)

Fill in the Blanks

Instructions: Complete the following statements with an appropriate word/term to be filled in the blank spaces.

1. Solid, liquid and gas are called the three ______________ of matter.
2. Among solid, liquid and gaseous states, ______________ state is least rigid.
3. The smell of perfume gradually spreads across a room due to ______________
4. ______________ have no definite volume and no definite shape.
5. Intermolecular space in gases is ______________ than that in liquids.
6. Liquid and ______________ states are known as fluid states.
7. Solids have a ______________ rate of diffusion than liquids.
8. The temperature at which a liquid changes into a gas/vapour is known as ______________
9. Change of solid state to liquid state is known as ______________.
10. The reaction in which energy is evolved is called an ______________ reaction.
11. Change of state directly from the solid to the gas without passing through the liquid state is known as ______________
12. The amount of heat needed to convert 1 kg of solid into liquid at its melting point is known as ______________
13. The interconversion of states without any change in composition and chemical nature of substance is known as a ______________ change.

True or False

Instructions: Read the following statements and write your answer as true or false.

1. The conversion of water into steam is a chemical change.
2. Intermolecular space in solids is less than that in liquids.
3. Rusting of iron is a chemical change.
4. The movement of pollen grains in water is very slow and in a zig zag manner.
5. The colour of an aqueous solution of $KMnO_4$ is purple due to diffusion.
6. The particles of matter attract each other.
7. Physical changes are easily reversible.
8. A solid exerts pressure only in the downward direction.
9. In the presence of impurities, the melting point always increases.
10. The rate of evaporation increases with humidity.
11. The boiling point of water on the kelvin scale is 373 K.
12. A process in which heat is released is an called endothermic reaction.

Very Short Answer Type Questions

1. Define interconversion of states of matter.
2. What are the two factors responsible for the change in the state of matter?
3. What is the role of intermolecular forces of attraction?

4. An aqueous solution of copper sulfate becomes blue. Which phenomenon it is due to?
5. Give reasons why a gas can fill the whole vessel in which it is enclosed.
6. Two liquids X and Y have boiling points 380 K and 390 K respectively. Which of the two has greater intermolecular forces of attraction?
7. Give reasons why solids cannot be compressed.
8. Which state of matter has definite shape and definite volume?
9. Name the physical state of matter
   (a) Which can be easily compressed
   (b) Which is most brittle
   (c) Which can flow but cannot fill the vessel completely
10. Name any three substances that can sublime.
11. Name the process for the following changes:
   (a) Liquid → Gas
   (b) Solid → Liquid
   (c) Gas → Liquid
12. Which will have more volume—ice or steam?
13. What is the impact of increase of surface area on the rate of evaporation?
14. If fish is being fried in a neighbouring home, you can smell it sitting in your house. Name the process that brings this smell to us.
15. In which change is more energy involved—physical or chemical?
16. Explain why liquids can flow.

Short Answer Type Questions

1. What are the characteristics of the particles of matter?
2. What are the three states of matter? Define them with two examples each.
3. Compare the three states of matter in terms of following in decreasing order:
   (a) Compressibility (b) Hardness
   (c) Energy of molecules (d) Intermolecular space
4. The smell of hot sizzling food reaches you several metres away, but to get the smell from cold food you have to go close. Why?
5. What is the physical state of water at
   (a) 250°C (b) 100°C?
6. What do you observe when barium chloride solution is mixed with a sodium sulfate solution?
7. Give reasons: when magnesium is burnt in air, there is an increase in mass after the reaction.
8. What do you understand by the term *latent heat*? What are the two types of latent heat?
9. What type of clothes should we wear in summer?
10. It is a hot summer day, Rahul and Ravi are wearing cotton and nylon clothes respectively. Who do you think would be more comfortable and why?
11. Give reasons for the following observations.
   (a) Naphthalene balls disappear with time without leaving any solid.
   (b) We can get the smell of perfume sitting several meters away.
12. For any substance, why does the temperature remain constant during the change of state?

Long Answer Type Questions

1. What happens to water if (a) it is kept in a deep freezer, (b) it is heated.
   Explain the phenomenon of change of state of water.
2. Comment on the following statements:
   (a) Evaporation produces cooling.
   (b) Rate of evaporation of an aqueous solution decreases with increase in humidity.
   (c) A sponge, though compressible, is a solid.
3. Explain the following:
   (a) Gases exert pressure
   (b) Evaporation causes cooling
   (c) Solids can be converted to liquids
   (d) Gases diffuse rapidly
4. Which characteristic of the particles of matter is illustrated by these observations?
   (a) When sugar is dissolved in water, there is no increase in the volume. Which characteristic of matter is illustrated by this observation?
   (b) A piece of chalk can be broken into small particles by hammering, but a piece of iron cannot be broken into small particles by hammering.
5. (a) Why does a desert cooler cool better on a hot dry day?
   (b) How does the water kept in an earthen pot (*matka*) become cool during summer?
   (c) Why does our palm feel cold when we put some acetone or petrol or perfume on it?
   (d) Why are we able to sip hot tea or milk faster from a saucer rather than a cup?
6. You are provided with a mixture of naphthalene and ammonium chloride by your teacher. Suggest an activity to separate them with a well labelled diagram.
7. Why does the temperature of a substance remain constant during its melting point or boiling point?
8. Classify the following as a physical or a chemical change:
   (a) Boiling of water (b) Manufacture of salt from seawater
   (c) Burning of paper (d) Drying of wet clothes
   (e) Butter getting rancid (f) Formation of clouds
   (g) Curdling of milk (h) Melting of wax
   (i) Making of a sugar solution, (j) Glowing of an electric tube light
   (k) Burning of coal (l) Photosynthesis

COMPETITION WINDOW (OBJECTIVE TYPE)

[For NEET, JEE (Main and Advanced), NTSE, KVPY and Olympiads]

Topic-wise MCQs

Matter and its types

1. Matter can have anything that
(a) has mass (b) is occupying space
(c) has both (a) and (b) (d) has none of these

2. The quantity of matter present in an object is known as its
(a) mass (b) density
(c) weight (d) gram

3. When we put some crystals of potassium permanganate in a beaker containing water, we observe that after sometime whole water has turned pink. This is due to
(a) melting of potassium permanganate crystals
(b) diffusion
(c) boiling
(d) sublimation of crystals

4. The state of matter having super energetic particles in the form of ionised gases is known as
(a) liquid state (b) gaseous state
(c) plasma state (d) Bose–Einstein condensate

5. The force that holds the particles of matter together is known as
(a) intermolecular force (b) nuclear force
(c) intermolecular space (d) bond

6. Which of the following is the densest state of matter?
(a) gas (b) solid
(c) plasma (d) liquid

7. The one in which interparticle forces are strongest, is
(a) ethane (b) carbon dioxide
(c) potassium chloride (d) hydrogen

8. On the basis of chemical classification of matter, which of the following is considered matter?
(a) element (b) compound
(c) mixture (d) all of these

9. Which of the following statements is correct?
(a) the particles of matter are very, very small
(b) the particles of matter are constantly moving
(c) the particles of matter attract each other
(d) all of these

Solid, liquid and gas

10. All liquids have different
(a) density (b) viscosity
(c) solubility (d) all of these

11. Which of the following have the highest compressibility?
(a) solids
(b) liquids
(c) gases
(d) all have the same compressibility

12. Which of the following describes the liquid phase?
(a) It has a definite volume but not a definite shape
(b) It has neither a definite shape nor a definite volume
(c) It has a definite shape and a definite volume
(d) It has a definite shape but not a definite volume

13. When a teaspoon of solid sugar is dissolved in a glass of liquid water, which phase or phases are present after mixing
(a) solid only (b) solid and liquid
(c) liquid only (d) none of these

14. The boiling point of alcohol is 78°C. What is this temperature in the kelvin scale?
(a) 273 K (b) 351 K
(c) 373 K (d) 88 K

15. A gas can be compressed to a fraction of its volume. The same volume of a gas can be spread all over a room. The reason for this is that
(a) gases consist of molecules having very large intermolecular spaces which can be reduced or increased under ordinary conditions
(b) gases consist of molecules which are in a state of random motion
(c) the volume occupied by molecules of a gas is negligible compared to the total volume of the gas
(d) none of these

16. Non-reacting gases have a tendency to mix with one another. This phenomenon is called
(a) diffusion (b) effusion
(c) explosion (d) chemical reaction

17. Which has the least energetic molecules?
(a) gases (b) plasmas
(c) solids (d) liquids

18. In which phase of matter would you expect alcohol to exist at room temperature?
(a) gas (b) plasma
(c) solid (d) liquid

19. The SI unit of temperature is
(a) fahrenheit (b) kelvin
(c) celcius (d) both (b) and (c)

Change of state of matter

20. At higher altitudes, generally
(a) boiling point of a liquid decreases
(b) boiling point of a liquid increases
(c) melting point of a solid increases
(d) no change in boiling point

21. In which phenomenon does water change into water vapour below its BP?
(a) boiling (b) condensation
(c) evaporation (d) both (a) and (c)

22. The liquid that has the highest rate of evaporation is
(a) water (b) alcohol
(c) petrol (d) nail-polish remover

23. The change of a liquid into vapour is called
(a) sublimation (b) vaporisation
(c) solidification (d) none of these

24. What is the term used to describe the phase change of a liquid to a gas?
(a) condensation (b) melting
(c) boiling (d) both (a) and (c)

25. What term is used to describe the phase change of a solid to a liquid?
(a) melting (b) boiling
(c) freezing (d) none of these

26. On leaving water in a glass, if some of its molecules turn into a gas, it is called ________
(a) extinction (b) egasoration
(c) solidification (d) evaporation

27. Select the best condition for liquefaction of a gas.
(a) decreasing the pressure
(b) increasing the temperature
(c) decreasing the pressure and increasing the temperature
(d) increasing the pressure and decreasing the temperature

28. Select the correct order (ascending) of evaporation for water, alcohol, petrol and kerosene.
(a) petrol > alcohol > kerosene > water
(b) petrol > alcohol > water > kerosene
(c) alcohol > petrol > water > kerosene
(d) water > alcohol > kerosene > petrol

29. Which of the following processes is known as fusion?
(a) change of liquid to vapour
(b) change of gaseous state to solid state
(c) change of liquid to solid
(d) change of solid to liquid

30. When a gas is compressed keeping temperature constant, it results in a/an
(a) increase in collisions among gaseous molecules
(b) increase in the speed of gaseous molecules
(c) decrease in the speed of gaseous molecules

Physical and chemical changes

31. Which of the following is a physical property?
(a) hardness (b) density
(c) rigidity (d) all of these

32. Which of the following is a physical change?
(a) crushing of a cool drink can
(b) mixing sand with water
(c) burning of coal
(d) both (a) and (b)

33. Which of the following is a chemical change?
(a) chopping of wood (b) sublimation of dry ice
(c) respiration (d) melting of ice

34. Which is a physical change?
(a) breaking of glass
(b) shredding of paper
(c) dissolving sugar in water
(d) all of these

35. A chemical change is characterised by
(a) formation of a new substance
(b) change of composition
(c) energy change
(d) all of these

36. Which of the following involves a chemical change?
(a) formation of biogas (b) blood clotting
(c) digestion of food (d) all of these

Miscellaneous

1. Which of the following is a characteristic feature of matter?
(a) mass and space (b) property of inertia
(c) influence of gravity (d) all of these

2. Which of the following represents intermolecular forces?
(a) dipole–dipole interaction
(b) ionic bond
(c) dispersion forces
(d) both (a) and (c)

3. Which of the following statement is correct regarding solids?
(a) solids have a fixed shape and a fixed volume
(b) solids have a high diffusion rate
(c) solids are quite hard
(d) both (a) and (c)

4. Which of the following are examples of amorphous solids?
(a) quartz (b) glass
(c) rubber (d) both (b) and (c)

5. Select the correct statement
(a) the spreading of purple colour of potassium permanganate into water is due to diffusion
(b) raisins swell in water due to osmosis
(c) water can exist only in the liquid state
(d) both (a) and (b)

6. Which of the following factors does not affect vapour pressure?
(a) temperature
(b) size of vessel
(c) intermolecular forces of attraction
(d) nature of solvent

7. Which one does not sublime?
(a) iodine (b) potassium iodide
(c) table salt (d) sugar

8. Dry ice is
(a) ice that has been dried
(b) solid carbon dioxide
(c) ice having no water of crystallisation
(d) none of these

9. Which of the following statement(s) about solids is correct?
(a) expand when the temperature rises
(b) have a low density
(c) change shape easily
(d) generally solids have a high density

10. The incorrect statement(s) amongst the following is/are
(a) diffusion of a liquid and a gas is known as intimate mixing
(b) diffusion also takes place in liquids
(c) gases diffuse at different rates
(d) some liquids diffuse at rates equal to that of gases

11. The property to flow is unique to fluids. Which one of the following statements is correct?
(a) only gases behave like fluids
(b) gases and solids behave like fluids
(c) gases and liquids behave like fluids
(d) only liquids are fluids

12. During summer, water kept in an earthen pot becomes cool because of the phenomenon of
(a) diffusion (b) transpiration
(c) osmosis (d) evaporation

13. Which of the following can exert pressure in all directions?
(a) solids (b) liquids
(c) gases (d) all of these

14. The state of matter can be changed by
(a) change of temperature
(b) change of pressure
(c) both of these
(d) none of these

15. Rate of evaporation depends upon
(a) temperature (b) pressure
(c) humidity (d) all of these

16. Which of the following is not correctly given here?
(a) gas into liquid — liquefaction
(b) liquid into solid — condensation
(c) liquid to vapour — evaporation
(d) solid to liquid — melting

17. Which of the following are examples of physical changes?
(a) melting of wax
(b) bending of a glass tube on heating
(c) making of ice cream
(d) all of these

Advanced and Olympiads

Single Choice

1. Which one of the following sets of phenomena would increase on raising the temperature?
(a) diffusion, evaporation, compression of gases
(b) evaporation, compression of gases, solubility
(c) evaporation, diffusion, expansion of gases
(d) evaporation, solubility, diffusion, compression of gases

2. Seema visited a Natural Gas Compressing Unit and found that the gas can be liquefied under specific conditions of temperature and pressure. While sharing her experience with friends she got confused. Help her to identify the correct set of conditions.
(a) low temperature, low pressure
(b) high temperature, low pressure
(c) low temperature, high pressure
(d) high temperature, high pressure

3. A few substances are arranged in the increasing order of 'forces of attraction' between their particles. Which one of the following represents a correct arrangement?
(a) water, air, wind (b) air, sugar, oil
(c) oxygen, water, sugar (d) salt, juice, air

4. Choose the correct statement of the following.
(a) Conversion of solid into vapours without passing through the liquid state is called vaporisation.
(b) Conversion of vapours into solid without passing through the liquid state is called sublimation.
(c) Conversion of vapours into solid without passing through the liquid state is called freezing.
(d) Conversion of solid into liquid is called sublimation.

5. Which condition out of the following will increase the evaporation of water?
(a) increase in the temperature of water
(b) decrease in the temperature of water
(c) less exposed surface area of water
(d) adding common salt to water

6. In which of the following conditions, will the distance between the molecules of hydrogen gas increase?
(i) Increasing pressure on hydrogen contained in a closed container.
(ii) Some hydrogen gas leaking out of the container.
(iii) Increasing the volume of the container of hydrogen gas.
(iv) Adding more hydrogen gas to the container without increasing the volume of the container.
(a) (i) and (iii) (b) (i) and (iv)
(c) (ii) and (iii) (d) (ii) and (iv)

7. Which of the following are correctly matched?
(i) Melting of wax — Physical change
(ii) A rock rolling down a hill — Physical change
(iii) Breaking of glass — Chemical change
(iv) Burning of oil — Chemical change
(a) (i) and (ii) (b) (ii) and (iii)
(c) (i), (ii) and (iv) (d) (i), (ii), (iii) and (iv)

8. Select the set showing the correct statements.
(i) The process of depositing a layer of zinc on iron prevents rusting of iron.
(ii) Crystallisation is a physical change.
(iii) Photosynthesis is an example of physical change.
(iv) During a chemical change, sound may be produced.
(a) (i), (ii) (b) (i), (iii)
(c) (i), (ii), (iii) (d) (i), (ii), (iv)

Multiple Choice

9. Which of these choices will change the state of matter?
(a) pressure (b) temperature
(c) heat (d) crushing a crystal

10. Select the one that is 'matter'
(a) humidity (b) feeling of heat
(c) water (d) smoke

11. Which is/are correct statement/s?
(a) Of the three states of matter, the one that is most compact is the solid state.
(b) Matter is continuous in nature.
(c) The density of a solid is generally more than that of a liquid.
(d) In the solid state, inter-particles space (that is, empty space) is minimum.

12. Which of the following properties are similar for solids, liquids and gases?
(a) density
(b) mass of the substance
(c) particle size of the substance
(d) movement of molecules

13. Which of these statement(s) is/are not true about gases?
(a) Gases have high density.
(b) Gases can be compressed more than solids.
(c) Gases have very specific shapes.
(d) Gases undergo diffusion fastest.

14. Which of the following statement(s) is/are true for gases?
(a) Volume of the gas is almost equal to the volume of the container confining the gas.
(b) A confined gas exerts uniform pressure on the walls of its container in all directions.
(c) Gases do not have a definite shape and volume.
(d) Mass of the gas cannot be determined by weighing a container in which it is enclosed.

15. Which of the following involve a chemical change?
(a) rusting of iron
(b) calcination of limestone
(c) metabolism
(d) chopping wood

Comprehension Type

Comprehension I: Matter is composed of tiny particles. it can be further divided into solid, liquid and gas on the basis of their physical state, and into elements, compounds and mixtures on the basis of their chemical composition. Solids, liquids, gases can be easily distinguished due to many differences in their structure and properties. Recently plasma and Bose–Einstein condensate have also been observed as states of matter. These forms can be interchanged into one another by changing the conditions of temperature, pressure, and so on.

16. Which of the following states of matter has the highest energy?
(a) solid (b) liquid
(c) gas (d) all have equal energies

17. Which of the following statements is incorrect?
(a) The strongest intermolecular force is present in solids.
(b) Gases diffuse fastest among solid, liquid and gas.
(c) Intermolecular space is maximum in solids.
(d) Gases are most compressible among solid, liquid and gas.

Comprehension II: Properties such as shape, size, colour and state of a substance are known as its physical properties. A change in which a substance undergoes a change in its physical properties is called a physical change. It is mostly reversible and in such a change no new substance is formed. For example, dissolution of sugar or salt in water. A change in which one or more new substances are formed is known as a chemical change. For example, digestion of food, ripening of fruits.

18. Which of the following represents characteristics of a physical change?
(a) no new substance is possible
(b) always involves exchange of heat
(c) a change of colour takes place
(d) a gas may be formed

19. Which of the following may accompany a chemical change?
(a) heat, light may be given off or absorbed
(b) a colour change may take place
(c) a sound may be produced
(d) all of these

20. Select the correct statements.
(a) Physical processes are mostly reversible.
(b) Fermentation of grapes is a chemical change.
(c) Digestion is a physical change.
(d) Both (a) and (b) are correct.

Matrix Matching

21. Match the following

Column I	Column II
(a) Solid	(p) No definite shape but a definite volume
(b) Liquid	(q) No definite shape and no definite volume
(c) Gas	(r) Definite shape and definite volume
(d) Plasma	(s) Fourth state of matter
	(t) Have intermolecular force of attraction

22. Match the following

Column I	Column II
(a) Solid into liquid	(p) sublimation
(b) Liquid into gas	(q) Liquification
(c) Gas into liquid	(r) Vaporisation
(d) Solid into gas	(s) Melting

Integer Type

23. Which of the following properties represent physical properties of matter?
Seeing, toxicity, hearing, feeling, heat of combustion, touch, smell, reactivity, measurement

24. Which of the following properties represent chemical properties of matter?
Toxicity, oxidation states, hearing, heat of combustion, touch, flammability, coordination number, reactivity, measure, enthalpy of formation

25. Write the number of processes involving a physical change:
Manufacture of salt from seawater, burning of paper, drying of wet clothes, butter getting rancid, formation of clouds, curdling of milk, melting of wax, making of a sugar solution, photosynthesis

26. Which of the following process can be used to obtain the gas or vapour state from solid or liquid state?
Evaporation, boiling, sublimation, condensation, liquefaction

27. The boiling point of a gas is 281 K. If it is X°C find the value of X.

28. The value of pressure in mm of mercury is 3800. If its value in atm is X atm, what is the value of X here?

29. If for *W* kg of water, the latent heat of vaporisation was found to be 90000 kJ/kg, what is the value of *W*?

Answer Keys

Fill in the Blanks

1. states 2. gas 3. diffusion
4. gas 5. more 6. gaseous
7. less 8. boiling point 9. melting
10. exothermic 11. sublimation
12. latent heat of fusion 13. physical

True or False

1. F 2. T 3. T 4. F 5. T
6. T 7. T 8. T 9. F 10. F
11. T 12. F

Very Short Answer Type Questions

1. A process in which things change from one state to another and get back to the original state, without any change in its chemical composition is called interconversion of state of matter.

2. Two factors responsible for the change of state of matter are, change in
   - temperature
   - pressure

3. Due to intermolecular forces of attraction, the molecules of matter are always in motion and attract each other with a force, due to which they are held together.

4. Aqueous solution of copper sulfate becomes blue. It is due to diffusion.

5. In gases, the molecules are free to move and they are not stuck to each other and the intermolecular forces of attraction are least. So, the gas almost fills the whole vessel in which it is enclosed.

6. Y
Y has higher boiling point than X so Y has stronger intermolecular force of attraction.

7. In solids, particles are closely packed so the intermolecular space is almost zero due to strong forces of attraction. Hence, the molecules are not free to move, it makes them hard and rigid. So solids cannot be compressed.

8. The solid state has definite shape and definite volume.

9. (a) gas (b) solid and (c) liquid

10. Camphor, ammonium chloride and naphthalene.

11. (a) vaporisation (b) melting (c) condensation

12. Steam has more volume but less density than ice.

13. On increasing surface area, the rate of evaporation increases.

14. It is due to diffusion.

15. In a chemical change, more energy is involved (absorbed or evolved).

16. Due to the weaker intermolecular forces in liquids, the particles are not closely packed and hence there are large intermolecular spaces. So, molecules in liquids can move randomly and hence liquids can flow easily.

Short Answer Type Questions

1. Characteristics of particles of matter are given below:
   - Particles of matter have space between them.
   - Particles of matter are continuously moving.
   - Particles of matter have an attraction force between them.
   - Particles of matter are very small in size.
2. The three states of matter are—solids, liquids and gases.
   - **Solids:** A solid has a definite volume and shape. Examples—wood, stone, iron, ice.
   - **Liquids:** A liquid has a definite volume but not a definite shape. Examples—water, juice, milk, oil.
   - **Gases:** A gas has neither definite shape nor a definite volume. Examples—air, hydrogen, oxygen.
3. (a) Compressibility: Gas > Liquid > Solid
   (b) Hardness: Solid > Liquid > Gas
   (c) Energy of molecules: Gas > Liquid > Solid
   (d) Intermolecular space: Gas > Liquid > Solid
4. Evaporation is directly proportional to temperature; therefore, hot food evaporates easily. Diffusion of hot food vapour with air becomes very fast and can reach a distant place within a very short time.
5. The boiling point of water is 100°C.
   (a) At 250°C, the state of water will be steam or water vapour—gaseous state.
   (b) At 100°C, there will be a transition of liquid state into the gaseous state. So, at this temperature, the state is/may be liquid as well as gaseous.
6. $\underset{(aq.)}{BaCl_2} + \underset{(aq.)}{Na_2SO_4} \rightarrow \underset{(ppt.)}{BaSO_4} + \underset{(aq.)}{2NaCl}$

   We can observe that a white insoluble solid (precipitate of barium sulfate) is formed along with a solution of sodium chloride. Then, wait for ten minutes for the reaction to be completed and the solid formed to settle down and then weigh the content again. Finally, note the reading.
   We will observe that the total mass of the apparatus + reactants = total mass of apparatus + products. Hence the law of conservation of mass is verified.
7. When magnesium ribbon is burnt in air, a white solid, magnesium oxide is formed. The mass of magnesium is more than the mass of magnesium oxide. This is because the mass of oxygen used is not taken. If that is considered, the total mass of the reactants and the products are found to be almost equal.

   $\underset{\text{(Magnesium)}}{2Mg} + \underset{\text{(Air)}}{O_2} \rightarrow \underset{\text{(Magnesium oxide)}}{2MgO}$
8. See text part.
9. We should wear light coloured cotton clothes in summer because,
   - cotton is a good absorber of water/sweat. It provides more surface area for the sweat to evaporate.
   - light colours absorb less heat.

   So, wearing light coloured cotton clothes helps us to feel cool and comfortable.
10. Rahul is wearing cotton clothes which are more comfortable in the summer because cotton absorbs the sweat which causes cooling on evaporation. Ravi is wearing nylon clothes which do not absorb sweat. Hence, Ravi will be uncomfortable.
11. (a) Naphthalene is a substance which directly changes from solid to gas on heating by the process of sublimation. So, the naphthalene balls disappear with time as they sublime due to the heat of the surroundings.
    (b) The smell (aroma) of perfume reaches several meters away due to the fast diffusion of the gaseous perfume particles through the air.
12. During the change of state of a substance, the temperature remains constant. This can be understood with the help of an example. When a solid is heated to its melting point, the temperature first rises and becomes constant when it reaches its melting point. Now, on further heating, the heat energy provided to the substance helps to break the force of attraction between the solid molecules. This heat is called latent heat. That is why, the temperature does not rise.

Long Answer Type Questions

1. (a) When water is placed in a deep freezer, it is cooled and changed into ice at 0°C ice.

   $\text{Water} \xrightarrow{\text{[freezer]}} \text{Ice (0°C)}$

   (b) Water on heating changes into steam at 100°C

   $\text{Water} \xrightarrow{\text{heating}} \text{Steam (100°C)}$

   The phenomenon of change of state of water: under ordinary conditions, water is in the liquid state but when it is placed in a deep freezer, it changes into ice at 0°C and when ice is kept at room temperature again, it changes back into liquid water.

   Likewise, when heating water, it changes into steam at 100°C, which on cooling changes back into liquid water. But there is no change in the chemical composition of water when its state changes from liquid to solid or liquid to the gaseous state.

   $\underset{\text{(Solid)}}{\text{Ice}} \underset{\text{Cool}}{\overset{\text{Heat}}{\rightleftharpoons}} \underset{\text{(Liquid)}}{\text{Water}} \underset{\text{Cool}}{\overset{\text{Heat}}{\rightleftharpoons}} \underset{\text{(Gas)}}{\text{Steam}}$
2. (a) Evaporation is a surface phenomenon. The particles from the surface of the liquid take energy from the surroundings and change into vapour which results in the decrease in energy of the surroundings. Hence, a cooling effect is produced during evaporation.
   (b) The amount of water present in the air is known as humidity. If the water vapour in the air is already present in a large amount, it is not able to take up more water through evaporation. Hence, the rate of evaporation of water will decrease. On a dry day, the air absorbs water more readily. Hence, the rate of evaporation is high on a dry day.

(c) A sponge is a solid but it has minute pores in which air is trapped. These pores make the sponge a soft material. When the sponge is pressed, the air present in the pores comes out and the sponge is compressed.

3. See text part.

4. See text part.

5. (a) A desert cooler functions on the basis of evaporation. The rate of evaporation increases with the increase in temperature and decrease in humidity. As evaporation increases when the day is hot and dry, the desert cooler functions better.

(b) Large number of tiny pores are present on the surface of the earthen pot. The water stored in the earthen pot evaporates faster through these pores due to the increased exposed surface area. As the process of evaporation causes cooling, the stored water inside the earthen pot becomes cool.

(c) Acetone, petrol and perfume, being volatile, evaporate very fast when exposed to larger surfaces. During the process they absorb the required latent heat of vaporisation from the palm (if kept on palm). So, the process causes cooling and the palm feels cool.

(d) A liquid has a larger surface area in a saucer than in a cup. Thus, it evaporates faster and cools faster in a saucer than in a cup. For this reason, we are able to sip hot tea or milk faster from a saucer than from a cup.

6. A mixture of naphthalene and ammonium chloride can be separated as follows:

Step 1: Put the mixture in a beaker and add water to it. Stir with a glass rod. Ammonium chloride being soluble in water gets dissolved leaving behind the insoluble naphthalene.

Step 2: Filter the solution. Naphthalene remains on the filter paper while ammonium chloride is obtained as filtrate.

Step 3: Evaporate the filtrate to get back ammonium chloride.

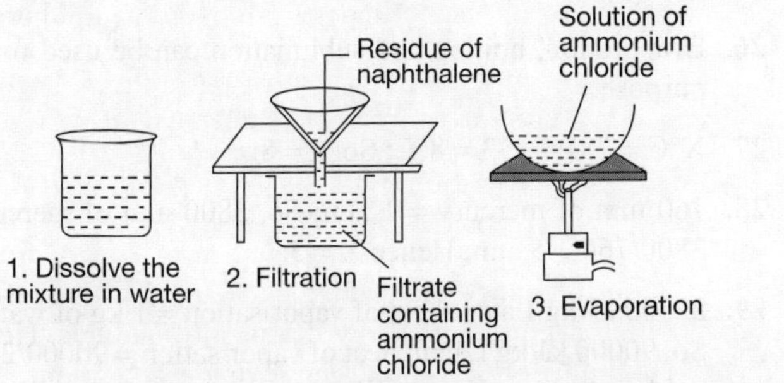

7. When a substance melts, it absorbs heat for the conversion of the solid into the liquid state. As we continue heating, the heat supplied is used up in converting the solid into the liquid state by overcoming the forces of attraction between the particles and there is no change in temperature till the whole solid is converted into liquid. This heat absorbed by the solid which resists the increase in temperature is called the latent heat of fusion.

When a liquid is heated, it starts getting converted into vapour. Further heat given to the liquid is used in changing the state and there is no increase in the temperature till the liquid starts boiling. This heat is known as latent heat of vaporisation. Hence, the temperature of a substance remains constant at its melting point or boiling point until all the substance melts or boils.

8. (a) Physical change (b) Physical change
(c) Chemical change (d) Physical change
(e) Chemical change (f) Physical change
(g) Chemical change (h) Physical change
(i) Physical change (j) Physical change
(k) Chemical change (l) Chemical change

Competition Window (Objective Type)

Topic-wise MCQs

1. (c)	**2.** (a)	**3.** (b)	**4.** (c)	**5.** (a)
6. (b)	**7.** (c)	**8.** (d)	**9.** (d)	**10.** (d)
11. (c)	**12.** (a)	**13.** (c)	**14.** (b)	**15.** (a)
16. (b)	**17.** (c)	**18.** (d)	**19.** (a)	**20.** (a)
21. (c)	**22.** (c)	**23.** (b)	**24.** (c)	**25.** (a)
26. (d)	**27.** (d)	**28.** (a)	**29.** (d)	**30.** (d)
31. (d)	**32.** (d)	**33.** (c)	**34.** (d)	**35.** (d)
36. (d)				

Miscellaneous

1. (d)	**2.** (d)	**3.** (d)	**4.** (d)	**5.** (d)
6. (b)	**7.** (d)	**8.** (b)	**9.** (d)	**10.** (a)
11. (c)	**12.** (d)	**13.** (c)	**14.** (c)	**15.** (d)
16. (b)	**17.** (d)			

Advanced and Olympiads

Single Choice

1. (c)	**2.** (c)	**3.** (c)	**4.** (b)	**5.** (a)
6. (c)	**7.** (c)	**8.** (d)		

Multiple Choice

9. (abc)	**10.** (acd)	**11.** (acd)	**12.** (bc)	**13.** (ac)
14. (abc)	**15.** (abc)			

Comprehension Type

16. (c)	**17.** (c)	**18.** (a)	**19.** (d)	**20.** (d)

Matrix Matching

21. (a)—(rt); (b)—(pt); (c)—(qt); (d)—(qst)

22. (a)—(s); (b)—(r); (c)—(q); (d)—(p)

Integer Type

23. 6	**24.** 7	**25.** 5	**26.** 3	**27.** 8
28. 5	**29.** 4			

Hints and Solutions

Competition Window (Objective Type)

Topic-wise MCQs

2. The mass of an object is a fundamental property of the object, a numerical measure of inertia, a fundamental measure of the amount of matter in the object.

3. This is due to diffusion of particles of potassium permanganate into the intermolecular spaces between particles of water.

10. Different liquids have different density, viscosity and solubility.

13. Sugar molecules occupy the intermolecular spaces present in water molecules. Thus, after dissolution only liquid is left.

14. K = 273 + t°C; K = 237 + 78 = 351 K

16. Effusion is the process in which individual molecules flow through a hole without collisions between molecules.

17. The order of energy for different states of matter is: Bose–Einstein < solids < liquids < gases < plasma condensate.

18. Alcohol, because of presence of intermediate magnitude of intermolecular forces, is a liquid at room temperature.

20. At higher altitudes, the atmospheric pressure decreases. Therefore, at lower temperature, the vapour pressure becomes equal to the atmospheric pressure. Thus, the boiling point of a liquid decreases.

21. Boiling point of water is 100°C whereas evaporation of water into water vapour occurs at room temperature.

22. Intermolecular forces are the least in the case of petrol. Thus, it has the highest rate of evaporation.

23. During vaporisation, energy is supplied to the liquid in the form of heat. The liquid reaches its boiling point and gets converted into gas (vapours).

Advanced and Olympiads

Single Choice

1. Rate of evaporation, diffusion and expansion of gases increase with an increase in temperature.

2. Gases can be compressed under low temperature and high pressure. Under these conditions the particles of gases come closer and liquefy.

3. Forces of attraction between the particles increase in the order, gases < liquids < solids. Hence, the correct arrangement is oxygen, water, sugar.

4. Conversion of vapours into solid without passing through the liquid state is called sublimation.

5. Rate of evaporation increases with an increase in the temperature of water.

6. (ii) If some hydrogen gas is leaked from the container the remaining gas will occupy the whole space and the distance between the molecules will increase.
   (iii) If the volume of the container is increased, the same number of molecules will occupy that space. Hence, the distance between the molecules will increase.

8. Photosynthesis is an example of a chemical change.

Multiple Choice

9. A crystal is a solid and even on crushing it you will get smaller crystals (solid).

13. Gases have very low intermolecular forces between their molecules thus they can be compressed and undergo rapid diffusion.

14. The mass of a gas can be determined by weighing the container, filled with gas and again weighing this container after removing the gas. The difference between the two weights gives the mass of the gas.

Integer Type

23. See, hear, feel, touch, smell and measure are physical properties of matter.

24. Toxicity, oxidation state, heat of combustion, flammability, coordination number, reactivity and enthalpy of formation are chemical properties of matter.

25. Manufacture of salt from seawater, drying of wet clothes, formation of clouds, melting of wax and making of a sugar solution are physical changes while the rest are chemical changes.

26. Evaporation, boiling and sublimation can be used for this purpose.

27. X°C = 281 – 273 = 8°C; So X = 8.

28. 760 mm of mercury = 1 atm; So, 3800 mm of mercury = 3800/760 = 5 atm; Hence X = 5.

29. 22500 kJ/kg Latent heat of vaporisation = 1 kg of water
   So, 90000 kJ/kg Latent heat of vaporisation = 90000/22500 = 4 kg of water; Hence, W = 4.

CHAPTER 2

Elements, Compounds and Mixtures

LEARNING OBJECTIVES

After studying this unit, you will be able to understand:

- Pure matter and pure substance
- Elements and their types
- Compounds and mixtures, and differentiate between them
- Various methods of separation of mixtures

Introduction

How do we judge whether milk, ghee, butter, salt, spices, mineral water and juice that we buy from the market are pure or not?

Fig. 2.1 Some consumables

For us, 'pure' signifies no adulteration. However, from a scientist's perspective, all these things are actually mixtures of different substances and hence they are not pure! For example, milk is actually a mixture of water, fat and proteins (Fig. 2.1). When a scientist claims that something is pure, it means that all the constituent particles of that substance are the same in their chemical nature. A pure substance consists of only a single type of species (particles). When we look around us, we see that most of the matter exists as mixtures of two or more pure components. For example, sea water, minerals and soil are all mixtures. All the matter around us is not pure. The matter around us is of two types—pure and mixture (Fig. 2.2).

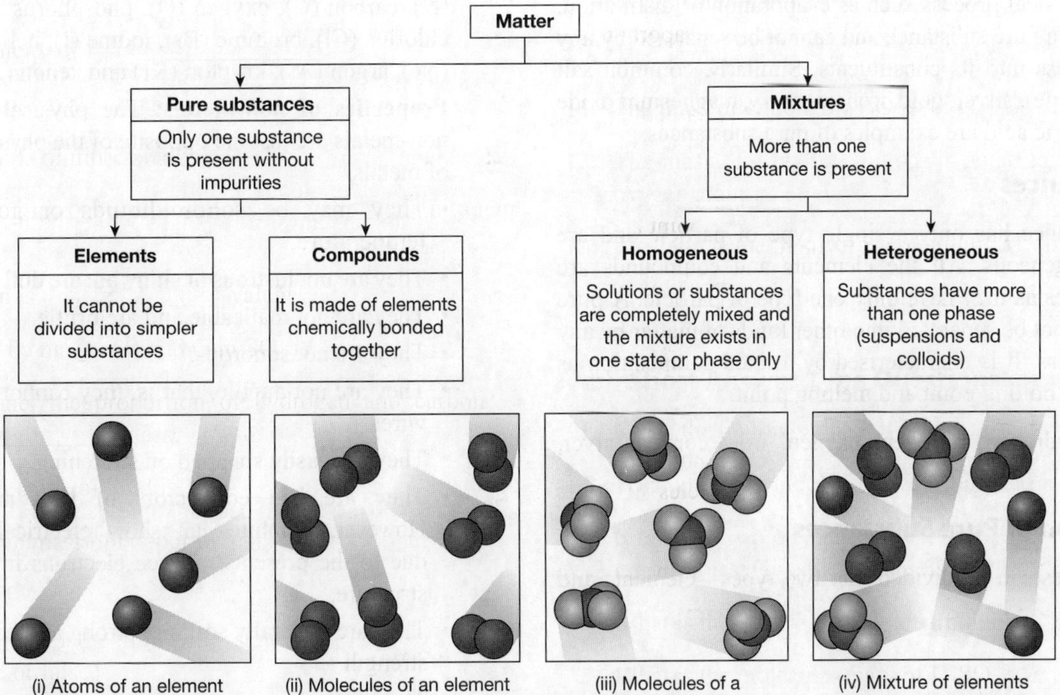

Fig. 2.2 Pure substances and mixtures

Table 2.1 Elements and their symbols

Element	Symbol	Element	Symbol	Element name in Latin	Symbol
Aluminium	Al	Hydrogen	H	Copper (Cuprum)	Cu
Arsenic	As	Iodine	I	Lead (Plumbum)	Pb
Barium	Ba	Fluorine	F	Potassium (Kalium)	K
Zinc	Zn	Chlorine	Cl	Antimony (Stibium)	Sb
Cadmium	Cd	Nitrogen	N	Silver (Argentum)	Ag
Calcium	Ca	Oxygen	O	Tin (Stannum)	Sn
Manganese	Mn	Phosphorus	P	Sodium (Natrium)	Na
Chromium	Cr	Sulfur	S	Gold (Aurum)	Au
Cobalt	Co	Bromine	Br	Mercury (Hydrogyrum)	Hg
Magnesium	Mg	Neon	Ne	Iron (Ferrum)	Fe

In Chapter 1, we learnt about the three states of matter (solid, liquid and gas). Before understanding the chemical nature of matter, it is important for us to understand the scientific meaning of the term *chemical substance.* In scientific terms, *a substance is a type of matter that cannot be separated into other kinds of matter by any physical process.* Then, that substance is a *pure form of matter* and not a mixture of many different kinds of matter. Most of the things that we use in our day-to-day life are in the form of mixtures. Pure substances are rare. For example, dissolved sugar can be separated from its solution by using a physical process such as evaporation or distillation. Thus, sugar is a pure substance and cannot be separated by any physical process into its constituents. Similarly, common salt (sodium chloride), silver, gold, iron, mercury, magnesium oxide and hydrochloric acid are examples of pure substances.

Pure Substances

A pure substance has only a single type of particle and are always homogeneous. All the elements and compounds are pure substances as they have only one type of particle. A pure substance cannot be separated into other kinds of matter by any physical process. It is characterised by a fixed composition as well as a fixed boiling point and melting point.

Examples, hydrogen, nitrogen, oxygen, iron, copper, silver, gold.

Classification of Pure Substances

Pure substances can be divided into two types—elements and compounds.

Element

An element is a substance that cannot be split up into two or more simpler substances by the usual chemical methods of applying heat, light and electric energy, because an element is made of only one type of atom. For example, oxygen is an element as it cannot be split up into two or more simpler substances by the usual methods of applying heat, light and electricity.

Types of Elements: Elements can be divided into three types—non-metals, metalloids and metals.

(i) Non-metals: A non-metal is an element that is neither malleable nor ductile and cannot conduct electricity [with the exception graphite (C)]. All the non-metals are gases or solids, except bromine which is a liquid non-metal at room temperature. Examples are, hydrogen (H), nitrogen (N), carbon (C), oxygen (O), phosphorus (P), sulfur (S), chlorine (Cl), bromine (Br), iodine (I), helium (He), neon (Ne), argon (Ar), krypton (Kr) and xenon (Xe).

Properties of non-metals: The physical properties of non-metals are just the opposite of the physical properties of metals.

- They may be solids, liquids or gases at room temperature.
- They are not lustrous or shiny but are dull in appearance.
- They are not malleable and are brittle.
- They are not sonorous.
- They are not ductile, that is, they cannot be drawn into wires.
- They are easily snapped on stretching.
- They are bad conductors of heat and electricity. However, graphite can show electrical conductance due to the presence of free electrons in its hexagonal structure.
- They are generally soft, not strong and have low tensile strength.
- They have comparatively low melting points, boiling points and densities.
- They may have many different colours.

(ii) **Metalloids:** The elements which show some properties of metals and some other properties of non-metals are known as metalloids. Their properties are intermediate between the properties of metals and non-metals. Metalloids are also known as semi-metals. Examples are, arsenic (As), selenium (Se), antimony (Sb), boron (B), silicon (Si) and germanium (Ge).

(iii) **Metals:** A metal is an element that is malleable and ductile and can conduct electricity. All the metals are solids except mercury, which is a liquid. Examples are, aluminium, iron, copper, silver, gold, zinc.

Properties of Metals

- Metals are solids at the room temperature. For example, iron, copper, aluminium, silver and gold are solids at room temperature. Only mercury is in the liquid state at room temperature.
- Metals are sonorous, so a metal makes a ringing sound when it is struck. For example, wires or strings for stringed musical instruments such as violins, guitars and sitars, and plate-type musical instruments such as cymbals.
- Metals are malleable so they can be beaten into thin sheets without breaking. For example, aluminium metal is quite malleable and can be converted into thin sheets (aluminium foil) which are used for packing food items like chocolates, medicines and cigarettes.
- Metals are ductile so they can be drawn or stretched into thin wires. However, all the metals are not equally ductile. For example, copper and aluminium metals are highly ductile and can be drawn into thin wires which are used in electrical wiring.
- Metals are good conductors of heat and electricity. Silver is the best conductor of heat and also has the highest thermal conductivity. Examples:
  - o Electric wires are made of copper and aluminium as they are very good conductors of electricity.
  - o The cooking utensils and water boilers are usually made of aluminium or copper as they are very good conductors of heat.
- Metals are shiny or lustrous and can be polished. This property of a metal containing a shining surface is known as metallic lustre. This property of metals makes them useful in making jewellery and decorative pieces. For example, gold, platinum and silver are used for making jewellery as they are bright and shiny. The shiny surface of metals also makes them good reflectors of light. For example, silver.
- Metals are generally hard; however, all the metals are not equally hard. The hardness varies from metal to metal and they cannot be cut with a knife. The exceptions are lithium, sodium and potassium which are soft metals.
- Metals are usually strong and have a high tensile strength so they can hold large weights without breaking. For example, iron (as steel) is very strong and has a high tensile strength so it can be used in the construction of railway lines, girders, machines, vehicles, bridges, buildings and chains
- Metals generally have high melting points and boiling points so most of the metals melt and vaporise at high temperatures. For example, iron has a melting point of 1808 K; solid iron melts and turns into liquid iron on heating to 1808 K. Tungsten has the highest melting point among all metals.
- Metals have high densities so they are heavy substances. For example, the density of iron metal is 7.8 g/cm3. Iridium (Ir) and osmium (Os) have the highest densities while lithium (Li) has the lowest density.
- Metals are usually silver or grey in colour with the exception of copper (reddish-brown) and gold (yellow).

Illustrations

1. Classify the given elements as non-metals, metals or metalloids.

Br, Ge, S, Sn, Pb, Al, B, Ga, Ar, Sb.

Solution: Non-metals: Br, S, Ar

Metals: Sn, Pb, Al, Ga

Metalloids: Ge, B, Sb

2. What is the effect of temperature on metallic conductance?

Solution: On increasing the temperature, metallic conductance decreases as mobile electrons are pushed away due to the vibration of slippery kernels.

3. (a) An element is sonorous and highly ductile. Under which category would you classify this element? What other characteristics do you expect the element to possess?

(b) What are the favourable qualities that gold acquires when it is alloyed with copper or silver for the purpose of making ornaments?

Solution:

(a) It is a metal. The element is expected to be lustrous, malleable and a good conductor of heat and electricity.

(b) Pure gold is highly malleable and soft. When it is alloyed with copper or silver it becomes hard and strong and can be moulded into various shapes.

Compounds

A compound is a substance made up of two or more elements chemically combined in a fixed proportion by mass. A compound is formed as a result of chemical reaction/s, between the constituent elements. It is important to note that the properties of compounds are different from the properties of the elements from which it is formed. For example:

- Methane (CH_4) is a compound made up of two elements, carbon and hydrogen, chemically combined in a fixed proportion of 3 : 1 by mass.
- Water (H_2O) is a compound made up of two elements, hydrogen and oxygen, chemically combined in a fixed proportion of 1 : 8 by mass.

Types of Compounds: On the basis of their properties, compounds can be further divided into three classes—acids, bases and salts. For example,

- Hydrochloric acid is an acid, potassium hydroxide is a base whereas potassium chloride is a salt.
- Sulfuric acid is an acid, sodium hydroxide is a base whereas sodium sulfate is a salt.

Features of a Compound

- A compound is always homogeneous in nature.
- In a compound, the constituents are present in definite proportions by mass.
- The properties of a compound are different from the properties of its constituents.
- A compound has a fixed melting point and boiling point.
- The constituents of a compound cannot be separated by simple physical processes.
- The formation of a compound is generally accompanied by an evolution of energy in the form of heat or light. For example, burning of a candle gives CO_2 and water, and liberates heat and light energy.

KNOWLEDGE BOOSTER

Stoichiometric and non-stoichiometric compounds

Compounds having a definite atomic composition (whole number ratio) are known as stoichiometric compounds. For example, H_2O, CO_2, NH_3, CH_4. Compounds having a variable atomic composition (non-whole number ratio) are called known as non-stoichiometric compounds. They are formed mainly by *d* and *f*-block elements. For example, $Fe_{0.93}O$, $Ni_{0.98}O$, $Cu_{1.7}S$.

Mixtures

Mixtures are constituted by more than one kind of pure form of matter, that is, a mixture is a substance which consists of two or more elements or compounds not chemically combined together. All the solutions are mixtures and the various substances present in a mixture are known as 'constituents or components of the mixture'. For example, lemonade (*nimbu pani*) is a mixture of water, lemon juice, sugar and salt.

A substance cannot be separated into other kinds of matter by a physical process. Sodium chloride is a substance and cannot be separated by any physical process into its chemical constituents. Similarly, sugar is a substance because it contains only one kind of pure matter and its composition is the same throughout.

A mixture has two or more different types of particles possessing different chemical nature. A mixture may be homogeneous or heterogeneous. All the mixtures are impure substances and a mixture does not have a fixed composition or a fixed melting point and boiling point.

Types of Mixtures

Depending upon the nature of the components that form a mixture, we can have different types of mixtures. Generally, mixtures are of two types—homogeneous mixtures and heterogeneous mixtures.

(i) Homogeneous mixtures: Those mixtures in which the substances are completely mixed together and are indistinguishable from one another and form only one phase are known as *homogeneous mixtures*. All the homogeneous mixtures are known as solutions. They may be further divided into three types:

- **Solid–liquid mixture:** A mixture of sugar in water is a homogeneous mixture as all the parts of the sugar solution have the same sugar–water composition and appear to be equally sweet and there is no visible boundary of separation between sugar and water particles in a sugar solution (that is, have one 1-phase system).
- **Solid–solid mixture:** Alloys (mixture of two or more metals) such as steel (80% Fe + 12% Cr + 8% Ni), brass (59.2% Cu + 36.4% Zn + 3.26% Pb) and bronze.
- **Liquid–liquid mixture:** Rectified sprit (alcohol + gasoline), mixture of benzene and toluene.

(ii) Heterogeneous mixtures: The mixtures in which all the substances remain separate and one substance is spread throughout the other substance as small particles, droplets or bubbles are known as heterogeneous mixtures (they have distinct phases). For example, a mixture of sugar and sand is a solid–solid heterogeneous mixture as different parts of this mixture will have different sugar–sand compositions and there is a visible boundary of separation between sugar and sand particles. The suspensions of solids in liquids are also examples of solid–liquid heterogeneous mixtures. A mixture having two or more immiscible liquids is also a case of liquid–liquid heterogeneous mixture.

Properties of Mixtures

- A mixture is usually heterogeneous.
- A mixture shows the properties of all the constituents present in it.
- A mixture can be separated into its constituents by physical processes.

- The composition of a mixture is variable; the constituents can be present in any proportion by mass.
- A mixture does not have a definite melting point and boiling point.
- As energy is usually neither given out nor absorbed in the preparation of a mixture, the formation of a mixture involves only a physical change.

Illustrations

1. Classify the given substances into compounds and mixtures.

Cloud, iodised table salt, steam, sucrose, steel, aerated drink, dry ice, milk, 22 carat gold, marsh gas, petrol, smoke

Solution:

Compounds: Cloud, steam, sucrose, dry ice, marsh gas

Mixtures: Iodised table salt, steel, aerated drink, petrol, smoke, milk, 22 carat gold

2. Classify the following into elements, compounds and mixtures.

Sodium, soil, sugar solution, silver, calcium carbonate, tin, silicon, coal, air, soap, methane, carbon dioxide, blood.

Solution:

Elements: Sodium, silver, tin and silicon

Compounds: Calcium carbonate, methane and carbon dioxide

Mixtures: Soil, sugar solution, coal, air, soap and blood

3. (a) How is sodium chloride different from its constituent elements?

(b) Why is iron sulfide a compound?

Solution:

(a) The properties of sodium and chlorine are completely different from those of sodium chloride. Sodium is a soft, highly reactive metal. Chlorine is a poisonous non-metallic gas while sodium chloride is a very useful non-poisonous compound which is added to our food to get minerals and also to add taste to it.

(b) Iron sulfide is a compound which can be broken into the elements iron and sulfur, both of which have different properties. The properties of the compound are entirely different from that of its constituent elements.

Separation of Mixtures

Most of the materials around us are mixtures and not pure substances. Such mixtures have two or more substances mixed in them. So, it may not be possible to use a mixture as such in homes and in industries. We might require only one or two separate constituents of a mixture for our uses. So, such mixtures have to be separated into their individual constituents to make them useful. Various methods of separation are used to get individual components from a mixture. Separation makes it possible to study and use the individual components of a mixture. Heterogeneous mixtures can be separated into their respective constituents by simple physical methods like handpicking, sieving and filtration. Some special techniques can also be used for the separation of the components of a mixture.

Mixture

Solid–solid
(i) Solvent extraction
(ii) Sublimation
(iii) Magnetic separation
(iv) Winnowing

Solid–liquid
(i) Filtration
(ii) Sedimentation
(iii) Centrifugation
(iv) Evaporation
(v) Crystallisation
(vi) Chromatography
(vii) Distillation

Liquid–liquid
(i) Fractional distillation
(ii) Separating funnel

(i) Separation of a mixture of two solids: All the mixtures having two solid substances can be separated by one of the following methods.

- **Separation by a suitable solvent:** In a few cases, one of the constituents of a mixture is soluble in a specific liquid solvent while the other constituent is insoluble. This difference in the solubilities of the constituents of a mixture can be used to separate them. For example, sugar is soluble in water while sand is insoluble, so a mixture of sugar and sand can be separated by using water as the solvent.

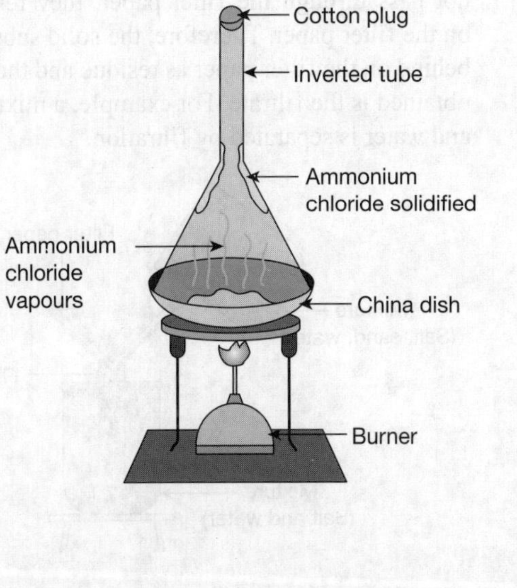

- **Separation by sublimation:** The process of sublimation can be used to separate those substances which sublime on heating, from a mixture. The solid substance obtained by cooling the vapour is called the sublimate. For example, iodine, naphthalene, anthracene, camphor and ammonium chloride can be separated from a mixture by sublimation.
- **Separation by a magnet:** Iron is attracted by a magnet; this tendency of iron is used to separate it from a mixture having iron as one of the constituents. For example, a mixture of iron filings and sulfur powder can be separated by using a magnet as iron filings are attracted by a magnet while sulfur is not.

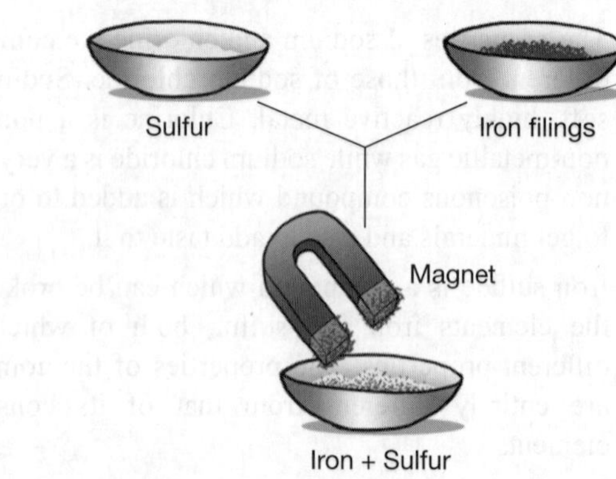

- **Winnowing:** This method is used to separate lighter particles from heavier particles by blowing air/by the force of wind. For example, separation of rice grain from husk. In its simplest form, it involves throwing the mixture into the air so that the wind blows away the lighter chaff while the heavier grains fall back.

(ii) Separation of a mixture of a solid and a liquid: All the mixtures having a solid and a liquid can be separated by any of the following processes.

- **Separation by filtration:** The process of removing insoluble solids from a liquid by using a filter paper is known as filtration. Filtration can be used for separating insoluble substances from a liquid. Here, the liquid passes through the filter paper and gets collected in the beaker kept below the funnel. As solid particles do not pass through the filter paper, they remain behind on the filter paper. Therefore, the solid substance is left behind on the filter paper as residue and the clear liquid obtained is the filtrate. For example, a mixture of chalk and water is separated by filtration.

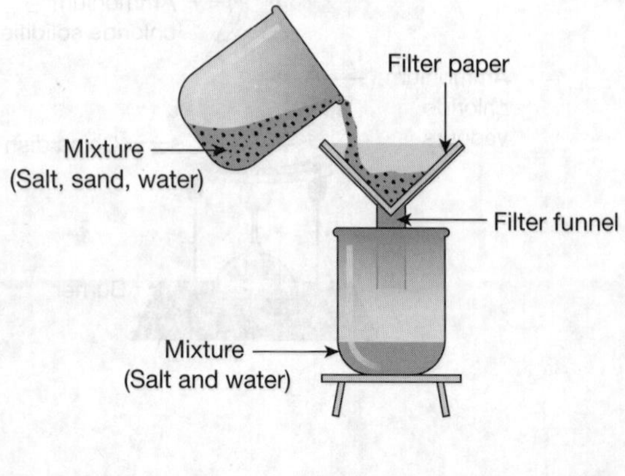

- **Sedimentation or decantation:** This method is applicable for a mixture having one solid and one liquid component and the solid must be insoluble in the liquid. In the sedimentation process, heavier components of the mixture settle at the bottom due to gravity. For example, settling of mud particles in water. Decantation follows sedimentation and this process involves pouring the clear, upper liquid out of the container, without disturbing the sediment. For example, after the tea leaves have settled down, the clear liquid tea from the top can be poured into a cup. Hence the transfer of clear tea is a case of decantation.

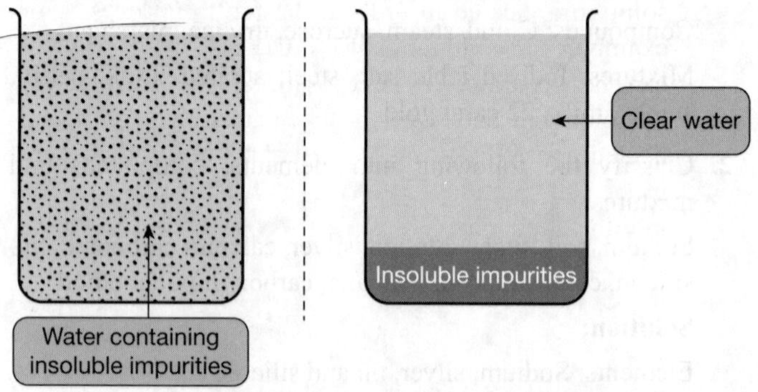

- **Separation by centrifugation:** the suspended particles of a substance in a liquid can be very rapidly separated by the centrifugation process with the help of a machine called a centrifuge. Here, the mixture of fine suspended particles in a liquid is taken in a test tube and the test tube is placed in a centrifuge machine and rotated rapidly for some time. As the mixture rotates around rapidly, a centrifugal force acts on the heavier suspended particles and brings them down to the bottom of the test tube. The clear liquid, being lighter, remains on top.

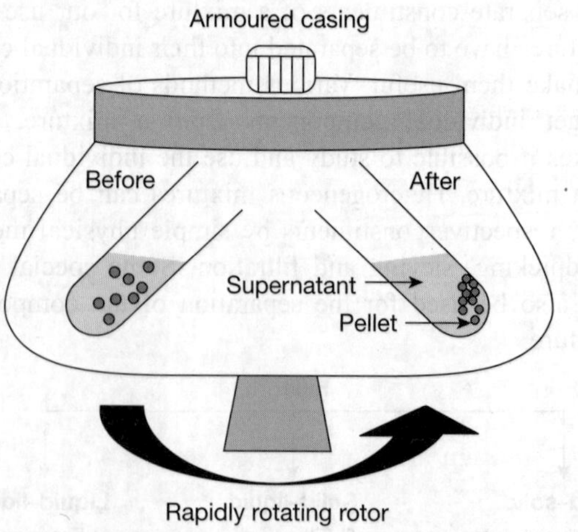

For example, the clay particles suspended in water can be very rapidly separated by centrifugation; the clay particles settle down at the bottom of the test tube and clear water remains at the top.

Uses: Centrifugation can be used in diagnostic laboratories for blood and urine tests, in dairies and at home to separate butter from cream.

A mixture of common salt and water cannot be separated by filtration or centrifugation as common salt is completely dissolved in water.

- **Separation by evaporation:** The process of changing of a liquid into vapour is called evaporation. It is used to separate a solid substance that has been dissolved in water or any other liquid solvent. The dissolved substance is left as a solid residue when all the water or liquid has been evaporated. This method is based on the fact that liquids vaporise easily whereas solids do not vaporise easily. Evaporation of a liquid can take place slowly even at room temperature, but the solution is heated in order to hasten the process. For example, (a) common salt (NaCl) dissolved in water can be separated by evaporation. The solution of common salt and water is taken in a china dish and heated gently by using a burner so the water present in the salt solution will form water vapour and escape into the atmosphere. When all the water present in the solution gets evaporated, common salt is left behind in the china dish as a white solid. The process of evaporation is used on a large scale to obtain common salt from sea water.
(b) Ink is a mixture of a dye in water and the dye can be separated from water by evaporation.

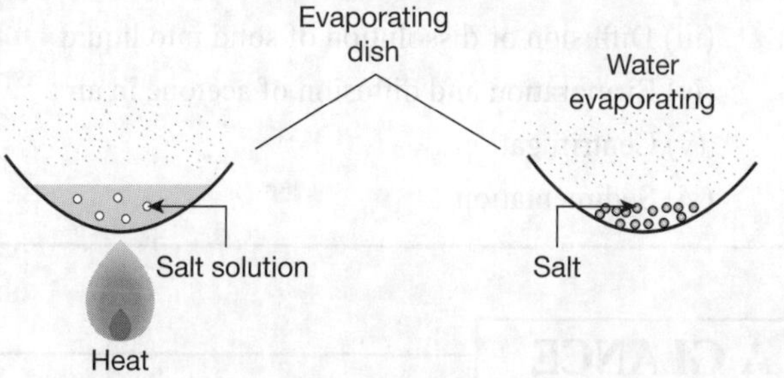

- **Purification by crystallisation:** The process of cooling a hot, concentrated solution of a substance to obtain crystals is known as crystallisation. It can be used for obtaining a pure solid substance from an impure sample. The process of crystallisation takes place as follows.
  - The impure solid substance is dissolved in a minimum amount of water to form a solution.
  - The solution is filtered to remove insoluble impurities.
  - The clear solution is heated gently on a water bath till a concentrated or saturated solution is obtained. It can be tested by dipping a glass rod in hot solution again and again. When small crystals form on the glass rod, the solution is saturated.
  - Now allow the hot, saturated solution to cool down slowly.
  - Crystals of pure solid are formed while impurities remain dissolved in solution.
  - Now separate the crystals of pure solid by filtration and dry them.

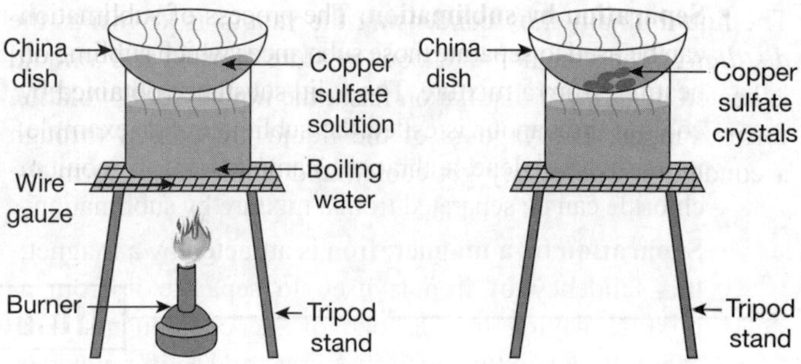

- **Separation by chromatography:** Chromatography is a technique of separating two or more dissolved solids which are present in a solution in extremely small quantities or those that cannot be separated by other methods. By using paper chromatography, we can separate two or more different substances present in the same solution. It is based on the fact that though two or more substances are soluble in the same solvent, their solubilities may be different. For example, (a) black ink is a mixture of several coloured substances which can be separated by paper chromatography. (b) To separate colours in a dye, pigments from natural colours, drugs from blood, chromatography is used.

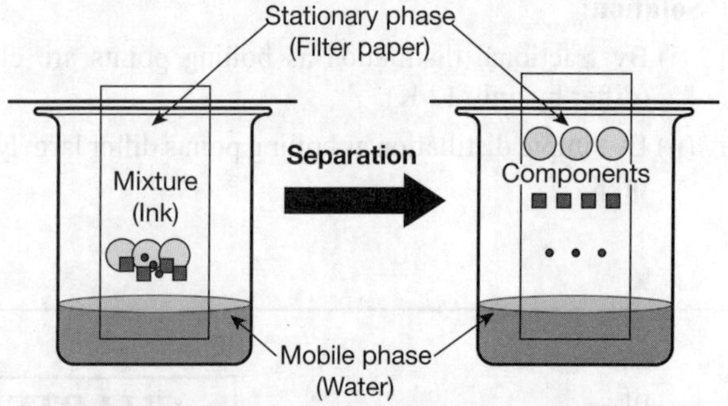

- **Separation by distillation:** Distillation is the process of heating a liquid to form vapour and then cooling the vapour to get back liquid. Distillation can be shown as:

$$\text{Liquid} \underset{\text{Cooling}}{\overset{\text{Heating}}{\rightleftharpoons}} \text{Vapour (or gas)}$$

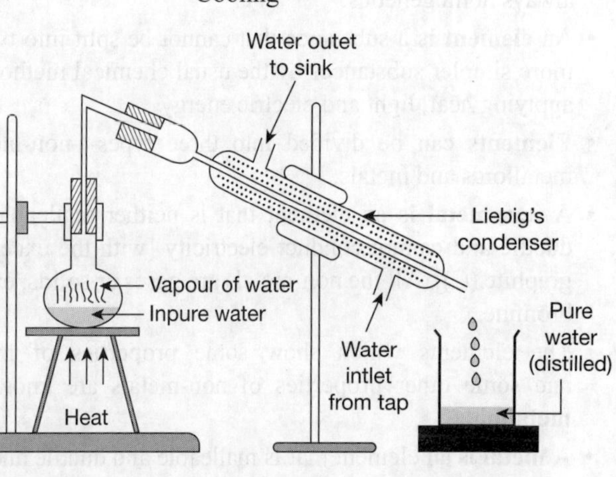

The liquid obtained by condensing the vapour is known as the *distillate*. When the homogeneous mixture of solid and a liquid is heated in a closed distillation flask, the liquid, being volatile, forms vapour. The vapours of the liquid are passed through a condenser where they get cooled and condense to form the pure liquid. This pure liquid is collected in a separate vessel. The solid, being non-volatile, remains behind in the distillation flask. For example, salt solution can be separated into salt and water, benzene and toluene can be separated, chlorobenzene and bromobenzene can be separated.

Illustrations

1. Name the process used to separate the given mixture.

(i) A mixture of methanol (338 K) and acetone (329 K)

(ii) Camphor having traces of common salt

(iii) Kerosene having water

Solution:

(i) By fractional distillation as boiling points are close

(ii) Sublimation

(iii) Solvent extraction using a separating funnel

2. Name the process used to separate the given mixture.

(i) A mixture of liquid X (366 K) and liquid Y (353 K).

(ii) A mixture of liquid P (355 K) and liquid Q (415 K).

Solution:

(i) By fractional distillation as boiling points are close (differ by only 12 K).

(ii) By simple distillation as boiling points differ largely by 60 K.

3. Name the process associated with the following:

(i) Dry ice is kept at room temperature and at 1 atm pressure.

(ii) A drop of ink placed on the surface of water contained in a glass spreads throughout the water.

(iii) A potassium permanganate crystal is in a beaker and water is poured into the beaker with stirring.

(iv) An acetone bottle is left open and the bottle becomes empty.

(v) Milk is churned to separate cream from it.

(vi) Settling of sand when a mixture of sand and water is left undisturbed for some time.

Solution:

(i) Sublimation of dry ice (solid) to CO_2 (gas)

(ii) Diffusion of ink into water

(iii) Diffusion or dissolution of solid into liquid

(iv) Evaporation and diffusion of acetone in air

(v) Centrifugation

(vi) Sedimentation

CHAPTER AT A GLANCE

- A **substance** is a type of matter that cannot be separated into other kinds of matter by any physical process. It is a pure form of matter.
- A **pure substance** comprises a single type of particle and is always homogeneous.
- An **element** is a substance that cannot be split into two or more simpler substances by the usual chemical methods of applying heat, light and electric energy.
- **Elements** can be divided into three types—non-metals, metalloids and metals.
- A **non-metal** is an element that is neither malleable nor ductile and cannot conduct electricity [with the exception graphite (C)]. All the non-metals are gases or solids, except bromine.
- The elements which show some properties of metals and some other properties of non-metals are known as **metalloids**.
- A **metal** is an element that is malleable and ductile and can conduct electricity. All the metals are solids except mercury, which is a liquid.
- A **compound** is a substance made up of two or more elements chemically combined in a fixed proportion by mass. It is formed as a result of chemical reaction/s, between the constituent elements.
- **Mixtures** are constituted by more than one kind of pure form of matter. That is, a mixture is a substance which consists of two or more elements or compounds not chemically combined together.
- Those mixtures in which the substances are completely mixed, are indistinguishable from one another and form a single phase are known as **homogeneous mixtures** (solutions). These can be of three types—solid–liquid mixture (mixture of sugar in water), solid–solid mixture (alloys) and liquid–liquid mixture.
- The mixtures in which all the substances remain separate and one substance is spread throughout the other substance as small particles, droplets or bubbles are known as **heterogeneous mixtures.**
- **Separation of mixtures:** Separation makes it possible to study and use the individual components of a mixture.

Mixture

Solid–solid	Solid–liquid	Liquid–liquid
(i) Solvent extraction	(i) Filtration	(i) Fractional distillation
(ii) Sublimation	(ii) Sedimentation	(ii) Separating funnel
(iii) Magnetic separation	(iii) Centrifugation	
(iv) Winnowing	(iv) Evaporation	
	(v) Crystallisation	
	(vi) Chromatography	
	(vii) Distillation	

- The process of **sublimation** can be used to separate those substances which sublime on heating from a mixture.
- A mixture of iron filings and sulfur powder can be separated by using a **magnet** as iron filings are attracted by a magnet while sulfur is not.
- **Winnowing** is used to separate lighter particles from heavier particles by blowing air/by the force of wind.
- The process of removing insoluble solids from a liquid using a filter paper is known as **filtration**.
- In the **sedimentation** process, heavier components of the mixture settle at the bottom due to gravity.
- **Decantation** follows sedimentation and it involves pouring clear, upper liquid out of the container, without disturbing the sediment.
- the suspended particles of a substance in a liquid can be very rapidly separated by using the centrifugation process with the help of a machine called a **centrifuge.**
- **Evaporation** is used to separate a solid substance that has been dissolved in water or any other liquid solvent.
- The process of cooling a hot, concentrated solution of a substance to obtain crystals is called **crystallisation**.
- To separate colours in a dye, pigments from natural colours, drugs from blood, **chromatography** is used.
- **Distillation** is the process of heating a liquid to form vapour and then cooling the vapour to get back liquid.

PRACTICE QUESTIONS

Analyse Your Concepts (School Exam Based)

Fill in the Blanks

Instructions: Complete the following statements with an appropriate word/term to be filled in the blank spaces.

1. ____________ is a pure substance.
2. An element is made up of only one kind of ____________
3. ____________ is a metal which exists in the liquid state.
4. Composition of a mixture is ____________.
5. Gases can be separated from air by ____________ method.
6. Brass is a mixture of ____________ and ____________.
7. Miscible liquids are separated by____________.
8. Immiscible liquids are separated by using a____________.
9. A heterogeneous mixture of a liquid and a solid is conveniently separated by____________.
10. If a mixture contains iron filings as one of the constituents, it can be separated by using a ____________.
11. Sublimation of ammonium chloride is a ____________ change.
12. A mixture of two or more miscible liquids, for which the difference in the boiling points is less than 25 K can be separated by a process called ____________.

True or False

Instructions: Read the following statements and write your answer as true or false.

1. Mercury, diamond and bromine are elements.
2. Sugar is a compound which contains the elements carbon, hydrogen, nitrogen and oxygen.
3. Wood is a mixture.
4. Fractional distillation is used when the substances differ substantially in their boiling points.
5. The separation of liquids by fractional distillation is based on the difference in their boiling points.
6. Solubility of a gas is directly proportional to temperature.
7. Chalk is a compound and gold is an element.
8. Gun powder is a compound.
9. Blood is a homogeneous mixture.
10. Mercury is a liquid metal.
11. Filtered tea is a heterogenous mixture.
12. Pt and Au are noble metals.

Match the Following

Instructions: Each question contains statements given in two columns which have to be matched. Statements (a, b, c, d) in column I have to be matched with statements (p, q, r, s) in column II.

1.

Column I	Column II
(a) Hydrargyrum	(p) Mixture
(b) Soap	(q) Pure substance
(c) Ozone	(r) Mercury
(d) Calcium	(s) Homonuclear molecule
(e) Urine	(t) Compound

Very Short Answer Type Questions

1. What is meant by a pure substance?
2. Name a method to check the purity of a liquid.
3. Give one example each of a homogeneous and a heterogeneous mixture.
4. Which of the two will scatter light, soap solution or sugar solution?
5. Name the process by which the coloured components can be obtained from blue ink.
6. Name the process used to separate a mixture of salt and ammonium chloride.
7. When we heat iron filings and sulfur till red hot, do we get a compound or a mixture?
8. Classify the following into elements and compounds:
   (i) H_2O (ii) He (iii) Cl_2 (iv) CO_2
   (v) Co (vi) Ar (vii) NH_3
9. Name the apparatus used to separate oil from water.
10. Name the process which is used in milk dairies to separate cream from milk.
11. 'Sea water' can be classified as a homogeneous as well as a heterogeneous mixture. Comment.
12. Can we separate alcohol dissolved in water by using a separating funnel?
    If yes, then describe the procedure. If not, explain.

Short Answer Type Questions

1. Try segregating the things around you as pure substances or mixtures. (a) Wood (b) Coal (c) Milk (d) Sugar (e) Common salt (f) Soap (g) Soil (h) Rubber.
2. Which of the following materials fall in the category of a 'pure substance'?
   (a) Ice (b) Milk (c) Iron (d) Hydrochloric acid (e) Calcium oxide (f) Mercury (g) Brick (h) Wood (i) Air.
3. Classify the substances given in the figure below into elements and compounds.

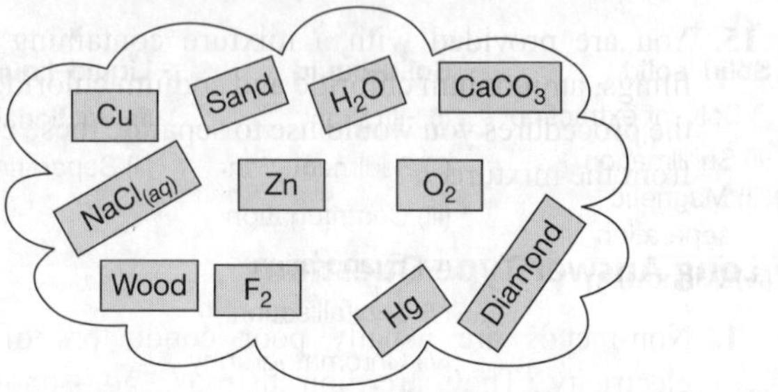

4. Give two examples for each of the following:
   (a) metals (b) non-metals
   (c) metalloids (d) inert gases
5. List the points of difference between homogeneous and heterogeneous mixtures.
6. Write the chemical name of the following and also give their molecular formulae:
   (a) baking soda (b) vinegar
   (c) marble (d) sand
7. Sucrose (sugar) crystals obtained from sugarcane and beetroot are mixed together. Will it be a pure substance or a mixture? Give reasons for the same.
8. While diluting a solution of salt in water, a student by mistake added acetone (boiling point 56°C). What technique can be employed to get back the acetone? Justify your choice.
9. Name: (a) a soft metal, (b) a metal which is brittle, (c) a non-metal which is lustrous, (d) a liquid metal, (e) a metal which is a poor conductor of electricity, (f) a non-metal which is a good conductor of electricity, (g) a liquid non-metal, (h) the hardest naturally occurring substance and (i) an inert gas.
10. Classify each of the following as a homogeneous or a heterogeneous mixture. Soda water, wood, air, soil, vinegar, filtered tea.
11. What types of mixtures are separated by the technique of crystallisation?
12. How will you separate a mixture containing kerosene and petrol (difference in their boiling points is more than 25°C) which are miscible with each other?
13. Name the technique used to separate
    (a) butter from curd (b) salt from sea water
    (c) camphor from salt
14. Give an example each for the mixture having the following characteristics. Suggest a suitable method to separate the components of these mixtures.
    (a) A volatile and a non-volatile component.
    (b) Two volatile components with an appreciable difference in boiling points.
    (c) Two immiscible liquids.

(d) One of the components changes directly from solid to gaseous state.

15. You are provided with a mixture containing sand, iron filings, ammonium chloride and sodium chloride. Describe the procedures you would use to separate these constituents from the mixture.

Long Answer Type Questions

1. Non-metals are usually poor conductors of heat and electricity. They are non-lustrous, non-sonorous, non-malleable and are coloured.
   (a) Name a lustrous non-metal.
   (b) Name a non-metal which exists as a liquid at room temperature.
   (c) The allotropic form of a non-metal is a good conductor of electricity. Name the allotrope.
   (d) Name a non-metal which is known to form the largest number of compounds.
   (e) Name a non-metal other than carbon which shows allotropy.
   (f) Name a non-metal which is required for combustion.
2. Which separation techniques will you apply for the separation of the following?
   (a) Sodium chloride from its aqueous solution.
   (b) Ammonium chloride from a mixture containing sodium chloride and ammonium chloride.
   (c) Small pieces of metal in the engine oil of a car.
   (d) Different pigments from an extract of flower petals.
   (e) Butter from curd.
   (f) Oil from water.
   (g) Tea leaves from tea.
   (h) Iron pins from sand.
   (i) Wheat grains from husk.
   (j) Fine mud particles suspended in water.
3. Fractional distillation is suitable for separation of miscible liquids with a boiling point difference of about 25 K or less. What part of fractional distillation apparatus makes it efficient and possess an advantage over the simple distillation process? Explain using a diagram.
4. Explain the following terms used in the separation of mixtures.
   (a) filtration
   (b) crystallisation
   (c) evaporation
   (d) sublimation
5. Suggest separation technique(s) one would need to employ to separate the following mixtures.
   (a) mercury and water
   (b) potassium chloride and ammonium chloride
   (c) common salt, water and sand
   (d) kerosene, water and salt

COMPETITION WINDOW (OBJECTIVE TYPE)

[For NEET, JEE (Main and Advanced), NTSE, KVPY and Olympiads]

Topic-wise MCQs

Substances

1. Antimony and polonium can be classified as
   (a) metalloids (b) non-metals
   (c) metals (d) cannot say
2. The most malleable metal here is
   (a) copper (b) iron
   (c) gold (d) magnesium
3. Which of the following is a compound?
   (a) diamond
   (b) stainless steel
   (c) bronze
   (d) hydrogen chloride
4. In which of the following may the constituents be present in any ratio?
   (a) solution
   (b) colloid
   (c) mixture
   (d) compound
5. Two substances P and Q when brought together form a substance R with the evolution of heat. The properties of R are entirely different from those of P and Q. The substance R is
   (a) a mixture (b) an element
   (c) a compound (d) any of these
6. Which one of the following will result in the formation of a mixture?
   (a) adding sodium metal to water
   (b) crushing of a marble tile into small particles
   (c) breaking of ice cubes into small pieces
   (d) adding milk in water
7. The chemical symbol P stands for
   (a) polonium (b) promethium
   (c) phosphorus (d) potassium
8. The purity of a solid substance can be checked by its
   (a) solubility in water
   (b) melting point
   (c) solubility in alcohol
   (d) boiling point

9. A sample has two substances with uniform properties. The sample contains
(a) an element
(b) a homogenous mixture
(c) a compound
(d) a heterogeneous mixture

10. Which of the following is considered to be a pure substance?
(a) muddy water (b) milk of magnesia
(c) granite (d) sodium bromide

11. Compounds
(a) can be separated by their physical properties
(b) contain only one type of element
(c) are the same as mixtures
(d) are different kinds of atoms chemically combined with each other

12. The elements present in ethane are
(a) H, O (b) O, P
(c) C, H (d) C, O

13. The element that has a symbol with two letters is
(a) boron (b) carbon
(c) uranium (d) tin

Homogeneous and heterogeneous mixtures, solutions

14. Which of the following is a case of a homogeneous substance?
(a) M and M candy (b) muddy water
(c) granite (d) calcium carbonate

15. Which of the following is an example of a heterogeneous substance?
(a) pieces of iron (b) candle
(c) bottled water (d) NaI

16. Assume air contains only nitrogen, oxygen and inert gases. Then air will be a
(a) heterogeneous mixture
(b) colloid
(c) compound
(d) homogeneous mixture

17. Which of the following is example of a homogenous mixture?
(a) alloy (b) blood
(c) milk (d) all of these

18. If addition of a solid in a liquid leads to a homogeneous mixture, then the solid will be
(a) partially soluble (b) insoluble
(c) soluble (d) all of these

19. Tincture of iodine has antiseptic properties. This solution is made by dissolving
(a) iodine in water
(b) iodine in potassium iodide
(c) iodine in alcohol
(d) iodine in vaseline

Methods of separation

20. Salt can be obtained from sea water by
(a) sublimation (b) filtration
(c) evaporation (d) decantation

21. A mixture of $ZnCl_2$ and $PbCl_2$ can be separated by
(a) sublimation (b) fractional distillation
(c) distillation (d) crystallisation

22. Camphor can be purified by
(a) sedimentation (b) sublimation
(c) distillation (d) filtration

23. The process that can used to separate oil and water is
(a) by using a separating funnel
(b) chromatography
(c) distillation
(d) sublimation

24. Filtration can be used to separate
(a) liquids from liquids
(b) liquids from gases
(c) solids from solids
(d) liquids from solids

25. Which of the following is based on difference in size?
(a) fractional distillation
(b) sublimation
(c) filtration
(d) solvent extraction

26. After evaporation, the residue left is
(a) soluble solid (b) solvent
(c) both (a) and (b) (d) none of these

27. A mixture of water and C_2H_5OH can be separated to a certain extent by
(a) fractional distillation (b) centrifugation
(c) sublimation (d) evaporation

28. Fractional distillation is based on the difference in the ________ of two liquids.
(a) solubility (b) melting point
(c) boiling point (d) none of these

29. A mixture of sand and sulfur may best be separated by
(a) fractional distillation
(b) dissolving in CS_2 and filtering
(c) fractional crystallisation from an aqueous solution
(d) magnetic method

30. Distillation is a good separation technique for
(a) solid alloys (b) gases
(c) solids (d) liquids

31. Magnetism can be beneficial for separating
(a) non-metallic solids and solids such as sulfur
(b) non-magnetic solids from non-magnetic liquids
(c) gases and non-metallic liquids
(d) magnetic solids and solids like sulfur

Miscellaneous

1. Which of the following sets represents correct statements given here?
   (i) Metals are ductile and malleable.
   (ii) Graphite can show electrical conductance.
   (iii) Non-metals show conductance.
   (iv) Metals are shiny while non-metals are dull (mostly).
   (a) (i), (iv) (b) (i), (ii), (iv)
   (c) (ii), (iii), (iv) (d) (i), (ii), (iii), (iv)

2. Which of the following sets of element contains only non-metals?
   (a) H, C, Hg, Ge (b) Fe, Co, Ni, P
   (c) S, Cl, Ar, Br (d) Al, Ga, Si, P

3. Rusting of an article made up of iron is called
   (a) corrosion and it is a physical as well as chemical change
   (b) dissolution and it is a physical change
   (c) corrosion and it is a chemical change
   (d) dissolution and it is a chemical change

4. The most common solvent on earth is ________
   (a) turpentine (b) water
   (c) petrol (d) gasoline

5. Which of the following statements is correct?
   (a) A homogeneous mixture must be uniform.
   (b) A heterogeneous mixture must contain at least three elements.
   (c) A mixture containing two compounds must be heterogeneous.
   (d) A pure substance must contain only one type of atom.

6. Tincture of iodine has antiseptic properties. This solution is made by dissolving
   (a) iodine in potassium iodide
   (b) iodine in vaseline
   (c) iodine in water
   (d) iodine in alcohol

7. A mixture of common salt, sulfur, sand and iron filings is shaken with carbon disulfide and filtered through a filter paper. The filtrate is evaporated to dryness in a china dish. What will be left in the dish after evaporation?
   (a) common salt (b) sulfur
   (c) sand (d) iron filings

8. A mixture of ethanol and water can be separated by
   (a) fractional distillation (b) sublimation
   (c) filtration (d) decantation

9. A mixture of methyl alcohol and acetone can be separated by
   (a) steam distillation
   (b) fractional distillation
   (c) distillation under reduced pressure
   (d) distillation

10. Melting points can separate materials because
    (a) heat causes molecules to disintegrate
    (b) many substances fuse at the melting point
    (c) substances melt at different temperatures
    (d) molecules vibrate rapidly when heated

11. Which of the following methods can be used for the separation of a liquid–liquid mixture?
    (a) separating funnel
    (b) solvent extraction
    (c) fractional distillation
    (d) both (a) and (c)

12. Which of the following sets of methods can be used for a mixture of a solid and solid type?
    (a) sublimation, filtration
    (b) evaporation, fractional distillation
    (c) magnetic separation, solvent extraction
    (d) sublimation, evaporation

13. Water droplets on the cover of the tea container represents.
    (a) fusion (b) sublimation
    (c) condensation (d) evaporation

14. Steam over tea represents the phenomenon of
    (a) fusion (b) boiling
    (c) evaporation (d) none of these

15. What type of mixtures are separated by crystallisation?
    (a) a mixture in which both the components are soluble in a solvent
    (b) a mixture in which both the components are insoluble in water
    (c) a mixture in which one component is soluble in a solvent
    (d) a mixture in which impurities are soluble in a solvent

Advanced and Olympiads

Single Choice

1. Which of the following statements are true for pure substances?
   (i) Pure substances contain only one kind of particle.
   (ii) Pure substances may be compounds or mixtures.
   (iii) Pure substances have the same composition throughout.
   (iv) Pure substances can be exemplified by all elements other than nickel.
   (a) (i) and (ii) (b) (i) and (iii)
   (c) (iii) and (iv) (d) (ii) and (iii)

2. Which of the following are homogeneous in nature?
   (i) ice (ii) wood
   (iii) soil (iv) air
   (a) (i) and (iii) (b) (ii) and (iv)
   (c) (i) and (iv) (d) (iii) and (iv)

3. Two chemical species X and Y combine together to form a product P which contains both X and Y
$X + Y \rightarrow P$
X and Y cannot be broken down into simpler substances by simple chemical reactions. Which of the following concerning the species X, Y and P are correct?
(i) P is a compound.
(ii) X and Y are compounds.
(iii) X and Y are elements.
(iv) P has a fixed composition.
(a) (i), (ii) and (iii)
(b) (i), (ii) and (iv)
(c) (ii), (iii) and (iv)
(d) (i), (iii) and (iv)

4. Two substances, A and B were made to react to form a third substance, A_2B according to the following reaction: $2A + B \rightarrow A_2B$
Which of the following statements concerning this reaction are incorrect?
(i) The product A_2B shows the properties of substances A and B.
(ii) The product will always have a fixed composition.
(iii) The product so formed cannot be classified as a compound.
(iv) The product so formed is an element.
(a) (i), (ii) and (iii)
(b) (ii), (iii) and (iv)
(c) (i), (iii) and (iv)
(d) (iii) and (iv)

5. In the separation of red ink from blue ink, a technique is used which is described below in the steps involved in the process. Which of the following set represents the proper sequence?
(i) The blue and red ink form spots at a certain distance on the paper.
(ii) Paper and solvent are taken as stationary and mobile phases, respectively.
(iii) Paper chromatography is used for this separation.
(iv) The paper is suspended in the closed jar by using a hook.
(v) A narrow strip of paper with a line drawn is cut and a mixture of red and blue ink is placed on the line marked on the paper, with the help of capillary.
(a) (ii) (iv) (iii) (v) (i)
(b) (iii) (ii) (v) (iv) (i)
(c) (ii) (iii) (iv) (v) (i)
(d) (iii) (ii) (iv) (v) (i)

Multiple Choice

6. Which of the following is an example of a homogeneous substance?
(a) Russian salad dressing
(b) aspirin
(c) sodium nitrate
(d) mixture of sand and rice grains

7. Which of the following is an example(s) of alloy?
(a) table salt
(b) gun metal
(c) bronze
(d) brass

8. Which of the following are compound(s)?
(a) calcium sulfate
(b) sugar
(c) diamond
(d) common salt

9. Which of the following is/are true for a compound?
(a) The constituents of a compound cannot be separated by simple physical methods.
(b) It is heterogenous in nature.
(c) Properties of a compound are entirely different from those of the elements present in it.
(d) A compound contains different elements in a fixed ratio.

10. Which of the following is/are monovalent?
(a) cesium (b) aluminium
(c) potassium (d) sodium

11. Which of the following is/are a mixture(s)?
(a) metal amalgam
(b) solution
(c) alloy
(d) phosphine

12. Which of the following are used for the separation of solid–solid mixtures?
(a) chromatography
(b) winnowing
(c) sublimation
(d) solvent extraction

13. What is correct on the basis of the phenomenon of evaporation?
(a) evaporation is a surface phenomenon
(b) it is favoured by a decrease in temperature
(c) it causes a cooling effect
(d) the rate of evaporation increases with an increase in the temperature

Comprehension Type

Comprehension I: An element is a substance which cannot be split into simpler substances by using a chemical method. An element can be a non-metal, a metalloid or a metal. Most of the elements are metals and are solid in nature, with a few liquid elements also. Metals and non-metals differ greatly in physical properties like hardness, density, melting point and boiling point.

14. Most of the elements are present mainly as
(a) non-metals in the liquid form
(b) non-metals in the solid state
(c) metals in the solid form
(d) metals in the liquid state

15. Which of the following set of elements contain liquid elements only?
(a) Cl, Br, Hg (b) Br, Hg, Ga
(c) Br, Hg, He (d) Br, Hg, Al

16. From the given elements select the elements with highest and lowest melting points—C, Fe, W, Hg, He, Na, Ca, Al
(a) W, C (b) Hg, W
(c) W, Hg (d) C, He

Matrix Matching

17. Match the following

Column I	Column II
(a) boron	(p) metal
(b) carbon	(q) non-metal
(c) potassium	(r) metalloid
(d) chlorine	(s) solid
	(t) gas

18. Match the following

Column I	Column II
(a) iron filings and common salt	(p) distillation
(b) salt and water	(q) magnetic separation
(c) benzene and toluene	(r) evaporation
(d) rice grains and alcohol	(s) filtration

Integer Type

19. How many of the following elements are liquid in nature?
F, Cl, Br, I, Hg, Ga, Fr, Cs, Mg, Ag

20. How many of these elements are metalloid elements?
As, Cl, Se, Sb, Al, B, Si, Hg, Ge, Al, Fe

21. How many of the given substances are mixtures?
Cloud, iodised table salt, ice, glucose, brass, aerated drink, dry ice, milk, 22 carat gold, marsh gas, diesel, smoke

22. How many of the following compounds can undergo sublimation?
Naphthalene, iodine, magnesium hydroxide, camphor, H_2SO_3, ammonium chloride, anthracene, CO_2

23. Which of the following are not compounds?
(a) chlorine gas (b) potassium iodide
(c) iron (d) iron sulfide
(e) aluminium (f) iodine
(g) carbon (h) carbon dioxide
(i) sulfur powder (j) methane

24. How many of the given methods can be used for the separation of a mixture of a solid and a liquid?
Filtration, sedimentation, separating funnel, evaporation, distillation, solvent extraction, centrifugation, chromatography, fractional distillation

Answer Keys

Fill in the Blanks

1. element
2. atoms
3. mercury
4. variable
5. fractional distillation
6. zinc, copper
7. fractional distillation
8. separating funnel
9. filtration
10. magnet
11. physical
12. fractional distillation

True or False

1. T	2. F	3. T	4. F	5. T
6. F	7. T	8. F	9. T	10. T
11. F	12. T			

Match the Following

1. (a)—(r); (b)—(t); (c)—(s); (d)—(q); (e)—(p)

Very Short Answer Type Questions

1. A substance having a single type of particle is known as a pure substance. For example, hydrogen and water are pure.
2. The purity of a liquid can be checked using the determination of boiling point method.
3. Homogeneous mixture is salt in water and heterogeneous mixture is salt and sulfur.
4. Soap solution can scatter light.
5. The coloured components can be obtained from blue ink by evaporation.
6. Sublimation is used to separate a mixture of salt and ammonium chloride.
7. When we heat iron filings and sulfur till red hot, we get a compound.

8. Elements are He, Ar, Cl_2 and Co; and compounds are H_2O, CO_2 and NH_3.

9. Separating funnel is used to separate oil from water.

10. Centrifugation is used in milk dairies to separate cream from milk.

11. Sea water can be classified as a homogeneous mixture because it contains salts dissolved in water. It can be classified as a heterogeneous mixture also since it contains mud, sand and decayed parts of plants.

12. No, a mixture of water and alcohol cannot be separated since both are miscible and they form a solution. Only immiscible liquids can be separated.

Short Answer Type Questions

1. (a) mixture (b) mixture (c) mixture (d) pure substance (e) pure substance (f) compound/mixture (g) mixture (h) pure substance

2. Ice, iron, calcium oxide, mercury are all pure substances as they have definite composition. Milk is a colloid, so it is a heterogeneous mixture. Hydrochloric acid is also a mixture of hydrogen chloride gas and water.

3. Elements: Cu, Zn, F_2, O_2, diamond (C), Hg
Compounds: $CaCO_3$, H_2O

4. (a) iron, silver (b) carbon, oxygen (c) antimony, silicon (d) helium, argon

5.

Homogeneous mixture	Heterogeneous mixture
1. Its constituent particles cannot be seen easily. 2. There are no visible boundaries of separation in a homogeneous mixture. 3. Its constituents cannot be easily separated. Examples: Alloys, solution of salt in water.	1. Its constituent particles can be seen easily. 2. Have visible boundaries of separation between the constituents. 3. Its constituents can be separated by simple methods. Examples: Mixture of sand and common salt, mixture of sand and water.

6. (a) Sodium bicarbonate (baking soda): $NaHCO_3$
(b) Acetic acid (vinegar): CH_3COOH
(c) Calcium carbonate (marble): $CaCO_3$
(d) Silicon dioxide (sand): SiO_2

7. The composition of the sugar (sucrose) will remain constant irrespective of the source of its preparation. Hence sugar or sucrose is a pure substance with fixed composition.

8. The mixture of acetone and salt solution in water can be separated by distillation since a difference in their boiling points is more than 25°C. Acetone will evaporate and get condensed first leaving behind the salt solution.

9. (a) gold (b) zinc (c) iodine (d) mercury (e) tungsten (f) graphite (g) bromine (h) diamond (i) neon or helium

10. Homogeneous mixtures: air, soda water, vinegar, filtered tea.
Heterogeneous mixtures: wood, soil.

11. Crystallisation method can be used for the purification of the mixtures that contain insoluble and/or soluble impurities. Have crystalline nature. Cannot be separated by filtration as some impurities are soluble.

12. Simple distillation can separate the mixture of kerosene and petrol (their boiling points differ by more than 25°C).
Method: In a distillation flask, a mixture of kerosene and petrol is taken as shown in figure.

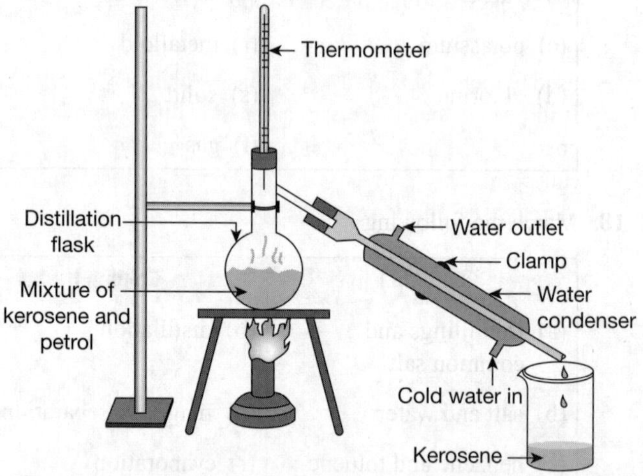

The mixture is heated slowly and the temperature is recorded with the help of thermometer. Petrol (bp = 70°C to 120°C) vaporises first and the temperature becomes constant for some time (till all petrol evaporates from the mixture). Vapours of petrol are condensed and collected in another container while the kerosene remains in the distillation flask. As soon as the temperature starts rising again, the heating is stopped and both the components are collected separately.

13. (a) By using the centrifugation method, butter can be separated from curd.
(b) By using the evaporation method, salt from sea water can be separated. Water vaporises on evaporation leaving behind the salt.
(c) Camphor from salt can be separated by the sublimation method. On subliming, camphor will be converted into vapour leaving behind the salt.

14. (a) Evaporation or distillation: Common salt solution
(b) Distillation: Acetone–water mixture
(c) Separation using separating funnel: Oil–water mixture
(d) Sublimation: Mixture of common salt and ammonium chloride

15. Main mixture: Sand + Fe + NH_4Cl + NaCl
Step I: Separate iron filings using a magnet.
Step II: Separate NH_4Cl by sublimation.
Step III: Add water, stir and filter to separate sand.
Step IV: Evaporate filtrate to get salt (NaCl).

Long Answer Type Questions

1. (a) Iodine (b) Bromine
(c) Graphite (d) Carbon
(e) Sulphur, phosphorus (f) Oxygen

2. (a) Evaporation (b) Sublimation
(c) Filtration (d) Chromatography
(e) Centrifugal machine or churning the curd by hand
(f) Decantation (g) Filtration
(h) Magnetic separation (i) Winnowing
(j) Coagulation and decantation

3. In fractional distillation, a fractionating column is used which is packed with glass beads or small plates. It increases the surface area for the vapours and they lose energy when they come in contact with beads or plates and can be quickly condensed. The length of the column will increase the efficiency of the process.

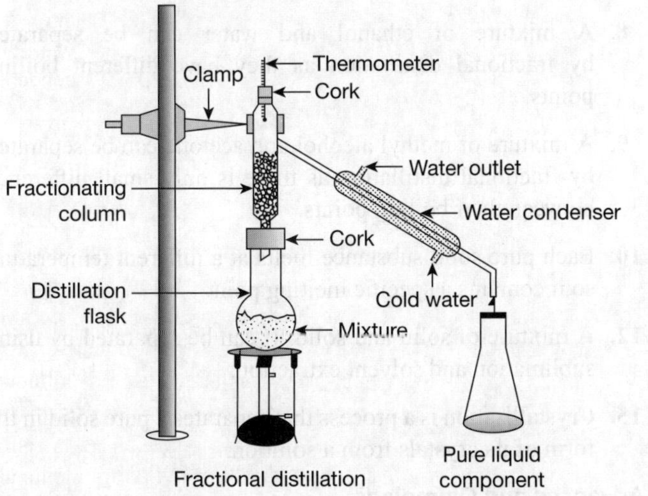

Fractional distillation

4. See text part.

5. (a) Mercury and water: Since there is a difference in the densities of mercury and water and both liquids are insoluble in each other, they can be separated using a separating funnel.
(b) Potassium chloride and ammonium chloride can be separated by sublimation. Ammonium chloride being volatile will be converted into vapour. KCl does not sublime.
(c) Filtration will separate the salt solution (salt and water) from the sand. Evaporation of the salt solution will separate the salt from the water.
(d) A separating funnel can be used to separate kerosene and salt solution. Evaporation of salt solution will separate salt and water.

Competition Window (Objective Type)

Topic-wise MCQs

1. (a)	2. (c)	3. (d)	4. (c)	5. (c)
6. (d)	7. (c)	8. (b)	9. (b)	10. (d)
11. (d)	12. (c)	13. (d)	14. (d)	15. (b)
16. (d)	17. (a)	18. (c)	19. (c)	20. (c)
21. (d)	22. (b)	23. (a)	24. (d)	25. (a)
26. (a)	27. (a)	28. (c)	29. (b)	30. (d)
31. (d)				

Miscellaneous

1. (b)	2. (c)	3. (c)	4. (b)	5. (a)
6. (d)	7. (b)	8. (a)	9. (b)	10. (c)
11. (d)	12. (c)	13. (c)	14. (b)	15. (c)

Advanced and Olympiads

Single Choice

1. (b)	2. (c)	3. (d)	4. (c)	5. (b)

Multiple Choice

6. (bc)	7. (bcd)	8. (abd)	9. (acd)	10. (acd)
11. (abc)	12. (bcd)	13. (acd)		

Comprehension Type

14. (c) **15.** (b) **16.** (d)

Matrix Matching

17. (a)—(rs); (b)—(qs); (c)—(ps); (d)—(qt)

18. (a)—(qs); (b)—(pr); (c)—(p); (d)—(s)

Integer Type

19. 5	20. 6	21. 7	22. 5	23. 6
24. 6				

Hints and Solutions

Competition Window (Objective Type)

Topic-wise MCQs

3. Hydrogen chloride (HCl) is a compound of hydrogen and chloride, stainless steel and bronze are alloys whereas diamond is the allotropic form of element carbon.
7. P stands for phosphorus.
8. Every solid has a fixed melting point so it can be used.
9. Homogeneous mixture is a solution having uniform composition and properties throughout.
10. Sodium iodide being a compound is a pure substance. Granite, muddy water and milk of magnesia all are mixtures.
12. The formula of ethane is C_2H_6. It contains the elements C, H.
13. The symbol of tin is Sn.
14. Calcium carbonate is a compound.
15. Candle is a heterogeneous mixture of wax and threads, iron is an element, bottled water and table salt (NaI) are compounds.
17. Alloys are homogenous mixture of metals.
18. Soluble solute leads to a homogenous mixture.
19. A solution of *tincture of iodine* is made by dissolving 2%–7% iodine in alcohol. Tincture solutions are characterised by the presence of alcohol since alcohol is a good solvent. Iodine does not dissolve in water. Tincture of iodine is used as an antiseptic.
20. Sea water is a solution of salt and water. During evaporation, water gets evaporated off and salt left as a residue.
22. Camphor being a substance that undergoes sublimation, can be purified by sublimation.
23. The mixture of oil and water can be separated using a separating funnel. Oil is less dense compared to water, so, it forms the upper layer.
25. Filtration is based on size difference.
26. Soluble solid is left behind while the solvent evaporates.
27. Both differ in their boiling point.
28. Due to difference in boiling points, each has a different tendency to form vapour.
29. Sulfur gets dissolved in CS_2 and separates out as a filtrate, while sand gets collected on the filter paper as residue.
30. Distillation is a separation technique used to separate miscible liquids having different boiling points.
31. Magnetism is useful for separation of magnetic and non-magnetic substances.

Miscellaneous

1. Non-metals do not show conductance; however, graphite shows conductance.
2. S, Cl, Ar, Br are non-metals.
3. Rusting of an article made up of iron is called corrosion. It is a chemical change because a new substance, hydrated iron oxide, called rust is formed.

   $$\text{Iron} + \text{oxygen} \xrightarrow{\text{Water}} \text{Hydrated iron oxide (rust)}$$

   $$4Fe + 3O_2 \rightarrow \underset{\text{(Rust)}}{2Fe_2O_3.xH_2O}$$

5. A homogeneous mixture must have uniform composition and properties.
6. Iodine dissolved in alcohol is known as tincture of iodine and has antiseptic properties.
7. Sulfur will be left behind. In the given mixture, only sulfur gets dissolved in carbon disulphide while the rest do not dissolve.
8. A mixture of ethanol and water can be separated by fractional distillation as they have different boiling points.
9. A mixture of methyl alcohol and acetone can be separated by fractional distillation as there is only small difference between their boiling points.
10. Each pure solid substance melts at a different temperature so it contains a specific melting point.
12. A mixture of solid and solid. It can be separated by using sublimation and solvent extraction.
15. Crystallisation is a process that separates a pure solid in the form of its crystals from a solution.

Advanced and Olympiads

Single Choice

1. Pure substances are made up of only one kind of particle and they have the same composition throughout. Mixtures are not pure substances. Only elements and compounds are pure substances.
2. Ice and air are homogeneous in nature since they have the same composition throughout and there are no visible boundaries between the components.
3. $X + Y \rightarrow P$

   X and Y are elements. Hence, they cannot be broken down into simpler substances. P is a compound; hence, it has a fixed composition.
4. $2A + B \rightarrow A_2B$

   The product A_2B is a new compound; hence, it does not show the properties of A and B. The product formed is a compound and not an element.

5. The correct sequence of steps is:
   (iii) Paper chromatography is used for the separation.
   (ii) Paper and solvent are taken as the stationary and mobile phases, respectively.
   (v) A narrow strip of paper with a line drawn is cut and a mixture of red and blue ink with the help of a capillary is placed on the line marked on the paper.
   (iv) The paper is suspended in the closed jar with the help of a hook.
   (i) The blue and red ink form spots at a certain distance on the paper.

Multiple Choice

7. Alloys are homogeneous mixtures of metals. Brass is an alloy of copper and zinc whereas bronze is an alloy of copper and tin.

8. Diamond is an allotrope of element carbon while all the rest are compounds.

9. Compounds are homogenous in nature.

10. Aluminium is trivalent.

11. Ammonia is a compound. Solution, alloy and amalgam are mixtures.

Comprehension Type

15. Br, Hg and Ga are liquid elements.

16. Carbon has the highest melting point while helium has the lowest melting point.

Integer Type

19. Br, Hg, Ga, Fr, Cs are liquid elements.

20. As, Se, Sb, B, Si and Ge are metalloids.

21. The given mixtures are iodised table salt, brass, aerated drink, milk, 22 carat gold, diesel, smoke.

22. Naphthalene, iodine, camphor, ammonium chloride and anthracene undergo sublimation.

23. Chlorine gas, iron, aluminium, iodine, carbon and sulfur powder are not compounds.

24. Filtration, sedimentation, evaporation, distillation, centrifugation and chromatography can be used for the separation of a mixture of a solid and a liquid.

CHAPTER

3 The Language of Chemistry

LEARNING OBJECTIVES

After studying this unit, you will be able to understand:

- Atoms, symbols of elements and their representation
- Atomic mass and its calculations
- Molecules, ions and radicals
- Valency and to write chemical formulae
- Laws of chemical combination

Introduction

Ancient Indian and Greek philosophers have always wondered about the unknown and unseen form of matter. The idea of divisibility of matter was considered long back in India, around 500 BC. An Indian philosopher Maharishi Kanad was one of the first persons to propose that matter (*padarth* in Hindi) is made up of very small particles called *parmanu* (John Dalton called these particles atoms). The word 'atom' means indivisible—something that cannot be divided further. Another Indian philosopher, Pakudha Katyayama, elaborated that the particles of matter (atoms) normally exist in a combined form and the combined form of atoms are known as *molecules.* All matter is made up of small particles known as atoms and molecules.

Atoms

The building blocks of all matter are *atoms*. An atom is the smallest particle of an element which may or may not exist independently but that can take part in a chemical reaction. Atoms of most elements are very reactive and do not exist in the free state. They exist in combination with the atoms of the same element or another element, as molecules. It means, the basic unit of matter in any state is considered as the *molecule* because a molecule is the smallest identity that can exist independently.

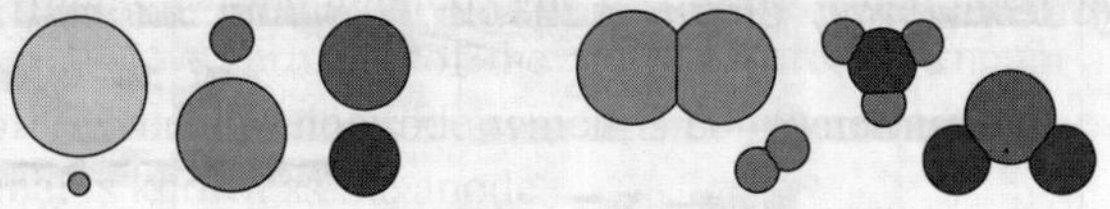

Atoms Molecules

Symbols of Elements

Dalton was the first scientist to use the symbols for elements. When he used a symbol for an element, he also meant a definite quantity of that element, that is, one atom of that element. Berzilius suggested that the symbols of elements can be made from one or two letters of the name of the element.

In the beginning, the names of elements were derived mainly from the names of the places where they were found for the first time. For example, the name copper was taken from Cyprus. Some names were also taken from some specific colours. For example, gold was taken from the Latin word meaning yellow.

Dalton's Symbols of Elements

Dalton was the first scientist to use symbols to represent elements. However, now it is just history.

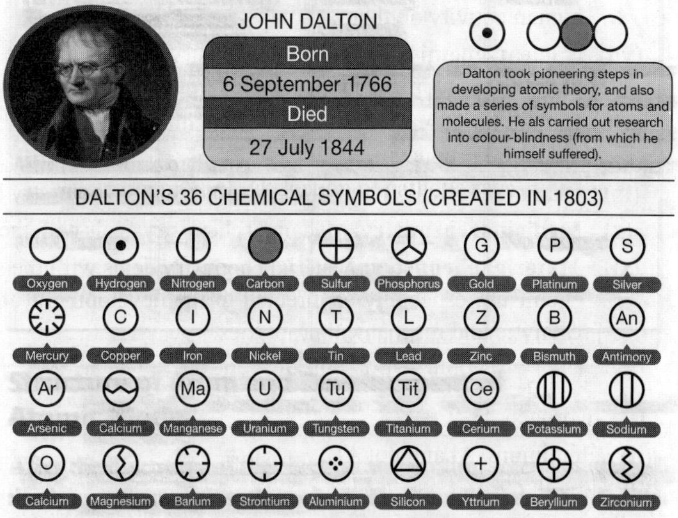

Symbol

In order to represent the elements, instead of using the entire name, scientists use abbreviated names; these are known as symbols. Hence, symbol may be defined as the abbreviation used for the naming of an element.

Modern Method or IUPAC System of Symbols

IUPAC (International Union of Pure and Applied Chemistry) approves names of elements. The symbols of elements are generally either the first letter or the first two letters or the first and the third letters of the name of an element. The first letter of a symbol is always written as a capital letter (upper case) and the second letter as a small letter (lower case). For example,

KNOWLEDGE BOOSTER

Chemistry and its branches

Chemistry is the science that deals with the properties, composition and structure of substances or matter (defined as elements and compounds), the transformations they undergo, and the energy that is released or absorbed during these processes.

For convenience, the study of chemistry is sub-divided into many branches. Some of the important ones are as follows:

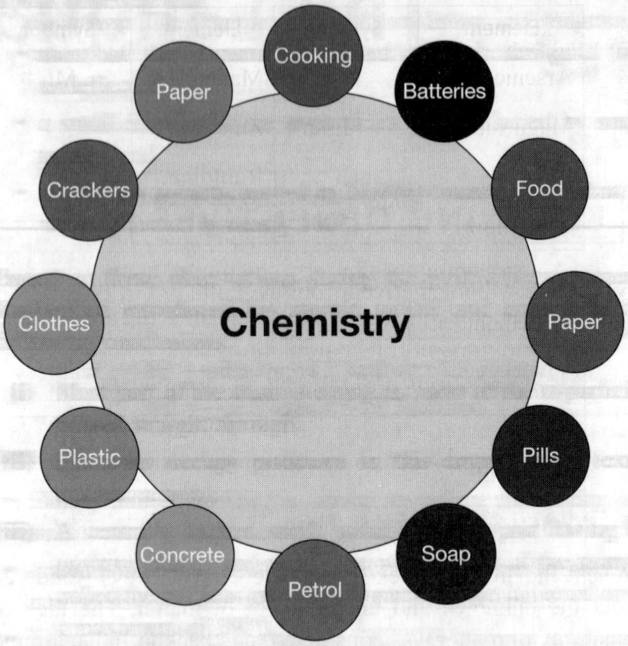

(i) **Inorganic chemistry:** Inorganic chemistry deals with the study of the compounds (except carbon compounds) obtained from minerals or inanimate sources. For example, HNO_3, H_2SO_4, CO_2.

(ii) **Organic chemistry:** Organic chemistry deals with the compounds of carbon and hydrogen (hydrocarbons and their derivatives) obtained from natural sources such as animal and plant sources. For example, carbohydrates, fats, proteins, vitamins, enzymes, polymers.

(iii) **Physical chemistry:** Physical chemistry deals with the physical principles and conditions that govern the chemical processes.

(iv) **Analytical chemistry:** Analytical chemistry deals with the study of the methods of detection and determination of elements and compounds.

(v) **Biochemistry:** Biochemistry deals with the study of chemical processes taking place in living organisms.

(vi) **Environmental chemistry:** Environmental chemistry deals with the study of the environment and changes occurring in the environment. That is, the chemical processes that occur in water, air, terrestrial and living environments, and the effects of human activity on them.

(vii) **Nuclear chemistry:** Nuclear chemistry deals with the study of the chemistry of radioactive substances and the energy changes taking place in the nucleus of the atom.

(viii) **Industrial chemistry:** Industrial chemistry deals with the study of chemical processes involved in the manufacture of industrial products.

(ix) **Pharmaceutical chemistry:** Pharmaceutical chemistry deals with pharmaceutical preparations and drug study.

(x) **Medicinal chemistry:** Medicinal chemistry deals with the study of structure–activity relationship, pharmacological activities.

(xi) **Material chemistry:** Material chemistry deals with the study of solid state chemistry, both inorganic and organic polymer chemistry, especially as directed to the development of materials with novel and/or useful optical, electrical, magnetic, catalytic and mechanical properties.

(i) hydrogen, H,
(ii) aluminium, Al and not AL,
(iii) cobalt, Co and not CO.

In the modern method, the symbols of elements can be classified as follows.

(i) Some symbols derived from the first letter of the names of the elements.

Element	Symbol	Element	Symbol
Hydrogen	H	Fluorine	F
Carbon	C	Phosphorus	P
Nitrogen	N	Sulfur	S
Oxygen	O	Iodine	I

(ii) Some symbols derived from the first two letters of the names of the elements.

Element	Symbol	Element	Symbol
Aluminium	Al	Silicon	Si
Barium	Ba	Argon	Ar
Lithium	Li	Calcium	Ca
Beryllium	Be	Nickel	Ni
Neon	Ne	Cobalt	Co

(iii) Some symbols derived from the first and the third letters of the names of the elements

Element	Symbol	Element	Symbol
Arsenic	As	Manganese	Mn
Magnesium	Mg	Zinc	Zn
Chlorine	Cl	Rubidium	Rb
Chromium	Cr		

(iv) Some symbols derived from the Latin or Greek names of the elements are given below

Element	Latin name	Symbol
Copper	Cuprum	Cu
Potassium	Kalium	K
Iron	Ferrum	Fe
Gold	Aurum	Au
Sodium	Natrium	Na
Silver	Argentum	Ag
Mercury	Hydragyrum	Hg
Tin	Stannum	Sn
Lead	Plumbum	Pb
Antimony	Stibium	Sb

Significance of the Symbol of an Element

A symbol represents:

- Name of the element.
- One atom of the element.
- A definite mass of the element (equal to atomic mass expressed in grams).
- One mole of atoms of the element. That is, the symbol also represents 6.022×10^{23} atoms of the elements.

Atomic Mass

In 1961, the International Union of Chemists universally accepted the carbon-12 isotope as the standard reference for measuring atomic masses in terms of atomic mass units. One atomic mass unit is a mass unit equal to exactly one-twelfth ($1/12^{th}$) the mass of one atom of carbon-12. The relative atomic masses of all elements have been found with respect to an atom of carbon-12. That is atomic mass of an element may be defined as the average relative mass of an atom of the element as compared with mass of an atom of carbon (C-12 isotope) taken as 12 amu.

$$\text{Atomic mass} = \frac{\text{Mass of 1 atom of the element}}{\frac{1}{12}\text{ of the mass of an atom of carbon-12}}$$

For example, the atomic mass of magnesium is 24 u which indicates that one atom of magnesium is 24 times heavier than $\frac{1}{12}$ of a carbon 12 atom. The atomic masses of some elements are given in Table 3.1.

Table 3.1 Atomic mass of some elements

S. No.	Element	Symbol	Atomic mass
1	Hydrogen	H	1 u
2	Carbon	C	12 u
3	Nitrogen	N	14 u
4	Oxygen	O	16 u
5	Sodium	Na	23 u
6	Magnesium	Mg	24 u
7	Aluminium	Al	27 u
8	Phosphorus	P	31 u
9	Sulfur	S	32 u
10	Chlorine	Cl	35.5 u
11	Potassium	K	39 u
12	Calcium	Ca	40 u
13	Iron	Fe	56 u
14	Copper	Cu	63.5 u

Gram Atomic Mass

Gram atomic mass of an element is defined as that quantity of the element whose mass expressed in grams is numerically equal to its atomic mass. To find the gram atomic mass we keep the numerical value the same as the atomic mass, but simply change the units from u to g. For example, the atomic mass of aluminium is 27 u. Its gram atomic mass is 27 g.

Illustrations

1. Which postulate of Dalton's atomic theory is the result of the law of conservation of mass?

   Solution: The given postulate of Dalton's atomic theory is the result of the law of conservation of mass. 'Atoms are indivisible particles, which cannot be created or destroyed in a chemical reaction.'

2. Which of the following symbols of elements are correct?

   (i) Cobalt: CO (ii) Carbon: c
   (iii) Aluminium: AL (iv) Helium: He
   (v) Sodium: So (vi) Iron: Fe
   (vii) Gold: Gd

Solution: Here only helium and iron have been given correct symbols. The correct symbols of these elements are as follows:

(i) Cobalt: Co (ii) Carbon: C
(iii) Aluminium: Al (iv) Helium: He
(v) Sodium: Na (vi) Iron: Fe
(vii) Gold: Au

3. The mass of one steel screw is 4.11 g. Find the mass of one mole of these steel screws. Compare this value with the mass of the earth (5.98×10^{24} kg). Which one of the two is heavier and by how many times?

Solution: Mass of 1 screw = 4.11 g

Mass of 1 mole of screws = $4.11 \times 6.022 \times 10^{23} = 2.475 \times 10^{24}$ g = 2.475×10^{21} kg

$$\frac{\text{Mass of earth}}{\text{Mass of 1 mole screws}} = \frac{5.98 \times 10^{24}\text{ kg}}{2.475 \times 10^{21}\text{ kg}} = 2416$$

It shows that mass of the earth is 2416 times more than the mass of one mole of screws.

How do atoms exist? Atoms of most of the elements are not able to exist independently. **Atoms usually exist in two ways**—molecules and ions. Atoms form molecules and ions. These molecules or ions aggregate in large numbers to form the matter that we can see, feel or touch.

Molecules

In general, a molecule is a group of two or more atoms that are chemically bonded together or tightly held together by attractive forces. A molecule is defined as the smallest particle of an element or a compound that is capable of independent existence and shows all the properties of that substance. It means, a molecule is the smallest identity that can exist individually. For example, hydrogen exists as H_2 (dihydrogen) and not as H. Atoms of the same element or of different elements can join together to form molecules. For example, H_2, N_2 and NH_3.

What is a molecule?

A molecule is an electrically neutral group of atoms joined together by chemical bonds

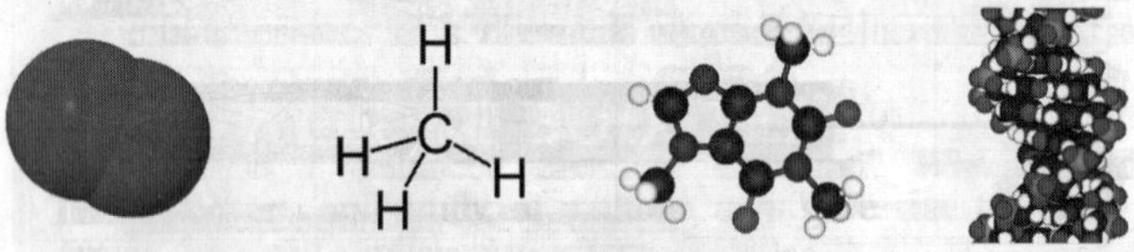

A molecule may consist of two atoms of the same element or many atoms of different elements

Types of Molecules

There are two types of molecules.

Molecules of Elements

The molecules of an element are constituted by the same type of atoms (homoatomic). These may be as follows:

(a) Monoatomic: Molecules of many elements like argon (Ar) and helium (He) are made up of only one atom of that element.

(b) Polyatomic: Molecules of most of the non-metals are made of two or more atoms.

- **Diatomic:** Such a molecule is formed by two atoms; for example, O_2, N_2, H_2, Cl_2.
- **Triatomic:** Such a molecule is formed by three atoms; for example, O_3 (ozone).

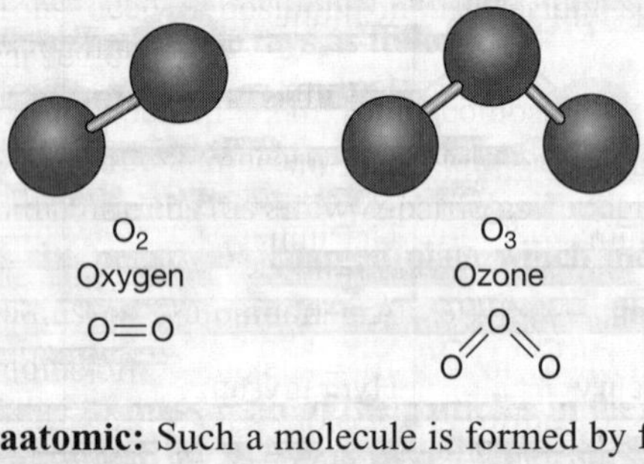

- **Tetraatomic:** Such a molecule is formed by four atoms; for example, P_4.
- **Octaatomic:** Such a molecule is formed by eight atoms; for example, S_8.

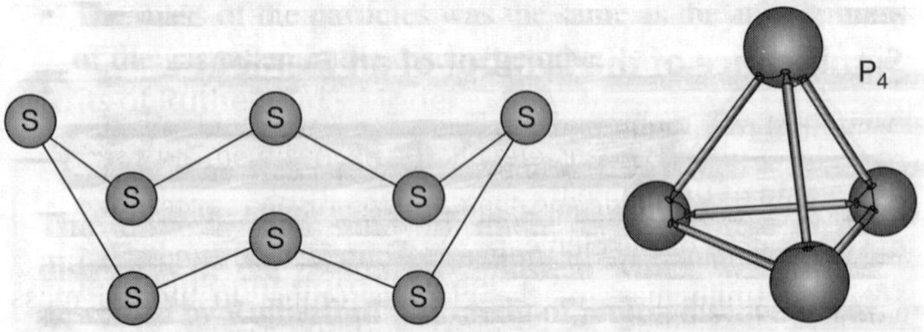

Molecules of Compounds

Atoms of different elements (heteroatomic) join together in definite proportions to form molecules of compounds. For example, hydrogen chloride is a compound. A molecule of hydrogen chloride (HCl) contains two different types of atoms: hydrogen (H) and chlorine (Cl) atoms.

Molecules of Some Compounds

Compound	Combining elements	Formula	Ratio by mass
Water	Hydrogen and oxygen	H_2O	1:8
Ammonia	Nitrogen and hydrogen	NH_3	14:3
Carbon dioxide	Carbon and oxygen	CO_2	3:8

Atomicity

The number of atoms constituting a molecule is known as its atomicity (Table 3.2).

Table 3.2 Atomicity of some common elements

Type of element	Name	Symbol	Atomicity
Non-metal	Helium	He	Monoatomic
	Argon	Ar	Monoatomic
	Neon	Ne	Monoatomic
	Hydrogen	H_2	Diatomic
	Chlorine	Cl_2	Diatomic
	Nitrogen	N_2	Diatomic
	Oxygen	O_2	Diatomic
	Phosphorus	P_4	Tetratomic
	Sulfur	S_8	Polyatomic
Metals	Sodium	Na	Monoatomic
	Iron	Fe	Monoatomic
	Aluminium	Al	Monoatomic
	Copper	Cu	Monoatomic

Ions

Compounds composed of metals and non-metals can contain charged species which are known as *ions*. An ion is a charged particle which can be negatively or positively charged. It can be an atom or group of atoms. It is formed by the loss or gain of electrons by an atom, so it contains an unequal number of electrons and protons.

Types of Ions

There are two types of ions—cations and anions.

Cation

A positively charged ion is known as a cation. It is formed by the loss of one or more electrons by an atom. The ions of all the metal elements are cations. For example, a sodium atom loses one electron to form a sodium ion, Na^+ (cation).

Na $\xrightarrow{-1\ \text{electron}}$	Na^+
Sodium atom	Sodium ion
Protons = 11 (+ charge)	Protons = 11 (+ charge)
Electrons = 11 (– charge)	Electrons = 10 (– charge)
Overall charge = 0	Overall charge = 1+

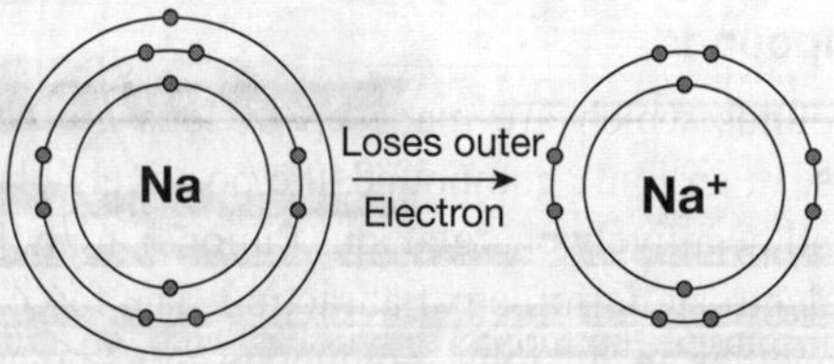

Sodium atom Sodium ion

Anion

A negatively charged ion is known as an anion. An anion is formed by the gain of one or more electrons by an atom. The ions of all the non-metal elements are anions. For example, a chlorine atom gains one electron to form a chloride ion, Cl^- (anion).

A normal or a neutral atom contains an equal number of protons and electrons but an anion contains more electrons than protons since it is formed by the addition of one or more electrons to an atom.

Cl $\xrightarrow{+1\ \text{electron}}$	Cl^-
Chlorine atom	Chlorine ion
Protons = 17 (+ charge)	Protons = 17 (+ charge)
Electrons = 17 (– charge)	Electrons = 18 (– charge)
Overall charge = 0	Overall charge = 1–

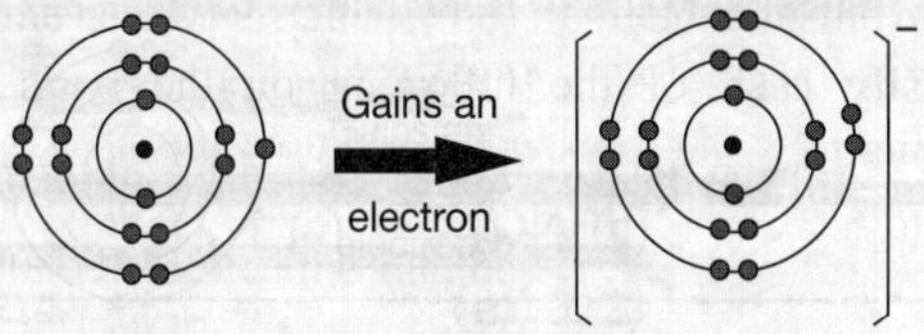

Chloride atom, Cl 2,8,7 Chloride ion, Cl^- [2,8,8]

Simple Ions

Such ions are formed from single atoms and are called *simple ions*. For example, the sodium ion Na^+, is a simple ion because it is formed from a single sodium atom, Na.

Compound Ions

Such ions are formed from groups of joined atoms and are called *compound ions*. For example, the ammonium ion NH_4^+, is a compound ion which is made up of two types of atoms joined together—nitrogen and hydrogen atoms.

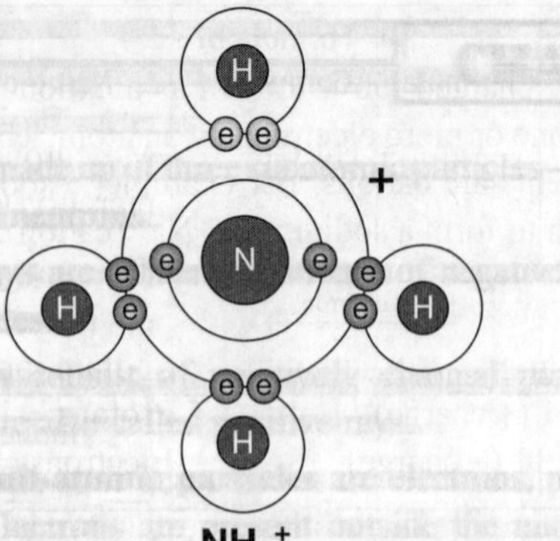

NH_4^+

For example (Table 3.3), sodium chloride (NaCl) is an ionic compound which is made up of an equal number of positively charged sodium ions (Na^+) and negatively charged chloride ions (Cl^-).

Table 3.3 Some ionic compounds

S. No.	Name	Formula	Ions present
1	Sodium chloride	NaCl	Na^+ and Cl^-
2	Potassium chloride	KCl	K^+ and Cl^-
3	Ammonium chloride	NH_4Cl	NH_4^+ and Cl^-
4	Magnesium chloride	$MgCl_2$	Mg^{2+} and Cl^-
5	Magnesium oxide	MgO	Mg^{2+} and O^{2-}
6	Aluminium oxide	Al_2O_3	Al^{3+} and O^{2-}
7	Sodium hydroxide	NaOH	Na^+ and OH^-
8	Copper sulfate	$CuSO_4$	Cu^{2+} and SO_4^{2-}

Ionic Compounds

The compounds which are made up of ions are called ionic compounds. In an ionic compound, the positively charged ions (cations) and negatively charged ions (anions) are held together by strong electrostatic forces of attraction. The forces that hold the ions together in an ionic compound are known as *ionic bonds* or *electrovalent bonds*. As an ionic compound consists of an equal number of positive ions and negative ions, the overall charge on an ionic compound is zero.

Illustrations

1. Write the cations and anions present (if any) in the following compounds:

(a) CH_3COONa (b) NaCl

(c) H_3 (d) NH_4NO_3

Solution:

(a) CH_3COONa: CH_3COO^-, Na^+

(b) NaCl: Na^+, Cl^-

(c) H_2: It is a covalent compound; hence, no ions are present in it.

(d) NH_4NO_3: NH_4^+, NO_3^-

2. Classify each of the following on the basis of their atomicity:

(a) F_2 (b) NO_2 (c) N_2O

(d) C_2H_6 (e) P_4 (f) H_2O_2

(g) P_4O_{10} (h) O_3 (i) HCl

(j) CH_4 (k) He (l) Ag

Solution: Monoatomic with atomicity (1): He, Ag

Diatomic with atomicity (2): F_2, HCl

Polyatomic with atomicity > 2

Atomicity (3): NO_2, N_2O, O_3

Atomicity (4): P_4, H_2O_2

Atomicity (5): CH_4

Atomicity (8): C_2H_6

Atomicity (14): P_4O_{10}

Valency

The word *valency* means, the power to combine. That is, the combining capacity of an element is known as its *valency*. It can be measured in terms of number of H-atoms or double the number of O-atoms that can combine with one atom of an element. As hydrogen has less combining capacity than any other element, its valency is taken as 1 as a standard value. For example, in CH_4 the valency of carbon is 4, in NH_3 the valency of nitrogen is 3 and in water (H_2O) the valency of oxygen is 2.

In the modern view, valency is equal to the number of electrons that an atom can share or lose or gain during a chemical reaction. The number of electrons present in the valence orbit of an atom are called valence electrons. In order to attain the octet state (8 valence electrons), every atom either loses or gains electrons or shares electrons.

Generally, elements having 1, 2, 3 valence electrons are metals and have a tendency to lose electrons and their valency is equal to the number of electrons lost. For group 1, group 2 and group 13, the valency is 1, 2 and 3 respectively.

For example,

$$\underset{2,\,8,\,1}{Na} - e^- \rightarrow \underset{2,\,8}{Na^+} \quad \text{(monovalent)}$$

$$\underset{2,\,8,\,2}{Mg} - 2e^- \rightarrow \underset{2,\,8}{Mg^{2+}} \quad \text{(divalent)}$$

$$\underset{2,\,8,\,3}{Al} - 3e^- \rightarrow \underset{2,\,8}{Al^{3+}} \quad \text{(trivalent)}$$

Elements having 4 valence electrons are non-metals like C and Si, and they undergo sharing of electrons. Their common valency is 4. For example, in CH_4 and CCl_4 carbon has 4 valency. Elements with 5, 6 and 7 electrons in their valence orbit are non-metals and have a tendency to gain electrons to complete their octet.

For example,

$$\underset{(2,\,5)}{N} + 3e^- \rightarrow \underset{2,\,8}{N^{3-}} \quad \text{(trivalent)}$$

$$\underset{(2,\,6)}{O} + 2e^- \rightarrow \underset{2,\,8}{O^{2-}} \quad \text{(divalent)}$$

$$\underset{(2,\,8,\,7)}{Cl} + e^- \rightarrow \underset{2,\,8,\,8}{Cl^-} \quad \text{(trivalent)}$$

In general, valency = number of valence electrons (up to 4). If the valence electrons are more than 4, the valency is given as, valency = valence electrons – 8.

Valency is not negative or positive it is just a number.

Number of electrons in outermost shell	1	2	3	4	5	6	7	8
Valency	1	2	3	4	3	2	1	0

Variable Valency

There are certain elements that exhibit more than one valency in their ions (or compounds). When an atom of an element loses not only its valence electrons but also loses electrons from its penultimate orbit, variable valency is observed (Table 3.4).

In case an element has two different positive valencies, then we use the suffix 'ous' for the lower valency state and the suffix 'ic' for the higher valency state. For example, iron can exist as Fe^{2+} (ferrous) or Fe^{3+} (ferric) in its compounds.

Nowadays, in place of 'ous' or 'ic', we can write the name of the compound by writing the valency in roman numbers within brackets. For example, $SnCl_2$ (stannous chloride) can be written as tin(II) chloride and $SnCl_4$ (stannic chloride) can be written as tin(IV) chloride.

Table 3.4 Some basic ions exhibiting variable valency and their compounds

Name of ions	Formula	Formula of Compound	Name of Compound
Cuprous	Cu^+	Cu_2O	Cuprous oxide or copper(I) oxide
Cupric	Cu^{2+}	CuO	Cupric oxide or copper(II) oxide

(Continued)

Name of ions	Formula	Formula of Compound	Name of Compound
Mercurous	Hg_2^{2+}	Hg_2Cl_2	Mercurous chloride or mercury(I) chloride
Mercuric	Hg^{2+}	$HgCl_2$	Mercuric chloride or mercury(II) chloride
Ferrous	Fe^{2+}	$FeCl_2$	Ferrous chloride or iron(II) chloride
Ferric	Fe^{3+}	$FeCl_3$	Ferric chloride or iron(III) chloride
Plumbous	Pb^{2+}	PbO	Plumbous oxide or lead(II) oxide
plumbic	Pb^{4+}	PbO_2	Plumbic oxide or lead(IV) oxide
Stannous	Sn^{2+}	SnO	Stannous oxide or tin(II) oxide
Stannic	Sn^{4+}	SnO_2	Stannic oxide or tin(IV) oxide
Aurous	Au^+	$AuCl$	Aurous chloride or gold(I) chloride
Auric	Au^{3+}	$AuCl_3$	Auric chloride or gold(III) chloride

Radical

The molecule of any compound is made up of two parts, each of which is called a radical. A radical is an atom or a group of atoms having the same or different elements, that behaves like a single unit with a positive or negative charge. For example, in NaCl, Na^+ and Cl^- are radicals.

A radical having only a single atom is called a simple radical while a radical having a group of two or more atoms (polyatomic) is called a compound radical. For example, Na^+ and Ca^{2+} are simple radicals while SO_4^{2-} and PO_4^{3-} are compound radicals.

Acidic and Basic Radicals

An acidic radical may be an atom or a group of atoms having a negative charge (Table 3.5). That is, anions or electronegative radicals are acidic radicals. A basic radical may be an atom or group of atoms having a positive charge. That is, cations or electropositive radicals are basic radicals. For example, in ammonium sulfate $(NH_4)_2SO_4$, the acidic radical is SO_4^{2-} with a combining power of 2 and the basic radical is NH_4^+ with a combining power of 1.

Table 3.5 Formulae and valencies of common ions

Electropositive ions (basic radicals)					
Monovalent		**Bivalent**		**Trivalent**	
Name	**Formula**	**Name**	**Formula**	**Name**	**Formula**
Potassium	K^+	Calcium	Ca^{2+}	Aluminium	Al^{3+}
Sodium	Na^+	Magnesium	Mg^{2+}	Chromium	Cr^{3+}
Copper(I) or cuprous	Cu^+	Manganese(II)	Mn^{2+}	Iron(III) or ferric	Fe^{3+}
Silver	Ag^+	Zinc	Zn^{2+}	Gold	Au^{3+}
Ammonium	NH_4^+	Iron(II) or ferrous	Fe^{2+}	**Tetravalent**	
Hydrogen	H^+	Nickel	Ni^{2+}	**Name**	**Formula**
Mercurous	Hg^+	Cobalt	Co^{2+}	Manganese(II)	Mn^{4+}
		Tin(II) or stannous	Sn^{2+}	Tin(IV) or stannic	Sn^{4+}
		Cadmium	Cd^{2+}	Lead(IV)	Pb^{4+}
		Lead(II)	Pb^{2+}	Platinum	Pt^{4+}
		Copper(II)	Cu^{2+}		

Electronegative ions (acidic radicals)					
Monovalent		**Bivalent**		**Trivalent**	
Name	**Formula**	**Name**	**Formula**	**Name**	**Formula**
Hydroxide	OH^-	Carbonate	CO_3^{2-}	Nitride	N^{3-}
Hydride	H^-	Chromate	CrO_4^{2-}	Phosphate	PO_4^{3-}
Fluoride	F^-	Dichromate	$Cr_2O_7^{2-}$	Phosphite	PO_3^{3-}
Chloride	Cl^-	Manganate	MnO_4^{2-}	Phosphide	P^{3-}
Bromide	Br^-	Permanganate	MnO_4^-	Borate	BO_3^{3-}
Iodide	I^-	Tetrathionate	$S_4O_6^{2-}$	Ferricyanide	$Fe(CN)_6^{3-}$
Bicarbonate	HCO_3^-	Sulfide	S^{2-}	Arsenite	AsO_3^{3-}
Bisulfite	HSO_3^-	Sulfite	SO_3^{2-}	**Tetravalent**	
Bisulfate	HSO_4^-	Sulfate	SO_4^{2-}	**Name**	**Formula**
Chlorate	ClO_3^-	Oxide	O^{2-}	Ferrocyanide	$Fe(CN)_6^{4-}$
Hypochlorite	ClO^-	Zincate	ZnO_2^{2-}	Carbide	C^{4-}
Nitrite	NO_2^-	Thiosulfate	$S_2O_3^{2-}$	Silicate	SiO_4^{4-}
Nitrate	NO_3^-	Oxalate	$C_2O_4^{2-}$		
Permanganate	MnO_4^-	Peroxide	O_2^{2-}		

Illustrations

1. Write any three metals that can show variable valencies along with their metal ions.

Solution: Iron (Fe) atom can show two valencies +2 (Fe^{2+}) and +3 (Fe^{3+}).

Copper (Cu) atom can show two valencies +1 (Cu^{+}) and +2 (Cu^{2+}).

Tin (Sn) atom can show two valencies +2 (Sn^{2+}) and +4 (Sn^{4+}).

2. Write the examples of any two acidic and basic radicals.

Solution: Acidic radicals: Carbonate (CO_3^{2-}), sulphate (SO_4^{2-})

Basic radicals: Ammonium ion (NH_4^{+}), aluminium ion (Al^{3+})

3. If we take basic radicals K^+ and Ca^{2+} and they combine with CO_3^{2-}, Cl^-, PO_4^{3-}. Write the formulae of the possible compounds formed.

Solution: K_2CO_3, KCl, K_3PO_4

$CaCO_3$, $CaCl_2$, $Ca_3(PO_4)_2$

Chemical Formulae

A compound is represented in the abbreviated form using a chemical formula. The chemical formula of a compound represents the composition of a molecule of the compound in terms of the symbols of the elements present in it. For example, water is a compound made up of 2 atoms of hydrogen and 1 atom of oxygen, so the formula of water can be written as H_2O. In the formula H_2O, the subscript 2 indicates 2 atoms of hydrogen. In the formula of water, oxygen O is written without a subscript and it indicates 1 atom of oxygen.

Formulae of Elements

The chemical formula of an element is a statement of the composition of its molecule in which the symbol tells us the element and the subscript tells us how many atoms are present in one molecule. For example, one molecule of hydrogen element contains two atoms of hydrogen, hence the formula of hydrogen is H_2.

Formulae of Compounds

The chemical formula of a compound is a statement of its composition in which the chemical symbols tell us which elements are present and the subscripts tell us how many atoms of each element are present in one molecule of the compound. For example, water is a compound whose molecule contains 2 atoms of hydrogen and 1 atom of oxygen. So, the formula of water is H_2O.

Writing the Formula of a Compound

The chemical formulae of different compounds can be written easily but for this we need to learn the symbols and combining capacity (valency) of the elements. Valency can be used to find out how the atoms of an element will combine with the atom(s) of another element to form a chemical compound.

(i) To write the chemical formula for a binary atomic compound: The rules that we have to follow while writing a chemical formula are as follows:

- We first write the symbols of all the elements which form the compound.
- Below the symbol of every element, we have to write down its valency. The valencies or charges on the ions must balance.
- Finally, we cross-over the valencies of the combining atoms. That is, with the first atom we write the valency of second atom (as a subscript); and with the second atom we write the valency of first atom (as subscript).

For example,

Element	H	Cl	Element	H	O	Element	H	S
Valency	1	1	Valency	1	2	Valency	1	2
Formula	HCl		Formula	H_2O		Formula	H_2S	
Element	N	H	Element	C	Cl	Element	P	Cl
Valency	3	1	Valency	4	1	Valency	5	1
Formula	NH_3		Formula	C_1Cl_4 or CCl_4		Formula	P_1Cl_5 or PCl_5	

(ii) To write the formula of a polyatomic compound: In the case of ionic compounds, follow these steps to write the formula.

Step 1: Write the symbols of formulae of the ions of the compound side by side with the positive ion on the left hand side and the negative ion on the right hand side.

Step 2: Enclose the polyatomic ion within brackets.

Step 3: Write the valency of each ion below its symbol.

Step 4: Reduce the valency numerals to a simple ratio by dividing with a common factor, if any.

Step 5: Cross the valencies. Do not write the charges positive or negative of the ions.

For example,

Formula of calcium nitrate.

Step 1: Writing the formula of the ions:

Ca^{2+} NO_3^-

Step 2: Ca^{2+} $(NO_3)^-$

Step 3: Ca^{2+} $(NO_3)^-$

2 1

Step 4: Not applicable, because the ratio is already simple.

Step 5: Ca^{2+} $(NO_3)^-$

2 1

Thus, the formula of calcium nitrate is $Ca(NO_3)_2$.

Formula of calcium phosphate.

Step 1: Writing the formula of the ions:

Ca^{2+} PO_4^{3-}

Step 2: Ca^{2+} $(PO_4)^{3-}$

Step 3: Ca^{2+} $(PO_4)^{3-}$

2 3

Step 4: Not applicable, because the ratio is already simple.

Step 5: Ca^{2+} $(PO_4)^3$

2 3

Thus, the formula of calcium phosphate is $Ca_3(PO_4)_2$.

Significance of the Formula of a Substance

The molecular formula of a compound has the following significance.

- It represents the name of the substance.
- It represents one molecule of the substance.
- It gives the names of all the elements present in the molecule.
- It gives the number of atoms of each element present in one molecule.
- It represents a definite mass of the substance.
- It also represents one mole of molecules of the substance. That is, the formula also represents 6.022×10^{23} molecules of the substance.

Molecular Mass and Formula Mass

Formula Unit Mass

The formula unit mass of a substance is the sum of the atomic masses of all atoms in a formula unit of the compound. Formula unit mass is calculated in the same manner as we calculate the molecular mass. The only difference is that we use the word formula unit for those substances whose constituent particles are ions. For example, sodium chloride as discussed above, has a formula unit NaCl. Its formula unit mass can be calculated as $1 \times 23 + 1 \times 35.5 = 58.5$ u.

KNOWLEDGE BOOSTER

Gram molecular mass

The molecular mass of a substance expressed in grams is known as its gram molecular mass.

For example, molecular mass of oxygen, $O_2 = 32$ u

So, gram molecular mass of oxygen, $O_2 = 32$ grams

For example, sulfuric acid has the formula H_2SO_4. It consists of two H, one S and four O atoms. The atomic masses of H, S and O are 1, 32 and 16 respectively. Therefore, the molecular mass of H_2SO_4 is:

Molecular mass of $H_2SO_4 = (2 \times \text{at. m. of H}) + (1 \times \text{at. m. of S}) + (4 \times \text{at. m. of O})$

$= (2 \times 1) + (1 \times 32) + (4 \times 16) = 98$ u

Illustrations

1. Write the molecular formulae for the following compounds:

(a) Copper(II) bromide (b) Aluminium(III) nitrate
(c) Calcium(II) phosphate (d) Iron(III) sulfide
(e) Mercury(II) chloride (f) Magnesium(II) acetate

Solution:

(a) Copper(II) bromide: $CuBr_2$
(b) Aluminium(III) nitrate: $Al(NO_3)_3$
(c) Calcium(II) phosphate: $Ca_3(PO_4)_2$
(d) Iron(III) sulfide: Fe_2S_3
(e) Mercury(II) chloride: $HgCl_2$
(f) Magnesium(II) acetate: $Mg(CH_3COO)_2$

2. Write the molecular formulae of all the compounds that can be formed by the combination of the following ions.

Cu^{2+}, Na^+, Fe^{3+}, Cl^-, SO_4^{2-}, PO_4^{3-}

Solution: Here the cations are Cu^{2+}, Na^+, Fe^{3+} and Cl^-, SO_4^{2-}, PO_4^{3-} are the anions. The possible compounds are:

(i) $CuCl_2$, $CuSO_4$, $Cu_3(PO_4)_2$
(ii) NaCl, Na_2SO_4, Na_3PO_4
(iii) $FeCl_3$, $Fe_2(SO_4)_3$, $FePO_4$

3. Give the chemical formulae for the following compounds and compute the ratio by mass of the combining elements in each one of them. You may use Appendix III.

(a) Ammonia (b) Carbon monoxide (c) Hydrogen chloride
(d) Aluminium fluoride (e) Magnesium sulfide

Solution:

Chemical formula		Ratio by mass
(a)	Ammonia (NH_3)	N(l): H(3) 14:3
(b)	Carbon monoxide (CO)	0(1): 0(1)12 :16 or 3 : 4
(c)	Hydrogen chloride (HCl)	H(l): Cl(l)1 : 35.5 or 2 : 71
(d)	Aluminium fluoride (AlF_3)	Al(l): F(3)27 :19 x 3 or 9 :19
(e)	Magnesium sulfide (MgS)	Mg(l): S(l)24 : 32 or 3 :4

KNOWLEDGE BOOSTER

Types of oxides

Oxides are mainly of the following types:

1. **Acidic oxide:** It reacts with water to give an acid.
   For example, $SO_3 + H_2O \rightarrow H_2SO_4$
2. **Basic oxide:** It reacts with water to give a base.
   For example, $Na_2O + H_2O \rightarrow 2NaOH$
   - An acidic oxide takes up oxygen while a basic oxide gives up oxygen. For example,
   $\underset{\text{Basic}}{CaO} + \underset{\text{Acidic}}{CO_2} \rightarrow CaCO_3$
   $6Na_2O + P_4O_{10} \rightarrow 4Na_3PO_4$
3. **Amphoteric oxide:** It reacts with both an acid and a base to form salts.
   For example, $Al_2O_3 + 6HCl \rightarrow 2AlCl_3 + 3H_2O$
   $Al_2O_3 + 2NaOH \rightarrow 2NaAlO_2 + H_2O$

Some acidic oxides: Cl_2O_7, P_4O_{10}, SO_2, CO_2, SiO_2, SO_3, N_2O_3, B_2O_3, CrO_3, WO_3, MoO_3, Mn_2O_7.

Some basic oxides: M_2O (here M = Group 1 metal), MO (Here M = Group 2 metal), Sc_2O_3, TiO_2, ZrO_2.

Some amphoteric oxides: BeO, CrO, Al_2O_3, Sb_2O_3, SnO, PbO, ZnO, Cr_2O_3.

Some neutral oxides: CO, N_2O, NO, H_2O.

Laws of Chemical Combination

Antoine L Lavoisier laid the foundation for the chemical sciences by establishing two important laws of chemical combination. These two laws of chemical combination played a significant role in the development of Dalton's atomic theory of matter. There are two important laws of chemical combination—(i) Law of conservation of mass and (ii) Law of constant proportions.

Law of Conservation of Mass

Law of conservation of mass was introduced by Lavoisier (father of chemistry) in 1774. According to this law of conservation of mass, matter can neither be created nor destroyed in a chemical reaction. The substances which combine together in a chemical reaction are called *reactants* whereas the new substances formed as a result of chemical reaction are called *products*. The law of conservation of mass means that in a chemical reaction, the total mass of products is equal to the total mass of reactants and that there is no change in mass during a chemical reaction.

For example, (i) Lavoisier observed that on heating mercuric oxide, it produced free mercury and oxygen. The sum of the masses of mercury and oxygen was found to be equal to the mass of mercuric oxide.

$$\underset{100\text{ g}}{\text{Mercuric oxide}} \rightarrow \underset{92.6\text{ g}}{\text{Mercury}} + \underset{7.4\text{ g}}{\text{Oxygen}}$$

Mass of the reactant = 100 g

Mass of the products = 92.6 + 7.4 g = 100.0 g = 100 g

Here, the total mass of the products formed is equal to the total mass of the reactants undergoing reaction, hence this data is in agreement with the law of conservation of mass.

(ii) A piece of ice (solid water) is taken in a small conical flask. It is well corked and weighed. The flask is now heated gently to melt the ice (solid) into water (liquid).

$$\text{Ice} \xrightarrow{\Delta} \text{Water}$$

The flask is again weighed. It is found that there is no change in the weight during this physical change also.

Law of Constant Proportions

Lavoisier, along with Joseph Proust and other scientists, observed that many compounds were composed of two or more elements and each such compound had the same elements in the same proportions, irrespective of where the compound came from and how it is prepared.

In a compound like water, the ratio of the mass of hydrogen to the mass of oxygen is always 1:8, whatever may be source (river, well) of water. It means if 18 g of water is decomposed, 2 g of hydrogen and 16 g of oxygen are always formed. Similarly in the case of ammonia, nitrogen and hydrogen are always present in the ratio 14:3 by mass, whatever be the method or the source from which it is obtained.

This led to the law of constant proportions which is also known as the *law of definite proportions*. According to this law, in a pure chemical substance or compound the elements are always present in definite proportions by mass.

For example, it is important to note that whatever sample of carbon dioxide is taken, it is observed that the ratio of carbon and oxygen is always same. That is, 12 : 32 or 3 : 8 in the methods given here.

(i) By burning carbon in oxygen: $C(s) + O_2(g) \rightarrow CO_2(g)\uparrow$

(ii) By heating sodium bicarbonate:

$$2NaHCO_3(s) \xrightarrow{\Delta} Na_2CO_3(s) + CO_2(g)\uparrow + H_2O(l)$$

(iii) By heating calcium carbonate:

$$CaCO_3(s) \xrightarrow{\Delta} CaO(s) + CO_2(g)\uparrow$$

(iv) By reacting calcium carbonate with hydrochloric acid:

$$CaCO_3(s) + 2HCl(aq) \rightarrow CaCl_2(aq) + CO_2(g)\uparrow + H_2O(l)$$

Illustrations

1. In a reaction, 5.3 g of sodium carbonate reacted with 6 g of ethanoic acid. The products were 2.2 g of carbon dioxide, 0.9 g water and 8.2 g of sodium ethanoate. Show that these observations are in agreement with the law of conservation of mass.

Sodium carbonate + ethanoic acid → sodium ethanoate + carbon + dioxide + water

Solution: According to the law of conservation of mass, mass can neither be created nor destroyed during a chemical reaction. It means that the mass remains the same. So, we add the masses of the reactants on the LHS and add the masses of all the products on the RHS.

LHS = 5.3 g + 6 g = 11.3 g

RHS = 8.2 g + 2.2 g + 0.9 g = 11.3 g

LHS = RHS

Hence these observations are in agreement with the law of conservation of mass.

2. Hydrogen and oxygen combine in the ratio of 1 : 8 by mass to form water. What mass of oxygen gas would be required to react completely with 3 g of hydrogen gas?

Solution: According to the law of constant proportions, the composition of a compound is always fixed. By applying this,

1 g of hydrogen gas combines with oxygen = 8 g

So, 3 g of hydrogen gas will combine with oxygen = $8 \times 3 = 24$ g

CHAPTER AT A GLANCE

- An **atom** is the smallest particle of an element which may or may not exist independently but but can take part in a chemical reaction. the basic unit of matter in any state is considered as the **molecule** because a molecule is the smallest identity to exist independently.
- **Symbol** is defined as the abbreviation used for the naming of an element.
- **Atomic mass** of an element may be defined as the average relative mass of an atom of the element as compared with mass of an atom of carbon (C-12 isotope) taken as 12 amu.
- **Gram atomic mass** is defined as that much quantity of the element whose mass expressed in grams is numerically equal to its atomic mass.
- A **molecule** is defined as the smallest particle of an element or a compound that is capable of an independent existence and shows all the properties of that substance.
- The number of atoms constituting a molecule is known as its **atomicity**.
- Compounds composed of metals and non-metals have charged species also and these charged species are known as **ions**.
- A positively charged ion is known as a **cation** while a negatively charged ion is known as an **anion**.
- The compounds that are made up of ions are called **ionic compounds**.
- The word valency means, power to combine. That is, the combining capacity of an element is known as its **valency**.
- In the modern view, **valency** is equal to number of electrons that an atom can share or lose or gain during a chemical reaction. The number of electrons present in the valence orbit of an atom are called **valence electrons**.
- The molecule of any compound is made of two parts which are separately called **radicals**.
- An acidic radical may be an atom or group of atoms having a negative charge. Anions or electronegative radicals are **acidic radicals**.

- A basic radical may be an atom or group of atoms having a positive charge. Cations or electropositive radicals are **basic radicals**.
- The **chemical formula** of a compound represents the composition of a molecule of the compound in terms of the symbols of the elements present in it.
- The **formula unit mass** of a substance is the sum of the atomic masses of all the atoms in a formula unit of the compound.
- **Acidic oxide:** It reacts with water to give an acid.
- **Basic oxide:** It reacts with water to give a base.
- **Amphoteric oxide:** It reacts with both acids and bases to form salts.
- **Some neutral oxides:** CO, N_2O, NO, H_2O.
- **Law of conservation of mass:** Matter can neither be created nor destroyed in a chemical reaction.
- **Law of constant proportions:** In a pure chemical substance or compound the elements are always present in definite proportions by mass.

PRACTICE QUESTIONS

Analyse Your Concepts (School Exam Based)

Fill in the Blanks

Instructions: Complete the following statements with an appropriate word/term to be filled in the blank spaces.

1. ● is a symbol for ________.
2. Ne, Ar are ________ molecules.
3. ________ is a pure substance which is made up of only one kind of atoms.
4. In SO_2, sulfur and oxygen are in the weight ratio of ________.
5. The formula of the oxide of Al^{3+} and O^{2-} is ________.
6. The atomicity of P_4 is ________.
7. N_2O_3 is known as nitrogen ________ oxide.
8. The formula of nitric acid is ________.
9. 1 mole of nitrogen atoms = ________ nitrogen atoms.
10. borate ion is ________ valent.
11. The ratio by mass of S and O in SO_3 is ________.
12. In methane, the proportion of hydrogen and carbon is ________ by mass.
13. The Latin name of gold is ________.
14. The atomic mass of copper is ________ u.

True or False

Instructions: Read the following statements and write your answer as true or false.

1. ⊗ this symbol represents the phosphorus atom.
2. S_8 is a polyatomic element.
3. The smallest particle of a compound is element.
4. 1 amu is equal to $1.66056 \times 10^{24-}$ g.
5. Mass of 6.022 atoms of an element is called its atomic mass.
6. In a pure chemical compound, the elements are always present in a definite proportion by mass.
7. The symbol of cobalt is CO.
8. Ions always have the same number of protons.
9. One mole of CO_2 and SO_2 contain the same number of oxygen atoms.
10. The formula mass of MgO is 28 amu.
11. The mass of 1 mole of N_2O is 44 g.
12. The mass of a hydrogen atom is equal to the mass of an oxygen atom.
13. The latin name of silver is argentum.
14. Both the chlorine atom and the chloride ion (Cl^-) have the same number of electrons.

Very Short Answer Type Questions

1. Name the building block of all matter.
2. What are the symbols for aluminium and iron?
3. Name any two elements whose symbols are derived from Latin names. Give their symbols.
4. What is the mass of 1 mole of ammonia (NH_3)?
5. What is the molecular mass of H_2SO_3?
6. Sodium carbonate ($Na_2CO_3.10H_2O$) is an important industrial compound. Calculate its formula mass.

7. Which of the following are triatomic molecules?
CO_2, NH_3, SO_2, CH_3OH, CH_4, H_2O

8. Helium gas consists of single atoms. What mass of helium contains 6.022×10^{23} atoms?

9. What is the ratio by mass of sulfur and hydrogen in H_2S?

10. Give two examples each of any bivalent metal ions and trivalent metal ions.

11. Which of the following has larger mass?
(a) 1 mole of phosphine (PH_3)
(b) 1 mole of ethane (C_2H_6)

12. How many moles of helium are present in 48 g of He?

13. What is the molar mass of phosphorus molecule (P_4)?

14. Write the structural formula of the acetate and nitrite ions.

Short Answer Type Questions

1. What are ionic and molecular compounds? Give examples.

2. Write down the formulae for the following compounds.
(a) zinc oxide (b) aluminium hydroxide (c) ferric sulfate

3. An element E has a valency of 3. Write the formula of its:
(a) bromide (b) oxide (c) sulfate (d) carbonate (e) nitrate

4. An element E shows valencies of 3 and 5. Write the formulae of its two oxides.

5. Give the formulae of the compounds formed from the following sets of elements:
(a) zinc and fluorine
(b) hydrogen and sulfur
(c) nitrogen and hydrogen
(d) tin and chlorine
(e) potassium and oxygen
(f) sulfur and oxygen

6. What are polyatomic ions? Give examples.

7. What do you mean by the term chemical formula and what does it represent?

8. How many atoms are present in a: (a) H_2S molecule (b) PO_4^{3-} ion?

9. Write down the names of compounds represented by the following formulae.
(a) $Al_2(SO_4)_3$ (b) $CaCl_2$
(c) K_2SO_4 (d) KNO_3
(e) $CaCO_3$

10. (a) In a chemical reaction, the sum of the masses of the reactants and products remains unchanged. This is called _____.
(b) A group of atoms carrying a fixed charge on them is called _____.
(c) The formula unit mass of $Ca_3(PO_4)_2$ is _____.
(d) Formula of sodium carbonate is _____ and that of ammonium sulfate is _____.

11. The valency of aluminium is 3. Write the valency of the other radicals present in the following compounds.
(a) aluminium chloride
(b) aluminium oxide
(c) aluminium nitride

12. What is variable valency? Give any two examples of elements exhibiting variable valency.

Long Answer Type Questions

1. Explain the following terms.
(a) Atomic mass
(b) Molecular mass
(c) Mole
(d) Avogadro constant
(e) Polyatomic ions

2. Calculate the molecular masses of H_2, O_2, Cl_2, CO_2, CH_4, C_2H_6, C_2H_4, NH_3, CH_3OH.

3. Raunak took 5 moles of carbon atoms in a container and Krish took 5 moles of sodium atoms in another container of the same weight.
(a) Whose container is heavier?
(b) Whose container has more number of atoms?

4. (a) Write the chemical formulae of the following compounds.
(i) magnesium chloride
(ii) calcium oxide
(iii) copper nitrate
(iv) aluminium chloride
(v) calcium carbonate
(b) Give the names of the elements present in the following compounds.
(i) quick lime
(ii) hydrogen bromide
(iii) baking powder
(iv) potassium sulfate

5. Which has more number of atoms—100 grams of sodium or 100 grams of iron (given, atomic mass of Na = 23 u, Fe = 56 u)?

6. (a) Why is it not possible to see an atom with the naked eye?
(b) If one mole of carbon atoms weighs 12 grams, what is the mass (in grams) of 1 atom of carbon?

COMPETITION WINDOW (OBJECTIVE TYPE)

[For NEET, JEE (Main and Advanced), NTSE, KVPY and Olympiads]

Topic-wise MCQs

Elements and symbols

1. Which one of the following information is conveyed by the symbol of an element?
(a) the atomic number of the element
(b) the atomic mass of the element
(c) the name of the element
(d) all of these

2. The chemical symbol for sodium is
(a) So (b) Sd
(c) NA (d) Na

3. The chemical symbol for bismuth is
(a) Be
(b) Bi
(c) B
(d) Ba

4. A group of atoms chemically bonded together is a (an)
(a) salt (b) element
(c) molecule (d) ion

5. Adding electrons to an atom will result in a (an)
(a) cation
(b) salt
(c) molecule
(d) anion

6. The correct symbol for gold is
(a) Ag (b) Si
(c) Au (d) Al

7. The total number of atoms represented by the compound $CuSO_4.5H_2O$ is
(a) 24 (b) 8
(c) 9 (d) 21

8. The chemical symbol for nitrogen gas is
(a) Ni (b) N_2
(c) N^+ (d) N

9. Which of the following symbols of elements is correct?
(a) Cobalt: CO
(b) Carbon: c
(c) Aluminium: AL
(d) Helium: He

10. How many of the following elements have been given correct symbol here:
(a) Nickel: NI
(b) Silver: Ag
(c) Vanadium: V
(d) Both (b) and (c)

11. To which of the following elements have symbols been given according to their Latin names?
(a) Mercury: Hg
(b) Lead: Pb
(c) Potassium: K
(d) All of these

Atomicity and valency

12. Atomicity of an element is
(a) number of atoms in a molecule
(b) number of neutrons in a metal
(c) number of electrons in a molecule
(d) number of ions in a molecule

13. Atomicity of phosphorus is
(a) 5 (b) 4
(c) 3 (d) 2

14. Valency of an atom or radical is
(a) its combining capacity
(b) size of atom
(c) its ionisation energy
(d) the electron affinity of the atom

15. The valency of gold in AuCl is
(a) 4 (b) 3
(c) 2 (d) 1

16. In MnO_2, valency of Mn is
(a) 4 (b) 6
(c) 2 (d) 1

17. Which of the following is triatomic?
(a) helium
(b) methane
(c) ammonia
(d) carbon dioxide

18. In P_2O_3 and P_2O_5, the valencies of phosphorus are respectively
(a) 2, 2 (b) 3, 5
(c) 5, 3 (d) 6, 10

19. Which of the following metal atoms can show variable valency?
(a) Ca
(b) Fe
(c) Cu
(d) both (b) and (c)

20. Which one will not show variable valency?
(a) Au (b) Sr
(c) Cu (d) Hg

Ions and radicals

21. Which of the following ions is not univalent?
(a) bicarbonate
(b) sulfite
(c) ammonium
(d) nitrite

22. Which of the following pairs are possible?
(a) Hg^+, Hg^{2+}
(b) Sn^{2+}, Sn^{4+}
(c) Na^+, Na^{2+}
(d) Both (a) and (b)

23. The polyatomic ion is
(a) sulfate
(b) nitride
(c) sulfide
(d) chloride

24. Which one is a bivalent ion?
(a) potassium
(b) sulfide
(c) magnesium
(d) both (b) and (c)

25. The charge on iron in ferrous sulfate is
(a) –1 (b) –2
(c) +1 (d) +2

26. Bicarbonate ion is
(a) HCO_3^- (b) HCO_3^+
(c) HCO^- (d) CO_3^{2-}

27. Phosphide and phosphate ions respectively are
(a) PO_4^{4-} and P^{4-}
(b) P^{4-} and PO_4^{4-}
(c) PO_4^{3-} and P^{3-}
(d) P^{3-} and PO_4^{3-}

28. Stannic and stannous ion are written respectively as
(a) Sn^{2+}, Sn^{4+} (b) Sn^{4+}, Sn^{2+}
(c) Sn^{3+}, Sn^{2+} (d) Sn^{2+}, Sn^{3+}

Molecular formula

29. A molecular positive radical (M^+), combines with a chromate radical. Now, identify the formula of the compound formed.
(a) $M_2Cr_2O_7$
(b) $M(Cr_2O_7)_2$
(c) $MCrO_4$
(d) M_2CrO_4

30. The formula of magnesium dihydrogen phosphate is
(a) $Mg(H_2PO_4)_2$
(b) $Mg_3(PO_4)_2$
(c) MgH_2PO_4
(d) $Mg(HPO_4)_2$

31. In which of the following cases, is the empirical formula the same as the molecular formula?
(a) C_2H_5COOH
(b) glucose
(c) sucrose
(d) C_6H_6

32. If the formula of a metallic nitrate is $A(NO_3)_3$, then what will be the formula of the nitride of that metal?
(a) AN_2 (b) AN
(c) A_2N (d) A_2N_3

33. The formula of a metal chloride is MCl_3. The formula of its phosphate is
(a) $M(PO_4)_2$ (b) MPO_4
(c) M_2PO_4 (d) M_3PO_4

34. The name of the compound $(NH_4)_2SO_4$ is
(a) ammonium sulfite
(b) ammonia sulfate
(c) ammonium sulfate
(d) ammonium sulfide

35. Write the formula of the compound formed by the combination of Ca^{2+} and N^{3-}.
(a) CaN (b) Ca_3N
(c) Ca_3N_2 (d) Ca_2N_3

Miscellaneous

1. In which of the following is the element not given the correct symbol as per as Dalton's theory?

	Element	Symbol
(a)	hydrogen	
(b)	oxygen	
(c)	sulfur	
(d)	gold	

2. The atomicity of O_3, P_4, S_6 are respectively
(a) 1, 1, 1 (b) 6, 4, 3
(c) 3, 4, 6 (d) 2, 4, 6

3. Write the formula of compound formed by combination of Li^+ and N^{3-}.
(a) LiN (b) Li_3N
(c) Li_3N_2 (d) Li_2N_3

4. The simplest formula of a compound containing 50% of an element P (atomic weight 10) and 50% of element Q (atomic weight 20) is
(a) PQ (b) P_2Q
(c) PQ_2 (d) P_2Q_3

5. An organic compound contains 4% sulfur. Its minimum molecular weight is
(a) 100 (b) 400
(c) 800 (d) 200

6. One molecule of sulfur contains
   (a) four atoms of sulfur
   (b) one atom of sulfur
   (c) two atoms of sulfur
   (d) eight atoms of sulfur

7. The law of definite proportions was given by
   (a) Proust (b) Michael Faraday
   (c) John Dalton (d) Humphry Davy

8. The percentage of copper and oxygen in samples of CuO obtained by different methods were found to be the same. This illustrates the law of
   (a) multiple proportions
   (b) reciprocal proportions
   (c) constant proportions
   (d) conservation of mass

9. The mass of sodium in 5.85 g of sodium chloride is
   (a) 2.3 g (b) 5.02 g
   (c) 6.8 g (d) 2.51 g

10. How many grams are contained in 1 g atom of Fe?
    (a) 28 g (b) 10 g
    (c) 1 g (d) 56 g

11. The formula of barium sulfate is
    (a) $Ba(SO_4)_2$ (b) $Ba_2(SO_4)_3$
    (c) Ba_2SO_4 (d) $BaSO_4$

12. Chromate and dichromate ions respectively are
    (a) CrO_4^{2-} and $Cr_2O_5^{2-}$
    (b) $Cr_2O_5^{2-}$ and CrO_4^{2-}
    (c) CrO_4^{2-} and $Cr_2O_7^{2-}$
    (d) $Cr_2O_7^{2-}$ and CrO_4^{2-}

13. The molecular weight of a compound which contains a total number of five radicals is 160, then it is
    (a) Al_2O_3 (b) Mg_3N_2
    (c) Ca_3P_2 (d) Fe_2O_3

14. The ratio of oxygen atoms present in one molecule of cupric nitrite and ferric sulfite is
    (a) 1 : 2 (b) 1 : 3
    (c) 4 : 9 (d) 2 : 3

15. Match the following

Column I	Column II
(i) sulfate	(p) HS^-
(ii) sulfite	(q) SO_4^{2-}
(iii) sulfide	(r) S^{2-}
(iv) bisulfide	(s) SO_3^{2-}

    (a) i—q, ii—s, iii—r, iv—p
    (b) i—s, ii—q, iii—r, iv—p
    (c) i—q, ii—s, iii—p, iv—r
    (d) i—q, ii—s, iii—r, iv—p

16. Select the set having correct statements.
    (i) CO_2, SO_2 are acidic oxides.
    (ii) CO, NO are neutral oxides.
    (iii) Na_2O, CaO are amphoteric oxides.
    (iv) BeO, Al_2O_3 are amphoteric oxides.
    (a) (i), (ii)
    (b) (ii), (iv)
    (c) (i), (ii), (iii)
    (d) (i), (ii), (iv)

17. Which set of ions represents basic radicals?
    (a) NH_4^+, NO_3^-
    (b) NH_4^+, Fe^{3+}
    (c) SO_4^{2-}, CO_3^{2-}
    (d) Fe^{3+}, S^{2-}

18. Which of the following formulae is not correctly given?
    (a) aluminium phosphate: $AlPO_4$
    (b) ferric sulfate: $FeSO_4$
    (c) calcium nitrate: $Ca(NO_3)_2$
    (d) mercurous chloride: Hg_2Cl_2

19. In which of the following pairs of ions does the name of the metal ion end with 'ous'?
    (a) Sn^{2+}, Fe^{3+} (b) Cu_2^{2+}, Hg_2^{2+}
    (c) Cu^{2+}, Hg_2^{2+} (d) Pb^{2+}, Pb^{4+}

20. Which of the following pairs can never show variable valencies?
    (a) Fe, Sn (b) Cu, Pb
    (c) Ca, K (d) Mn, Cr

Advanced and Olympiads

Single Choice

1. In which of the following is the element given the correct symbol as per Dalton's theory?

Element	Symbol
(i) silver	●
(ii) lead	Ⓛ
(iii) mercury	
(iv) gold	Ⓖ

   (a) (i), (ii), (iii) (b) (i), (iii), (iv)
   (c) (ii), (iii), (iv) (d) (i), (ii), (iii), (iv)

2. In which pair are both the symbols incorrectly given?
   (a) iron (Fe) and silver (Ag)
   (b) cobalt (CO) and aluminium (AL)
   (c) nickel (ni) and gold (Au)
   (d) phosphorus (Ph) and argon (Ar)

3. The ratio of the number of molecules present in a given mass of oxygen and sulfur trioxide is
   (a) 2 : 5 (b) 3 : 2
   (c) 2 : 3 (d) 5 : 2

4. The mass of one C atom is
   (a) 12 g (b) 1.99×10^{-23} g
   (c) 6.02×10^{23} g (d) 6 g

5. How many atoms of each type of element are present in the chemical formula $(CH_3)_3CCH_2CCHO$?

	Carbon	Hydrogen	Oxygen
(a)	6	9	2
(b)	6	12	1
(c)	4	3	1
(d)	4	6	1

6. Match the following

Column I	Column II
(i) monovalent	(p) borate
(ii) bivalent	(q) cuprous
(iii) trivalent	(r) carbide
(iv) tetravalent	(s) stannous

(a) i—r, ii—p, iii—q, iv—s
(b) i—s, ii—q, iii—p, iv—r
(c) i—q, ii—s, iii—p, iv—r
(d) i—q, ii—s, iii—r, iv—p

Multiple Choice

7. Which of the following statements is/are true about an atom?
(a) Atoms are not able to exist independently.
(b) Atoms are the basic units from which molecules and ions are formed.
(c) Atoms are always neutral in nature.
(d) Atoms aggregate in large numbers to form the matter that we can see, feel or touch.

8. Which of the following is/are the correct formula/e?
(a) SiO_2 (b) BaCl
(c) H_2S (d) $NaHSO_4$

9. Which of the following are the best examples of the law of conservation of mass?
(a) 2 g of hydrogen combines with 16 g of oxygen to form 18 g of water
(b) A sample of air increases in volume when heated at constant pressure but its mass remains unaltered
(c) When 6 g of carbon is heated in a vacuum there is no change in mass
(d) 12 g of carbon combines with 32 g of oxygen to form 44 g of CO_2

10. Which of the following represents a polyatomic ion?
(a) phosphate (b) nitride
(c) sulfite (d) sulfate

11. Which of the following element(s) has a symbol having two letters?
(a) aluminium (b) lead
(c) oxygen (d) tin

12. Chemical formula gives information about
(a) total number of atoms of each element present in the compound
(b) nature of the compound
(c) physical properties of the compound
(d) which elements are present in the compound

13. Which of the following ion is/are divalent?
(a) Cu^{2+} (b) BO_4^{3-}
(c) SO_4^{2-} (d) Zn^{2+}

14. Which set includes all species with same magnitude of charge?
(a) zinc ion, calcium ion, cupric ion
(b) ferrous ion, ferric ion, stannic ion
(c) sulfite, sulfate, carbonate
(d) phosphate, borate, nitride

Comprehension Type

Comprehension I: In the modern view, valency is equal to the number of electrons that an atom can share or lose or gain during a chemical reaction. The number of electrons present in the valence orbit of an atom are called valence electrons. In order to attain the octet state (8 valence electrons), every atom either loses or gains or shares electrons. There are certain elements that exhibit more than one valency in their ions (or compounds). When an atom of an element not only loses its valence electrons but also loses electrons from its penultimate orbit, variable valency is observed.

15. In which of the following pairs do both the metals show variable valency?
(a) Na, Mg (b) Na, Pb
(c) Sn, Pb (d) Zn, Pb

16. If mercury can exist in 1 and 2 valency states as Hg^+ and Hg^{2+}, the possible pair of compounds with Cl^- and O^{2-} may be
(a) Hg_2Cl_2, $HgCl_2$ (b) Hg_2O, HgO
(c) HgO, HgO_2 (d) both (a) and (b)

Comprehension II: The formula of a binary compound, that is, a compound formed by two elements only, can be obtained by transposing their valencies. Suppose an element P has a valency *n* and element Q has valency *m*, the compound formed between P and Q usually has the formula P_nQ_m. The subscripts should be divided by a common factor, if any.

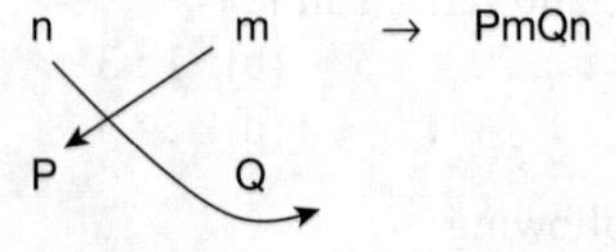

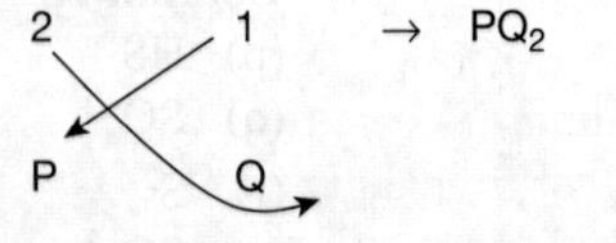

17. The formula formed when calcium ion reacts with carbonate ion is
(a) $Ca(CO_3)_2$ (b) $Ca_3(CO_3)_2$
(c) Ca_2CO_3 (d) $CaCO_3$

18. The ions Mg^{2+} and PO_4^{3-} combine to form
(a) $Mg_2(PO_4)_3$ (b) $Mg_3(PO_4)_3$
(c) $Mg_3(PO_4)_2$ (d) $Mg_2(PO_4)_2$

Matrix Matching

19. Match the following

Column I	Column II
(a) CO_3^{2-}	(p) $PO_4^{3-} = 1P + 4(O) = 5$
(b) PO_4^{3-}	(q) $CO = 1C + 1(O) = 2$
(c) P_2O_5	(r) $CO_3^{2-} = 1C + 3(O) = 4$
(d) CO	(s) $P_2O_5 = 2P + 5(O) = 7$

20. Match the following

Column I	Column II
(a) sodium	(p) plumbum
(b) gold	(q) stannum
(c) tin	(r) aurum
(d) lead	(s) natrium

Integer Type

21. What is the valency of sulfur in sulfur trioxide (SO_3)?

22. Find the value of n in Na_nBO_3?

23. Calculate the sum of the valencies of nitrogen, helium and sodium.

24. How many of the following ions possess either +2 or –2 charge? Consider only magnitude and not sign.
Sulfate, nitrate, carbonate, cupric ion, mercuric ion, potassium ion, stannous ion, phosphate, zinc ion

25. How many of the following metals can show variable valency?
Ca, Zn, Fe, Sn, Pb, Cr, Mn, Cu, Hg, Na

26. How many of the following are acidic radicals?
S^{2-}, NO^{2-}, NH^{4+}, CH_3COO^-, CO_3^{2-}, Fe^{3+}, SO_4^{2-}, Cu^{2+}, PO_4^{3-}

Answer Keys

Fill in the Blanks

1. carbon 2. monoatomic 3. element
4. 1 : 1 5. Al_2O_3 6. 4
7. tri 8. HNO_3 9. 6.022×10^{23}
10. tri 11. 2 : 3 12. 1 : 3
13. aurum 14. 63.5

True or False

1. T 2. T 3. F 4. T 5. F
6. T 7. F 8. F 9. T 10. F
11. T 12. F 13. T 14. F

Very Short Answer Type Questions

1. Atoms are the building blocks of all matter.
2. The symbols of aluminium and iron are Al and Fe respectively.
3. (i) Na for sodium from natrium
(ii) Ag for silver from argentum.
4. The mass of 1 mole of ammonia is 17 g.
$NH_3 = 14 + 3 \times 1 = 17$
5. The molecular mass of H_2SO_3 is 82 u (2 + 32 + 48 = 82 u).
6. The formula mass of ($Na_2CO_3.10H_2O$) is 286 u as follows
Na = 23 u, C = 12 u, O = 16 u, H = 1 u
So, formula mass of $Na_2CO_3.10H_2O = 2 \times 23 + 12 + 48 + 10 \times 18 = 286$ u.
7. CO_2, SO_2 and H_2O are triatomic molecules.
8. As number of atoms = 6.022×10^{23} = 1 mol of helium = 4 u of helium.
9. The ratio by mass of sulfur and hydrogen in hydrogen sulfide (H_2S) is 32 : 2 or 16 : 1.
10. Mg^{2+} and Ca^{2+} are bivalent metal ions and Fe^{+3} and Cr^{+3} are trivalent metal ions (cation).
11. (a) 1 mole of phosphine = 31 + 3 = 34 g
(b) 1 mole of ethane = 24 + 6 = 30 g
So 1 mole of phosphine has more mass than 1 mole ethane.
12. Number of moles of helium

$$= \frac{\text{weight in g}}{\text{atomic weight of helium}} = \frac{48}{4} = 12$$

13. The molar mass of phosphorus molecule (P_4) = 31 × 4 = 124 g.
14. Acetate ion is CH_3COO^- and nitrite is NO_2^-.

Short Answer Type Questions

1. An ionic compound is made up of ions, or, it contains a cation which is a positive ion and an anion which is a negative ion. For example, potassium chloride is an ionic compound made up of K^+ and Cl^- ions.
A molecular compound is made up of molecules. For example, ammonia (NH_3), carbon dioxide (CO_2).
2. (a) ZnO (b) $Al(OH)_3$ (c) $Fe_2(SO_4)_3$

3. (a) EBr_3 (b) E_2O_3 (c) $E_2(SO_4)_3$
(d) $E_2(CO_3)_3$ (e) $E(NO_3)_3$

4. Here E is a phosphorus-like element and its oxides are E_2O_3 and E_2O_5, that is, P_2O_3 and P_2O_5.

5. The possible compounds and their formulae are as follows:
   (a) Zinc fluoride: ZnF_2
   (b) Hydrogen sulfide: H_2S
   (c) Ammonia: NH_3
   (d) Stannous chloride: $SnCl_2$ and stannic chloride: $SnCl_4$
   (e) Potassium oxide: K_2O
   (f) Sulfur dioxide: SO_2 and sulfur trioxide: SO_3

6. Polyatomic ions are a group of atoms which carry a fixed charge (either positive or negative) on them and behave as ions. For example, ammonium ion (NH_4^+), sulfate ion (SO_4^{2-}), carbonate ion (CO_3^{2-}).

7. Chemical formula of a compound or element is the symbolic representation of its composition. It represents
   - the number and kind of atoms present per molecule of the compound,
   - molar mass of the compound,
   - one mole of the compound.

8. In H_2S there are three atoms:
   (a) 2 atoms of hydrogen + 1 atom of sulfur = three (3) atoms
   In phosphate there are five atoms:
   (b) 1 atom of phosphorus + 4 atoms of oxygen = five (5) atoms

9. (a) $Al_2(SO_4)_3$ is aluminium sulfate
   (b) $CaCl_2$ is calcium chloride
   (c) K_2SO_4 is potassium sulfate
   (d) KNO_3 is potassium nitrate
   (e) $CaCO_3$ is calcium carbonate

10. (a) Law of conservation of mass
   (b) Polyatomic ion
   (c) The formula unit mass of $Ca_3(PO_4)_2$ =

$$310 = (3\times Ca) + (2\times P) + (8\times O)$$
$$= (3\times 40) + (2\times 31) + (8\times 16)$$
$$= (120 + 62 + 128) = 310 \text{ u}$$

   (d) Na_2CO_3, $(NH_4)_2SO_4$

11. (a) The molecular formula of aluminium chloride: $AlCl_3$. Valency of Al is 3.
   Another radical present is chloride (Cl^-) and its valency is 1.
   (b) The molecular formula of aluminium oxide: Al_2O_3. Valency of Al is 3.
   Another radical present is oxide (O^{2-}) and its valency is 2.
   (c) The molecular formula of aluminium nitride: AlN. Valency of aluminium is 3.
   Another radical present is nitride (N^{3-}) and its valency is 3.

12. Some elements have multiple valencies, which means that they show variable valency.
   Ferrous is written as iron (II) and ferric is written as iron (III).

Metal	Radicals	Valency
Iron	Ferrous [iron (II)] Ferric [iron (III)]	2 3
Copper	Cuprous [copper (I)] Cupric [copper (II)]	1 2

Long Answer Type Questions

1. See text part.

2. (i) Molecular mass of H_2 = Atomic mass of hydrogen × 2 = 1 × 2 = 2 u
   (ii) Molecular mass of O_2 = Atomic mass of oxygen × 2 = 16 × 2 = 32 u
   (iii) Molecular mass of Cl_2 = Atomic mass of chlorine × 2 = 35.5 × 2 = 71 u
   (iv) Molecular mass of CO_2 = (Atomic mass of carbon × 1) + (Atomic mass of oxygen × 2)
   = 12 + (16 × 2) = 12 + 32 = 44 u
   (v) Molecular mass of CH_4 = (Atomic mass of carbon × 1) + (Atomic mass of hydrogen × 4)
   = 12 + (1 × 4) = 12 + 4 = 16 u.
   (vi) Molecular mass of C_2H_6 = (Atomic mass of carbon × 2) + (Atomic mass of hydrogen × 6)
   = (12 × 2) + (1 × 6) = 24 + 6 = 30 u
   (vii) Molecular mass of C_2H_4 = (Atomic mass of carbon × 2) + (Atomic mass of hydrogen × 4)
   = (12 × 2) + (1 × 4) = 24 + 4 = 28 u
   (viii) Molecular mass of NH_3 = (Atomic mass of nitrogen × 1) + (Atomic mass of hydrogen × 3)
   = (14 × 1) + (1 × 3) = 14 + 3 = 17 u
   (ix) Molecular mass of CH_3OH = (Atomic mass of carbon × 1) + (Atomic mass of hydrogen × 3) + (Atomic mass of oxygen × 1) + (Atomic mass of hydrogen × 1)
   = 12 + 3 + 16 + 1 = 32 u

3. (a) Mass of container containing 5 mol of C atoms = 5 × 12 = 60 g
   Mass of container containing 5 mol of Na atoms = 5 × 23 = 115 g
   The container that Krish has is heavier.
   (b) Both containers have the same number of atoms as they contain the same number of moles.

4. (a) (i) Magnesium chloride: $MgCl_2$
   (ii) Calcium oxide: CaO
   (iii) Copper nitrate: $Cu(NO_3)_2$
   (iv) Aluminium chloride: $AlCl_3$
   (v) Calcium carbonate: $CaCO_3$

(b)

Compounds	Elements
(i) Quick lime: Calcium oxide, CaO	Calcium, oxygen
(ii) Hydrogen bromide: HBr	Hydrogen, bromine
(iii) Baking powder: Sodium hydrogen carbonate, $NaHCO_3$	Sodium, hydrogen, carbon, oxygen
(iv) Potassium sulfate: K_2SO_4	Potassium, sulfur, oxygen

5. Molar mass of sodium = 23 g

1 mole atom = 6.022×10^{23} atoms

23 g sodium contains = 6.022×10^{23} atoms

1 g sodium contains = 6.022×10^{23} atoms

100 g sodium contains = 6.022×10^{23} atoms = 2.618×10^{24} atoms

By using the formula, we find number of atoms in 100 g Fe.

Number of atoms of an element in a given mass

$$= \frac{\text{Given mass}}{\text{Gram atomic mass 100 g}} \times \text{Avogadro's number}$$

$$= \frac{100\text{ g}}{56\text{ g}} \times 6.022 \times 10^{23} = 1.075 \times 10^{24} \text{ atoms}$$

It means 100 g of sodium has more number of atoms compared to 100 g of iron.

6. (a) It is not possible to see it with the naked eye as an atom is extremely small in size. Generally, the radius of an atom is of the order of nanometres. For example, atomic radius of the hydrogen atom is 10^{-10} m.

(b) 1 mol of carbon atoms = 6.022×10^{23} atoms

Molar atomic mass = 12 g

6.022×10^{23} carbon atoms weigh = 12 g

1 carbon atom weighs

$$= \frac{\text{Weight of carbon}}{\text{number of carbon atom}}$$

$$= \frac{12}{6.022 \times 10^{23}} = 1.99 \times 10^{-23} \text{ g}$$

Competition Window (Objective Type)

Topic-wise MCQs

1. (c)	**2.** (d)	**3.** (b)	**4.** (c)	**5.** (d)
6. (c)	**7.** (d)	**8.** (b)	**9.** (a)	**10.** (d)
11. (d)	**12.** (a)	**13.** (b)	**14.** (a)	**15.** (d)
16. (a)	**17.** (d)	**18.** (b)	**19.** (d)	**20.** (b)
21. (b)	**22.** (d)	**23.** (a)	**24.** (d)	**25.** (d)
26. (a)	**27.** (d)	**28.** (b)	**29.** (d)	**30.** (a)
31. (c)	**32.** (b)	**33.** (b)	**34.** (c)	**35.** (c)

Miscellaneous

1. (c)	**2.** (c)	**3.** (b)	**4.** (b)	**5.** (c)
6. (d)	**7.** (a)	**8.** (c)	**9.** (d)	**10.** (d)
11. (d)	**12.** (c)	**13.** (d)	**14.** (c)	**15.** (d)
16. (d)	**17.** (b)	**18.** (b)	**19.** (b)	**20.** (c)

Advanced and Olympiads

Single Choice

1. (c) **2.** (b) **3.** (d) **4.** (b) **5.** (b) **6.** (c)

Multiple Choice

7. (bcd) **8.** (acd) **9.** (ad) **10.** (acd) **11.** (abd) **12.** (ad) **13.** (acd) **14.** (acd)

Comprehension Type

15. (c) **16.** (d) **17.** (d) **18.** (c)

Matrix Matching

19. (a)—(r); (b)—(p); (c)—(s); (d)—(q)

20. (a)—(s), (b)—(r), (c)—(q), (d)—(p)

Integer Type

21. 6 **22.** 3 **23.** 6 **24.** 6 **25.** 6 **26.** 6

Hints and Solutions

Competition Window (Objective Type)

Topic-wise MCQs

7. 1 atom of Cu + 1 atom of sulfur + 9 atoms of O + 10 atoms of H. Hence, total number of atoms in this compound is 21.

8. Nitrogen gas exists as a diatomic molecule hence, its symbol is N_2.

9. The correct symbols of these elements are as follows:

(a) cobalt: Co (b) carbon: C

(c) aluminium: Al (d) helium: He

Hence, only helium is given the correct symbol.

10. The correct symbol of nickel is Ni.

12. The number of atoms present in the molecule of an element is called its atomicity.

13. The atomicity of phosphorus is 4, that is, P_4.

17. Carbon dioxide (CO_2) consists of one atom of carbon and two atoms of oxygen.

18. In P_2O_3 and P_2O_5 the valencies of phosphorus are 3, 5 respectively.

19. Calcium can have a valency of 1, copper can have valencies of 1 and 2, and iron can have valencies of 2 and 3.

20. Sr can show only one valency of 2.

21. Sulfide is bivalent, that is, SO_3^{2-}.

22. Sodium can have only Na^+ and not Na^{2+} hence this pair is not possible.

23. Sulfate is a polyatomic ion consisting of sulfur and oxygen, that is, SO_4^{2-}.

25. The formula of ferrous sulfate is $FeSO_4$. Fe $\rightarrow$ +2 charge; SO_4 $\rightarrow$ –2 charge.

28. Stannic and stannous ions are written respectively as Sn^{4+}, Sn^{2+}.

32. $A(NO_3)_3 \Rightarrow$ valency of A is + 3

Metal nitride $A^{+3}\ N^{-3} \Rightarrow AN$

35.

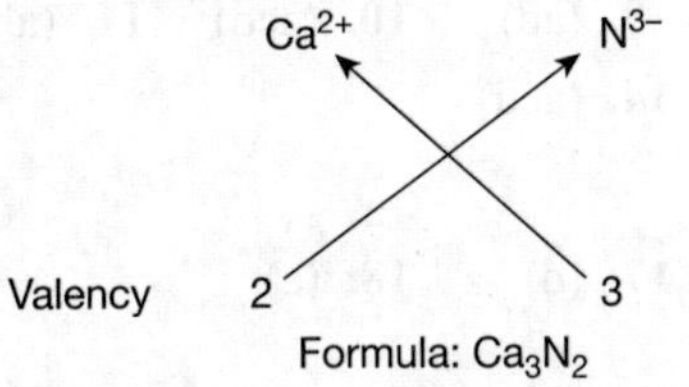

Miscellaneous

1. The correct symbol of sulfur is ⊕. The used symbol ⊗ is of phosphorus.

3.

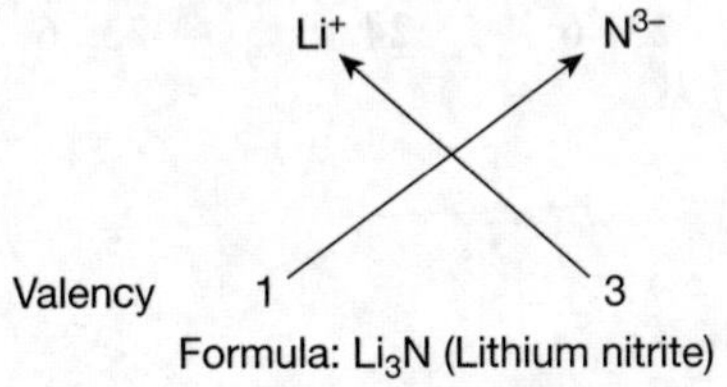

4. P: 50%, Q: 50%; P : Q = 5 : 2.5 = 2 : 1; Hence, P_2Q.

5. 4 g sulfur is in 100 g of compound, hence 32 g sulfur is in

$$= \left(\frac{100}{4} \times 32\right) = 80 \text{ g of compound.}$$

8. Law of constant proportions, according to which a pure chemical compound always contains the same elements combined together in the same definite proportion of weight.

9. 58.5 g of sodium chloride contains = 23 g of sodium

So, 5.85 g of sodium chloride contains

$$= \frac{23}{58.5} \times 5.85 = 2.3 \text{ g}$$

10. 1 g atom = 1 mol

11.

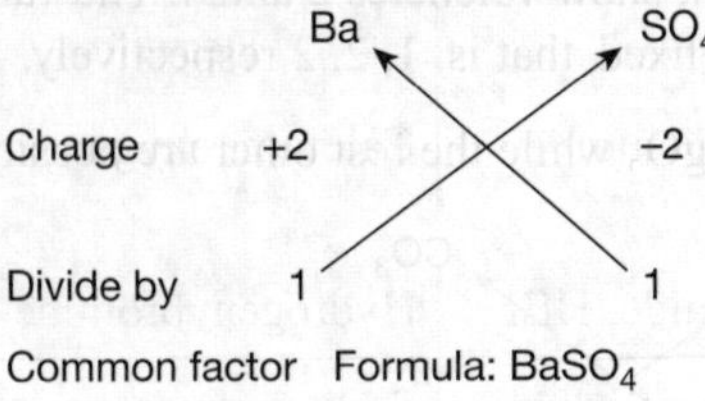

16. Both Na_2O and CaO are basic oxides so this is not a correct statement. The rest are correct.

17. NH_4^+, Fe^{3+} are basic radicals while the rest are acidic radicals.

18. Ferric sulfate is $Fe_2(SO_4)_3$ and $FeSO_4$ is ferrous sulfate.

19. In the case of lower charge, the name ends with 'ous'; as in Cu_2^{2+} (cuprous), Hg_2^{2+} (mercurous). In the case of higher charge, the name ends with 'ic'; as in Fe^{3+} (ferric).

20. The elements of group 1 (K) and group 2 (Ca) have fixed valencies, 1 and 2 respectively. The rest of the element exhibit variable valencies.

Advanced and Olympiads

Single Choice

1. The correct symbol of silver is ⓢ. The used symbol ● is of carbon. The rest other are correct.

2. Cobalt is Co and aluminium is Al. In option (b) both the symbols are incorrect. Nickel is Ni and phosphorus is P.

3. 32 g of oxygen contains X molecules

32 g of SO_3 contains $\frac{80}{32}$ X molecules

Ratio of number of molecules present in 32 g of O_2 and

$$SO_3 = X : \frac{80}{32} X = 2.5 : 1 = 5 : 2.$$

4. The mass of 6.023×10^{23} atoms of carbon = 12 g

So, the mass of 6.023×10^{23} atoms of carbon

$$= \frac{12}{6.023 \times 10^{23}} \text{ g} = 1.99 \times 10^{-23} \text{ g}$$

Multiple Choice

7. Atoms of inert gases exist in monoatomic or independent form.

8. The valency of barium is 2 and that of chlorine is 1. Thus, the correct formula is $BaCl_2$.

10. Sulfite (SO_3^{2-}) sulfate (SO_4^{2-}) and phosphate (PO_4^{3-}) ions are polyatomic as they contain more than one type of atom. N^{3-} is monoatomic.

11. The symbols of tin, aluminium and lead are Sn, Al and Pb respectively.

14. Zinc ion, calcium ion, cupric ion: +2 charge; sulfite, sulfate, carbonate: –2 charge; phosphate, borate, nitride: –3 charge; ferrous, ferric, stannic ions have +2, +3 and +4 charge respectively.

Comprehension Type

15. Both Sn and Pb can show valencies 2 and 4. The valencies of Na, Mg, Zn are fixed, that is, 1, 2, 2 respectively.

16. Hg cannot form HgO_2 while the rest other are possible.

17.

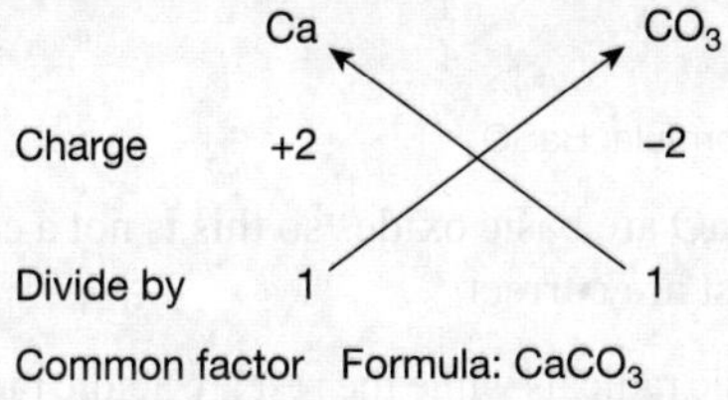

18.

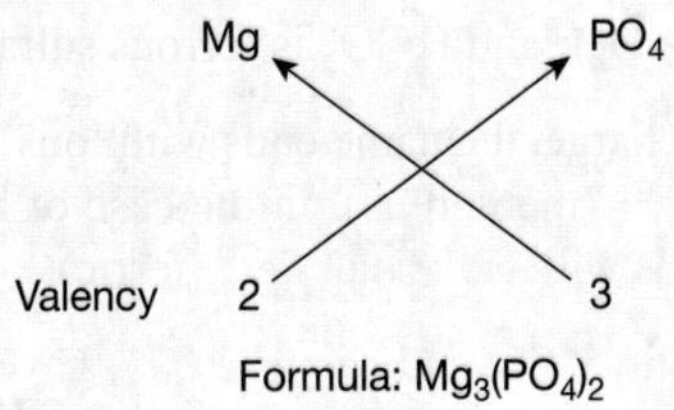

Integer Type

21. Sulfur has valency 6 in sulfur trioxide (SO_2).

22. The formula of sodium borate is Na_3BO_3. So, $n = 3$.

23. N (7) = 2, 5; valency = 5. He (2) = 2; valency = 0. Na (11) = 2, 8, 1; valency = 0;

So, sum of valencies = 5 + 0 + 1 = 6

24. Sulfate, carbonate: –2 charge; cupric, mercuric, stannous, zinc ions: +2 charge; nitrate (–1), potassium ion (+1) and phosphate (–3).

25. Fe, Sn, Pb, Cr, Mn, Cu and Hg can show variable valencies while Ca (2), Zn (2) and Na (1) have fixed valency.

26. S^{2-}, NO_2^-, CH_3COO^-, CO_3^{2-}, SO_4^{2-} and PO_4^{3-} are acidic radicals.

CHAPTER

4 Chemical Reactions

LEARNING OBJECTIVES

After studying this unit, you will be able to understand:

- Chemical reactions, reactants, products
- How to define a chemical equation and to write it
- Combination reactions, decomposition reactions, displacement reactions, double displacement reactions, oxidation and reduction reactions and differentiate between them

Introduction

Chemical reactions are the processes by which new substances with new properties are formed. Chemical reactions involve chemical changes. During a chemical reaction, a rearrangement of atoms takes place between the reacting species to form new species having entirely different properties. Chemical reactions involve breaking of old chemical bonds which exist between the atoms of the reacting substances, and then forming of new chemical bonds between the rearranged atoms of the new substances. During a chemical reaction, atoms of one element do not change into those of another element however only a rearrangement of atoms takes place in a chemical reaction. Let us now discuss reactants and products of a chemical reaction.

Chemical reactions in everyday life

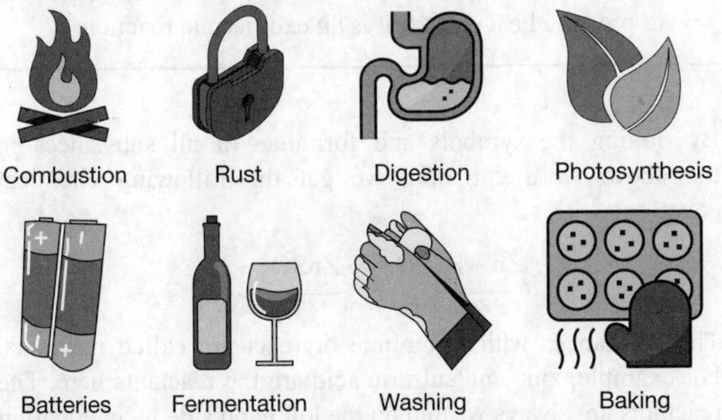

Reactant: The substances which take part in a chemical reaction are known as reactants.

For example, in the breaking up of calcium carbonate into calcium oxide and carbon dioxide, calcium carbonate is the reactant. Similarly, sodium and water are the reactants when they react.

$$\underset{\text{Calcium carbonate}}{CaCO_3} \rightarrow \underset{\text{Calcium oxide}}{CaO} + \underset{\text{Carbon dioxide}}{CO_2}$$

Products: The new substances produced as a result of a chemical reaction are known as products.

For example, hydrogen and sodium hydroxide are the products of the reaction between sodium and water.

$$\underset{\text{Sodium}}{Na} + \underset{\text{Water}}{H_2O} \rightarrow \underset{\text{Sodium hydroxide}}{NaOH} + \underset{\text{Hydrogen}}{H_2}$$

When a magnesium ribbon is heated, it burns in air with a dazzling white flame to form a white powder called magnesium oxide.

$$\underset{\text{(As ribbon)}}{\text{Magnesium}} + \underset{\text{(From air)}}{\text{Oxygen}} \xrightarrow{\text{Heat}} \underset{\text{(White powder)}}{\text{Magnesium oxide}}$$

The burning of magnesium in air to form magnesium oxide is an example of a chemical reaction.

The magnesium ribbon used here usually has a coating of magnesium oxide on its surface, which is formed by the slow action of oxygen of air on it. So, before burning in air, the magnesium ribbon is cleaned by rubbing with a sand paper. This is done to remove the protective layer of magnesium oxide from the surface of the magnesium ribbon so that it may readily combine with the oxygen of air (on heating). The dazzling (very bright) white light given out during the burning of a magnesium ribbon is harmful to the eyes. So, the magnesium ribbon should be burned by keeping it as far as possible from the eyes.

Characteristics or Tests of Chemical Reactions

The conversion of reactants into products in a chemical reaction is often accompanied by some features (changes) which can be observed easily. These easily observable features which take place as a result of chemical reactions are called *characteristics* of chemical reactions. The important characteristics of chemical reactions are:

- change in state,
- formation of a precipitate,
- change in temperature,
- change in colour, and
- evolution of a gas

Change in State

Many chemical reactions are characterised by a change in state—the physical state changes from solid to liquid and gas. For example, the combustion reaction of candle wax is characterised by a change in state from solid to liquid and gas (because wax is a solid, combustion of wax leads to a liquid at room temperature, carbon dioxide produced by the combustion of wax is a gas).

Formation of a Precipitate

Some chemical reactions are characterised by the formation of a precipitate. A precipitate is a solid product which separates out from the solution during a chemical reaction. For example, when potassium iodide solution is added to a solution of lead nitrate, a yellow precipitate of lead iodide is formed.

Change in Temperature

Many chemical reactions are characterised by a change in temperature. For example, when quicklime reacts with water, slaked lime is formed and a huge amount of heat energy is produced. When barium hydroxide [$Ba(OH)_2$] is added to ammonium chloride (NH_4Cl) taken in a test tube and mixed with a glass rod, barium chloride, ammonia and water are formed. A lot of heat energy is absorbed.

Change in Colour

Some chemical reactions are characterised by a change in colour. For example, when citric acid reacts with potassium permanganate solution, the purple colour of potassium permanganate solution disappears.

Evolution of a Gas

Many chemical reactions are characterised by the evolution of a gas. For example, when zinc granules react with dilute sulfuric acid, bubbles of hydrogen gas are produced.

Illustrations

1. Why should a magnesium ribbon be cleaned before burning in air?

   Solution: Magnesium ribbon is cleaned or rubbed with a sand paper before burning to remove the layer of magnesium oxide formed on its surface (due to exposure to air). This layer interferes in the further reaction of the pure metal.

2. Which one is a chemical change—rusting of iron or melting of iron?

   Solution: Rusting of iron is a chemical change.

3. Why is respiration considered as an exothermic reaction?

   Solution: As respiration results in the oxidation of glucose to produce heat energy, it is an exothermic reaction.

Chemical Equations

All chemical reactions are represented by chemical equations. The method of representing a chemical reaction with the help of symbols and formulae of the substances involved in it, is called a chemical equation. For example, zinc metal reacts with dilute sulfuric acid to give zinc sulfate and hydrogen gas. This is known as the *word equation.*

KNOWLEDGE BOOSTER

The amount of product actually obtained in a reaction is known as the actual yield.

Balancing of a chemical equation refers to establishing a mathematical relationship between the quantity of reactants and products. These quantities are expressed as grams or moles.

By putting the symbols and formulae of all substances in the above word equation, we get the following chemical equation:

$$\underbrace{Zn + H_2SO_4}_{\text{Reactants}} \rightarrow \underbrace{ZnSO_4 + H_2}_{\text{Products}}$$

The substances which combine or react are called reactants. For example, zinc and sulfuric acid are the reactants here. The reactants are always written on the left hand side in an equation with a plus sign (+) between them.

The new substances produced in a reaction are called products. Zinc sulfate and hydrogen are the products in the case. The products are always written on the right hand side in an equation with a plus sign (+) between them.

When potassium nitrate is heated, it gives potassium nitrite and oxygen. This reaction may be represented in the form of a chemical equation as follows.

$$\underset{\text{Potassium nitrate}}{KNO_3} \rightarrow \underset{\text{Potassium nitrite}}{KNO_2} + \underset{\text{Oxygen}}{O_2}$$

Rules for Writing a Chemical Equation

These rules have to be followed while writing a chemical equation.

(i) The reactants taking part in the reaction are written in terms of their symbols or molecular formulae on the left hand side of the equation.

(ii) A plus (+) sign is added between the formulae of the reactants.

(iii) The products of the reaction are written in terms of their symbols or molecular formulae on the right hand side of the equation.

(iv) A plus (+) sign is added between the formulae of the products.

(v) In between the reactants and the products, an arrow sign (→) is inserted to show which way the reaction is occurring.

$$P + Q \rightarrow R + S$$

In this chemical equation, P and Q are the reactants, and R and S are the products. The arrow indicates that the reaction proceeds towards the formation of R and S.

Balanced and Unbalanced Chemical Equations

Balanced Equation

A balanced chemical equation has an equal number of atoms of different elements in the reactants and products. For example,

$$Zn + H_2SO_4 \rightarrow ZnSO_4 + H_2$$

Let us count the number of atoms of all the elements in the reactants and products separately.

	In reactants	In products
Number of Zn atoms	1	1
Number of H atoms	2	2
Number of S atoms	1	1
Number of O atoms	4	4

Unbalanced Equation

An unbalanced chemical equation has an unequal number of atoms of one or more elements in the reactants and products. For example,

$$H_2 + O_2 \rightarrow H_2O$$

Let us count the number of hydrogen atoms and oxygen atoms in the reactants as well as products.

	In reactants	In products
Number of H atoms	2	2
Number of O atoms	2	1

Balancing of Chemical Equations

The process of making the number of different types of atoms equal on both the sides of an equation is known as *balancing of the equation*. Simple equations can be balanced by the hit and trial method in following steps.

Step 1: Examine the number of atoms of different elements present in the unbalanced equation.

For example, hydrogen burns in oxygen to form water. This reaction can be written in an equation form as:

$$H_2 + O_2 \rightarrow H_2O$$

Let us count the number of hydrogen atoms and oxygen atoms in the reactants and products.

	In reactants	In product
Number of H atoms	2	2
Number of O atoms	2	1

Step 2: Balancing of number of atoms on both sides.

The number of hydrogen atoms is equal on both the sides (2 each), but the number of oxygen atoms is unequal. There are 2 oxygen atoms on the left side but only 1 oxygen atom on the right side. To have 2 oxygen atoms on the right side, we multiply H_2O by 2 and write $2H_2O$.

$$H_2 + O_2 \rightarrow 2H_2O$$

Let us count the number of various atoms on both the sides again.

	In reactants	In product
Number of H atoms	2	4
Number of O atoms	2	2

Though the number of oxygen atoms has become equal (2 on both sides), the number of hydrogen atoms has now become unequal. There are 2 hydrogen atoms on the left side but 4 hydrogen atoms on the right side. To have 4 hydrogen atoms on the left side, we multiply H_2 by 2 and write $2H_2$.

$$2H_2 + O_2 \rightarrow 2H_2O$$

Let us count the number of various atoms on both the sides once again.

	In reactants	In product
Number of H atoms	4	4
Number of O atoms	2	2

This chemical equation contains an equal number of atoms of hydrogen and oxygen on both the sides, so this is a balanced equation.

Balancing of Ionic Equations

In these equations, charge balancing of atoms/ions on both sides of the equation has to be done. For example,

Initial $Cu^{2+}(aq) + H_2S \rightarrow CuS(s) + H^{+}(aq)$

Balanced $Cu^{2+}(aq) + H_2S \rightarrow CuS(s) + 2H^{+}(aq)$

We have balanced the charges. It was +2 on the LHS and we have made it +2 on the RHS. The number of hydrogen, copper and sulfur atoms are also balanced on both sides.

Compound	Formula	Ions involved
Sodium chloride	$NaCl$	Na^+ and Cl^-
Potassium nitrate	KNO_3	K^+ and NO_3^-
Magnesium chloride	$MgCl_2$	Mg^{2+} and Cl^-
Magnesium oxide	MgO	Mg^{2+} and O^{2-}
Calcium chloride	$CaCl_2$	Ca^{2+} and Cl^-
Calcium oxide	CaO	Ca^{2+} and O^{2-}
Ammonium chloride	NH_4Cl	NH_4^+ and Cl-
Ammonium sulfate	$(NH_4)_2SO_4$	NH_4^+ and SO_4^{2-}
Cupric sulfate	$CuSO_4$	Cu^{2+} and SO_4^{2+}
Cupric chloride	$CuCl_2$	Cu^{2+} and Cl^-

Making Equations More Informative

The equation which gives more information about the chemical reaction is known as a more informative or information giving equation. A chemical equation can be made more informative by these three ways:

- by indicating the 'physical states' of the reactants and products,
- by indicating the 'heat changes' taking place in the reaction, and
- by indicating the 'conditions' under which the reaction takes place.

We will discuss these three points in detail one by one.

1. **Physical states of reactants and products**: There can be four physical states for the reactants and products of a chemical reaction: solid, liquid, aqueous solution and gas.
   - Solid state is indicated by the symbol (*s*).
   - Liquid state is indicated by the symbol (*l*).
   - Aqueous solution (solution made in water) is indicated by the symbol (*aq*).
   - Gaseous state is indicated by the symbol (*g*).

   The physical states of the reactants and products are shown by putting the above *state symbols* just after their symbols or formulae in an equation.

   For example, $Zn(s) + H_2SO_4(aq) \rightarrow ZnSO_4(aq) + H_2(g)$

2. **Heat changes in an equation**: There are two types of reactions on the basis of heat changes involved—exothermic reactions and endothermic reactions.

   (i) Exothermic reaction: Those reactions in which heat is evolved are known as exothermic reactions. For example,
   - Any combustion process is always exothermic.
   $$C + O_2 \rightarrow CO_2 + \text{Heat} + 94.2 \text{ kcal}$$
   - Any neutralisation reaction is always exothermic
   $$HCl + NaOH \rightarrow NaCl + H_2O + \text{Heat} + 13.7 \text{ kcal}$$
   - When heat is evolved during product formation.
   $$N_2 + 3H_2 \rightarrow 2NH_3 + 22.4 \text{ kcal}$$

   (ii) Endothermic reaction: A reaction in which heat or energy is absorbed during the course of the reaction is known as an endothermic reaction. For example,
   - Any dissociation reaction is always endothermic.
   $$CaCO_3 \xrightarrow{\Delta} CaO + CO_2 - \text{Heat}$$
   - Salt hydrolysis reaction is always endothermic.
   $$FeCl_3 + 3H_2O \rightarrow 3HCl + Fe(OH)_3 - \text{Heat}$$
   - Formation of some compounds is endothermic.
   $$N_2 + O_2 \rightarrow 2NO - 42 \text{ kcal}$$
   - The electrolysis of water to form hydrogen and oxygen is also an endothermic reaction.

3. **Conditions under which the reaction takes place**: If heat is needed for a reaction to occur, then the heat sign delta (Δ) is put over the arrow of the equation. If the reaction takes place in the presence of a catalyst, then the symbol or formula of the catalyst is also written above or below the arrow sign in the equation. Presence of catalyst make the reaction possible even at lower temperature with a faster rate. For example,

$$\underset{\text{Potassium chlorate}}{2KClO_3(s)} \xrightarrow[MnO_2]{\Delta} \underset{\text{Potassium chloride}}{2KCl(s)} + \underset{\text{Oxygen}}{3O_2(g)}$$

$$\underset{\text{Carbon monoxide}}{CO(g)} + \underset{\text{Hydrogen}}{2H_2(g)} \xrightarrow[ZnO+CrO_3]{300 \text{ atm; } 300°C} \underset{\substack{\text{Methanol} \\ \text{(Methyl alcohol)}}}{CH_3OH(l)}$$

Illustrations

1. Write a balanced equation for the reaction between magnesium and hydrochloric acid. Name the product obtained and identify the type of reaction.

   Solution: $Mg(s) + 2HCl(\text{dil.}) \rightarrow MgCl_2(aq) + H_2(g)$

   The products obtained are magnesium chloride and hydrogen gas. The reaction is a displacement reaction.

2. Predict whether the following chemical equations present the correct balanced chemical equation or not.

   (i) $Cu(s) + 2H_2SO_4 \rightarrow CuSO_4 + SO_2 + 2H_2O$

   (ii) $\underset{\text{Zinc nitrate}}{Zn(NO_3)_2} \rightarrow 2ZnO + 4NO_2 + O_2$

Solution:

(i) No, the equation is not correctly balanced as it is not specified whether the acid is dilute or concentrated. Moreover the state of reactants and products is not mentioned here. Hence the correct balanced equation will be as follows,

$$\underset{\text{Copper}}{Cu(s)} + \underset{\text{(Conc.)}}{2H_2SO_4} \rightarrow \underset{\text{Copper sulphate}}{CuSO_4(s)} + \underset{\text{Sulphur dioxide}}{SO_2(g)} \uparrow + \underset{\text{Water}}{2H_2O(l)}$$

(ii) No, the equation is not correctly balanced as atoms of different elements are not the same in the reactants and products. Moreover, the states of the reactants and products are not specified here. So the correct balanced equation will be as follows,

$$\underset{\text{Zinc nitrate}}{2Zn(NO_3)_2(s)} \rightarrow \underset{\text{Zinc oxide}}{2ZnO(s)} + 4NO_2(g) + O_2(g)$$

3. Write the balanced chemical equations from the following word equations. Also include the physical state of each element or compound involved in the reaction.

(i) Sodium metal reacts with water to give hydrogen gas and an aqueous hydroxide solution.

(ii) Potassium chlorate on heating gives potassium chloride and oxygen gas.

Solution:

(i) $2Na(s) + 2H_2O(l) \rightarrow H_2(g) + 2NaOH(aq)$

(ii) $2KClO_3(s) \rightarrow 2KCl(s) + 3O_2(g)$

Types of Chemical Reactions

Chemical reactions are mainly of these types:

- combination reactions,
- decomposition reactions,
- displacement reactions,
- double displacement reactions and
- oxidation and reduction reactions.

Combination Reactions

The reactions in which two or more substances combine to form a single compound, are known as combination reactions. In a combination reaction, two or more elements can combine to form a compound, two or more compounds can combine to form a new compound, or an element and a compound can combine to form a new compound. For example,

$CaO + H_2O \rightarrow Ca(OH)_2$

(i) Magnesium and oxygen combine when heated, to form magnesium oxide.

$$\underset{\text{Magnesium}}{2Mg(s)} + \underset{\text{Oxygen}}{O_2(g)} \xrightarrow{\text{Combination}} \underset{\text{Magnesium oxide}}{2MgO(s)}$$

(ii) Hydrogen burns in oxygen to form water.

$$\underset{\text{Hydrogen}}{2H_2(g)} + \underset{\text{Oxygen}}{O_2(g)} \xrightarrow{\text{Combination}} \underset{\text{Water}}{2H_2O(l)}$$

(iii) Carbon (coal) burns in air to form carbon dioxide.

$$\underset{\text{Carbon (coal)}}{C(s)} + \underset{\text{Oxygen (From air)}}{O_2(g)} \xrightarrow{\text{Combination}} \underset{\text{Carbon dioxide}}{CO_2(g)}$$

Illustrations

1. Write balanced chemical equations for the following reactions:

(i) Sodium metal reacts with water to form sodium hydroxide and hydrogen gas.

(ii) Magnesium reacts with dil. HCl.

Solution:

The reactions for these equations are

(i) $$\underset{\text{Sodium}}{2Na(s)} + \underset{\text{Water}}{2H_2O(l)} \rightarrow \underset{\text{Sodium hydroxide}}{2NaOH(aq)} + \underset{\text{Hydrogen}}{H_2(g)}$$

(ii) $Mg + 2HCl \rightarrow MgCl_2 + H_2$

2. Write the chemical equations for the reactions taking place when:

(i) Iron reacts with steam

(ii) Copper is heated in air

Solution:

(i) $3Fe(s) + 4H_2O(g) \rightarrow Fe_3O_4(s) + 4H_2(g)$

(ii) $2Cu + O_2 \xrightarrow{\text{Heat}} 2CuO(s)$

Decomposition Reactions

The reactions in which a compound splits up into two or more simpler substances is called a decomposition reaction. Decomposition reactions are carried out by applying heat, light or electricity. Heat, light or electricity provides energy which breaks a compound into two or more simpler compounds. A decomposition reaction is the opposite of a combination reaction. Here are some examples of decomposition reactions.

$$2Pb(NO_3)_2 \xrightarrow{\text{Heat}} 2PbO + 4NO_2\uparrow + O_2\uparrow$$

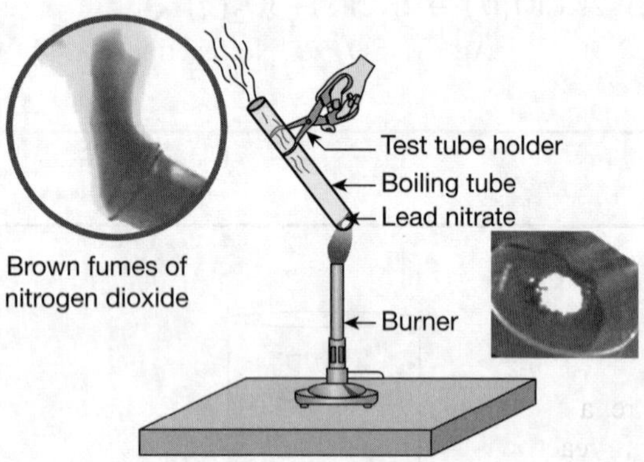

(i) When calcium carbonate is heated, it decomposes to give calcium oxide and carbon dioxide.

$$\underset{\text{Calcium carbonate (Limestone)}}{CaCO_3(s)} \xrightarrow[\text{(Decomposition)}]{\text{Heat}} \underset{\text{Calcium oxide (Lime)}}{CaO(s)} + \underset{\text{Carbon dioxide}}{CO_2(g)}$$

(ii) When lead nitrate is heated strongly, it breaks down to form simpler substances like lead monoxide, nitrogen dioxide and oxygen.

$$\underset{\text{Lead nitrate (Colourless)}}{2Pb(NO_3)_2(s)} \xrightarrow[\text{(Decomposition)}]{\text{Heat}} \underset{\text{Lead monoxide (Yellow)}}{2PbO(s)} + \underset{\text{Nitrogen dioxide (Brown fumes)}}{4NO_2(g)} + \underset{\text{Oxygen}}{O_2(g)}$$

(iii) When electric current is passed through molten sodium chloride, it decomposes into sodium metal and chlorine gas.

$$\underset{\text{Sodium chloride (Molten)}}{2NaCl(l)} \xrightarrow[\text{(Decomposition)}]{\text{Electricity}} \underset{\text{Sodium metal}}{2Na(s)} + \underset{\text{Chlorine gas}}{Cl_2(g)}$$

Illustrations

1. Why are decomposition reactions called the opposite of combination reactions? Write equations for these reactions.

Solution: A decomposition reaction is one in which a compound is broken down into simpler elements or compounds. For example,

$$\underset{\text{Calcium carbonate}}{CaCO_3(s)} \xrightarrow{\Delta} CaO(s) + CO_2(g)$$

It is just the opposite of a combination reaction in which two or more reactants combine to form a single product.

$$CaO(s) + CO_2(g) \rightarrow \underset{\text{Calcium carbonate}}{CaCO_3(s)}$$

2. Why do we store silver chloride in dark-coloured bottles?

Solution: We store silver chloride in dark-coloured bottles to avoid its decomposition in the sunlight.

$$CaO(s) + H_2O(l) \rightarrow \underset{\text{Calcium hydroxide}}{Ca(OH)_2(aq)}$$

Reactivity Series of Metals

On the basis of reaction with water, metals can be arranged in the decreasing order of reactivity. This arrangement of metals in the decreasing order of reactivity in the form of a series is known as the activity or reactivity series of metals.

K, Na, Ca, Mg, Al, Zn, Fe, Pb [H], Cu, Hg, Ag, Au ⟶

Potassium being the most reactive metal is placed at the top of the list and the least reactive metal being gold is placed at the bottom of the list.

Although hydrogen is a non-metal, it is included in this series because it can form a positive ion. It would occupy a position based on the formation of its positive ion.

The ability of metals to reduce water to hydrogen decreases on going down the series.

$$K > Na > Ca > Mg > Al > Zn \ldots\ldots\ldots\ldots$$

Some metals are so reactive that they will displace hydrogen from water. This reaction is called *hydrolysis*.

$$2K + 2H_2O \rightarrow 2KOH + H_2$$

Lead and the metals that are further below in the activity series show no reaction with water, even when the metal is hot and steam is used.

Illustrations

1. Can we store $CuSO_4$ solution in an iron vessel?

Solution: As iron is above copper in the reactivity series, it is more reactive than copper. Hence, iron displaces copper from copper sulfate solution. As a result, the liberated copper appears in the form of a red precipitate. Therefore, $CuSO_4$ solution cannot be stored in an iron vessel.

$$Fe + CuSO_4(g) \rightarrow FeSO_4 + Cu$$

2. Zinc can produce hydrogen on reacting with acids but silver cannot. Explain.

Solution: As zinc is above hydrogen in reactivity series, it is more reactive. It can displace hydrogen from acids such as HCl and H_2SO_4.

$$Zn + \underset{\text{(dil.)}}{H_2SO_4} \rightarrow ZnSO_4 + H_2$$

As Ag is below hydrogen and less reactive than hydrogen, it cannot displace hydrogen from acid.

$$Ag + H_2SO_4 \rightarrow \text{No reaction}$$

Displacement Reactions

A reaction in which one element takes the place of another element in a compound is called a *displacement reaction*. In general, a more reactive element displaces a less reactive element from its compound. Here are few examples.

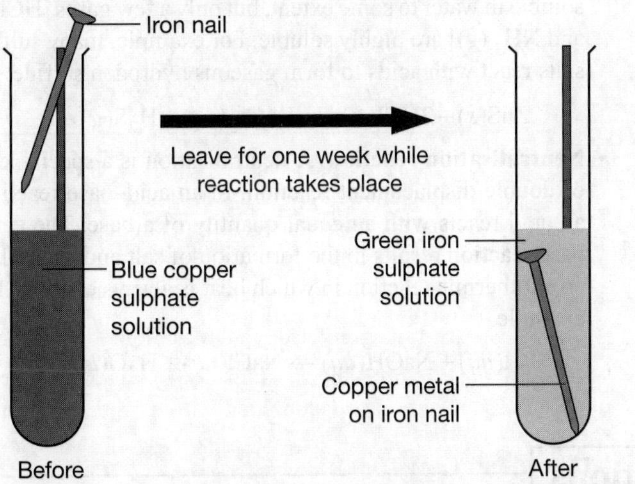

(i) When a strip of zinc metal is placed in copper sulfate solution, then zinc sulfate solution and copper are obtained.

$$\underset{\substack{\text{Copper sulphate}\\\text{(Blue solution)}}}{CuSO_4(aq)} + \underset{\substack{\text{Zinc}\\\text{(Silvery-white)}}}{Zn(s)} \rightarrow \underset{\substack{\text{Zinc sulphate}\\\text{(Colourless solution)}}}{ZnSO_4(aq)} + \underset{\substack{\text{Copper}\\\text{(Red-brown)}}}{Cu(s)}$$

Here, a displacement reaction takes place because zinc is more reactive than copper.

(ii) When a piece of magnesium metal is placed in copper sulfate solution, then magnesium sulfate solution and copper metal are formed.

$$\underset{\substack{\text{Copper sulphate}\\\text{(Blue solution)}}}{CuSO_4(aq)} + \underset{\substack{\text{Magnesium}\\\text{(Silvery white)}}}{Mg(s)} \rightarrow \underset{\substack{\text{Magnesium sulphate}\\\text{(Colourless solution)}}}{MgSO_4(aq)} + \underset{\substack{\text{Copper}\\\text{(Red-brown)}}}{Cu(s)}$$

(iii) When a piece of iron metal (say, an iron nail) is placed in copper sulfate solution, then iron sulfate solution and copper metal are formed.

$$\underset{\substack{\text{Copper sulphate}\\\text{(Blue solution)}}}{CuSO_4(aq)} + \underset{\substack{\text{Iron}\\\text{(Grey)}}}{Fe(s)} \rightarrow \underset{\substack{\text{Iron sulphate}\\\text{(Greenish solution)}}}{FeSO_4(aq)} + \underset{\substack{\text{Copper}\\\text{(Red-brown)}}}{Cu(s)}$$

Illustrations

1. Why does the colour of copper sulfate solution change when an iron nail is dipped in it? Write two observations.

Solution: The blue colour of copper sulfate solution changes due to the displacement reaction taking place. Iron displaces copper from copper sulfate solution to give the following observations:

- Pale green coloured solution of $FeSO_4$ is formed.
- Reddish-brown copper metal gets deposited.

$$Fe(s) + CuSO_4(aq) \rightarrow FeSO_4(aq) + Cu(s)$$

Double Displacement Reactions

The reactions in which two compounds react by an exchange of ions to form two new compounds is known as a double displacement reaction. Such a reaction usually occurs in solution and one of the products, being insoluble, precipitates out (separates as a solid). Some of the examples of double displacement reactions are given below.

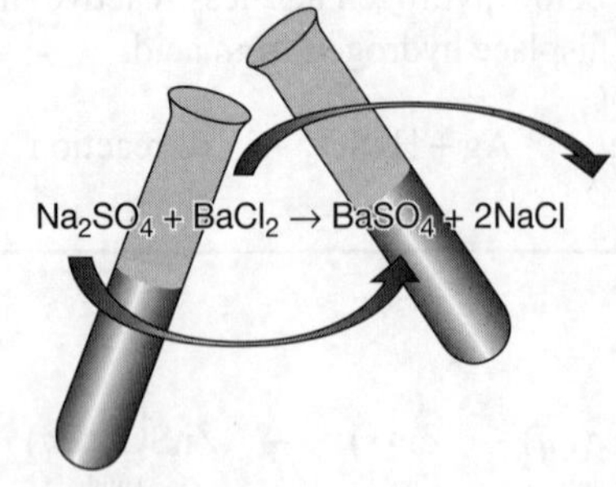

(i) When silver nitrate solution is added to sodium chloride, a white precipitate of silver chloride is formed along with sodium nitrate solution.

$$\underset{\text{Silver nitrate}}{AgNO_3(aq)} + \underset{\text{Sodium chloride}}{NaCl(aq)} \rightarrow \underset{\text{Silver chloride (White ppt.)}}{AgCl(s)} + \underset{\text{Sodium nitrate}}{NaNO_3(aq)}$$

(ii) When barium chloride solution is added to sodium sulfate solution, a white precipitate of barium sulfate is formed along with sodium chloride solution.

$$\underset{\text{Barium chloride}}{BaCl_2(aq)} + \underset{\text{Sodium sulphate}}{Na_2SO_4(aq)} \rightarrow \underset{\text{Barium sulphate (white ppt.)}}{BaSO_4(s)} + \underset{\text{Sodium chloride}}{2NaCl(aq)}$$

Types of Double Displacement Reactions

Double displacement reactions can be further classified as precipitation, gas formation and acid–base neutralisation reactions.

- **Precipitation reaction:** A precipitation reaction is possible when two solutions are mixed and a solid separates from the solution. The solid part that forms and separates from the solution is called the *precipitate*. For example,

$$Pb(NO_3)_2(aq) + 2KI(aq) \rightarrow PbI_2(s)\downarrow + 2KNO_3(aq)$$

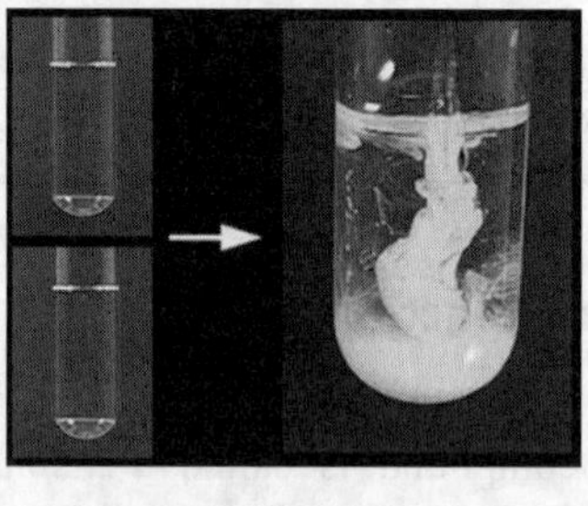

- **Gas-formation reaction:** A double displacement reaction is possible when an insoluble gas is formed. All gases are soluble in water to some extent, but only a few gases [$HCl(g)$ and NH_3 (g)] are highly soluble. For example, many sulfide salts react with acids to form gaseous hydrogen sulfide:

$$ZnS(s) + 2HCl(aq) \rightarrow ZnCl_2(aq) + H_2S(g)$$

- **Neutralisation reaction:** A neutralisation is a special kind of double displacement reaction. In an acid–base reaction, an acid reacts with an equal quantity of a base. The acid–base reaction results in the formation of salt and water. It is an exothermic reaction in which heat is always evolved. For example,

$$HCl(aq) + NaOH(aq) \rightarrow NaCl(aq) + H_2O(l) + \text{Heat}$$

Illustrations

1. Translate the following statement into a chemical equation and then balance it. Barium chloride reacts with aluminium sulfate to give aluminium chloride and a precipitate of barium sulfate. State how this reaction can be classified (two types).

Solution:

Chemical equation: $\underset{\text{Barium chloride}}{3BaCl_2(aq)} + \underset{\text{Aluminium sulphate}}{Al_2(SO_4)_3(aq)} \rightarrow$

$$\underset{\text{Barium sulphate}}{3BaSO_4(s)} + \underset{\text{Aluminium chloride}}{2AlCl_3(aq)}$$

This reaction can be classified as a double displacement reaction and a precipitation reaction.

2. What is observed when a solution of potassium iodide is added to a solution of lead nitrate? Name the type of reaction. Write a balanced chemical equation to represent the above chemical reaction.

Solution: A yellow precipitate of lead iodide is formed. It is a double displacement reaction and also a precipitation reaction.

$$Pb(NO_3)_2(aq) + 2KI(aq) \rightarrow PbI_2(s) + 2KNO_3(aq)$$

3. A solution of potassium chloride when mixed with a solution of silver nitrate, forms an insoluble white substance. Write the chemical reaction involved and also mention the type of chemical reaction.

Solution:

$$KCl(aq) + AgNO_3(aq) \rightarrow \underset{\text{Silver chloride (white ppt.)}}{AgCl(s)} + \underset{\text{Potassium nitrate}}{KNO_3(aq)}$$

The reaction is a double displacement reaction. It is also called a precipitation reaction due to the formation of the white precipitate of AgCl.

Oxidation and Reduction Reactions

The earlier concept of oxidation and reduction is based on the addition or removal of oxygen or hydrogen. So, in terms of oxygen or hydrogen, oxidation and reduction reactions can be defined as follows.

Oxidation: It is defined as:
Addition of oxygen: The addition of oxygen to a substance is called oxidation. For example,

$$2Mg + O_2 \rightarrow 2MgO$$

Removal of hydrogen: The removal of hydrogen from a substance is also called oxidation. For example,

$$4HCl + MnO_2 \rightarrow MnCl_2 + Cl_2 + 2H_2O$$

Oxidation also involves following processes:

(i) Addition of an electronegative element or increase in its ratio.

$$2SnCl_2 + Cl_2 \rightarrow 2SnCl_4$$

(ii) Removal of an electropositive element or decrease in its ratio.

$$Hg_2Cl_2 \rightarrow HgCl_2 + Hg$$

(iii) Loss of electrons by an atom, ion or molecule—de-electronation.

$$Na - 1e^- \rightarrow Na^+$$

$$Fe^{2+} - 1e^- \rightarrow Fe^{3+}$$

Reduction: It is defined as:
Removal of oxygen: The removal of oxygen from a substance is also called reduction.

$$PbO + C \rightarrow Pb + CO$$

Addition of hydrogen: The addition of hydrogen to a substance is called reduction.

$$H_2 + Cl_2 \rightarrow 2HCl$$

Reduction also involves following processes:

(i) Removal of an electronegative element or decrease in its ratio.

$$2FeCl_3 \rightarrow 2FeCl_2 + Cl_2$$

(ii) Addition of an electropositive element or increase in its ratio

$$2HgCl_2 \rightarrow Hg_2Cl_2 + Cl_2$$

(iii) Gain of electrons by atom or ion—electronation.

$$Cu^{2+} + e^- \rightarrow Cu^+$$

Oxidising agent: It oxidises the other substance and in the process gets reduced.

(i) The substance which gives oxygen for oxidation is called an oxidising agent.

(ii) The substance which removes hydrogen is also called an oxidising agent. For example,

$$PbS + \underset{\text{Oxidising agent}}{4O_3} \rightarrow PbSO_4 + 4O_2$$

Some common oxidants

(i) Molecules of most electronegative elements. For example, O_3, O_2, F_2, Cl_2.

(ii) Elements in their highest oxidation states in compounds. For example,

$$K\overset{+7}{Mn}O_4,\ K_2\overset{+6}{Cr_2}O_7,\ H_2\overset{+6}{S}O_4,\ H\overset{+5}{N}O_3,\ \overset{+6}{S}O_3,\ \overset{+4}{C}O_2,\ \overset{+3}{Fe}Cl_3$$

Some oxides like MgO, CaO, CrO_3, H_2O_2.

Reducing agent: It reduces the other substance and, in the process it gets oxidised.

(i) The substance which gives hydrogen for reduction is called a reducing agent.

(ii) The substance which removes oxygen is also called a reducing agent.

Some examples of reductants

- Metals from group 1 (Li, Na) and Group 2 (Mg, Ca).
- Non-metals such as C, S, P, H.
- HX such as HI, HBr, HCl.
- Metal hydrides (MH) such as LiH, CaH_2.

Redox Reactions

The oxidation and reduction reactions are also called *redox reactions* (the term *red* stands for reduction and *ox* stands for oxidation). Here, oxidation and reduction take place simultaneously, that is, one species gets oxidised while another one gets reduced. For example,

$$A_1 + A_2^{n+} \rightarrow A_1^{n+} + A_2$$

Here A_1 is oxidised into A_1^{n+} while A_2^{n+} is reduced into A_2.

(i) $2\overset{+3}{Fe}Cl_3 + \overset{+2}{Sn}Cl_2 \rightarrow 2FeCl_2 + \overset{+4}{Sn}Cl_4$

Here $SnCl_2$ is oxidised into $SnCl_4$ while $FeCl_3$ is reduced into $FeCl_2$.

(ii) When copper oxide is heated with hydrogen, copper metal and water are formed.

$$\underset{\text{Copper oxide}}{CuO} + \underset{\text{Hydrogen}}{H_2} \xrightarrow{\text{Heat}} \underset{\text{Copper}}{Cu} + \underset{\text{Water}}{H_2O}$$

Here, copper oxide is being reduced to copper while hydrogen is being oxidised to water.

Illustrations

1. What is a redox reaction? Identify the substance oxidised and the substance reduced in the following reactions.

(i) $2PbO + C \rightarrow 2Pb + CO_2$

(ii) $2MnO_2 + 4HCl \rightarrow MnCl_2 + 2H_2O + Cl_2$

Solution: The reaction in which oxidation and reduction take place simultaneously is called a redox reaction.

(i) Substance oxidised: C
Substance reduced: PbO

(ii) Substance oxidised: HCl
Substance reduced: MnO_2

Neutralisation Reactions

Reaction between an acid and a base to form a salt and water, liberating heat in the process is called a neutralisation reaction.

$$\text{Acid} + \text{Base} \rightarrow \text{Salt} + \text{Water} + \text{Heat}$$

$$HCl + NaOH \rightarrow NaCl + H_2O + \text{Heat}$$

$$HNO_3 + KOH \rightarrow KNO_3 + 2H_2O + \text{Heat}$$

$$H_2SO_4 + Ca(OH)_2 \rightarrow CaSO_4 + 2H_2O + \text{Heat}$$

Heat of Neutralisation

The enthalpy change during the neutralisation of 1 g-mole of an acid and a base is called heat of neutralisation. It is also the heat of formation of water. It is always exothermic, $\Delta H = -\text{ve}$. The heat of neutralisation of a strong acid and a strong base is always 13.7 kcal/mol or 57.3 kJ/mol. It is independent of the nature of the strong acid or strong base. For example,

$$H^+ Cl^- + Na^+ OH^- \rightarrow NaCl + H_2O + 13.7 \text{ kcal/mol or } 57.3 \text{ kJ/mol}$$

$$H^+ + OH^- \rightarrow H_2O + 13.7 \text{ kcal or } 57.3 \text{ kJ/mol}$$

$$\Delta H = -13.7 \text{ kcal or } -57.3 \text{ kJ/mol}$$

Uses

- Slaked lime $[Ca(OH)_2]$ is added to reduce the acidity of soil.
- Antacid tablets with $Mg(OH)_2$ or cold milk can be used to treat acidity by neutralising the HCl in the stomach.
- Stings of ants/bees containing HCOOH (formic acid) can be neutralised by rubbing soap (NaOH).
- Stings of wasps contain an alkali and can be treated by neutralising with CH_3COOH.
- Lithium hydroxide is used to neutralise CO_2 exhaled by astronauts in spaceships or submariners in submarines.

CHAPTER AT A GLANCE

- **Chemical reactions** are the processes by which new substances with new properties are formed.
- The substances which take part in a chemical reaction are known as **reactants**.
- The new substances produced as a result of chemical reaction are known as **products**.
- The important characteristics of chemical reactions are change in state, formation of precipitate, change in temperature, change in colour and evolution of a gas.
- The method of representing a chemical reaction with the help of symbols and formulae of the substances involved in it is called a **chemical equation**.
- A **balanced chemical equation** has an equal number of atoms of different elements in the reactants and products.
- An **unbalanced chemical equation** has an unequal number of atoms of one or more elements in the reactants and products.
- A reaction in which heat or energy is released during the course of a reaction is known as an **exothermic** reaction.
- A reaction in which heat or energy is absorbed during the course of a reaction is known as an **endothermic** reaction.
- A **catalyst** is a substance which is usually used to increase the rate of a reaction. However, it does not take part in the reaction and remains unchanged in mass and properties.
- A reaction in which two or more substances combine to form a single compound, is known as a **combination reaction.**
- A reaction in which a compound splits up into two or more simpler substances is called a **decomposition reaction.**
- On the basis of their reaction with water, metals can be arranged in the decreasing order of reactivity. This arrangement of metals in the decreasing order of reactivity in the form of a series is known as the **activity** or **reactivity series** of metals.

  $$\xrightarrow{\text{K, Na, Ca, Mg, Al, Zn, Fe, Pb [H], Cu, Hg, Ag, Au}}$$

- A reaction in which one element takes the place of another element in a compound is called a **displacement reaction.**
- A reaction in which two compounds react by an exchange of ions to form two new compounds is known as a **double displacement reaction.**

- A precipitation reaction is possible when two solutions are mixed together and a solid separates from the solution. The solid part that forms and separates from the solutions is called the **precipitate**.
- A **neutralisation reaction** is an acid–base reaction in which an acid reacts with an equal quantity of a base. This results in the formation of salt and water, and is an exothermic reaction. It is an exothermic reaction in which heat is always evolved.
- **Oxidation** involves the addition of oxygen or removal of hydrogen or loss of electrons.
- **Reduction** involves removal of oxygen or addition of hydrogen or addition of electrons.
- An **oxidising agent** oxidises others but itself gets reduced.
- A **reducing agent** reduces others but itself gets oxidised.
- In a **redox reaction**, oxidation and reduction take place simultaneously—one species gets oxidised while another one gets reduced.

PRACTICE QUESTIONS

Analyse Your Concepts (School Exam Based)

Fill in the Blanks

Instructions: Complete the following statements with an appropriate word/term to be filled in the blank spaces.

1. When new substances are formed in a process, the process is called ______.
2. The new substances produced in a reaction are known as ______.
3. A reaction in which heat is given out along with the products is called ______ reaction.
4. A reaction in which energy is absorbed is known as ______ reaction.
5. Two different atoms or groups of atoms (ions) are exchanged during ______ reactions.
6. The reaction of an acid and a base to form salt and water is known as ______ reaction.
7. The uptake of oxygen by a substance is known as ______.
8. Calcium carbonate on heating gives CaO and CO_2. It is a case of ______ reaction.
9. Potassium on reaction with water displaces hydrogen from water. This reaction is called ______ reaction.
10. The reaction of $BaCl_2$ and Na_2SO_4 is an example of ______ reaction.
11. The reaction of zinc with copper sulfate is an example of ______ reaction.
12. Between K, Fe and Au, the most reactive element is ______.
13. A ______ is a substance which changes the rate of a chemical reaction without itself undergoing a chemical change.
14. The chemical reaction between hydrogen and chlorine is a ______ reaction.

True or False

Instructions: Read the following statements and write if they are true or false.

1. A chemical reaction is characterised by a change of temperature, colour, and so on.
2. The number of atoms of each element is conserved during a chemical reaction.
3. Reactants and products are written on the left and right-hand side respectively in a reaction.
4. Rusting is a double decomposition reaction.
5. The reaction of CaO with water to give calcium hydroxide is known as a combination reaction.
6. The formation of gas bubbles in a liquid during a reaction is called effervescence.
7. The amount of product actually formed in a reaction is called the actual yield.
8. $PbS + 4H_2O_2 \rightarrow PbSO_4 + 4H_2O$. Here H_2O_2 acts as a reductant.
9. A neutralisation reaction is always exothermic in nature.
10. Reaction between liquid O_2 and liquid hydrogen to form water is a case of exothermic reaction.
11. A neutralisation reaction is endothermic in nature.
12. When a piece of copper is added to the silver nitrate solution, it turns blue in colour.

Very Short Answer Type Questions

1. What happens during a chemical reaction?
2. Define chemical energy.
3. Write the example of an endothermic reaction.
4. Write the balanced chemical equation with state symbols for the following reaction.

Solutions of barium chloride and sodium sulfate in water react to give insoluble barium sulfate and a solution of sodium chloride.

5. Give an example of a double displacement reaction.
6. Write the chemical equation involved when calcium metal reacts with aqueous hydrochloric acid to produce a solution of calcium chloride and hydrogen gas.
7. What happens when sodium reacts with water?
8. Write the reaction that occurs when potassium chlorate ($KClO_3$) is heated?
9. Why should a magnesium ribbon be cleaned before burning it in air?
10. Complete and balance the following equations:
(a) $KOH + \_\_\_\_\_\_\_\_ \rightarrow K_2SO_4 + H_2O$
(b) $Mg(OH)_2 + \_\_\_\_\_\_\_\_ \rightarrow MgCO_3 + H_2O$
11. Why is silver nitrate solution stored in coloured bottles?
12. Why does iron turn the solution green on reaction with aqueous copper sulfate?
13. Write the reaction when hydrogen peroxide is exposed to sunlight.
14. Ferrous sulfate decomposes with the evolution of a gas having a characteristic odour of burning sulfur. Write the chemical reaction involved and identify the type of reaction.
15. What happens when water is added to quick lime? Write the chemical equation.

Short Answer Type Questions

1. Give one example each of which illustrates the following characteristics of a chemical reaction: (a) evolution of a gas, (b) change of colour.
2. What is a balanced chemical equation? Why should chemical equations be balanced?
3. Consider the following chemical reaction:
$$X + \text{Barium chloride} \rightarrow \underset{\text{White ppt.}}{Y} + \text{Sodium chloride}$$
(a) Identify X and Y.
(b) Identify the type of reaction.
4. A solution of potassium chloride when mixed with silver nitrate solution, forms an insoluble white substance. Write the chemical reaction involved and also mention the type of the chemical reaction.
5. Zinc liberates hydrogen gas when reacted with dilute hydrochloric acid, whereas copper does not. Explain why.
6. Why are decomposition reactions called the opposite of combination reactions? Write equations for these reactions.
7. What type of chemical reactions take place when
(a) limestone is heated?
(b) a magnesium wire is burnt in air?
(c) electricity is passed through water?
(d) ammonia and hydrogen chloride are mixed?
(e) silver bromide is exposed to sunlight?
8. Name the reducing agent in the following reaction:
$$3MnO_2 + 4Al \rightarrow 3Mn + 2Al_2O_3$$
State which is more reactive, Mn or Al and why.
9. State the type of chemical reactions with chemical equations that take place when ammonia and hydrogen chloride gases are mixed.
10. Write the chemical equation of a reaction in which the formation of precipitate occurs.
11. Name the type of chemical reaction represented by the following equations:
(a) $CaO + H_2O \rightarrow Ca(OH)_2$
(b) $3BaCl_2 + Al_2(SO_4)_3 \rightarrow 2AlCl_3 + 3BaSO_4$
(c) $2FeSO_4 \rightarrow Fe_2O_3 + SO_2 + SO_3$
12. Define the term decomposition reaction. Give one example each of thermal decomposition and electrolytic decomposition.
13. Silver chloride turns grey when exposed to sunlight. Give the chemical equation to explain it.
14. What would you observe on adding zinc granules to freshly prepared ferrous sulfate solution? Give reasons for your answer.
15. In an industrial process used for the manufacture of sodium hydroxide, a gas A is formed as a byproduct. The gas A reacts with lime water to give a compound B which is used as a bleaching agent in the chemical industry. Identify A and B. Also give the chemical equations of the reactions involved.
16. Why is respiration considered an exothermic reaction? Explain.

Long Answer Type Questions

1. How do the following help in bringing about a chemical change?
(a) Pressure (b) Light (c) Catalyst
2. Discuss the importance of the decomposition reaction in the metal industry with three points.
3. 2 g of ferrous sulfate crystals are heated in a dry boiling tube.
(a) List any two observations.
(b) Name the type of chemical reaction taking place.
(c) Write the balanced chemical equation for the reaction and name the products formed.
4. Write the balanced chemical equations for the following reactions.
(a) Calcium hydroxide + Carbon dioxide → Calcium carbonate + Water

(b) Zinc + Silver nitrate → Zinc nitrate + Silver
(c) Aluminium + Copper chloride → Aluminium chloride + Copper
(d) Barium chloride + Potassium sulfate → Barium sulfate + Potassium chloride

5. You might have noted that when copper powder is heated in a china dish, the reddish-brown surface of the copper powder becomes coated with a black substance.
(a) Why has this black substance formed?
(b) What is this black substance?
(c) Write the chemical equation of the reaction that takes place.
(d) How can the black coating on the surface be turned reddish-brown?

6. You are provided with two containers made of copper and aluminium. You are also provided with solutions of (a) dilute HCl, (b) dilute HNO_3, (c) $ZnCl_2$ and (d) H_2O. In which of the above containers can these solutions be kept?

COMPETITION WINDOW (OBJECTIVE TYPE)

[For NEET, JEE (Main and Advanced), NTSE, KVPY and Olympiads]

Topic-wise MCQs

Chemical reactions and chemical equations

1. A chemical equation is balanced to satisfy one of the following laws in chemical reactions. This law is called:
(a) law of conservation of magnetism
(b) law of conservation of mass
(c) law of conservation of motion
(d) law of conservation of momentum

2. One of the following does not happen during a chemical reaction. This is
(a) formation of new substances with completely different properties
(b) breaking of old chemical bonds and formation of new chemical bonds
(c) atoms of one element change into those of another element to give new products
(d) rearrangement of atoms occurs to form new products

3. Which one of the following processes involve chemical reactions?
(a) storing of oxygen gas under pressure in a gas cylinder
(b) liquefaction of air
(c) keeping petrol in a china dish in the open
(d) heating copper wire in the presence of air at a high temperature

4. Which of the following is not a characteristic of a chemical change?
(a) it is irreversible
(b) no net energy change is involved
(c) a new substance is formed
(d) it involves absorption or liberation of energy

5. The chemical reaction between quicklime and water is characterised by:
(a) formation of slaked lime precipitate
(b) evolution of hydrogen gas
(c) change in colour of the product
(d) change in temperature of mixture

6. In which of the following chemical equations do the abbreviations represent the correct states of the reactants and products involved at reaction temperature?
(a) $2H_2(l) + O_2(l) \rightarrow 2H_2O(g)$
(b) $2H_2(g) + O_2(l) \rightarrow 2H_2O(l)$
(c) $2H_2(g) + O_2(g) \rightarrow 2H_2O(l)$
(d) $2H_2(g) + O_2(g) \rightarrow 2H_2O(g)$

7. In the equation, $KOH + HNO_3 \rightarrow KNO_3 + H_2O$ nitric acid behaves as
(a) a dehydrating agent
(b) a nitrating
(c) an acid
(d) an oxidising agent

8. A student added dilute HCl to a test tube containing zinc granules and made following observation:
(a) the solution remained colourless
(b) the solution becomes green in colour
(c) the zinc surface became dull and black
(d) a gas evolved which burnt with a pop sound

9. $xFe + yH_2O \rightarrow Fe_3O_4 + H_2$
What are the values of x and y respectively?
(a) 3, 3 (b) 4, 4
(c) 3, 4 (d) 3, 2

10. What will be the value of a, b and c in following equation?
$H_2C_2O_4 + aH_2O_2 \rightarrow bCO_2 + cH_2O$
(a) a = 1, b = 2, c = 2
(b) a = 2, b = 2, c = 2
(c) a = 1, b = 2, c = 4
(d) a = 2, b = 2, c = 1

Types of chemical reactions

11. Which of the following is not an endothermic reaction?
(a) $CaCO_3 \rightarrow CaO + CO_2$
(b) $C_6H_{12}O_6 + 6O_2 \rightarrow 6CO_2 + 6H_2O$
(c) $6CO_2 + 6H_2O \rightarrow C_6H_{12}O_6 + 6O_2$
(d) $2H_2O \rightarrow 2H_2 + O_2$

12. Electrolysis of water is a decomposition reaction. The mole ratio of hydrogen and oxygen gases liberated during electrolysis of water is
(a) 1:1
(b) 2:1
(c) 4:1
(d) 1:2

13. Which of the following reactions involves the combination of two elements?
(a) $CaO + CO_2 \rightarrow CaCO_3$
(b) $4K + O_2 \rightarrow 2K_2O$
(c) $NH_3 + HCl \rightarrow NH_4Cl$
(d) $SO_2 + \frac{1}{2}O_2 \rightarrow SO_3$

14. Which of the following is not a decomposition reaction?
(a) $N_2 + 3H_2 \rightarrow 2NH_3$
(b) $2H_2O \xrightarrow{\text{Electrolysis}} H_2 + O_2$
(c) $CaCO_3 \xrightarrow{\text{Heat}} CaO + CO_2$
(d) $2HgO \xrightarrow{\text{Heat}} 2Hg + O_2$

15. Thermal decomposition of potassium carbonate will produce
(a) carbon dioxide
(b) oxygen
(c) potassium hydroxide
(d) no other product

16. Consider the reaction
$NaBr(aq) + AgNO_3(aq) \rightarrow NaNO_3(aq) + AgBr(s)$
(a) double displacement reaction
(b) displacement reaction
(c) decomposition reaction
(d) combination reaction

17. One of the following is an exothermic reaction. This is
(a) process of photosynthesis
(b) electrolysis of water
(c) conversion of limestone into quicklime
(d) process of respiration

18. The following reaction is an example of a
$4NH_3(g) + 5O_2(g) \rightarrow 4NO(g) + 6H_2O(g)$
(i) displacement reaction
(ii) combination reaction
(iii) redox reaction
(iv) neutralisation reaction
(a) (i) and (iv)
(b) (ii) and (iii)
(c) (i) and (iii)
(d) (iii) and (iv)

19. Barium chloride on reacting with ammonium sulfate forms barium sulfate and ammonium chloride. Which of the following correctly represents the type of the reaction involved?
(i) displacement reaction
(ii) precipitation reaction
(iii) combination reaction
(iv) double displacement reaction
(a) (i) only
(b) (ii) only
(c) (iv) only
(d) (ii) and (iv)

20. Which among the following is(are) double displacement reaction(s)?
(i) $Pb + CuCl_2 \rightarrow PbCl_2 + Cu$
(ii) $Na_2SO_4 + BaCl_2 \rightarrow BaSO_4 + 2NaCl$
(iii) $C + O_2 \rightarrow CO_2$
(iv) $CH_4 + 2O_2 \rightarrow CO_2 + 2H_2O$
(a) (i) and (iv)
(b) (ii) only
(c) (i) and (ii)
(d) (iii) and (iv)

21. Which of the following is(are) an endothermic process(es)?
(i) dilution of sulfuric acid
(ii) sublimation of dry ice
(iii) condensation of water vapours
(iv) evaporation of water
(a) (i) and (iii)
(b) (ii) only
(c) (iii) only
(d) (ii) and (iv)

22. In the double displacement reaction between aqueous potassium iodide and aqueous lead nitrate, a yellow precipitate of lead iodide is formed. While performing the activity if lead nitrate is not available, which of the following can be used in place of lead nitrate?
(a) lead sulfate (insoluble)
(b) lead acetate
(c) ammonium nitrate
(d) potassium sulfate

23. The following reaction is used for the preparation of oxygen gas in the laboratory
$2KClO_3(s) \xrightarrow[\text{Catalyst}]{\text{Heat}} 2KCl(s) + 3O_2(g)$
Which of the following statement(s) is(are) correct about the reaction?
(a) It is a decomposition reaction and endothermic in nature.
(b) It is a combination reaction.
(c) It is a decomposition reaction and is accompanied by the release of heat.
(d) It is a photochemical decomposition reaction and is exothermic in nature.

Reactivity and precipitation

24. A white precipitate will be formed if we add common salt solution to:
(a) $Ba(NO_3)_2$ solution
(b) KNO_3 solution
(c) $AgNO_3$ solution
(d) $Mg(NO_3)_2$ solution

25. What happens when a copper rod is dipped in iron sulfate solution?
(a) reaction is exothermic
(b) copper displaces iron
(c) no reaction takes place
(d) blue colour of copper sulfate solution is obtained

26. An acid that can decolourise purple coloured potassium permanganate solution is
(a) carbonic acid
(b) hydrochloric acid
(c) sulfuric acid
(d) citric acid

27. When copper turnings are added to silver nitrate solution, a blue-coloured solution is formed after some time. It is because copper:
(a) is oxidised to Cu^{2+}
(b) is reduced to Cu^{2+}
(c) forms a blue-coloured complex with $AgNO_3$
(d) displaces silver from the solution

28. A dilute solution of sodium carbonate was added to two test tubes—one containing dilute HCl (x) and the other containing dilute NaOH (y). The correct observation was
(a) a colourless gas liberated in test tube x
(b) a colourless gas liberated in test tube y
(c) a brown-coloured gas liberated in test tube x
(d) a brown-coloured gas liberated in test tube y

Oxidation and reduction

29. The removal of oxygen from a substance is known as
(a) corrosion
(b) oxidation
(c) rancidity
(d) reduction

30. The process of respiration is:
(a) a combination reaction which is endothermic
(b) an oxidation reaction which is exothermic
(c) an oxidation reaction which is endothermic
(d) a reduction reaction which is exothermic

31. The chemical reaction involved in the corrosion of iron metal is that of
(a) oxidation as well as combination
(b) reduction as well as displacement
(c) oxidation as well as displacement
(d) reduction as well as combination

32. Hydrogen sulfide (H_2S) is a strong reducing agent. Which of the following reactions shows its reducing action?
(a) $CuSO_4 + H_2S \rightarrow CuS + H_2SO_4$
(b) $Cd(NO_3)_2 + H_2S \rightarrow CdS + 2HNO_3$
(c) $Pb(NO_3)_2 + H_2S \rightarrow PbS + 2CH_3COOH$
(d) $2FeCl_3 + H_2S \rightarrow 2FeCl_2 + 2HCl + S$

33. In which of the following reactions does H_2O_2 act as a reducing agent?
(a) $Ag_2O + H_2O_2 \rightarrow 2Ag + H_2O + O_2$
(b) $PbS + 4H_2O_2 \rightarrow PbSO_4 + 4H_2O$
(c) $2FeSO_4 + H_2SO_4 + H_2O_2 \rightarrow Fe_2(SO_4)_3 + 2H_2O$
(d) $2KI + H_2O_2 \rightarrow 2KOH + I_2$

34. When the gases sulfur dioxide and hydrogen sulfide mix in the presence of water, the reaction is $SO_2 + 2H_2S \rightarrow 2H_2O + 3S$. Here hydrogen sulfide is acting as:
(a) an oxidising agent
(b) a reducing agent
(c) a catalyst
(d) a dehydrating agent

35. The following reaction describes the rusting of iron $4Fe + 3O_2 \rightarrow Fe_2O_3$. Which one of the following statements is(are) correct?
(a) metallic iron is oxidised to Fe_2O_3
(b) this is an example of an oxidation reaction
(c) Fe is reduced to Fe_2O_3
(d) both (a) and (b)

Miscellaneous

1. When copper powder is heated, it gets coated with
(a) red copper oxide
(b) black copper oxide
(c) reduction
(d) none of these

2. A reaction of a type PQ + RS → PS + RQ, involves
(a) no chemical change
(b) decomposition of PQ and RS
(c) exchange of ions of PQ and RS
(d) combination of PQ and RS

3. In which of the following reactions are gaseous species involved?
(a) decomposition of calcium carbonate
(b) reaction of zinc with dilute H_2SO_4
(c) decomposition of $KClO_3$
(d) all of these

4. Consider the following equation of the chemical reaction of a metal A:

$$4A + 3O_2 \rightarrow 2A_2O_3$$

This equation represents:

(a) combination reaction as well as an oxidation reaction
(b) oxidation reaction as well as a displacement reaction
(c) decomposition reaction as well as an oxidation reaction
(d) combination reaction as well as a reduction reaction

5. Which of the following can be decomposed by the action of light?
(a) $PbCl_2$ (b) KCl
(c) CuCl (d) AgCl

6. A dilute ferrous sulfate solution was gradually added to a beaker containing acidified permanganate solution. The light purple colour of the solution fades and finally disappears. Which of the following is the correct explanation for the observation?
(a) $KMnO_4$ is an oxidising agent, it oxidises $FeSO_4$
(b) $FeSO_4$ acts as an oxidising agent and oxidises $KMnO_4$
(c) the colour disappears due to dilution; no reaction is involved
(d) $KMnO_4$ is an unstable compound and decomposes in the presence of $FeSO_4$ to a colourless compound.

7. Which of the following are exothermic processes?
(i) reaction of water with quicklime
(ii) dilution of an acid
(iii) evaporation of water
(iv) sublimation of camphor (crystals)
(a) (i) and (ii)
(b) (ii) and (iii)
(c) (i) and (iv)
(d) (iii) and (iv)

8. Which of the following processes is not endothermic?
(a) the formation of CS_2 from carbon and sulfur
(b) the formation of NO from N_2 and O_2
(c) the decomposition of calcium carbonate
(d) the formation of ammonia from nitrogen and hydrogen

9. Which among the following statement(s) is(are) true? Exposure of silver chloride to sunlight for a long duration turns grey due to
(i) the formation of silver by decomposition of silver chloride
(ii) sublimation of silver chloride
(iii) decomposition of chlorine gas from silver chloride
(iv) oxidation of silver chloride
(a) (i) only
(b) (i) and (iii)
(c) (ii) and (iii)
(d) (iv) only

10. Which of the following are the characteristics of chemical reactions?
(i) evolution of a gas
(ii) change in physical state
(iii) formation of a precipitate
(iv) constant temperature
(a) (i), (ii) and (iii)
(b) (ii) and (iii)
(c) (i) and (iv)
(d) (i), (ii), (iii) and (iv)

11. Which of the following gases can be used for storage of a fresh sample of an oil for a long time?
(a) carbon dioxide or oxygen
(b) nitrogen or oxygen
(c) carbon dioxide or helium
(d) helium or nitrogen

12. Which of the following are combination reactions?
(i) $2KClO_3 \xrightarrow{\text{Heat}} 2KCl + 3O_2$
(ii) $MgO + H_2O \rightarrow Mg(OH)_2$
(iii) $4Al + 3O_2 \rightarrow 2Al_2O_3$
(iv) $Zn + FeSO_4 \rightarrow ZnSO_4 + Fe$
(a) (i) and (iii)
(b) (iii) and (iv)
(c) (ii) and (iv)
(d) (ii) and (iii)

13. When ferrous sulfate is heated strongly, decomposition occurs to give ferric oxide, SO_2 and SO_3. During this reaction, the colour changes from reactant to product is
(a) brown to green
(b) green to brown
(c) yellow to green
(d) green to blue

14. A shining brown coloured element M on heating in air becomes black in colour. This element M and the black-coloured compound are respectively
(a) Zn, ZnO (b) Cu, CuO
(c) Ni, NiO (d) Fe, FeO

15. The vapour density of fully dissociated ammonium chloride would be
(a) slightly less than half of that of ammonium chloride
(b) half of that of ammonium chloride
(c) double of that of ammonium chloride
(d) none of these

16. The pressure inside a closed glass vessel in which hydrogen iodide is heated would
(a) increase with temperature
(b) decrease with temperature
(c) remain constant
(d) first decrease and then increase

Advanced and Olympiads

Single Choice

1. Which of the following are examples of an electrochemical reaction?
(a) dissociation of fused potassium chloride
(b) combustion of glucose

(c) dissociation of acidulated water
(d) both (a) and (c)

2. Which of the following sets represents the correct statements?
(i) Heat of neutralisation is always exothermic
(ii) Heat of salt hydrolysis is exothermic
(iii) Thermal decomposition is an endothermic reaction
(iv) Heat of formation may be endothermic or exothermic
(a) (i), (ii)
(b) (i), (iii)
(c) (i), (iii), (iv)
(d) (i), (ii), (iii), (iv)

3. Select the set showing correct statements.
(i) Mixing, heat, light, electricity and pressure are needed for a reaction.
(ii) A reaction is characterised by a change of state, change of colour, and so on.
(iii) In an exothermic reaction, heat is evolved.
(iv) In an endothermic reaction, heat is evolved.
(v) In endothermic reaction, heat is absorbed.
(a) (i), (ii), (iii), (iv)
(b) (i), (ii), (iii), (v)
(c) (ii), (iii), (iv), (v)
(d) (i), (ii), (iii)

4. In a balanced equation $H_2SO_4 + aHI \rightarrow H_2S + bI_2 + cH_2O$, the values of a, b, c are
(a) a = 3, b = 5, c = 2
(b) a = 4, b = 8, c = 5
(c) a = 8, b = 4, c = 4
(d) a = 5, b = 3, c = 4

5. Which of the following statements about the given reaction are correct?
$3Fe(s) + 4H_2O(g) \rightarrow Fe_3O_4(s) + 4H_2(g)$
(i) Iron metal gets oxidised.
(ii) Water gets reduced.
(iii) Water acts as a reducing agent.
(iv) Water acts as an oxidising agent.
(a) (i), (ii) and (iii)
(b) (iii) and (iv)
(c) (i), (ii) and (iv)
(d) (ii) and (iv)

6. Solid calcium oxide reacts vigorously with water to form calcium hydroxide accompanied by the liberation of heat. This process is called slaking of lime. Calcium hydroxide dissolves in water to form its solution (called lime water). Which among the following is(are) true about slaking of lime and the solution formed?
(i) It is an endothermic reaction.
(ii) It is an exothermic reaction.
(iii) The pH of the resulting solution will be more than 7.
(iv) The pH of the resulting solution will be less than 7.
(a) (i) and (ii)
(b) (ii) and (iii)
(c) (i) and (iv)
(d) (iii) and (iv)

7. Three beakers labelled as A, B and C, each containing 25 mL of water were taken. A small amount of NaOH, anhydrous $CuSO_4$ and NaCl were added to the beakers A, B and C respectively. It was observed that there was an increase in the temperature of the solutions contained in beakers A and B, whereas in case of beaker C, the temperature of the solution falls. Which one of the following statement(s) is(are) correct?
(i) In beakers A and B, exothermic processes have occurred.
(ii) In beakers A and B, endothermic processes have occurred.
(iii) In beaker C, an exothermic process has occurred.
(iv) In beaker C, an endothermic process has occurred.
(a) (i) only
(b) (ii) only
(c) (i) and (iv)
(d) (ii) and (iii)

Multiple Choice

8. In the reaction $ZnO + C \rightarrow Zn + CO$
(a) C acts as an oxidising agent
(b) ZnO is reduced
(c) this reaction does not represent a redox reaction
(d) C acts as a reducing agent

9. Which of the following involves a chemical reaction?
(a) process of respiration
(b) digestion of food in our body
(c) melting of candle wax on heating
(d) burning of candle wax when heated

10. Which of the following is(are) an example of a displacement reaction?
(a) $ZnCO_3 \rightarrow ZnO + CO_2$
(b) $MgO + CO_2 \rightarrow MgCO_3$
(c) $CuSO_4 + H_2S \rightarrow CuS + H_2SO_4$
(d) $Cu + 2AgNO_3 \rightarrow CuNO_3 + 2Ag$

11. $2Ca + O_2 \rightarrow 2CaO$. This reaction is an example of a/an
(a) displacement reaction
(b) decomposition reaction
(c) oxidation reaction
(d) combination reaction

12. Which of the following process is/are endothermic?
(a) formation of CS_2 from carbon and sulfur
(b) formation of NO from N_2 and O_2
(c) decomposition of lithium carbonate
(d) formation of SO_3 from SO_2 and O_2

13. Double decomposition reactions are
(a) $Zn + CuSO_4 \rightarrow ZnSO_4 + Cu$
(b) $BaCl_2 + H_2SO_4 \rightarrow BaSO_4 + 2HCl$

(c) $Na_2S + 2HCl \rightarrow 2NaCl + H_2S$
(d) $NaOH + HNO_3 \rightarrow H_2O + NaNO_3$

14. Which of the following reactions is/are not redox reactions?
(a) $2S_2O_7^{2-} + 2H_2O \rightarrow 4SO_4^{2-} + 4H^+$
(b) $BaCl_2 + MgSO_4 \rightarrow BaSO_4 + MgCl_2$
(c) $Ca(OH)_2 + 2HCl \rightarrow CaCl_2 + 2H_2O$
(d) $Cu_2S + 2FeO \rightarrow 2Cu + 2Fe + SO_2$

15. Which of the following is/are not redox reactions?
(a) $CH_3COOH + C_2H_5OH \rightarrow CH_3COOC_2H_5 + H_2O$
(b) $H_2SO_4 + 2NaOH \rightarrow Na_2SO_4 + 2H_2O$
(c) $BaCl_2 + H_2SO_4 \rightarrow BaSO_4 + 2HCl$
(d) $2FeCl_3 + SnCl_2 \rightarrow 2FeCl_2 + SnCl_4$

Comprehension Type

Comprehension I: A chemical reaction is a process in which a new substance is formed. It is characterised by a change of state, colour, temperature, composition, and so on. chemical reactions can be of many types, such as, combination, decomposition, displacement, double displacement and redox. If in a chemical reaction heat is evolved, it is called an exothermic reaction and if heat is absorbed, it is called an endothermic reaction.

16. Which of the following is not a characteristic of a chemical reaction?
(a) change of composition
(b) change in temperature
(c) formation of a precipitate
(d) change in physical state only without change of formula

17. In which case is the reaction type not correctly matched?
(a) $2SO_2 + O_2 \rightarrow 2SO_3$ — Combination reaction
(b) $2HI \rightarrow H_2 + I_2$ — decomposition reaction
(c) $2HgCl_2 + SnCl_2 \rightarrow Hg_2Cl_2 + SnCl_4$ — Displacement reaction
(d) $H_2 + Cl_2 \xrightarrow{\text{Sunlight}} 2HCl$ — Photochemical reaction

Comprehension II: Certain substances such as hydroiodic acid and ammonium chloride are found to split up into simpler molecules on heating, which reunite on cooling to give the original substances. This type of process is referred to as thermal dissociation and differs from simple decomposition in being reversible. The vapour density of a gas, which decreases as dissociation proceeds, serves as a measure of the degree of dissociation at a particular temperature.

18. Thermal dissociation is a process in which
(a) an acid is neutralised by a base
(b) a compound decomposes on heating
(c) a compound ionises on heating
(d) the compound decomposes to simpler molecules which reunite

19. Which one of the following reactions involves thermal dissociation?
(a) $H_2 + Br_2 \rightarrow 2HBr$
(b) $PCl_3 + Cl_2 \rightarrow PCl_5$
(c) $2Zn(NO_3)_2 \rightarrow 2ZnO + 4NO_2 + O_2$
(d) $2Ca + O_2 \rightarrow 2CaO$

20. During the thermal dissociation of a gas, the vapour density
(a) increases
(b) decreases
(c) remains constant
(d) increases in some cases and decreases in others

Matrix Matching

21. Match the following

Column I	Column II
(a) $C + O_2 \rightarrow CO_2$	(p) Displacement
(b) $AgBr \xrightarrow{\text{light}} Ag + Br$	(q) Combination
(c) $Zn + CuCl_2 \rightarrow ZnCl_2 + Cu$	(r) Decomposition
(d) $CH_3CH_2OH \xrightarrow{\text{Cu}} CH_3CHO + H_2$	(s) Oxidation

22. Match the following

Column I	Column II
(a) Displacement reaction	(p) $AgCl(s) \xrightarrow{\text{Sunlight}} 2Ag(s) + Cl_2(g)$
(b) Double displacement reaction	(q) $Zn(s) + CuSO_4(aq) \rightarrow ZnSO_4(aq) + Cu(s)$
(c) Thermal decomposition reaction	(r) $CaCO_3(s) \xrightarrow{\text{Heat}} CaO(s) + CO_2(g)$
(d) Photolytic decomposition reaction	(s) $AgNO_3(aq) + NaCl(aq) \rightarrow AgCl(s) + NaNO_3(aq)$
	(t) $Pb(NO_3)_2(aq) + 2KI(aq) \rightarrow PbI_2(s) + 2KNO_3(aq)$

Integer Type

23. How many of the following may be the important characteristics of chemical reactions?
(i) Change in state (ii) Formation of a precipitate (iii) Change in temperature (iv) Change in colour (v) Evolution of a gas and (vi) Formation of gaseous products only.

24. How many of these metals are less reactive than hydrogen?
Na, Ca, Mg, Al, Cu, Hg, Ag, Fe, Pt

25. Consider the following reaction
$PbS + nO_3 \rightarrow PbSO_4 + O_2$
What is the value of coefficient '*n*' in the above equation?

26. How many of the following compounds can behave like an oxidant?
CO_2, C, H_2CO_3, H_2S, HNO_3, H_2O_2, HX, SO_3, H_3PO_4

27. How many of the following reactions are mostly exothermic?
Neutralisation, decomposition, salt hydrolysis, combination, combustion, photochemical.

Answer Keys

Fill in the Blanks

1. chemical reaction
2. product
3. exothermic
4. endothermic
5. double displacement
6. neutralisation
7. oxidation
8. thermal decomposition
9. hydrolysis
10. double decomposition
11. redox
12. K
13. catalyst
14. combination

True or False

1. T	2. T	3. T	4. F	5. T
6. T	7. T	8. F	9. T	10. T
11. F	12. T			

Very Short Answer Type Questions

1. A chemical reaction involves the breaking of chemical bonds between the atoms or groups of atoms of reacting substances and rearrangement of atoms making new bonds to form new substances.

2. Every substance has a fixed amount of stored energy, which is in the form of potential energy and is referred to as its chemical energy.

3. Here is an example of an endothermic reaction:
$N_2 + O_2 \xrightarrow[3000°C]{\Delta} 2NO$

4. $BaCl_2(aq) + Na_2SO_4(aq) \rightarrow BaSO_4(s) + 2NaCl(aq)$

5. An example of a double displacement reaction is the reaction of lead nitrate and hydrochloric acid.
$Pb(NO_3)_2(aq)$(Lead nitrate) + $2HCl(aq)$(Hydrochloric acid) $\rightarrow PbCl_2(s)$(Lead chloride) + $2HNO_3(aq)$(Nitric acid)

6. $Ca(s) + 2HCl(aq) \rightarrow CaCl_2(aq) + H_2(g)$

7. Sodium being more reactive than hydrogen, displaces hydrogen from water.
$2Na(s) + H_2O(l) \rightarrow 2NaOH(aq) + H_2(g)$

8. $2KClO_3(s) \rightarrow 2KCl(s) + 3O_2(g)$

9. Magnesium ribbon should be cleaned so that the oxide layer over it is removed and magnesium metal is available for the reaction.

10. (a) $2KOH + H_2SO_4 \rightarrow K_2SO_4 + 2H_2O$
(b) $Mg(OH)_2 + CO_2 \rightarrow MgCO_3 + H_2O$

11. Silver nitrate solution is kept in a coloured bottle to avoid its decomposition in the presence of sunlight.

12. Iron reacts with copper sulfate solution to give a green solution of ferrous sulfate.
$Fe + CuSO_4 \rightarrow \underset{\text{Green}}{FeSO_4} + Cu$

13. In the presence of sunlight, it decomposes into water and oxygen as follows:
$2H_2O_2 \rightarrow 2H_2O + O_2$

14. $2FeSO_4(s) \xrightarrow{\text{Heat}} Fe_2O_3(s) + SO_2(g) + SO_3(g)$
It is a thermal decomposition reaction.

15. Slaked lime is formed with a hissing sound and a large amount of heat is evolved.
$CaO + H_2O \rightarrow Ca(OH)_2 + \text{Heat}$

Short Answer Type Questions

1. (a) When zinc reacts with dil. sulfuric acid, hydrogen gas is evolved, with an effervescence.
$\underset{\text{Zinc}}{Zn} + \underset{\text{dil. Sulfuric acid}}{H_2SO_4} \rightarrow \underset{\text{Zinc sulfate}}{ZnSO_4} + \underset{\text{Hydrogen}}{H_2}$

(b) When blue coloured copper sulfate solution reacts with hydrogen sulfide gas, a black coloured solid substance, copper sulfide is formed.
$\underset{\text{Copper sulfate}}{CuSO_4} + \underset{\text{Hydrogen sulfide}}{H_2S} \rightarrow \underset{\text{Copper sulfide}}{CuS} + \underset{\text{Sulfuric acid}}{H_2SO_4}$

2. A chemical reaction in which the number of atoms of each element is equal on both, the reactants side and products side, is called a balanced chemical equation. An example

of a balanced chemical reaction is given below:

$$N_2(g) + 3H_2(g) \rightarrow 2NH_3(g)$$

As we can see, there are 2 atoms of N and 6 atoms of H on both the sides. Thus, this reaction is called a balanced chemical equation.

A reaction should always be written in the balanced form as an unbalanced reaction violates the fundamental principle of law of conservation of mass. Hence, in a chemical reaction, the total mass of the reactants should be equal to the total mass of the products.

3. (i) X is Na_2SO_4 and Y is $BaSO_4$.

(ii) $\underset{\text{Sodium sulphate}}{Na_2SO_4(aq)} + BaCl_2(aq) \rightarrow \underset{\text{White ppt.}}{BaSO_4 \downarrow} + 2NaCl(aq)$

The reaction is a double displacement reaction. It is also called a precipitation reaction.

4. $KCl(aq) + AgNO_3(aq) \rightarrow AgCl(s) + KNO_3(aq)$
It is a double displacement and precipitation reaction.

5. Zinc is above hydrogen whereas copper is below hydrogen in the activity series of metals. That is why zinc displaces hydrogen from dilute hydrochloric acid, while copper does not.
$Zn + HCl \rightarrow ZnCl_2 + H_2$
$Cu + HCl \rightarrow$ No reaction

6. Decomposition reactions are those in which a compound breaks down to form two or more substances. In a combination reaction, a single product is formed from two or more reactants. Thus, a decomposition reaction is the exact opposite of a combination reaction.
For example, $2H_2O(l) \rightarrow 2H_2(g) + O_2(g)$
$2H_2(g) + O_2(g) \rightarrow 2H_2O(l) + \text{energy}$

7. (a) Decomposition (b) Combination (c) Decomposition (d) Combination (e) Decomposition

8. Al is a reducing agent. Al is more reactive than Mn because Al displaces Mn from its oxide.

9. Ammonia and hydrogen chloride gases are mixed to form ammonium chloride.

$NH_3 + HCl \rightarrow NH_4Cl$
The reaction is a combination reaction.

10. $\underset{\text{Lead nitrate}}{Pb(NO_3)_2(aq)} + 2KI(aq) \rightarrow \underset{\substack{\text{Lead iodide} \\ \text{(Yellow ppt.)}}}{PbI_2(s)} + 2KNO_3(aq)$

A yellow precipitate of PbI_2 is formed.

11. The type of chemical reactions are as follows:
(a) Combination reaction
(b) Double displacement reaction
(c) Decomposition reaction

12. **Decomposition reaction:** A reaction in which a compound is broken down into simpler elements or simple compounds is known as a decomposition reaction.

Thermal decomposition: $CaCO_3(s) \xrightarrow{\Delta} CaO(s) + CO_2(g)$

Electrolytic decomposition: $2H_2O(l) \xrightarrow{\text{Electricity}} 2H_2(g) + O_2(g)$

13. The large amount of energy present in sunlight makes silver chloride (AgCl) decompose into silver (Ag^+) and chloride (Cl^-) ions. The metallic silver appears as small flakes which are essentially black. These tiny black spots evenly distributed in the mass of white AgCl makes it look grey.

$$2AgCl(s) \rightarrow 2Ag(s) + Cl_2(g)$$

14. The solution turns
(a) green to colourless
(b) black coating is formed on zinc.
Reason: Zinc is more reactive than iron so it displaces the iron from its salt solution.

15. $A \rightarrow Cl_2$ (Chlorine gas)
$B \rightarrow CaOCl_2$ (Calcium oxychloride)
$2NaCl + 2H_2O \rightarrow 2NaOH + Cl_2 + H_2$
$Cl_2 + Ca(OH)_2 \rightarrow CaOCl_2 + H_2O$

16. The carbohydrates are broken down to form glucose. This glucose combines with oxygen in the cells of our body and releases energy; so, it is an exothermic reaction.
$C_6H_{12}O_6(aq)$(Glucose) + $6O_2(g) \rightarrow 6CO_2(g) + 6H_2O(l)$ + energy

Long Answer Type Questions

1. (a) Some chemical reactions can take place when reactants are subjected to high pressure. For example, nitrogen and hydrogen produce ammonia gas when reactants are subjected to high pressure.

$$N_2 + 3H_2 \underset{200\text{ atm.}}{\overset{450°C}{\rightleftharpoons}} 2NH_3$$

(b) Some chemical reactions can take place in the presence of light. For example, photosynthesis.

$$6CO_2 + 6H_2O \underset{\text{cholorphyll}}{\overset{\text{Sunlight}}{\rightleftharpoons}} C_6H_{12}O_6 + 6O_2$$

(c) A catalyst can increase or decrease a chemical reaction's rate. Some chemical reactions desperately need a catalyst to change the rate of the reaction, in case it is too slow or too fast.

Positive catalyst: Finely divided iron is used as a positive catalyst to increase the rate of reaction in the manufacturing of ammonia from hydrogen and nitrogen.

$$N_2 + 3H_2 \underset{450°C}{\overset{\substack{\text{Iron (catalyst)} \\ 200\text{–}900}}{\rightleftharpoons}} 2NH_3 + \text{Heat}$$

Negative catalyst: When a catalyst decreases the rate of reaction, it is called a negative catalyst or a retardant. For example, phosphoric acid is used to decrease the rate of the decomposition of hydrogen peroxide.

$$2H_2O_2 \xrightarrow{H_3PO_4} 2H_2O + O_2$$

2. (i) A balanced chemical equation consists of an equal number of atoms of different elements in the reactants and products.

According to the law of conservation of mass, matter can neither be created nor destroyed in a chemical reaction. So, to follow the law, an equation should be balanced.

(ii) (a) $P_4(s)+10Cl_2(g) \rightarrow 4PCl_5(s)$

(b) $CH_4(g)+2O_2(g) \rightarrow CO_2(g)+2H_2O(l)+$Heat energy

(c) $C_6H_{12}O_6(aq)+6O_2(g) \rightarrow 6CO_2(aq)+6H_2O(l)+$ Heat energy

3. (a) Observations:

Colour changes from green to white

Formation of reddish-brown ferric oxide (Fe_2O_3)/ evolution of SO_2/SO_3 gas.

(b) Decomposition reaction

(c) $2FeSO_4 \xrightarrow{Heat} \underset{\text{Ferric oxide}}{Fe_2O_3} + \underset{\text{Sulfur dioxide}}{SO_2} + \underset{\text{Sulfur trioxide}}{SO_3}$

4. (a) $Ca(OH)_2 + CO_2 \rightarrow CaCO_3 + H_2O$

(b) $Zn + 2AgNO_3 \rightarrow Zn(NO_3)_2 + 2Ag$

(c) $2Al + 3CuCl_2 \rightarrow AlCl_3 + 3Cu$

(d) $BaCl_2 + K_2SO_4 \rightarrow BaSO_4 + 2KCl$

5. (a) When copper is heated in air, oxidation takes place

(b) CuO/Copper oxide

(c) $2Cu + O_2 \rightarrow 2CuO$

(d) On passing hydrogen gas over the heated material

6. (i) When solutions are stored in a copper container,

(a) Dilute HCl: Copper does not react with dilute HCl. Therefore, it can be stored.

(b) Dilute HNO_3: Nitric acid acts as a strong oxidising agent and reacts with the copper vessel, therefore it cannot be stored.

(c) $ZnCl_2$: Zinc is more reactive than copper (Cu) therefore, no displacement reaction occurs and hence it can be stored.

(d) H_2O: Copper does not react with water. Therefore, it can be stored.

(ii) When solutions are stored in an aluminium container,

(a) Dilute HCl: Aluminium reacts with dilute HCl to form its salt and hydrogen is evolved. Therefore, it cannot be stored.

$$2\,Al + 6HCl \rightarrow 2\,AlCl_3 + 3\,H_2$$

(b) Dilute HNO_3: Aluminium gets oxidised by dilute HNO_3 to form a layer of Al_2O_3 and can be stored.

(c) $ZnCl_2$: Aluminium being more reactive than zinc can displace zinc ion from the solution. Therefore, the solution cannot be stored.

$$2\,Al + 3\,ZnCl_2 \rightarrow 2\,AlCl_3 + 3Zn$$

(d) H_2O: Aluminium does not react with cold or hot water. Therefore, water can be stored.

Aluminium is attacked by steam to form aluminium oxide and hydrogen.

$$2Al(s) + 3H_2O(g) \rightarrow Al_2O3(s) + 3H_2(g)$$

Competition Window (Objective Type)

Topic-wise MCQs

1. (b)	2. (c)	3. (d)	4. (b)	5. (d)
6. (d)	7. (c)	8. (d)	9. (c)	10. (c)
11. (b)	12. (b)	13. (b)	14. (a)	15. (d)
16. (a)	17. (d)	18. (c)	19. (d)	20. (b)
21. (d)	22. (b)	23. (a)	24. (c)	25. (c)
26. (d)	27. (c)	28. (a)	29. (d)	30. (b)
31. (a)	32. (d)	33. (a)	34. (b)	35. (d)

Miscellaneous

1. (b)	2. (c)	3. (d)	4. (a)	5. (d)
6. (a)	7. (a)	8. (d)	9. (a)	10. (a)
11. (d)	12. (d)	13. (b)	14. (b)	15. (b)
16. (c)				

Advanced and Olympiads

Single Choice

1. (d)	2. (c)	3. (b)	4. (c)	5. (c)
6. (b)	7. (c)			

Multiple Choice

8. (ad)	9. (ab)	10. (cd)	11. (cd)	12. (abc)
13. (bcd)	14. (abc)	15. (abc)		

Comprehension Type

16. (d)	17. (c)	18. (b)	19. (b)	20. (c)

Matrix Matching

21. (a)—(q); (b)—(r); (c)—(p); (d)—(s)

22. (a)—(q); (b)—(st); (c)—(r); (d)—(p)

Integer Type

23. 5	24. 4	25. 4	26. 5	27. 3

Hints and Solutions

Competition Window (Objective Type)

Topic-wise MCQs

4. Heat change is possible during a reaction.

7. The reaction represents a neutralisation reaction in which a base (KOH) reacts with an acid (HNO_3) to form a salt (KNO_3) and water (H_2O).

8. $Zn + 2HCl \rightarrow ZnCl_2 + H_2$

Hydrogen gas burns with a pop sound.

9. The balanced equation is $3Fe + 4H_2O \rightarrow Fe_3O_4 + H_2$

So, x = 3, y = 4

13. Except (b) all other reactions involve compounds.

15. Potassium carbonate is thermally stable and does not decompose under ordinary conditions.

22. Lead sulfate being insoluble will not dissociate into Pb^{2+} ions.

25. Iron is more reactive than copper, hence Cu will not displace iron from iron sulfate, hence no reaction will take place.

27. Cu is more reactive than Ag

$Cu + 2AgNO_3 \rightarrow CuNO_3 + 2Ag$ reaction occurs.

28. $Na_2CO_3 + 2HCl \rightarrow 2NaCl + H_2O + CO_2$

$Na_2CO_3 + NaOH \rightarrow$ No reaction

32. $FeCl_3 + H_2S \xrightarrow[\text{Oxidation}]{\text{Reduction}} FeCl_2 + HCl + S$

In the given reaction, H_2S undergoes oxidation and hence behaves as a reducing agent.]

33. In (d), (b) and (c) H_2O_2 acts as an oxidising agent.

Miscellaneous

1. $2Cu + O_2 \xrightarrow{\Delta} \underset{\text{Black}}{2CuO}$

3. Decomposition of calcium carbonate and $KClO_3$ gives CO_2 and O_2 respectively. Zinc on reaction with H_2SO_4 gives H_2. For reactions see text part.

8. The reactions in option a, b and c are endothermic while the formation of ammonia is an exothermic reaction. See text part.

13. $\underset{\substack{\text{Ferrous sulphate}\\ \text{(Green colour)}}}{2FeSO_4(s)} \xrightarrow[\text{(Decomposition)}]{\text{Heat}} \underset{\substack{\text{Ferric oxide}\\ \text{(Brown colour)}}}{Fe_2O_3(s)} + SO_2(g) + SO_3(g)$

14. $\underset{\text{Brown}}{Cu} + \frac{1}{2}O_2 \xrightarrow{\Delta} \underset{\text{Black}}{CuO}$

15. Ammonium chloride on dissociation gives ammonia and HCl.

$NH_4Cl \rightarrow NH_3 + HCl$

Molecular weight = 2 × vapour density

The molecular weights of $NH_4Cl.NH_3$ and HCl are 53.5 g/mol, 17 g/mol and 36.5 g/mol respectively. The average molecular weight of fully dissociated NH_4Cl will be $\frac{17+36.5}{2} = 26.75$ g/mol which is one half of 53.5 g/mol (the molecular weight of pure undissociated NH_4Cl).

16. $2HI \rightarrow H_2 + I_2$

Volume is constant after dissociation. Thus, using Boyle's law, pressure will also be constant since PV = Constant.

Advanced and Olympiads

Single Choice

1. Dissociation of fused potassium chloride and dissociation of acidulated water are examples of electrochemical reactions while combustion of glucose is a case of exothermic reaction and is known as respiration.

2. Heat of salt hydrolysis is always endothermic, so statement (ii) is incorrect while the rest are correct.

3. Heat is absorbed in an endothermic reaction.

4. The values of a, b, c are 8, 4, 4 respectively. Hence, the reaction is

$H_2SO_4 + 8HI \rightarrow H_2S + 4I_2 + 4H_2O$

5. The substance which oxidises the other substances in a chemical reaction is known as an oxidising agent. Likewise, the substance which reduces the other substance in a chemical reaction is known as reducing agent.

Multiple Choice

8. $ZnO + C \xrightarrow[\text{Oxidation}]{\text{Reduction}} Zn + CO$

C oxidises itself and reduces ZnO; hence it acts as a reducing agent.

10. (a) is decomposition and (b) is combination.

12. Reactions in options (a), (b) and (c) are endothermic while formation of SO_3 is an exothermic reaction. See text part.

14. In redox reactions, oxidation and reduction take place side by side.

$Cu_2S + 2FeO \rightarrow 2Cu + 2Fe + SO_2$

The oxidation number of Cu changes from +1 to 0 (reduction) and the oxidation number of S changes from –2 to +4 (oxidation). Hence, (d) is a redox reaction.

15. The reaction in which both oxidation and reduction take place simultaneously is called a redox reaction.

$-2e^-$ oxidised

$$\overset{+3}{2FeCl_3} + \overset{+2}{SnCl_2} \rightarrow \overset{+2}{2FeCl_2} + \overset{+4}{SnCl_4}$$

+e reduced

Therefore, only (d) is a redox reaction.

Comprehension Type

16. Change in physical state only without a change of formula is a case of physical change and not a chemical reaction.

17. $2HgCl_2 + SnCl_2 \rightarrow Hg_2Cl_2 + SnCl_4$ is a redox reaction and not a displacement reaction.

Integer Type

23. Statement (i) to (v) are features of a chemical reaction. It is not necessary that only gaseous products are formed during a chemical reaction.

24. Na, Ca, Mg, Al and Fe are more reactive than hydrogen while Cu, Hg, Pt and Ag are less reactive.

25. The balanced equation is $PbS + 4O_3 \rightarrow PbSO_4 + 4O_2$

26. CO_2, H_2CO_3, SO_3, H_3PO_4 and HNO_3 can act like oxidants while H_2S, C, HX are reductants.

27. Neutralisation, combination and combustion are exothermic while decomposition, salt hydrolysis and photochemical reactions are endothermic.

CHAPTER 5 Atomic Structure

LEARNING OBJECTIVES

After studying this unit, you will be able to understand:

- The characteristics, properties and distinguishing properties of cathode rays and anode rays
- The characteristics, properties and distinguishing properties of the electron, proton and neutron
- The structure of the atom in terms of Thomson's model, Rutherford's model and Bohr's model
- To differentiate between atomic number and mass number

Introduction

Atoms are made up of three subatomic particles electrons, protons and neutrons. An electron has a negative charge, a proton has a positive charge, whereas a neutron has no charge, it is neutral. Protons and neutrons are present in a small nucleus at the centre of the atom. Electrons are outside the nucleus. The atoms of different elements differ in the number of electrons, protons and neutrons.

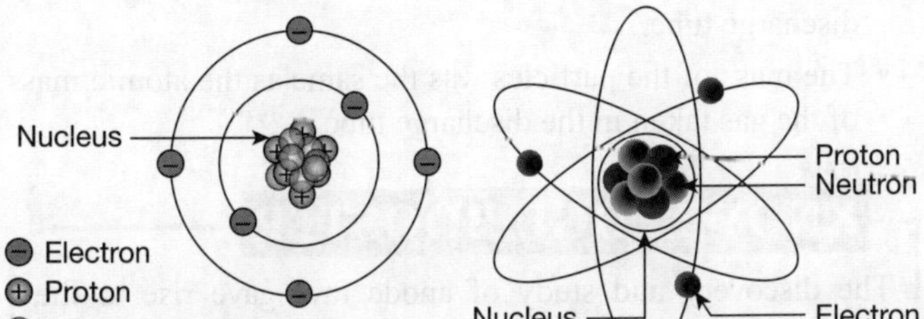

Particles in an atom

Cathode Rays

Sir William Crook was first to design a discharge tube, so it is also called a Crook's discharge tube. A discharge tube is a long glass tube fitted with metal electrodes on either end across which high voltage can be applied. The electrode which is connected to the negative terminal of the power source is known as the cathode while the electrode which is connected to the positive terminal is known as the anode.

J J Thomson modified it as follows. The discharge tube is connected to a vacuum pump for controlling the pressure of the gas inside the discharge tube. When the pressure of the gas is reduced to nearly 10^{-2} atm and a potential difference of nearly 10000 volts is applied to the electrodes, an electric current flows and at the same time, light is emitted by the gas. The fluorescence (green coloured) is caused due to the bombardment of the walls of the tube by rays emanating from the cathode. These rays are known as cathode rays. Actually, these rays are a beam or stream of negatively charged small particles.

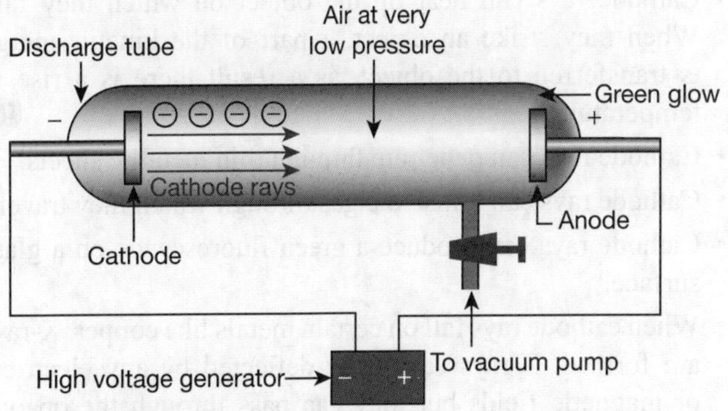

Fig. 5.1 Production of cathode rays

Features of Cathode Rays

J J Thomson described cathode rays as follows.

- Cathode rays always travel in a straight line as they create a shadow on the opposite side of the cathode.

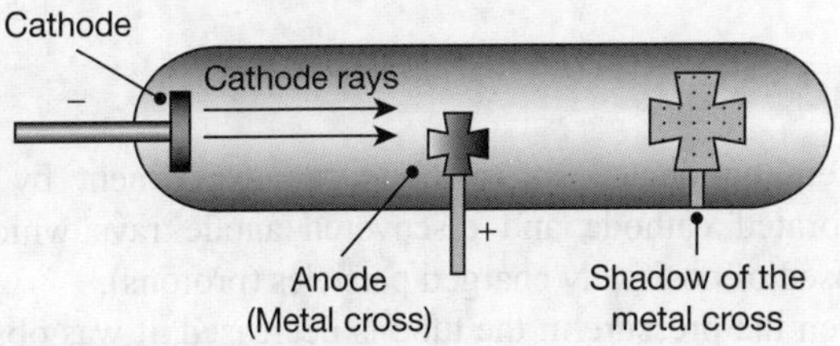

Cathode rays cast shadows of the objects placed in their path

- Cathode rays are composed of negatively charged particles. We know this because they bend towards the positively charged plate when subjected to electric and magnetic fields.

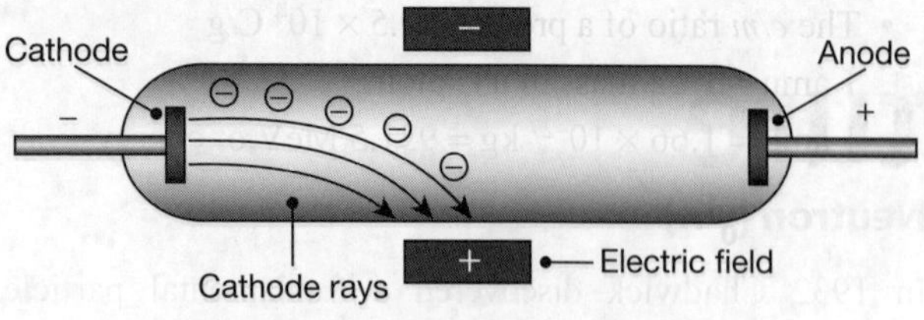

Effects of electric field on cathode rays

- Cathode rays are composed of material particles and possess energy, so they can produce mechanical effects. For example, they can rotate a light paddle wheel placed in their way.

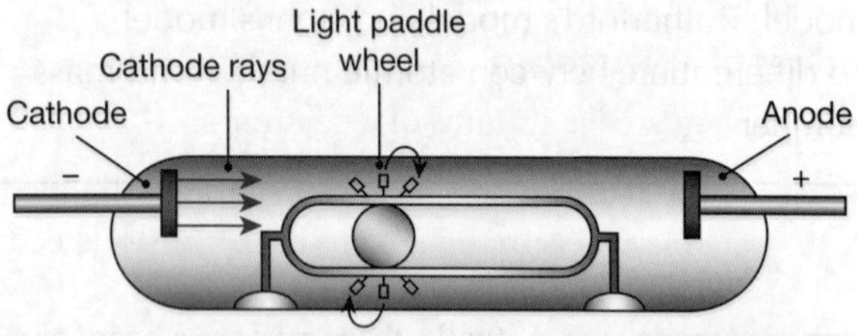

Cathode rays can rotate a light paddle wheel placed in their path

- Cathode rays can heat up the object on which they fall. When they strike an object, a part of the kinetic energy is transferred to the object, as a result there is a rise in temperature.
- Cathode rays can penetrate through thin metallic sheets.
- Cathode rays can ionise the gas through which they travel.
- Cathode rays can produce a green fluorescence on a glass surface.
- When cathode rays fall on certain metals like copper, X-rays are formed. X-rays cannot be deflected by any electrical or magnetic fields but they can pass through the opaque material and can be stopped by only solid objects such as bones.

KNOWLEDGE BOOSTER

The discovery and study of cathode rays gave rise to the discovery of the electron by J J Thomson.

Anode Rays

Goldstein conducted the cathode ray experiment by using a perforated cathode and discovered anode rays which are composed of positively charged particles (protons).

When the pressure in the tube is decreased, it was observed that in addition to cathode rays, a new kind of rays are also found which came through the perforations (holes) of the cathode. These rays travelled in the opposite direction of the cathode rays and passed through the holes of the cathode and struck the other end of the discharge tube. When these radiations struck the end of the tube, the fluorescent radiations were seen. Such rays were called *canal rays* as they are passed through the holes or canals in the cathode. These rays were also called anode rays as they move from the anode toward the cathode. It was observed that the anode rays consist of positively charged particles. So, these rays were also called *positive rays*.

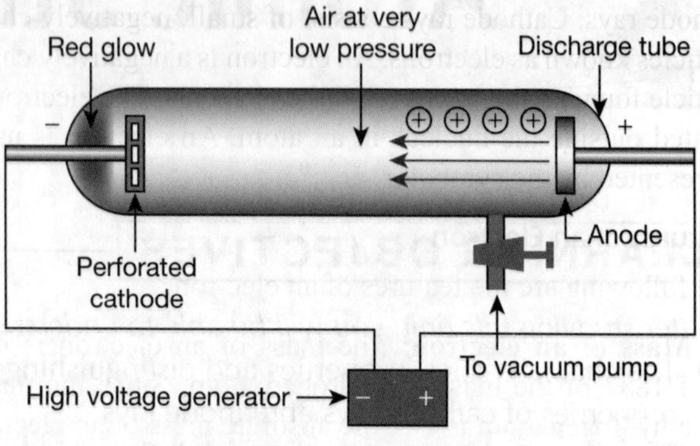

Production of anode rays

Features of Anode Rays

Goldstein described anode rays as follows.

- Anode rays travel in straight lines.
- Anode rays consist of material particles.
- Anode rays are deflected by electric and magnetic fields towards the negatively charged plate which indicates that they are positively charged or composed of positively charged particles.
- The charge to mass ratio of the particles in the anode rays was determined by W Wien using Thomson's technique. Charge to mass ratio of the particles in the anode rays depends upon the nature and mass of the gas taken in the discharge tube.
- The mass of the particles was the same as the atomic mass of the gas taken in the discharge tube.

KNOWLEDGE BOOSTER

The discovery and study of anode rays gave rise to the discovery of the proton by Goldstein which was further described by Rutherford (the credit of proton discovery was given to Rutherford).

Sub-atomic Particles

The main sub-atomic particles are the electron, proton and neutron. Electrons are present outside the nucleus while the protons and neutrons are present inside the nucleus.

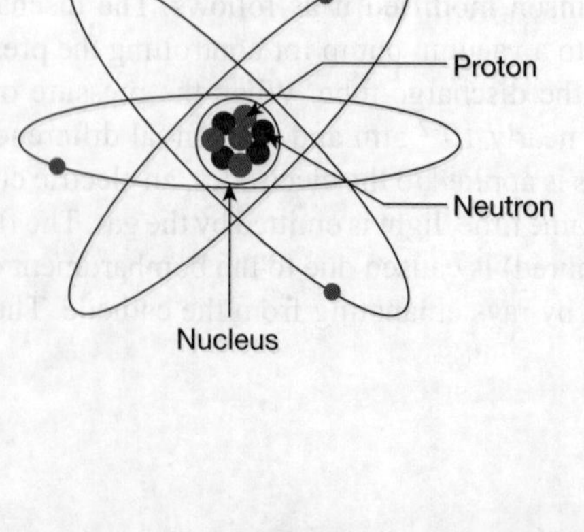

Electron ($_{-1}^{0}e$)

The electron was discovered by J J Thomson during the study of cathode rays. Cathode rays consist of small, negatively charged particles known as electrons. An electron is a negatively charged particle found in the atoms of all the elements. The electrons are located outside the nucleus in an atom. An electron is usually represented by the symbol e^- or $_{-1}^{0}e$.

Features of an Electron

The following are the features of an electron.

- **Mass of an electron:** The mass of an electron is nearly 1/1837 of the mass of hydrogen atom. Since the mass of a hydrogen atom is 1 u, the absolute mass of an electron is 0.0005488 amu or 9.1×10^{-28} grams or 9.1×10^{-31} kg.
- **Charge of an electron:** The charge on an electron is confirmed by Milliken's oil drop experiment and its value is 1.6×10^{-19} Q or -4.8×10^{-10} esu.
- **Specific charge ratio (*e/m* ratio):** Thomson gave the *e/m* ratio or specific charge ratio for the electron and the value is 1.76×10^8 c/g. Cathode rays are a beam or stream of electrons. The *e/m* ratio for cathode rays is constant and independent of the nature of the gas in the discharge tube and the nature of the cathode or electrodes in the tube. That is, the electron is an universal fundamental particle.

  e/m ratio or specific charge ratio or Thomson ratio order:

  $$_{-1}e^- > {}_1p^1 > {}_2\alpha^4 > {}_0n^1$$
- An electron can leave its orbit for a maximum of 10^{-8} s.
- The particle nature of the electron is confirmed by the scintillation effect on the ZnS screen.
- Photoelectric effect, emission of β-particles, thermal emission or heating of a metal filament confirm the fundamental particle nature of the electron.
- The density of the electron is 2.17×10^{17} g/cc.

Proton ($_1^1p$)

The proton was discovered by Goldstein and described by Rutherford. A proton is a positively charged particle found in the atoms of all the elements. The protons are located in the nucleus of an atom. The hydrogen atom contains only one proton in its nucleus, the atoms of all other elements contain more than one proton. A proton is usually represented by the symbol p^+ or $_1^1p$.

Features of a Proton

The following are the features of a proton.

- **Mass of proton:** The mass of a proton has been found to be equal to 1.673×10^{-27} kg. This is almost equal to that of an atom of hydrogen. Since the mass of a hydrogen atom is 1 amu, the relative mass of a proton is nearly 1.0072 amu.
- **Charge of proton:** The charge present on a proton is equal and opposite to the charge present on an electron. The value of charge on a proton is 1.602×10^{-19} coulomb or $+4.8 \times 10^{-10}$ esu of positive charge.
- The *e/m* ratio of a proton is 9.5×10^4 C/g

  1 amu = 1/12 mass of a C atom

  1 amu = 1.66×10^{-27} kg = 931.5 MeV of energy

Neutron ($_0^1n$)

In 1932 Chadwick discovered a fundamental particle, the neutron. The neutron is a neutral particle found in the nucleus of an atom. The subatomic particle not present in a hydrogen atom is the neutron. A hydrogen atom contains only one proton and one electron. A neutron is represented by the symbol $_0^1n$. It was discovered by bombarding Be atoms with α-particles:

$$_4Be^9 + {}_2He^4 \rightarrow {}_6C^{12} + {}_0n^1$$

Features of a Neutron

The following are the features of a neutron:

- **Mass of a neutron:** The mass of neutron is slightly higher than the mass of a proton. The relative mass of a neutron is 1 u. The absolute mass of a neutron is 1.674×10^{-27} kg or 1.0086 amu.
- **Charge of a neutron:** A neutron has no charge and it is electrically neutral.

Table 5.1 Comparison between electron, proton and neutron

	Electron	Proton	Neutron
(i) Symbol	e/e^-	p/p^+	n
(ii) Nature	Negatively charged	Positively charged	Neutral
(iii) Relative mass	1/1840 of a H atom	Equal to H atom	Equal to H atom
(iv) Actual mass	9.1×10^{-28} g	1.673×10^{-24} g	1.674×10^{-24} g
(v) Charge	(–1) (1.602×10^{-19} C)	(+1) (1.602×10^{-19} C)	No charge

Structure of Atom and Development of Atomic Model

After the discovery of the electron and proton, scientists started thinking of arranging these particles in an atom. This led to the development of the concept of atomic models. An atomic model depicts the arrangement of fundamental particles in an atom. J J Thomson was the first scientist to propose a model for the structure of the atom. Mainly, there are three atomic models.

(i) Thomson's atomic model

(ii) Rutherford's atomic model

(iii) Bohr's atomic model

Thomson's Atomic Model

Thomson was the first to propose a detailed model of the atom just after the discovery of electrons. He proposed that an atom consists of a uniform sphere of positive electricity in which the electrons are distributed or embedded more or less uniformly. The negative and the positive charges are equal in magnitude. Thus, the atom as a whole is electrically neutral. This model of atom is known as the *plum pudding model* or *watermelon model*.

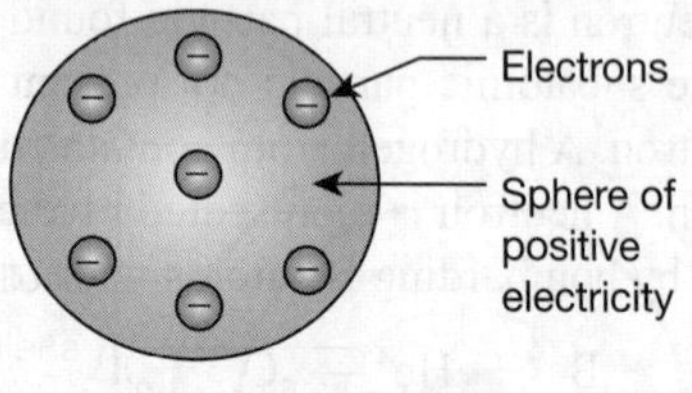

The salient features of Thomson's atomic model are:

- An atom consists of a positively charged sphere and the electrons are embedded in it.
- The negative and positive charges are equal in magnitude. So, the atom as a whole is electrically neutral.

Although Thomson's model could explain that atoms are electrically neutral, it cannot explain how the positively charged particles are shielded from the negatively charged electrons without getting neutralised. The results of experiments carried out by other scientists (Rutherford and others) could not be explained by this model so it was rejected.

Rutherford's Atomic Model

Rutherford and his students (Hans Geiger and Ernest Marsden) bombarded a very thin gold foil with α-particles (helium nuclei). This famous α-particle scattering experiment is represented in Fig. 5.2.

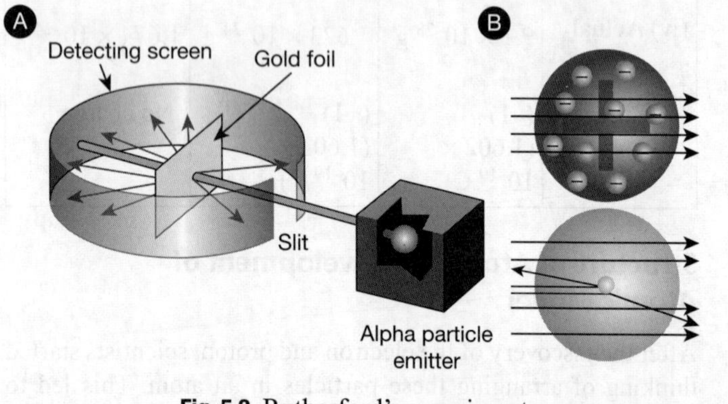

Fig. 5.2 Rutherford's experiment

A stream of high energy α-particles from a radioactive source was directed at a thin foil (thickness <100 nm) of gold metal. The thin gold foil had a circular fluorescent zinc sulfide screen around it. Whenever α-particles struck the screen, a tiny flash of light was produced at that point.

The results of the scattering experiment were quite unexpected. According to Thomson's model of the atom, the mass of each gold atom in the foil should have been spread evenly over the entire atom, and α-particles had enough energy to pass directly through such a uniform distribution of mass. It was expected that the particles would slow down and change directions only by small angles as they passed through the foil. It was observed that:

- most of the α-particles passed through the gold foil undeflected,
- a small fraction of the α-particles was deflected by small angles, and
- a very few α-particles (~1 in 20,000) bounced back, that is, were deflected by nearly 180°.

Based on these observations during the gold foil experiment, Rutherford introduced his atomic model and arrived at the following conclusions.

(i) Most part of the atom is empty as most of the α-particles passed straight through.

(ii) Electrons occupy positions in this empty space (extra nuclear region).

(iii) A centrally located small, solid, compact part having all positive charge and nearly the whole mass of the atom is called the nucleus (as a few α-particles get defected up to a maximum of 180°).

(iv) The radius of an atom is 10^{-10} m or 10^{-8} cm.
The radius of the nucleus = 10^{-15} m or 10^{-13} cm.
Atomic radius > nucleus radius by 10^5 times.
Density of nucleus is 10^{17} kg/m^3 or 10^{14} g/cm^3 and the volume of the nucleus is 10^{-39} cm^3.

(v) Centrifugal force develops between electrons and the nucleus. So, the electrons revolve around the nucleus as stars move around the sun (planetary or solar model).

Merits of Rutherford's Model

- Explains the discovery of the nucleus.
- Circulatory rotation of the electrons around the nucleus.

Demerits of Rutherford's Model

There are some draw backs in Rutherford's atomic model.

(i) Cannot explain stability of atom: This model was unable to explain the stability of an atom. According to Rutherford's postulate, electrons revolve at a very high speed around the nucleus of an atom in a fixed orbit. However, Maxwell explained that accelerated charged particles release electromagnetic radiations. Hence, electrons revolving around the nucleus will release electromagnetic radiation. The electromagnetic radiation will possess energy from the electronic motion as a result of which the orbits will gradually shrink. Finally, the orbits will shrink and collapse into the nucleus of the atom. According to the calculations, if Maxwell's explanation is followed, Rutherford's model will collapse within 10^{-8} seconds. Therefore, Rutherford's atomic model was not following Maxwell's theory and it was unable to explain an atom's stability.

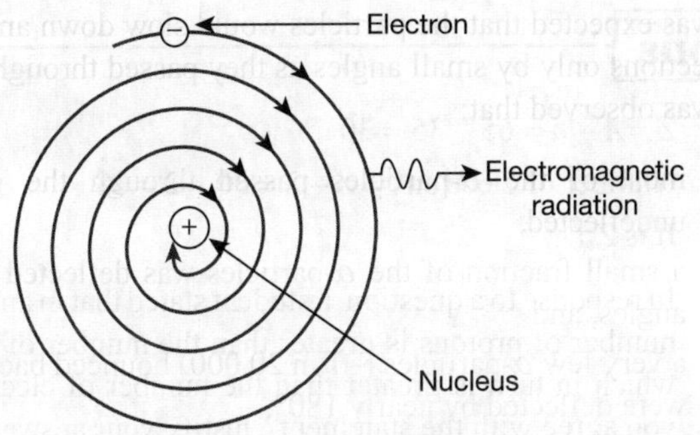

(ii) Rutherford's theory was incomplete because it did not mention anything about the arrangement of electrons in the orbit. This was one of the major drawbacks of Rutherford's atomic model.

(iii) It cannot explain the number and velocity of electrons.

(iv) According to him, the atomic spectrum is continuous and non-linear. However, it is seen to be linear and discontinuous.

KNOWLEDGE BOOSTER

Bohr's atomic model

Bohr modified and removed all the demerits of Rutherford's model and introduced the circular orbit concept in which both energy and angular momentum are quantised (fixed values).

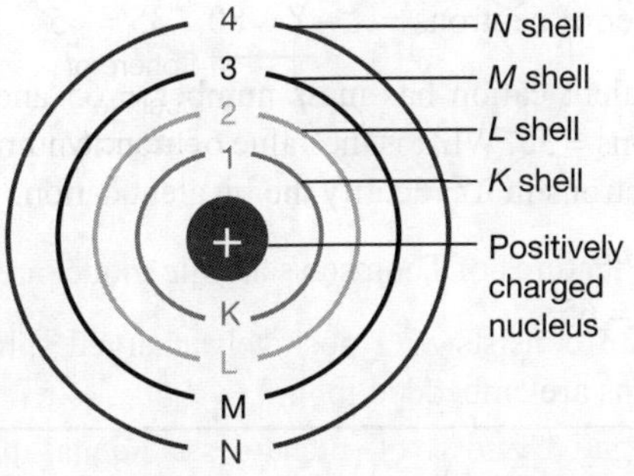

Illustrations

1. What are the limitations of J J Thomson's model of the atom?

Solution: The limitations of J J Thomson's model of the atom are as follows. It could not explain the results of the scattering experiment performed by Rutherford.

It did not have any experimental support.

2. What are the limitations of the Rutherford's model of the atom?

Solution: The limitations of Rutherford's model of the atom are as follows. It failed to explain the stability of an atom. It doesn't explain the spectrum of hydrogen and other atoms.

Composition of Nucleus

Atomic Number (*Z*)

Moseley postulated that the frequency of the X-rays was related to the charge present on the nucleus of the atom of the element used as a cathode and found that $\sqrt{v} = a(Z - b)$ where v is the frequency, Z is the nuclear charge and a and b are constants. a = probability constant, b = a constant having the same values for all lines of the X-ray spectrum.

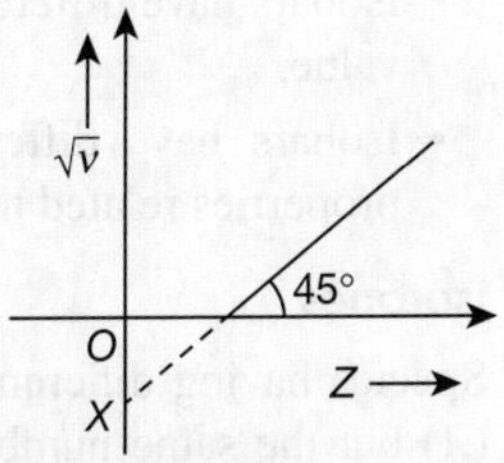

The number of unit positive charges carried by the nucleus of an atom is called the atomic number of the element, that is,

$$Z = p = e \text{ (from atoms)}$$
$$Z = p \text{ (for ions)}$$

Mass number (*A*): It is equal to the sum of protons and neutrons (or atomic number and neutrons) in an atom. Mass number is nearly equal to atomic weight.

Relationship Between Mass Number and Atomic Number

Mass number = No. of protons + No. of neutrons
Mass number = Atomic number + No. of neutrons

$$A = p + n = Z + n$$

For example, helium atom has 2 protons, 2 neutrons and 2 electrons. Its mass number is equal to 2 + 2 = 4.

Oxygen has 8 protons, 8 neutrons and 8 electrons. its mass number is 16.

- In some cases, particularly heavier elements, the number of neutrons is more than the number of protons. For example, mercury has an atomic number equal to 80. So, it has 80 protons and 80 electrons. But the mass number of mercury is 200. Therefore, the number of neutrons in mercury is 200 – 80 = 120.

Representation of an Element or Atom

Generally, an atom is represented by its symbol. Atomic number is written on the lower side of the symbol and the mass number is written on the upper side.

$$^{A}_{Z}X$$

A = mass number, Z = atomic number, X = symbol of element. For example, $^{7}_{3}Li$ indicates that lithium has an atomic number equal to 3 and a mass number equal to 7.

Illustrations

1. Calculate the number of protons, neutrons and electrons in $_{35}Br^{80}$.

   Solution: In this case, $_{35}Br^{80}$, $Z = 35$, $A = 80$, species is neutral

   Number of protons = number of electrons = $Z = 35$

   Number of neutrons = $A - Z = 80 - 35 = 45$

2. A bivalent cation has mass number = 65 and number of neutrons = 35. What is the value of atomic number, number of electrons in it? Identify the bivalent cation.

   Solution:

   As $A = 65$,

   $n = 35$

   $Z = A - n = 65 - 35 = 30$

   Number of electrons = $Z - 2 = 30 - 2 = 28$

   It is Zn^{+2}.

3. In response to a question, a student stated that in an atom, the number of protons is greater than the number of neutrons, which in turn is greater than the number of electrons. Do you agree with the statement? Justify your answer.

   Solution: The given statement is not correct as the number of protons can never be greater than the number of neutrons. The number of neutrons can be equal to or greater than the number of protons but the number of protons is always equal to the number of electrons for an atom as it is neutral.

Concept of Isotopes, Isobars, Isoelectronics and Isotones

Isotopes

They were discovered by Soddy. Atoms of the same element with the same atomic number (Z) but different mass number (A) are called isotopes.

Isotopes of hydrogen: It has three isotopes. These three isotopes are commonly known as hydrogen (H) or protium, deuterium (D) and tritium (T) respectively. As the atomic number is the same for all the three, they all have one electron and one proton but different number of neutrons.

	Protium	Deuterium	Tritium
Atomic number, Z	1	1	1
Mass number, A	1	2	3
Number of protons	1	1	1
Number of electrons	1	1	1
Number of neutrons	0	1	2
Electronic configuration	K 1	K 1	K 1

Isotopes of carbon: Carbon has the following two isotopes.

	$^{12}_{6}C$	$^{14}_{6}C$
Z =	6	6
p =	6	6
n =	6	8
A =	12	14

Features of Isotopes

- Isotopes have the same chemical properties.
- Isotopes have the same number of protons and electrons.

Isotopes differ in atomic weight, number of neutrons and physical properties.

Isobars

Atoms of different elements with the same mass number (A) but different atomic number (Z) are called isobars. For example,

Argon (atomic number 18) and calcium (atomic number 20) are isobars because they have the same mass number 40.

Argon ($^{20}_{18}Ar$)	Calcium ($^{40}_{20}Ca$)
At. no. = 18	At. no. = 20
Mass no. = 40	Mass no. = 40
No. of electrons = 18	No. of electrons = 20
No. of protons = 18	No. of protons = 20
No. of neutrons = 22	No. of neutrons = 20

Features of Isobars

- Isobars have different number of p, n but the same $(p + n)$ value.
- Isobars have different chemical properties but physical properties related to mass are the same.

Isotones

Species having different atomic number (Z) and mass number (A) but the same number of neutrons are called isotones. The atoms have different number of protons of atomic number. Isotones belong to two or more different elements. They have (A – Z) the same, and A and Z are different. For example,

- $_{6}C^{14}$ $\quad$ $_{7}N^{15}$ $\quad$ $_{8}O^{16}$

  n: 14 – 6 $\quad$ 15 – 7 $\quad$ 16 – 8

  = 8 $\quad$ = 8 $\quad$ = 8

  $_{14}Si^{30}$ $\quad$ $_{15}P^{31}$ $\quad$ $_{16}S^{32}$

Isoelectronics

Atoms, ions or molecules having the same number of electrons are called isoelectronic. For example, CH_4, Ne, Na^+, NH_4^+, Mg^{2+}, F^-, O^{2-} have 10 electrons, so they are isoelectronic.

Illustrations

1. From the given species, select the isotopes, isobars and isotones.

$^{12}_{6}C, ^{14}_{7}N, ^{14}_{6}C, ^{15}_{7}N, ^{16}_{8}O$

Solution:

Isotopes: $^{12}_{6}C$, $^{14}_{6}C$ are isotopes as they have the same atomic number. $^{14}_{7}N$, $^{15}_{7}N$ are isotopes as they have the same atomic number.

Isobars: $^{14}_{7}N, ^{14}_{6}C$ are isobars as they have a mass number 14.

Isotones: $^{14}_{7}N, ^{14}_{6}C$ are isotones as they have 8 neutrons each.

2. What is the difference in physical and chemical properties of isotopes?

Solution: Isotopes have the same chemical properties as their atomic numbers are the same. Their physical properties are different as they depend on mass.

KNOWLEDGE BOOSTER

The distribution of electrons in different orbits or shells is governed by a scheme known as the *Bohr Bury scheme*. The arrangement of electrons in various energy levels or orbits of an atom is known as the electronic configuration of the atom.

According to this scheme:

- The electrons are arranged around the nucleus in different energy levels or energy shells. The electrons first occupy the shell with the lowest energy, that is, closest to the nucleus (lower *n* value).
- The first or the innermost energy shell (*K* or $n = 1$) can have only two electrons.
- The second shell (*L* or $n = 2$) can have up to 8 electrons.
- From the third shell (*M* or $n = 3$) onwards, the shells become larger and the third shell can have a maximum of 18 electrons.

In general, the maximum number of electrons that can be present in any orbit is given by $2n^2$ where *n* is the number of the energy shell.

Maximum number of electrons in different orbits

Orbit	Value of *n*	Maximum number of electrons in the orbit
K	1	$2 \times 1^2 = 2$
L	2	$2 \times 2^2 = 8$
M	3	$2 \times 3^2 = 18$
N	4	$2 \times 4^2 = 32$

- The outermost shell or orbit of an atom cannot have more than 8 electrons and the shell next to the outermost shell cannot have more than 18 electrons.

Valence shell and valence electrons: The outermost orbit of an atom is known as its valence shell and the electrons present in the outermost orbit are known as valence electrons.

Valency: The valency of an element is the combining capacity of the atoms of the element with atoms of the same or different elements. Valency is equal to the number of valence electrons. For example,

(i) The electronic configuration of lithium is 2, 1 or it may be represented as:

K L

2 1

The outermost orbit is the *L* shell and the number of electrons in its valence shell is one.

> Only the valence electrons of an atom take part in chemical reactions because they have more energy than all the inner electrons of the atom.

(ii) The electronic configuration of beryllium is 2, 2 or it may be represented as:

K L

2 2

The outermost orbit is the *L* shell and the number of electrons in its valence shell is two.

CHAPTER AT A GLANCE

- **Atoms** are made up of three subatomic particles—electrons, protons and neutrons.
- **Cathode** rays are a beam or stream of negatively charged small particles.
- **Anode** rays consist of positively charged particles. So, these rays are also called positive rays.
- The main **sub-atomic particles** are electrons, protons and neutrons. Electrons are present outside the nucleus while protons and neutrons are present inside the nucleus.
- An **electron** is a negatively charged particle found in the atoms of all the elements. The electrons are located outside the nucleus in an atom. An electron is usually represented by the symbol e^- or ${}_{-1}^{0}e$.
- The absolute **mass of an electron** is 0.0005488 amu or 9.1×10^{-28} g or 9.1×10^{-31} kg.
- **Charge on an electron (e^-)** was confirmed by Milliken's oil drop experiment and its value is 1.6×10^{-19} Q or -4.8×10^{-10} esu.
- *e/m* ratio or specific charge ratio or Thomson ratio order: ${}_{-1}e^- > {}_{1}p^1 > {}_{2}\alpha^4 > {}_{0}n^1$
- A **proton (${}_{1}^{1}p$)** is a positively charged particle found in the atoms of all the elements. The protons are located in the nucleus of an atom.
- The **mass of proton** has been found to be equal to 1.673×10^{-27} kg or 1.0072 amu.
- The value of the **charge on a proton** is 1.602×10^{-19} coulomb or $+4.8 \times 10^{-10}$ esu of positive charge.
- The **neutron (${}_{0}^{1}n$)** is a neutral particle found in the nucleus of an atom.
- The **absolute mass** of a neutron is 1.674×10^{-27} kg or 1.0086 amu.
- **Thomson's atomic model:** an atom consists of a uniform sphere of positive electricity in which the electrons are distributed or embedded more or less uniformly. The negative and the positive charge are equal in magnitude. Thus, the atom as a whole is electrically neutral.
- **Rutherford's atomic model:** Rutherford bombarded a very thin gold foil with α-particles (helium nuclei) called the Rutherford's famous α-particle scattering experiment.
- The centrally located small, solid, compact part having all the positive charge and nearly the whole mass is called the **nucleus**.
- **Radius of atom** is = 10^{-10} m or 10^{-8} cm. Radius of nucleus = 10^{-15} m or 10^{-13} cm. Atomic radius > nucleus radius by 10^5 times.
- **Rutherford's model** cannot explain (i) stability of atom, (ii) number and velocity of electrons, (iii) linear nature of spectrum.
- **Bohr** modified and removed all the demerits of Rutherford's model and introduced the circular orbit concept in which both energy and angular momentum are quantised (fixed values).
- The number of unit positive charges carried by the nucleus of an atom is called the **atomic number** of the element. $Z = p = e$ (from atoms), $Z = p$ (for ions).
- **Mass number (*A*)** is equal to the sum of protons and neutrons, or atomic number and neutrons. Mass number is nearly equal to atomic weight.
  Mass number = Atomic number + No. of neutrons ($A = p + n = Z + n$)
- Generally, an atom is represented by its **chemical symbol**. Atomic number is written on the lower side of the symbol and the mass number is written on the upper side.
- Atoms of same element with same atomic number (*Z*) but different mass number (*A*) are called **isotopes**.
- Atoms of different elements with the same mass number but different atomic number are called **isobars**.
- Species having different atomic number and mass number but the same number of neutrons are called **isotones**.
- Atoms, ions or molecules having the same number of electrons are called **isoelectronic**.
- The arrangement of electrons in various energy levels or orbits of an atom is known as the **electronic configuration** of the atom.
- In general, the maximum number of electrons that can be present in any orbit is given by $\mathbf{2n^2}$ where *n* is the number of the energy shell.
- **Valency** of an element is the combining capacity of the atoms of the element with atoms of the same or different elements. Valency is equal to number of valence electrons.

PRACTICE QUESTIONS

Analyse Your Concepts (School Exam Based)

Fill in the Blanks

Instructions: Complete the following statements with an appropriate word/term to be filled in the blank spaces.

1. Cathode rays are a beam or stream of ____________ charged particles.
2. Cathode rays on passing through an electric field deflect towards the ____________ plate.
3. Thomson's atomic model is also known as ____________ model.
4. The subatomic particle not present in a hydrogen atom is ____________.
5. The number of protons in the nucleus of an atom is equal to its ____________.
6. The total number of protons and neutrons in the nucleus of an atom is equal to its ____________.
7. According to Thomson, electrons are uniformly distributed in the ____________ charged sphere.
8. An atom has atomic mass number 10 and atomic number 20. The atom has ____________ electrons.
9. Na^+ and Mg^{2+} are known as ____________.
10. An atom of an element has 12 protons, 12 electrons and 12 neutrons. The atomic mass of the atom is ____________.
11. When cathode rays fall on a metal surface like Cu ____________ are formed.
12. Rutherford used a thin film of ____________ metal.
13. The ____________ of an atom is very hard and dense.
14. Neutrons are ____________ particles having a mass equal to that of protons.

True or False

Instructions: Read the following statements and write your answer as true or false.

1. Protons are present in all matter and have a positive charge.
2. In F^-, the number of electrons is more than the number of protons.
3. The number of neutrons in the atom of an element is equal to its atomic number.
4. Isotopes of an element differ in the number of neutrons.
5. The *e/m* ratio of a neutron is less than that of an electron.
6. The name 'proton' was given by Chadwick.
7. Matter is electrical and negative in nature.
8. Mass of an electron is about 1840 times that of a proton.
9. Neutrons were discovered by Henry Becquerel.
10. Anode rays are made of protons.
11. J J Thomson proposed that the nucleus of an atom contains only nucleons.
12. A neutron is formed by an electron and a proton combining together. Therefore, it is neutral.
13. Ar-40 and Ca-40 are isobars.
14. Na^+, Mg^{2+}, O^{2-} are isotones.

Very Short Answer Type Questions

1. What are canal rays?
2. What is the nature of charge on cathode and anode rays respectively?
3. By how many times is a proton heavier than an electron?
4. The atomic number of oxygen and sulfur are 8 and 16 respectively. What will be the number of electrons in these?
5. Out of O-16 and O-18 isotopes, which one has a higher number of neutrons?
6. The mass number and atomic number of an element are 24 and 12 respectively. What is the number of nucleons present in it?
7. An element has an atomic number 19 and mass number 39. How many electrons and protons are present in its uni-positive ion?
8. Who determined the charge of the electron for the first time?
9. The nucleus of an element contains 17 protons and 18 neutrons. What are its atomic number and mass number?
10. Write the decreasing order of *e/m* ratio of *e*, *p*, *n*, α -particles?
11. The radius of an atom is larger than the radius of a nucleus by how many times?
12. Is it possible for the atom of an element to have one electron, one proton and no neutron? If so, name the element.
13. Calculate the number of neutrons present in the nucleus of an element X which is represented ${}^{31}_{15}X$.

Short Answer Type Questions

1. Give the nuclear compositions of $^{18}_{8}O$ and $^{15}_{7}N$.
2. An element may be represented as $^{40}_{18}E$. Find the number of electrons, protons and neutrons in this atom. Give its electronic configuration.
3. What are the characteristic features of Thomson's model of the atom and why was the theory rejected?
4. Isotopes have the same chemical properties but different physical properties. Explain.
5. The atomic nucleus of an element has a mass number of 23 and 12 neutrons. What is the atomic number of the element?
6. Why is the atom neutral even though it contains charged particles?
7. Write any two observations which support the fact that atoms are divisible.
8. Will ^{35}Cl and ^{37}Cl have different valencies? Justify your answer.
9. Why did Rutherford select a gold foil in his α-ray scattering experiment?
10. What information do you get from Fig. 4.4 about the atomic number, mass number and valency of atoms X, Y and Z? Give your answer in a tabular form.

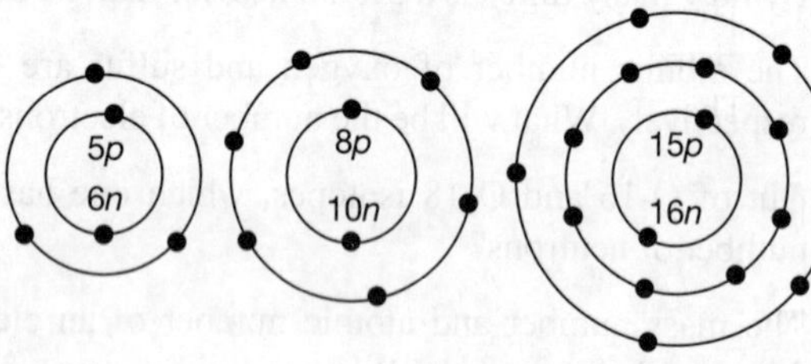

(x) (y) (z)

11. Composition of the nuclei of two atomic species X and Y are given as:

	X	Y
Protons =	6	6
Neutrons =	6	8

Give the mass numbers of X and Y. What is the relation between the two species?

Long Answer Type Questions

1. What is Thomson's model of the atom? Why was it rejected?
2. Hydrogen has three isotopes written as $^1_1H, ^2_1H : ^3_1H$. Explain why,
   (a) These isotopes have almost identical chemical properties.
   (b) They are electrically neutral.
3. What observations in the scattering experiment led Rutherford to make the following conclusions?
   (a) Most of the space in an atom is empty.
   (b) The whole mass of an atom is present in the centre of the atom.
   (c) The nucleus has a positive charge.
4. Explain the following:
   (a) If an atom contains one electron and one proton, will it carry any charge or not?
   (b) On the basis of Thomson's model of an atom, explain how the atom is neutral as a whole.
   (c) On the basis of Rutherford's model of an atom, which subatomic particle is present in the nucleus of an atom?
5. Match the names of the scientists given in column A with their contributions towards the understanding of the atomic structure as given in column B

Column (A)	Column (B)
(a) Ernest Rutherford	(i) Indivisibility of atoms
(b) J J Thomson	(ii) Stationary orbits
(c) Dalton	(iii) Concept of nucleus
(d) Neils Bohr	(iv) Discovery of electrons
(e) James Chadwick	(v) Atomic number
(f) E Goldstein	(vi) Neutron
(g) Mosley	(vii) Canal rays

6. Fill in the blanks in the following statements.
   (a) Rutherford's α-particle scattering experiment led to the discovery of the______.
   (b) Isotopes have the same ______ but different ______.
   (c) Neon and chlorine have atomic numbers 10 and 17 respectively. Their valencies will be______ and ______ respectively.
   (d) The electronic configuration of silicon is ______ and that of sulfur is______.
7. Complete the following table.

Atomic number	Mass number	Number of Neutrons	Number of protons	Number of electrons	Name of the atomic species
9	–	10	–	–	–
16	32	–	–	–	Sulfur
–	24	–	12	–	–
–	2	–	1	–	–
–	1	0	1	0	–

COMPETITION WINDOW (OBJECTIVE TYPE)

[For NEET, JEE (Main and Advanced), NTSE, KVPY and Olympiads]

Topic-wise MCQs

Sub-atomic particles

1. Which of the following is never true for cathode rays?
(a) they are electromagnetic waves
(b) they possess kinetic energy
(c) they produce mechanical pressure
(d) they produce heat

2. The anode ray experiment lead to the discovery of ______.
(a) electrons (b) nucleus
(c) protons (d) neutrons

3. Which of the following is not a basic particle of an element?
(a) a molecule (b) an atom
(c) an ion (d) none of these

4. What is an electron?
(a) either a wave or a particle, depending on how it is observed
(b) a particle
(c) a wave
(d) neither of these

5. Proton is
(a) ionised hydrogen molecules
(b) ionised hydrogen atom
(c) nucleus of deuterium
(d) an α-particle

6. Neutron was discovered by
(a) Rutherford (b) Bohr
(c) Chadwick (d) Thomson

7. The charge on an electron is
(a) unit negative (b) -1.6×10^{-19} C
(c) -4.8×10^{-10} esu (d) all

8. The mass of an electron is
(a) 9.1×10^{-10} g (b) 9.1×10^{-28} g
(c) 9.1×10^{-18} g (d) 9.1×10^{-25} g

9. The heaviest particle is
(a) electron (b) neutron
(c) proton (d) all are equally massive

10. The mass of a neutron is of the order of
(a) 10^{-27} kg (b) 10^{-26} kg
(c) 10^{-24} kg (d) 10^{-23} kg

11. The *e/m* ratio is highest for
(a) neutron (b) proton
(c) electron (d) He^{2+}

12. Which of these pairs has almost similar masses?
(a) proton–electron (b) neutron–$_1H^1$
(c) electron–$_1H^1$ (d) neutron–electron

Atomic models

13. The first model of an atom was given by
(a) Dalton (b) E Goldstein
(c) Rutherford (d) J J Thomson

14. Thomson atomic model is known as
(a) plum pudding model
(b) planetary model
(c) watermelon model
(d) both (a) and (c)

15. Rutherford's alpha (α)-particles scattering experiment resulted in the discovery of the
(a) electron (b) proton
(c) nucleus in the atom (d) atomic mass

16. When α-particles are sent through a thin metal foil, most of them go straight through the foil because
(a) α-particles are much heavier than electrons
(b) most part of the atom is empty
(c) α-particles move with high velocity
(d) α-particles are positively charged

17. In the Thomson's model of the atom, which of the following statements are correct?
(i) The mass of the atom is assumed to be uniformly distributed over the atom.
(ii) The positive charge is assumed to be uniformly distributed over the atom.
(iii) The electrons are uniformly distributed in the positively charged sphere.
(iv) The electrons attract each other to stabilise the atom.
(a) (i), (ii) and (iii) (b) (i) and (iii)
(c) (i) and (iv) (d) (i), (iii) and (iv)

18. In an atom, the constituent electrons
(a) move around the nucleus in fixed energy levels
(b) move around the nucleus in a random way
(c) do not move
(d) are uniformly distributed

19. When α-particles are sent through a thin metal foil only one out of ten thousand of them rebounded. This observation led to the conclusion that
(a) only a unit positive charge is present in an atom
(b) the number of electrons revolving around the nucleus of the atom is more
(c) positively charged particles are concentrated at the centre of the atom
(d) a massive sphere with more negative charge and unit sphere with negative charge and unit positive charge is present at the centre of the atom

Atomic number, mass number

20. Which of the following statements is always correct?
(a) An atom has an equal number of electrons and protons
(b) An atom has an equal number of electrons and neutrons
(c) An atom has an equal number of protons and neutrons
(d) An atom has an equal number of electrons, protons and neutrons

21. Which of the following are true for an element?
(i) Atomic number = number of protons + number of electrons
(ii) Mass number = number of protons + number of neutrons
(iii) Atomic mass = number of protons = number of neutrons
(iv) Atomic number = number of protons = number of electrons
(a) (i) and (ii) (b) (i) and (iii)
(c) (ii) and (iii) (d) (ii) and (iv)

22. A neutral atom of an element contains 17 electrons. Its nucleus has 20 neutrons. Its mass number is
(a) 37 (b) 3
(c) 17 (d) 20

23. The number of electrons in an element X is 16 and the number of neutrons is 16. Which of the following is the correct representation of the element?
(a) ${}^{32}_{15}X$ (b) ${}^{32}_{16}X$
(c) ${}^{16}_{16}X$ (d) ${}^{15}_{16}X$

24. The mass number of an atom whose bi-positive ion has 10 electrons and 12 neutrons is
(a) 21 (b) 23
(c) 22 (d) 24

25. The atomic number of an element is 13 and its mass number is 27. The correct order representing the number of electrons, proton and neutrons respectively in this atom is
(a) 13, 14, 13 (b) 27, 27, 13
(c) 13, 13, 14 (d) 13, 13, 13

Isotopes, isobars...

26. Isotopes of an element have
(a) the same physical properties
(b) different chemical properties
(c) different number of neutrons
(d) different atomic numbers

27. Which one of the following pairs represents isobars?
(a) ${}_{20}Ca^{40}$ and ${}_{19}K^{39}$ (b) ${}_{20}Ca^{40}$ and ${}_{18}Ar^{40}$
(c) ${}_{2}He^{3}$ and ${}_{2}He^{4}$ (d) ${}_{12}Mg^{24}$ and ${}_{12}Mg^{25}$

28. An isotone of ${}^{76}_{32}Ge$ is
(a) ${}^{77}_{34}Se$ (b) ${}^{78}_{34}Se$
(c) ${}^{77}_{32}Ge$ (d) ${}^{87}_{33}As$

29. Which one of the following groupings represents a collection of isoelectronic species? (Atomic numbers Cs: 55, Br: 35)
(a) K^+, Ca^{2+}, Mg^{2+} (b) O^{2-}, F^-, Na^+
(c) Be, Al^{3+}, I^- (d) Ca^{2+}, Cs^+, Br^-

30. Which among the following are isobars?
(a) ${}_{b}X^{a}$, ${}_{b+1}Y^{a}$ (b) ${}_{b}X^{a}$, ${}_{c}Y^{b}$
(c) ${}_{b}X^{a}$, ${}_{b}Y^{a+1}$ (d) ${}_{b}X^{a}$, ${}_{b-1}Y^{a-1}$

31. Which one of the following sets of ions represents a collection of isoelectronic species?
(a) K^+, Ca^{2+}, Sc^{3+}, S^{2-}
(b) Na^+, Ca^{2+}, Sc^{3+}, Cl^-
(c) K^+, Cl^-, Mg^{2+}, Sc^{3+}
(d) Na^+, Mg^{2+}, Al^{3+}, S^{2-}

32. Which of the following are isoelectronic?
(a) Ne and S (b) K^+ and Ca
(c) K^+ and Ar (d) Na^+ and K^+

33. If two neutrons are added to an element P, then it will get converted to its
(a) isotone (b) isobar
(c) isotope (d) none of the above

34. Which of the following pairs are isotopes?
(a) carbon monoxide and carbon dioxide
(b) C-12 and C-14
(c) oxygen and ozone
(d) water and steam

Electronic configuration, valency

35. In an atom, valence electrons are present in
(a) any one of its orbits (b) first orbit
(c) outermost orbit (d) next to outermost orbit

36. The maximum number of electrons that can be accommodated in the second shell ($n = 2$) is
(a) 2 (b) 10
(c) 18 (d) 8

37. The number of valence electrons in the P^{3-} ion is:
(a) 5 (b) 8
(c) 15 (d) 18

38. Which one of the following is the correct electronic configuration of magnesium?
(a) 2, 2, 8 (b) 8, 2, 2
(c) 2, 2, 8 (d) 2, 8, 2

39. The maximum number of electrons which a shell (designated by n) can accommodate is
(a) $2n^2$ (b) n^2
(c) $2n$ (d) $2n+1$

40. The total number of shells (orbits), valence electrons in Fe are
(a) 3, 2 (b) 4, 2
(c) 4, 6 (d) 3, 6

41. The ion of an element has 3 positive charges. Mass number of the atom is 27 and the number of neutrons is 14. What is the number of electrons in the ion?
(a) 13 (b) 10
(c) 14 (d) 16

42. The core charge on oxygen is equal to
(a) –6e (b) +8e
(c) +2e (d) –2e

43. Elements with valency 1 are
(a) always non-metals
(b) always metalloids
(c) either metals or non-metals
(d) always metals

44. An atom with 6 protons and 6 neutrons will have a valency of
(a) 3 (b) 7
(c) 1 (d) 4

45. The electron distribution in a Silicon atom ($Z = 14$) is
(a) 2,8,4 (b) 2,8,2
(c) 8,2,4 (d) 2,4,8

Miscellaneous

1. A tripositive ion has 15 protons. What is the number of electrons in its tetrapositive ion?
(a) 11 (b) 10
(c) 15 (d) 12

2. In which of the following pairs of shells, is the energy difference between two adjacent orbits minimum?
(a) *M*, *N* (b) *N*, *O*
(c) *K*, *L* (d) *L*, *M*

3. Which of the following correctly represents the electronic distribution in the Na atom?
(a) 3,8 (b) 2,8,1
(c) 1,8,2 (d) 8,2,1

4. In a sample of methyl ethanoate (CH_3COOCH_3) the two oxygen atoms have the same number of electrons but different number of neutrons. Which of the following is the correct reason for it?
(a) one of the oxygen atoms has gained two neutrons
(b) one of the oxygen atoms has gained electrons
(c) the two oxygen atoms are isotopes
(d) the two oxygen atoms are isobars

5. Select the set having correct statements.
(i) According to Thomson's model, the positive charge is assumed to be uniformly distributed over the atom.
(ii) According to Rutherford, the nucleus is present at the centre of the atom
(iii) Atomic radius is 10000 times larger than that the nuclear radius
(iv) Bohr introduced the circular orbit concept
(a) (i), (ii) and (iii) (b) (i) and (iii)
(c) (i) and (iv) (d) (i), (ii) and (iv)

6. Select the correct statements.
(i) The electron is a negatively charged particle while the proton is positively charged
(ii) A neutron is heavier than a proton
(iii) The neutrons are present outside the nucleus
(iv) *e*/*m* ratio is maximum for the electron
(a) (i), (ii) and (iv) (b) (i) and (iv)
(c) (i) and (iii) (d) (i), (ii) and (iii)

7. Which among the following pairs have different number of valence electrons?
(a) Mg^{2+}, Ar (b) O^{2-}, F^-
(c) Na^+, Al^{3+} (d) P^{3-}, Ar

8. An element has two isotopes with mass numbers 16 and 18. The average atomic weight is 16.5. The percentage abundance of these isotopes is ________ and ________, respectively.
(a) 50, 50 (b) 25, 75
(c) 75, 25 (d) 33.33, 66.67

9. Some of the elements have fractional atomic masses. The reason for this could be
(a) the nuclear reactions
(b) the presence of neutrons in the nucleus
(c) the existence of isobars
(d) the existence of isotopes

10. The number of electrons present in the valence shell of an atom with atomic number 20 is
(a) 1 (b) 8
(c) 2 (d) 0

11. Which of the following arrangements of electrons represent magnesium (Mg)?
(a) 2, 8, 1 (b) 2, 8, 4
(c) 2, 8, 3 (d) 2, 8, 2

12. Which of the following have an equal number of neutrons and protons?
(a) oxygen (b) chlorine
(c) deuterium (d) both (a) and (c)

13. The number of electrons in an element with atomic number *m* and atomic mass *n* will be
(a) $(m + n)$ (b) m
(c) $(m - n)$ (d) $(n - m)$

14. The relative atomic masses of many elements are not whole numbers because
(a) of the existence of isotopes
(b) of the presence of impurities
(c) they cannot be determined accurately
(d) the atoms ionise during the determination of their masses

15. Which of the following describes an isotope with a mass number of 39 that contains 20 neutrons in its nucleus?
(a) $^{39}_{19}K$ (b) $^{20}_{19}K$
(c) $^{39}_{20}Ca$ (d) $^{39}_{18}Ar$

16. Which of the following isotopes is used as the standard for atomic mass?
(a) ^{14}C (b) ^{16}O
(c) ^{12}C (d) ^{14}N

17. Members of which of the following have similar chemical properties?
(a) allotropes
(b) isobars
(c) isotopes
(d) both isotopes and allotropes

18. While performing cathode ray experiments, it was observed that there was no passage of electric current under normal conditions. Which of the following can account for this observation?
(a) air is a poor conductor of electricity under normal conditions
(b) carbon dioxide is present in air
(c) dust particles are present in air
(d) none of the above

19. The fluorescence on the walls of a discharge tube is due to
(a) canal rays (b) anode rays
(c) cathode rays (d) none of the above

20. Which of the following electronic configurations is/are incorrect?
(a) Cl (17): 2, 7, 8 (b) P (15): 2, 8, 5
(c) B (5): 2, 3 (d) F (9): 2, 7

Advanced and Olympiads

Single Choice

1. Select the set having correct statements.
(i) Atom is divisible into sub-atomic particles
(ii) Atom is neutral in nature
(iii) Hydrogen atom has 1 neutron
(iv) Atomic number is always equal to the number of protons
(a) (i), (ii) and (iii) (b) (i), (iii) and (iv)
(c) (ii), (iii) and (iv) (d) (i), (ii) and (iv)

2. Which of the following statements about Rutherford's model of atom are correct?
(i) considered the nucleus as positively charged
(ii) established that α-particles are four times as heavy as a hydrogen atom
(iii) can be compared to solar system
(iv) was in agreement with Thomson's model
(a) (i) and (iii) (b) (ii) and (iii)
(c) (i) and (iv) (d) only (i)

3. In the Thomson's model of the atom, which of the following statements are correct?
(i) the mass of the atom is assumed to be uniformly distributed over the atom
(ii) the positive charge is assumed to be uniformly distributed over the atom
(iii) the electrons are uniformly distributed in the positively charged sphere
(iv) the electrons attract each other to stabilise the atom
(a) (i) and (iv)
(b) (i), (iii) and (iv)
(c) (i), (ii) and (iii)
(d) (i) and (iii)

4. The atomic masses of two isotopes of O are 15.9936 and 17.0036. Select the incorrect statement
(a) total number of electrons in each atom is 16
(b) total mass number in each atom is 33
(c) total number of neutrons in each atom is 17
(d) total number of protons in each atom is 17

5. Which of the following is not correctly matched here?
(a) C-14, N-15 — Isotones
(b) Sc^{3+}, Ti^{4+} — Isoelectronic
(c) B-10, B-12 — Isobars
(d) Ca-40, K-40 — Isobars

6. Which of the following pairs have the same number of valence electron(s)?
(a) N, P, S (b) F, Cl, Ar
(c) Li, Na, K (d) Be, Mg, Ar

7. Rutherford's α-particle scattering experiment showed that
(i) electrons have negative charge
(ii) the mass and positive charge of the atom is concentrated in the nucleus
(iii) neutrons exist in the nucleus
(iv) most of the space in the atom is empty
Which of the above statements are correct?
(a) (i) and (iii) (b) (ii) and (iv)
(c) (i) and (iv) (d) (iii) and (iv)

8. Atomic models have been improved over the years. Arrange the following atomic models in their chronological order
(i) Rutherford's atomic model
(ii) Thomson's atomic model
(iii) Bohr's atomic model
(a) (i), (ii) and (iii) (b) (ii), (iii) and (i)
(c) (ii), (i) and (iii) (d) (iii), (ii) and (i)

Multiple Choice

9. Which are correct amongst the following statements about cathode rays?
(a) particles forming them possess the same mass and same charge
(b) *e/m* ratio of the particles forming them depends upon the gas filled in the discharge tube
(c) they are deflected by electric and magnetic fields
(d) *e/m* ratio is constant, independent of the gas filled and the cathode material

10. Which of the following atoms has two neutrons in its nucleus?
(a) protium (b) tritium
(c) helium (d) lithium

11. Which of the following statements are correct?
(a) ${}^{13}_{7}N$ and ${}^{14}_{7}N$ are isotopes
(b) ${}^{14}_{7}N$ and ${}^{16}_{8}O$ are isotones
(c) ${}^{14}_{7}N$ and ${}^{14}_{6}C$ are isobars
(d) ${}^{14}_{7}N$ and ${}^{17}_{8}O$ are isotones

12. Which of the following concepts was considered in Rutherford's atomic model?
(a) the electrical neutrality of the atom
(b) existence of nuclear forces of attraction on the electron
(c) electrons revolve around the nucleus at very high speeds
(d) the quantisation of energy and angular momentum

13. Which of the following statements are true?
(a) The total number of electrons and protons is always equal in a neutral atom.
(b) Most of the space in an atom is empty.
(c) The total number of electrons in any energy level can be calculated by the formula $2n^2$.
(d) The total number of electrons and protons in an atom or ion is always equal.

14. In Bohr's model, which of the following are correct?
(a) energy is quantised in each orbit
(b) angular momentum is quantised in each orbit
(c) it is applicable for multi-electronic species also
(d) by using its radius, the velocity in any orbit can be found out

15. Which pairs are correctly matched here?
(a) CO_2 and N_2O are isoelectronic
(b) ${}^{15}_{7}N$ and ${}^{16}_{8}O$ are isotones
(c) CO and NO are isoelectronic
(d) C-13 and N-15 are isoneutronic

Comprehension Type

Comprehension I: Atomic number is equal to the number of protons or electrons present in any atom. It is denoted by *Z*. In the case of ions, the number of electrons differ from those in the atom due to uptake or losing of electrons; thus, *Z* cannot be equal to electrons for ions. Mass number (*A*) is the sum of protons + neutrons present in an atom.

16. A dipositive ion has 22 protons. What is the number of electrons in its tetrapositive ion?
(a) 22 (b) 16
(c) 18 (d) 20

17. If the atomic weights of N and Si are 14 and 28, respectively, then what is the ratio of the number of neutrons in them?
(a) 2 : 1 (b) 2 : 3
(c) 5 : 3 (d) 1 : 2

18. Two nuclides P and Q are isoneutronic. Their mass number are 31 and 32, respectively. If the atomic number of P is 15, then the atomic number of Q will be
(a) 14 (b) 15
(c) 16 (d) 18

Matrix Matching

19. Match the following

Column I	Column II
(a) cathode rays	(p) Millikan
(b) anode rays	(q) Goldstein
(c) neutron	(r) Thomson
(d) charge on electron	(s) Chadwick

20. Match the following

Column I	Column II
(a) Isotopes	(p) have same number of atoms and same electrons
(b) Isobar	(q) have same number of neutrons
(c) Isotones	(r) have same sum of $p + n$
(d) Isosteres	(s) have same number of protons
	(t) have same mass number

Integer Type

21. The *e/m* ratio of a proton (${}_{1}{}^{1}H$) is x times more than that of the α -particle (${}^{4}_{2}\alpha$)

22. The number of neutrons present in ${}^{15}_{7}N$ is *n*. What is the value of *n*?

23. The radius of the atom is 10^n times higher than that of the radius of the nucleus. What is the value of *n*?

24. How many of the following species are isoelectronic?
Na, K^+, Mg^{2+}, Sc^{3+}, S^{2-}, Ar, P^{3-}, Cl^-, P, Ti^{4+}

25. In Millikan's experiment, the static electric charge on the oil drops was obtained by shining X-rays. If the static electric charge on the oil drop is -9.6132×10^{-19} C, calculate the number of electrons present in it.

26. How many of the following are isotones?
${}^{12}_{6}C$, ${}^{14}_{6}C$, ${}^{14}_{7}N$, ${}^{15}_{7}N$, ${}^{16}_{8}O$, ${}^{17}_{8}O$, ${}^{18}_{8}O$

Answer Keys

Fill in the Blanks

1. negative
2. positive
3. plum pudding model
4. neutron
5. atomic number
6. mass number
7. positively
8. 10
9. isoelectronic
10. 24
11. X-rays
12. gold metal
13. nucleus
14. neutral

True or False

1. T 2. T 3. F 4. T 5. T
6. F 7. T 8. T 9. F 10. T
11. F 12. F 13. T 14. F

Very Short Answer Type Questions

1. Canal rays are positively charged radiations that can pass through a perforated cathode plate. These rays are made of positively charged particles known as protons.
2. The nature of charge present on cathode and anode rays are negative and positive respectively.
3. A proton is heavier than an electron. It is 1840 times heavier than an electron.
4. For a neutral atom, the number of electrons = atomic number.
   So, the number of electrons in oxygen = 8
   So, the number of electrons in sulfur = 16
5. O-18 has more neutrons than O-16.
   Number of neutrons = $A - Z$
   In O-18, number of neutrons = 18 – 8 = 10
   In O-16, number of neutrons = 16 – 8 = 8
6. Number of nucleons = Mass number = $p + n$ = 12 + (24 – 12) = 24
7. Atomic number (Z) = 19
   Number of protons = Z = 19
   Number of electrons = Z – positive charge = 19 – 1 = 18
8. Charge on electron was confirmed for the first time by Millikan through the oil drop experiment.
9. The atomic number and mass number are 17 and 35 respectively.
10. The decreasing order of the e/m ratio is $e > p > \alpha > n$.
11. The radius of the atom is larger than the radius of the nucleus by 100000 times.
12. Yes, it is possible. For example, hydrogen has one electron, one proton and no neutron. It is represented as ${}^{1}_{1}H$.
13. In ${}^{31}_{15}X$, number of protons = number of electrons = 15
    Number of neutrons =Mass number (A) – no. of protons (p) = 31 – 15 = 16

Short Answer Type Questions

1. In ${}^{17}_{8}O$, the number of protons = 8
   Number of neutrons = 17 – 8 – 9
   In ${}^{15}_{5}N$, the number of protons = 7
   Number of neutrons = 15 – 7 – 8
2. In ${}^{40}_{18}E$, Z = 18, A = 40
   Number of electrons or protons = atomic number (Z)
   So, number of electrons = 18
   Number of protons = 18
   Number of neutrons = $A - Z$ = 40 – 18 = 22
3. See text part.
4. Isotopes have the same chemical properties as they have the same atomic number and chemical properties are decided by the atomic number. Isotopes have different atomic masses so mass related physical properties like boiling point and density differ.
5. We know that
   Mass number = No. of protons + No. of neutrons
   23 = No. of protons + 12
   ∴ No. of protons = 23 – 12 = 11
   Now, Atomic number = No. of protons = 11.
6. An atom is neutral in spite of having charged particles because the number of positive species or particles (protons) are equal to the number of negative species (electrons). For example, in Mg there are 12 protons and 12 electrons.
7. Atoms are divisible. With the discovery of electrons and protons it was established that the atom is divisible and is made up of negatively charged electrons and positively charged protons along with some neutral particles called neutrons.
8. No, both ^{35}Cl and ^{37}Cl will have the same valencies, as ^{35}Cl and ^{37}Cl are isotopes. As the isotopes have the same number of electrons and protons and they differ only in the number of neutrons, their electron distribution (2, 8, 7) will be the same.
9. A light metal cannot be used. On being hit by fast moving α-particles, the atoms of a light metal will be simply pushed forward and no scattering can occur. So, Rutherford selected the gold foil as gold is a heavy metal with a high mass number, highly malleable and can be beaten to get very thin foils.
10.

Element	Atomic number	Mass number	Valency
X	5	11	3
Y	8	18	2
Z	15	31	3,5

11. Mass number of X = Number of protons + Number of neutrons = 6 + 6 = 12
Mass number of Y = Number of protons + Number of neutrons = 6 + 8 = 14
As these two atomic species X and Y have the same atomic number, but different mass numbers, they are isotopes.

Long Answer Type Questions

1. See text part.
2. See text part.
3. See text part in Rutherford's atomic model.
4. (a) An electron is a negatively charged particle, whereas a proton is a positively charged particle and the magnitude of their charges is equal. So, an atom containing one electron and one proton will not carry any charge. Thus, it will be a neutral atom.
   (b) As per Thomson's model of the atom, an atom consists of both negative and positive charges which are equal in number and magnitude hence, they balance each other as a result of which atom as a whole is electrically neutral in nature.
   (c) On the basis of Rutherford's model of an atom, protons are present in the nucleus of an atom.
5. (a)—(iii), (b)—(iv), (c)—(i), (d)—(ii), (e)—(vi), (f)—(vii) and (g)—(v).
6. (a) nucleus, (b) atomic number, mass number, (c) 0 and 1 and (d) silicon: 2, 8, 4, sulfur: 2, 8, 6.
7.

Atomic number	Mass number	Number of neutrons	Number of protons	Number of electrons	Name of the atomic species
9	19	10	9	9	Fluorine
16	32	16	16	16	Sulfur
12	24	12	12	12	Magnesium
1	2	1	1	1	Deuterium
1	1	0	1	0	Hydrogen ion

Competition Window (Objective Type)

Topic-wise MCQs

1. (a) 2. (c) 3. (a) 4. (c) 5. (b)
6. (c) 7. (d) 8. (b) 9. (b) 10. (a)
11. (c) 12. (b) 13. (d) 14. (d) 15. (c)
16. (b) 17. (a) 18. (a) 19. (c) 20. (a)
21. (d) 22. (a) 23. (b) 24. (d) 25. (c)
26. (c) 27. (b) 28. (b) 29. (b) 30. (a)
31. (a) 32. (c) 33. (c) 34. (b) 35. (c)
36. (d) 37. (b) 38. (d) 39. (a) 40. (b)
41. (b) 42. (b) 43. (c) 44. (d) 45. (a)

Miscellaneous

1. (a) 2. (b) 3. (b) 4. (c) 5. (d)
6. (a) 7. (a) 8. (c) 9. (d) 10. (c)
11. (d) 12. (d) 13. (b) 14. (a) 15. (a)
16. (c) 17. (c) 18. (a) 19. (c) 20. (a)

Advanced and Olympiads

Single Choice

1. (d) 2. (a) 3. (c) 4. (d) 5. (c)
6. (c) 7. (b) 8. (c)

Multiple Choice

9. (acd) 10. (bc) 11. (ac) 12. (abc) 13. (abc)
14. (abd) 15. (abc)

Comprehension Type

16. (c) 17. (d) 18. (c)

Matrix Matching

19. (a)—(r); (b)—(q); (c)—(s); (d)—(p)
20. (a)—(s); (b)—(rt); (c)—(q); (d)—(p)

Integer Type

21. 2 22. 8 23. 5 24. 7 25. 6
26. 3

Hints and Solutions

Competition Window (Objective Type)

Topic-wise MCQs

1. Cathode rays are never electromagnetic waves.
6. The neutron was discovered by Chadwick.

$$_4Be^9 + _2\alpha^4 \rightarrow _6C^{12} + _0n^1$$

11. Due to low mass.
13. J J Thomson proposed the first model of the atom.
15. Rutherford's alpha (α)-particles scattering experiment resulted in the discovery of the nucleus in the atom. A large number of particles went straight through the atom while a very small number of particles were deflected back showing the presence of a dense positively charged nucleus.

17. Thomson's model could be compared with the raisin pudding model according to which the mass of the atom is uniformly distributed over the atom in the form of positive charge and electrons are uniformly distributed over the atom. Electrons do not attract each other.

20. An atom has an equal number of electrons and protons since it is neutral.

21. Mass number = $p + n$, Atomic number = $p = e$.

23. Number of electrons = 16, Number of neutrons = 16. Hence, atomic number of element X is 16 and atomic mass is 32. The element is represented as ${}_{16}^{32}X$.

24. Mass number = 12 + 12 = 24.

27. Isobars have the same atomic mass.

28. Isotones have the same number of neutrons.

29. O^{2-}, F^- and Na^+ have 10 electrons each.

31. These have 18 electrons.

32. Contain the same number of electrons (18).

33. The numbers of neutrons are different in the isotopes of the same element.

36. Maximum number of electrons = $2n^2 = 2 \times 2^2 = 8$.

40. Fe (26) = 2, 8, 14, 2; Number of shells = 4; Valence electrons = 2

41. Mass number of the atom = 27; Number of neutrons = 14; Number of protons = 27 – 14 = 13; Number of electrons in the atom = 13
Number of electrons in the ion with 3 positive charges = 13 – 3 = 10]

42. Core charge = + Ze (Z = atomic number) = +8e

43. Element with valency 1 can be both metals or non-metals.

44. For the atom n = 6, p = 6, hence e = 6; Distribution of electrons = 2,4.
So, valency = 4.

45. Silicon has 14 electrons. Its electron distribution is 2, 8, 4.

Miscellaneous

1. X^{3+} has 15 protons, then in X, there are 15 protons and 15 electrons; in X^{3+}, there are 15 protons and 12 electrons; in X^{4+}, there are 15 protons and 11 electrons.

3. ${}_{11}Na - \underset{2}{K} \quad \underset{8}{L} \quad \underset{1}{M}$

4. Two isotopes have the same number of electrons and protons but different number of neutrons.

5. As the atomic radius is 100000 times larger than the radius of the nucleus, statement (iii) is incorrect while the rest are correct.

10. Z = 20, electronic configuration = 2, 8, 8, 8, 2. Therefore, it has two valence electrons.

12. In oxygen, the number of protons = number of neutron = 8. In deuterium, the number of protons = number of neutrons = 1.

20. The correct configuration of Cl (17) = 2, 8, 7.

Advanced and Olympiads

Single Choice

1. The hydrogen atom has no neutron so statement (iii) is incorrect while the rest are correct.

2. Rutherford's model of the atom explained the presence of the nucleus in the centre of the atom and electrons around the nucleus revolving in round orbitals like the solar system.

4.

	First isotope of O	Second isotope of O
Atomic mass	15.9936	17.0036
(a) Number of neutrons	16 – 8 = 8	17 – 8 = 9
(b) Number of protons (Mass number – Number of neutrons)	16 – 8 = 8	17 – 9 = 8
(c) Number of electrons	8	8
(d) Mass number	16	17

5. B-10, B-12 are isotopes as they have same atomic number 5.

6. As Li, Na, K all have 1 valence electron.

7. Two important observations of Rutherford's α-particle scattering experiment were—(i) the mass and positive charge of the atom is concentrated in the nucleus, (ii) most of the space in the atom is empty.

8. Thomson's atomic model, followed by Rutherford's model which is followed by Bohr's model.

Multiple Choice

10. Tritium (${}_1^3H$) and helium (${}_2^4He$) have 2 neutrons each.

12. According to Rutherford's theory, an atom is electrically neutral and electrons revolve around the nucleus. Nuclear forces of attraction exist between the nucleus and electrons. The only assumption that Rutherford did not consider was the quantisation of energy and angular momentum.

13. The number of electrons and protons are equal in atoms but not in ions.

15. CO and NO are not isoelectronic as they have different number of electrons (14 and 15 respectively). All the rest are correct.

Comprehension Type

16. Let the element be X; then X^{2+} has 22 protons; then in X there are 22 protons and 22 electrons; in X^{2+} there are 22 protons and 20 electrons; in X^{4+} there are 22 protons and 18 electrons.

17. Number of neutrons in ${}_7^{14}N = 14 - 7 = 7$; Number of neutrons in ${}_{14}^{28}Si = 28 - 14 = 14$

The ratio of the number of neutrons in N and Si is 7 : 14 or 1 : 2.

18.

$^{31}_{15}P$	$^{32}_{z}Q$
Protons = 15	Protons + Neutrons = 32
Protons + Neutrons = 31	Neutrons = 16
Neutrons = 16	Protons = 32 – 16 = 16

Integer Type

21. The charge on the α-particle is twice that of the proton and the mass is 4 times that of the proton. So, it will have half the *e/m* ratio of the protons; that is, the proton has twice the *e/m* ratio of the α-particle. So $x = 2$.

22. Number of neutrons = A – Z = 15 – 7 = 8. So, $n = 8$.

23. The radius of the atom is 10^5 times higher than that of the radius of the nucleus, so the value of n is 5.

24. K^+, Sc^{3+}, S^{2-}, Ar, P^{3-}, Cl^- and Ti^{4+} are isoelectronic as they all have 18 electrons.

25. $$\text{Total electrons present} = \frac{\text{Total charge}}{\text{Charge on 1 electron}}$$

$$= \frac{-9.6132 \times 10^{-19}\,\text{C}}{-1.6022 \times 10^{-19}\,\text{C}} = 6$$

26. $^{14}_{6}C$, $^{15}_{7}N$, $^{16}_{8}O$ are isotones as they have 8 neutrons.

CHAPTER

6 Hydrogen

LEARNING OBJECTIVES

After studying this unit, you will be able to understand:

- The occurrence and position of hydrogen
- The various methods of preparation
- Its physio-chemical properties

Introduction

It is the first, lightest, smallest, non-metallic element in the periodic table. It does not have any neutrons. It has maximum specific heat. It exists in the H^+ (hydronium) and H^- (hydride) states.

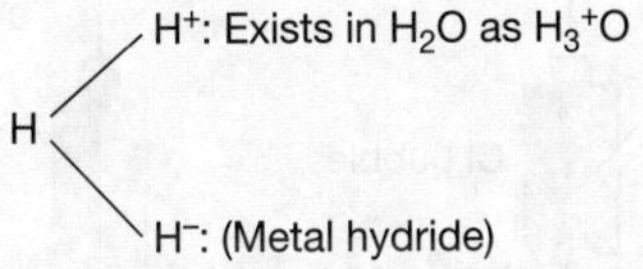

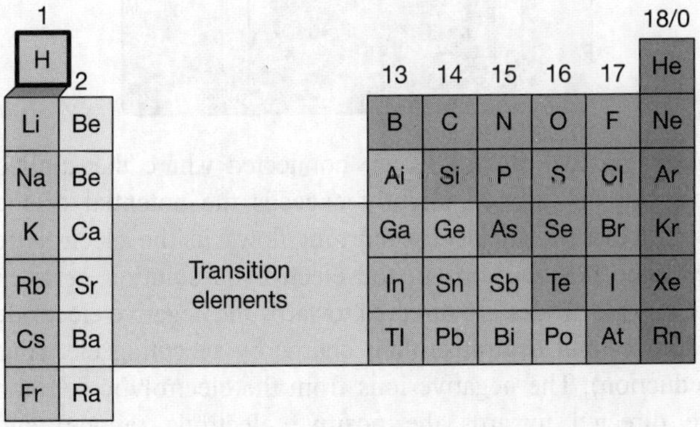

The position of hydrogen in the periodic table

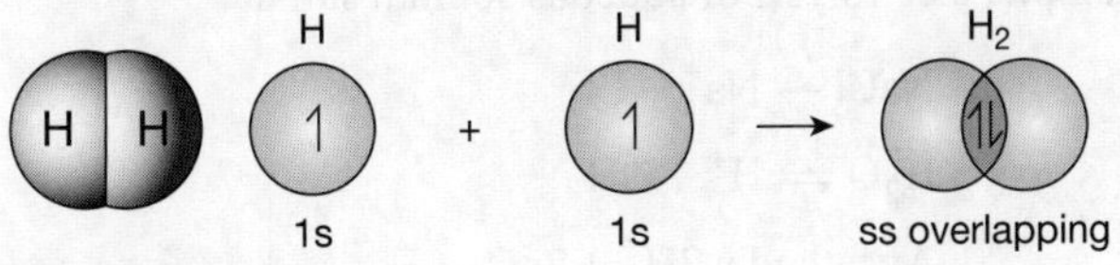

Hydrogen molecule
Symbol = H
Relative atomic mass = 1
Molecular formula = H_2
Relative molecular mass = 2
Valency = 1
Atomic number = 1
Electronic configuration: 1 (K shell)
Position in periodic table: 1st period; group I

Discovery

Hydrogen was discovered by Henry Cavendish and he prepared it by the reaction of iron with dilute acid. He also introduced its elementary nature and said that it gives water on burning in air. He called it inflammable air due to its combustible nature. Lavoisier named it hydrogen. ('Hydro' means water and 'gen' means creating.)

Occurrence

It is the most abundant element in the universe (70%) and the third most abundant element on the surface of the earth. It is found in traces in the earth's crust (0.98%) in the free state and in the atmosphere (0.01%). Volcanic gases also contain 0.025% of hydrogen. In the combined state, it is present in acids, alkalis, hydrocarbons, petroleum, and so on. For example, in water it is 11.1% by weight.

Position in the Periodic Table

Hydrogen is the first element in the periodic table with an atomic number of 1. As it has only 1 valence electron, it is placed in the first group and the first period of the periodic table. Since an atom of hydrogen contains only one electron in its *K*-orbit, it has an equal tendency to lose or to gain one electron to get a stable configuration. Therefore, hydrogen belongs neither to the first group nor to the seventh group and its position is not certain even now.

Methods of Preparation of Dihydrogen

Electrolysis of Acidulated Water

Highly pure hydrogen can be prepared by the electrolysis of water having a small amount of an acid or a base.

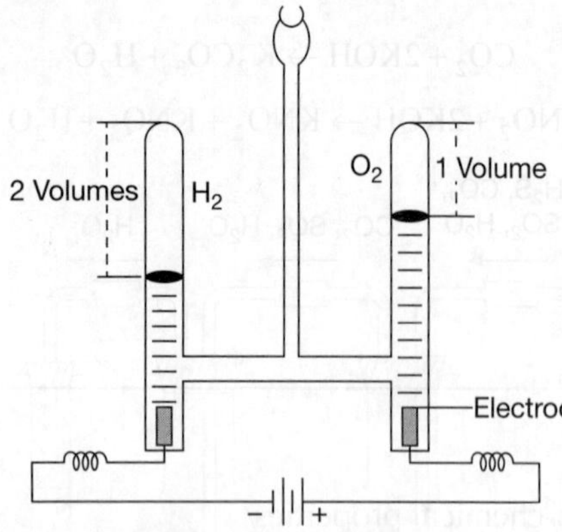

Hofmann's voltameter

$$2H_2O \xrightarrow{\text{Electrolysis}} 2H_2 + O_2$$

$$H_2O \rightleftharpoons H^+ + OH^-$$

At the cathode: $H^+ + e^- \rightarrow H$

$$H + H \rightarrow H_2(g)$$

At the anode: $4OH^- \rightarrow 4OH + 4e^-$

$$4OH \rightarrow 2H_2O + O_2\ (g)$$

Pure hydrogen is formed at the cathode and pure oxygen is formed at the anode respectively.

KNOWLEDGE BOOSTER

Electrolysis

Electrolysis is the passing of a direct electric current through an electrolyte producing chemical reactions at the electrodes which causes decomposition of the materials. The main components required to achieve electrolysis are an electrolyte, electrodes and an external power source.

In chemistry and manufacturing, electrolysis is a technique that uses direct electric current to drive an otherwise non-spontaneous chemical reaction. Electrolysis is commercially important as it is a stage in the separation of elements from naturally occurring sources (such as ores) using an electrolytic cell.

Electrolyte

An electrolyte is a substance that produces an electrically conducting solution when dissolved in a polar solvent, such as water. The dissolved electrolyte separates into cations and anions, which disperse uniformly through the solvent. Electrically, such a solution is neutral. For example, NaCl, $CuSO_4$, H_2SO_4.

Electrode

An electrode is a solid electric conductor that carries electric current into non-metallic solids, liquids, gases, plasmas, vacuums. In an electrochemical cell, reduction and oxidation reactions take place at the electrodes. The electrode at which reduction takes place is known as the cathode and the electrode at which oxidation takes place is known as the anode.

Electrolytic cell

An electrolytic cell uses electrical energy to drive a non-spontaneous redox reaction. An electrolytic cell is a kind of electrochemical cell. It is often used to decompose chemical compounds in a process called electrolysis.

One of the electrodes is connected to the positive pole of the battery to become the positively charged electrode (anode) at which oxidation reactions occur. The other electrode is connected to the negative pole of the battery at which reduction reactions occur (cathode). Electrolytes which are used as ionic conductors in these cells differ from electronic conductors (metals).

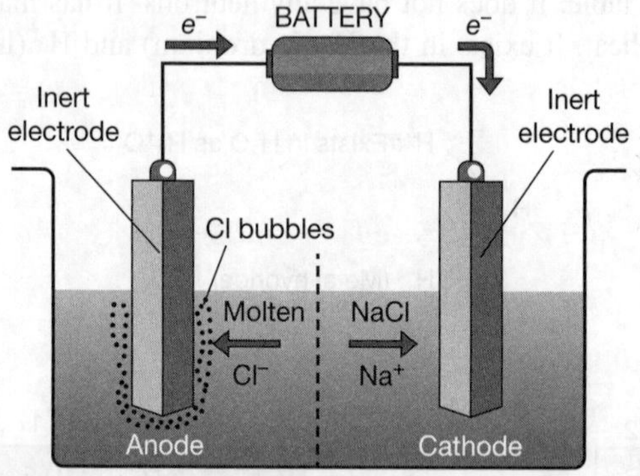

When the two electrodes are connected where the applied potential on the cell slightly exceeds the potential of the reversible cell, an electric current flows in the electrolytic cell. The positive ions in the electrolytic solution between the two electrodes are directed towards the negative electrode (cathode) and neutralise their charge by accepting electrons (reduction). The negative ions from the electrolytic solution are directed towards the positive electrode (anode) and neutralise their charge by losing electrons (oxidation). For example, electrolysis of aqueous sodium sulfate.

$$NaCl \rightarrow Na^+ + Cl^-$$

$$H_2O \rightleftharpoons H^+ + OH^-$$

At cathode: $2H^+ + 2e^- \rightarrow H_2$

At anode: $2OH^- \rightarrow \frac{1}{2}O_2 + H_2O + 2e^-$

$$Na_2SO_4 \rightarrow 2Na^+ + SO_4^{2-}$$

$$H_2O \rightleftharpoons H^+ + OH^-$$

At cathode: $2H^+ + 2e^- \rightarrow H_2$

At anode: $2OH^- \rightarrow \frac{1}{2}O_2 + H_2O + 2e^-$

Laboratory Preparation of Hydrogen

Reactants

The reactants used are granulated zinc [having impurity of copper (catalyst)], dilute hydrochloric acid or dilute sulfuric acid.

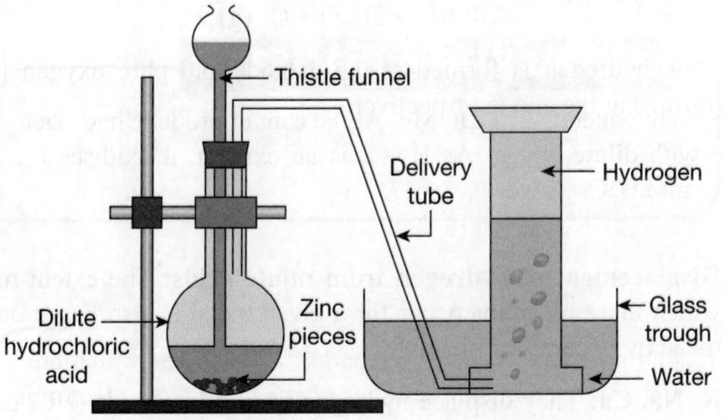

Fig. 6.1 Preparation of hydrogen gas

Procedure

Take some pieces of granulated zinc in a flask (Fig. 6.1) and pour dilute HCl or H_2SO_4 into it. After the reaction, hydrogen is released.

$$Zn + 2HCl(dil) \rightarrow ZnCl_2 + H_2 \uparrow$$

$$Zn + 2H_2SO_4(dil) \rightarrow ZnSO_4 + H_2 \uparrow$$

Observation

Reaction will gradually start in the form of effervescence and evolution of hydrogen gas. When all the air from the apparatus has been expelled, collect the hydrogen gas over water.

Collection of Hydrogen

Hydrogen is collected by the downward displacement of water.

- It is virtually insoluble in water. About 20 mL of hydrogen dissolves in 1 litre of water under normal conditions.
- It forms an explosive mixture with air and so it cannot be collected by downward displacement of air even though it is lighter than air.

Impurities Present with Hydrogen

It has impurities of hydrogen sulfide (H_2S), sulfur dioxide (SO_2), oxides of nitrogen, phosphine (PH_3), arsine (AsH_3), carbon dioxide and water vapour. These impurities can be removed from hydrogen by passing it through the following.

(i) Silver nitrate solution (for removing arsine and phosphine).

$$AsH_3 + 6AgNO_3 \rightarrow Ag_3As + 3AgNO_3 + 3HNO_3$$

$$PH_3 + 6AgNO_3 \rightarrow Ag_3P + 3AgNO_3 + 3HNO_3$$

(ii) Lead nitrate solution (for removing hydrogen sulfide).

$$Pb(NO_3)_2 + H_2S \rightarrow PbS + 2HNO_3$$

(iii) Caustic potash solution (for removing sulfur dioxide, carbon dioxide and oxides of nitrogen).

$$SO_2 + 2KOH \rightarrow K_2SO_3 + H_2O$$

$$CO_2 + 2KOH \rightarrow K_2CO_3 + H_2O$$

$$2NO_2 + 2KOH \rightarrow KNO_2 + KNO_3 + H_2O$$

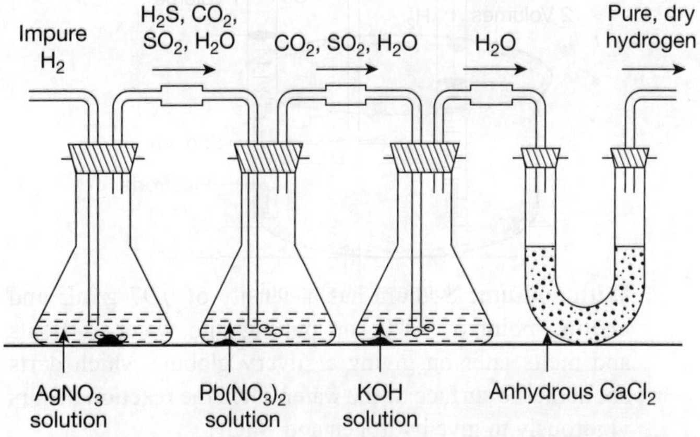

(iv) Drying agents like fused $CaCl_2$ or P_2O_5 (for removing water vapour). Pure and dried hydrogen gas is collected over mercury as there is no reaction between them.

- Here concentrated sulfuric acid is not used as a drying agent as it reacts with hydrogen.

$$H_2SO_4 + H_2 \rightarrow 2H_2O + SO_2$$

Precautions

A mixture of hydrogen and air explodes violently on bringing it near a flame. So, the following precautions are necessary.

- The apparatus must be airtight in order to avoid any leakage.
- There must not be any flame burning near the apparatus.
- Hydrogen gas must be collected only after the escape of all air from the apparatus.
- The end of the thistle funnel must be dipped under an acid in order to prevent the gas from escaping from the funnel.

KNOWLEDGE BOOSTER

- Pure zinc is not used as it is non-porous in nature. So, impure granulated zinc is used which becomes porous due to the presence of some impurities. Conc. sulfuric acid cannot be used as it gives SO_2 and not hydrogen.

$$Zn + conc.\ H_2SO_4 \rightarrow ZnSO_4 + SO_2 + 2H_2O$$

- In the laboratory method, these metals are not used because of the following disadvantages.

Metal		Disadvantage
Na, K	–	React violently
Ca, Mg	–	Very expensive
Fe	–	Impurities of H_2S and SO_2
Al	–	Protective coating of Al_2O_3
Cu, Hg	–	Low reactivity

With Metals

Many metals react with H_2O at different temperatures to release hydrogen.

(i) **With very active metals:** Metals such as Na, K and Ca react with cold water to form hydrogen and their hydroxides. These reactions are exothermic in nature.

With sodium: Sodium has a density of 0.97 g/mL and melting point of 97°C; on putting into water, it floats and melts later on giving a silvery globule which darts about on the surface of the water. Then the reaction occurs vigorously to give hydrogen and NaOH.

$$2Na + 2H_2O \rightarrow 2NaOH + H_2$$

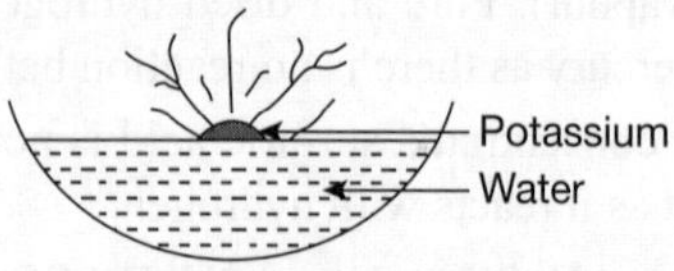

With potassium: Potassium has a density of 0.86 g/mL and melting point of 62°C; on putting into water, it floats and melts later on giving a silvery globule which darts about on the surface of the water. Then the reaction occurs vigorously to give hydrogen and KOH.

$$2K + 2H_2O \rightarrow 2KOH + H_2$$

- Reaction with potassium is more vigorous and exothermic than with sodium. It is better to use the amalgam of these metals to avoid any accident and for the safe preparation of H_2.
- Pure H_2 burns with a pale blue flame but here, the colour of the flame is lilac due to the presence of traces of potassium vapour.
- Due to high reactivity of Na, K with air, they are kept in kerosene.

(ii) **With less active metals:** Metals like Mg, Al and Zn, react with H_2O on heating to form hydrogen.

$$Mg + H_2O \xrightarrow{\Delta} MgO + H_2$$

Magnesium burns in steam with intense white light to give H_2 and white ash or MgO. MgO crumbles down due to heating.

(iii) When steam is passed over metals like Al, Fe, Co, Ni and Sn, hydrogen is formed.

$$3Fe + \underset{\text{Steam}}{4H_2O} \xrightleftharpoons{\text{High temp}} Fe_3O_4 + 4H_2$$

$$2Al + 3H_2O \xrightarrow{800°C} Al_2O_3 + 3H_2$$

Here the reaction stops after sometime as the oxides of these metals stick to the surface of the respective metals and no steam can come in contact with the rest of the metal.

From Acids

When dilute acids are treated with those metals which are above hydrogen in the electrochemical series, hydrogen is formed.

For example, $Zn + H_2SO_4(aq) \rightarrow ZnSO_4 + H_2(g)$

$$Fe + 2HCl(aq) \rightarrow FeCl_2 + H_2(g)$$

KNOWLEDGE BOOSTER

Active metals like Zn, Mg, Al, Fe cannot produce hydrogen with dilute HNO_3. As HNO_3 is an oxidant, it oxidises H_2 into H_2O.

Displacement of hydrogen from dilute acids: The extent to which these reactions occur for a given metal is also based on the activity series of metals.

K, Na, Ca: They displace hydrogen from dilute acids (HCl or H_2SO_4) with explosive violence.

Mg, Al, Zn: They displace hydrogen from dilute acids (HCl or H_2SO_4) vigorously but not violently forming their respective salts.

Fe, Ni, Sn, Pb: They displace hydrogen from dilute acids (HCl or H_2SO_4) gently forming their respective salts.

[H] Cu, Hg, Ag, Au: They do not displace hydrogen from dilute acids.

Industrial Methods

Bosch and Lane methods are used to prepare hydrogen.

Bosch Method

This method includes two steps.

(i) In the first step, steam is passed over hot coke around 1000°C in a converter (furnace) to get water gas.

$$C + H_2O \xrightarrow{1000°C} \underbrace{CO + H_2}_{\text{Water gas}}$$

(ii) In the second step, water gas mixed with steam is passed over the catalyst heated up to 450°C to get hydrogen.

$$\underbrace{CO + H_2}_{\text{Water gas}} + H_2O \xrightarrow[450°C]{Fe_3O_4 + Cr_2O_3} 2H_2 + CO_2 \uparrow$$

The first step is endothermic and the second step is exothermic.

Removal of CO_2

CO_2 can be separated from hydrogen by passing the mixture through cold water under 30 atm pressure where CO_2 dissolves leaving behind H_2.

Or

When the mixture is passed through caustic potash solution, CO_2 is removed as K_2CO_3.

$$2KOH + CO_2 \rightarrow K_2CO_3 + H_2O$$

Removal of CO

To remove CO, the mixture is passed through ammoniacal cuprous chloride solution where CO is removed leaving behind H_2.

$$CuCl + CO + 2H_2O \rightarrow CuCl.CO.2H_2O$$

Lane Method

Super-heated steam is passed over iron at a high temperature to get hydrogen. Iron can be regenerated by passing water gas over the iron oxide.

$$3Fe + 4H_2O \xrightarrow[1000^\circ]{\Delta} Fe_3O_4 + 4H_2\uparrow$$

This process is called gassing and it is continued for nearly 10 minutes till the surface of iron is covered by a layer of the iron oxide and the reaction begins to slow down. This iron oxide can be further reduced into iron if hydrogen supply is cut off and water gas is circulated. The circulation of water gas is called the reduction period and it is nearly for 20 minutes.

$$Fe_3O_4 + H_2 \rightarrow 3FeO + H_2O$$
$$FeO + H_2 \rightarrow Fe + H_2O$$
$$Fe_3O_4 + CO \rightarrow 3FeO + CO_2$$
$$FeO + CO \rightarrow Fe + CO_2$$

Water Gas Shift Reaction

$$CO + H_2O \xrightarrow[\text{Fe and Co oxides}]{673\text{ K}} CO_2 + H_2$$

Test for Hydrogen

(i) It is combustible. However, it does not support combustion.

(ii) It burns silently in air or O_2 with a pale blue flame to give water.

$$2H_2 + O_2 \rightarrow 2H_2O$$

- If H_2 is premixed with air or oxygen, it explodes with a pop sound.

Illustrations

1. (a) Why is nitric acid not used in the preparation of hydrogen?

(b) Why is hot concentrated sulfuric acid not used in the preparation of hydrogen?

Solution:

(a) Nitric acid is not used in the preparation of hydrogen because it is a powerful oxidising agent, and the oxygen formed due to its decomposition oxidises hydrogen to give water, thus defeating the purpose of the reaction.

$$3Zn + 8HNO_3 \rightarrow 3Zn(NO_3)_2 + 4H_2O + 2NO$$

(b) Hot concentrated sulfuric acid is not used in the preparation of hydrogen because it will produce sulfur dioxide.

$$Zn + 2H_2SO_4 \rightarrow ZnSO_4 + SO_2 + 2H_2O$$

2. Thin strips of magnesium, copper and iron are taken.

(a) Write down what happens when these metals are treated as follows:

(i) Heated in presence of air.

(ii) Heated with dil. HCl and added to an aqueous solution of zinc sulfate.

(iii) Added to an aqueous solution of zinc sulfate.

(b) Arrange these metals in the increasing order of reactivity.

Solution:

(a) (i) When we heat magnesium, copper and iron in the presence of air they form their respective oxides.

(ii) Magnesium and iron react with HCl liberating hydrogen and forming their respective salts.

(iii) Magnesium will displace zinc from zinc sulfate solution as magnesium is more reactive than zinc in the activity series of metals.

(b) Cu < Fe < Mg

Physical Properties

(i) It is a colourless, tasteless and odourless gas.

(ii) It is the lightest among all the known substances and has a density of 0.0899 g/cm^3.

(iii) It is less soluble in water.

(iv) It is highly inflammable and should be handled with care.

(v) Its melting and boiling points are 13.957 K and 20.39 K respectively.

Property	H_2
Molecular mass	2.016
Boiling point (K)	20.39
Enthalpy of dissociation (kJ mol^{-1}) (298.2 K)	435.88
Enthalpy of fusion (kJ mol^{-1})	0.117
Enthalpy of vaporisation (kJ mol^{-1})	0.904
Internuclear distance (pm)	74.14
Melting point (K)	13.957

Chemical Properties

It is not quite reactive at ordinary temperatures due to its high bond dissociation energy (436 kJ mol^{-1}) but at higher temperatures it becomes more reactive.

(i) Combustion: Hydrogen undergoes combustion in air or O_2 (with an almost invisible pale blue flame) giving water through an exothermic reaction.

$$2H_2 + O_2 \rightarrow 2H_2O + 485 \text{ kJ}$$

(ii) Reaction with halogens: It reacts with halogens and forms HX.

$$H_2(g) + Cl_2(g) \xrightarrow{\text{light}} 2HCl(g)$$

$$H_2(g) + Br_2(g) \xrightarrow[\text{catalyst}]{620 \text{ K}} 2HBr(g)$$

(iii) Reaction with nitrogen: Hydrogen combines with nitrogen in a 3 : 1 ratio at 490°C and 200 atm pressure in the presence of iron as catalyst to produce ammonia.

$$N_2(g) + 3H_2(g) \rightleftharpoons 2NH_3(g)$$

(iv) Reaction with non-metals of oxygen family: It reacts with oxygen, sulfur, selenium and tellurium to form their corresponding compounds under favourable conditions.

$$2H_2 + O_2 \xrightarrow{654^\circ C} 2H_2O$$

$$H_2 + S \xrightarrow[\text{Pressure}]{427^\circ C} H_2S$$

(v) Reaction with carbon: It combines with carbon at 1147°C to form methane.

$$C(s) + 2H_2(g) \xrightarrow{1147^\circ C} CH_4(g)$$

However, in the presence of an electric arc, it gives acetylene.

$$2C(s) + H_2(g) \xrightarrow[3027^\circ C]{\text{Electric arc}} C_2H_2(g)$$

(vi) Reaction with carbon monoxide: It reacts with carbon monoxide at 427°C and 200 atm. pressure in the presence of the catalyst ZnO/Cr_2O_3 to give methanol.

$$CO + 2H_2 \xrightarrow[700 \text{ K, } 200 \text{ atm}]{ZnO/Cr_2O_3} CH_3OH$$

(vii) Reaction with metals: It reacts with many metals at high temperature to form metal hydrides.

For example, $2Li + H_2 \rightarrow 2LiH$

$$Ca + H_2 \rightarrow CaH_2$$

(viii) Reducing properties: It can reduce metal oxides into metals.

For example, $Fe_3O_4 + 4H_2 \rightarrow 3Fe + 4H_2O$

$$Ag_2O + H_2 \rightarrow 2Ag + H_2O$$

(ix) Hydrogenation reactions: Hydrogen is used for these reactions in the presence of catalysts like Ni, Pd and Pt.

For example, $\text{Vegetable oil} + H_2 \xrightarrow{\Delta, \text{ Ni}} \text{Vegetable ghee}$

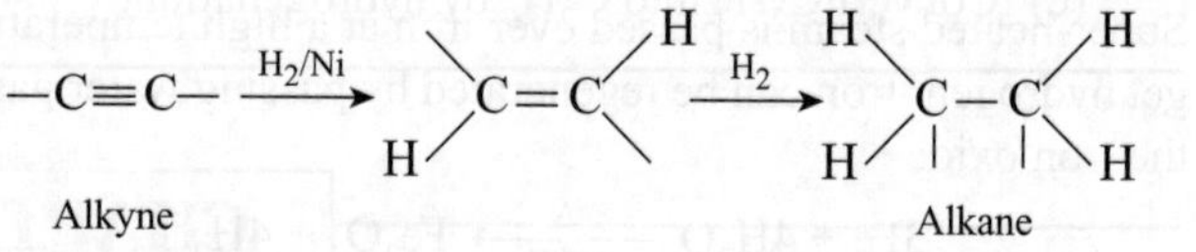

Uses of Hydrogen

(i) In airships, balloons: In airships and balloons, a mixture of 15% hydrogen and 85% helium is used as hydrogen is highly inflammable.

(ii) In the preparation of compounds: It is used to prepare many useful compounds such as methyl alcohol and ammonia.

(iii) As a cryogenic fluid: It is used to produce low temperatures—as a cryogenic fluid.

(iv) In the manufacture of synthetic petrol: Hydrogen is used in the manufacture of synthetic petrol by heating with coal and heavy oils under very high pressure in the presence of a catalyst.

(v) Oxygen–hydrogen torch: A mixture of H_2 and O_2 when burnt in a specially designed apparatus is known as an oxygen–hydrogen torch. It produces a very high temperature of nearly 2500°C. This flame is used for cutting, welding metals, melting platinum and quartz, and also for fusing alumina to get rubies and sapphires (gems).

(vi) Atomic hydrogen torch: It creates a very high temperature up to 2800°C which is used for welding, for alloys having metals like W, Mn, Cr, and so on.

(vii) As a fuel: Hydrogen has a high heat of combustion; so, it can be used as a fuel in the form of coal gas, water gas. Liquid hydrogen when mixed with liquid oxygen can be used as a fuel for rockets.

Illustrations

1. What do you observe when hydrogen gas is passed through a soap solution?

Solution: When hydrogen gas is passed through a soap solution, soap bubbles filled with hydrogen fly high and burst. This behaviour proves that hydrogen is lighter than air.

2. Under what conditions can hydrogen be made to combine with

(a) nitrogen and (b) chlorine?

Name the products in each case and write the equation for each.

Solution:

(a) Three volumes of hydrogen and one volume of nitrogen react at 450–500°C and 200–900 atm in the presence of finely divided iron catalyst with molybdenum as a promoter to give ammonia.

$$N_2 + 3H_2 \rightleftharpoons 2NH_3$$

(b) Equal volumes of hydrogen and chlorine react slowly in diffused sunlight to form hydrogen chloride.

$$H_2 + Cl_2 \rightarrow 2HCl$$

3. (a) Write the reaction between carbon monoxide and hydrogen.

(b) Write the reaction between carbon and hydrogen under electric arc.

(c) Convert C_2H_2 into C_2H_6 by hydrogenation.

Solution:

(a) $CO + 2H_2 \xrightarrow[700\ K,\ 200\ atm]{ZnO/Cr_2O_3} CH_3OH$

(b) $2C(s) + H_2(g) \xrightarrow[3027°C]{Electric\ arc} C_2H_2(g)$

(c) $HC \equiv CH + H_2 \xrightarrow[Ni]{\Delta} H_2C = CH_2 \xrightarrow[Ni]{\Delta} CH_3 - CH_3$

CHAPTER AT A GLANCE

- **Hydrogen** is the first, lightest, smallest, non-metallic element in the periodic table.
- It is the most abundant element in the universe (70%) and the third most abundant element on the surface of the globe.
- as hydrogen has only 1 valence electron, it is placed in the first group and the first period of the **periodic table**.
- **Highly pure hydrogen** can be prepared by the electrolysis of water having small amount of an acid or a base.
- **Electrolysis** is the passing of a direct electric current through an electrolyte producing chemical reactions at the electrodes and decomposition of the materials.
- An **electrolyte** is a substance that produces an electrically conducting solution when dissolved in a polar solvent, such as water.
- An **electrode** is a solid electric conductor that carries electric current into non-metallic solids, liquids, gases, plasmas or vacuums.
- An **electrolytic cell** uses electrical energy to drive a non-spontaneous redox reaction.
- In the laboratory, **hydrogen** is prepared by treating zinc with dil. HCl or H_2SO_4. Concentrated sulfuric acid is not used as a drying agent as it reacts with hydrogen. Pure zinc is not used as it is non-porous, so impure granulated zinc is used which becomes porous due to the presence of some impurities.
- Metals like Na, K, Ca react with cold water to form hydrogen and their hydroxides. These reactions are exothermic in nature.
- Pure H_2 burns with a pale blue flame.
- Due to the high reactivity of Na, K with air they are kept in kerosene.
- Active metals like Zn, Mg, Al, Fe cannot produce hydrogen with dilute HNO_3 since HNO_3 oxidises H_2 into H_2O.
- Bosch method and Lane method are used to prepare hydrogen in industries.
- Hydrogen burns silently in air or O_2 with a pale blue flame to give water.
- In airships and balloons, a mixture of 15% hydrogen and 85% helium is used as hydrogen is highly inflammable.
- Hydrogen is used in the manufacture of synthetic petrol by heating with coal and heavy oils under very high pressure in the presence of a catalyst.
- A mixture of H_2 and O_2 when burnt in a specially designed apparatus is known as an oxygen–hydrogen torch. It produces a very high temperature of nearly 2500°C. This flame is used for cutting and welding metals.

PRACTICE QUESTIONS

Analyse Your Concepts (School Exam Based)

Fill in the Blanks

Instructions: Complete the following statements with an appropriate word/term to be filled in the blank spaces.

1. Hydrogen was discovered by ________.
2. Hydrogen is a ________ atomic gaseous molecule.
3. Hydrogen has a ________ value of ionisation energy.
4. Sodium hydroxide + zinc → hydrogen + ________.
5. Hydrogen is ________ soluble in water.
6. Hydrogen has a ________ specific heat value.
7. ________ metal is preferred for collecting hydrogen from hot water.
8. ________ metal is preferred for collecting hydrogen from steam.
9. Calcium + Water → Calcium hydroxide + ________.
10. ________ metal is preferred for collecting hydrogen from cold water.
11. Hydrogen reacts with nitrogen at 490°C and 200 atm pressure to give ________.

True or False

Instructions: Read the following statements and write your answer as true or false.

1. Hydrogen is the most abundant element in the universe.
2. Water has 2% by weight of hydrogen.
3. Hydrogen is a good oxidising agent.
4. Sodium amalgam reacts smoothly with cold water.
5. Hydrogen adsorbed on the surface of Pt is known as occluded hydrogen.
6. Ag and Au can easily form hydrogen on reaction with dilute acids.
7. Potassium reacts less violently than sodium with water.
8. Hydrogen in the presence of potassium vapour burns with a lilac-coloured flame.
9. Zn and Mg can produce hydrogen easily with HNO_3.
10. Hydrogen is a non-inflammable gas.
11. Hydrogen is used to convert vegetable oil into ghee.

Very Short Answer Type Questions

1. How did the name hydrogen originate?
2. Write some sources of hydrogen.
3. Why is hydrogen called 'inflammable air'?
4. How will you justify the position of hydrogen in the periodic table?
5. How can you say hydrogen shows dual nature?
6. Which test should be done before collecting hydrogen in a gas jar?
7. Give equations to express the reaction between
   (a) calcium and water
   (b) steam and red-hot iron
8. Hydrogen can be prepared from zinc by using water. Write the equation.
9. Hydrogen can be prepared from zinc by using alkali. Write the equation.
10. Why is purified and dry hydrogen collected over mercury?
11. Why should the end of the thistle funnel be dipped under acid?
12. Based on the reactions of water on metals, arrange these metals in the increasing order of reactivity. Iron, sodium magnesium, zinc, calcium.
13. If the following are kept in closed vessels at over 400°C, what will happen to them?
    (a) Iron filings and steam
    (b) Hydrogen and magnetic oxide of iron
14. Why is helium used in airships and balloons?

Short Answer Type Questions

1. Give the general group study of hydrogen with reference to
   (a) valence electrons (b) burning
   (c) reducing power.
2. Compare hydrogen with alkali metals on the basis of:
   (a) reaction with oxygen (b) oxide formation
   (c) ion formation (d) reducing-power
3. Compare hydrogen and halogens on the basis of:
   (a) physical state (b) ion formation
   (c) valency (d) reaction with oxygen
4. Which metal is preferred for the preparation of hydrogen
   (a) from water? (b) from acid?

5. Answer the following:
   (a) Write the reaction of steam with red hot iron.
   (b) Why is this reaction considered a reversible reaction?
   (c) How can the reaction proceed continuously?

6. Hydrogen can be manufactured by the Bosch process.
   (a) Give the equations involved with conditions.
   (b) How can you obtain hydrogen from a mixture of hydrogen and carbon monoxide?

7. A small piece of calcium metal is put into a small trough containing water. There is effervescence and white turbidity is formed.
   (a) What gas is formed in the reaction and how will you test this gas?
   (b) Write a balanced equation for the reaction.
   (c) What will you observe when a few drops of red litmus solution are added to the turbid solution?

8. By applying activity series, name a metal which shows the following properties.
   (a) That reacts readily with cold water.
   (b) That displaces hydrogen from hot water.
   (c) That forms a base which is insoluble in water.
   (d) That displaces hydrogen from dilute HCl.

9. Under what conditions can hydrogen be made to combine with (a) sulfur and (b) oxygen? Name the products in each case and write the equation for each.

10. What is an oxygen–hydrogen torch and for what is used?

Long Answer Type Questions

1. (a) Why are zinc and aluminium considered to have a unique nature? Give balanced equations to support your answer.
   (b) Write balanced equations for the following.
      (i) Iron reacts with dil. HCl
      (ii) Aluminium reacts with fused sodium hydroxide
      (iii) Lead reacts with potassium hydroxide
      (iv) Zinc reacts with caustic soda solution

2. For the laboratory preparation of hydrogen, give the following:
   (a) materials used
   (b) chemical equation
   (c) method of collection
   (d) fully labelled diagram

3. Give reasons for the following.
   (a) Hydrogen is collected by the downward displacement of water and not of air, even though it is lighter than air.
   (b) A candle brought near the mouth of a jar containing hydrogen gas starts burning but is extinguished when pushed inside the jar.
   (c) Apparatus for the laboratory preparation of hydrogen should be airtight and away from a naked flame.
   (d) Potassium and sodium are not used for reaction with dilute hydrochloric acid or dilute sulfuric acid in the laboratory preparation of hydrogen.

4. Write the names of impurities present in hydrogen prepared in the laboratory and how can these impurities be removed from hydrogen?

5. When hydrogen is passed over a black solid compound P, the products are a colourless liquid and a reddish-brown metal Q. Substance Q is divided into two parts, each placed in separate test tubes. Dilute HCl is added to one part of substance Q and dilute HNO_3 to the other.
   (a) Name the substances P and Q.
   (b) Give two tests for the colourless liquid formed in the experiment.
   (c) What happens to substance P when it reacts with hydrogen?
   (d) Write an equation for the reaction between hydrogen and substance P.
   (e) Is there any reaction between substance Q and dilute hydrochloric acid?
   Give proper reasons for your answer.

COMPETITION WINDOW (OBJECTIVE TYPE)

[For NEET, JEE (Main and Advanced), NTSE, KVPY and Olympiads]

Topic-wise MCQs

Preparation

1. Which of the following is incorrect regarding hydrogen?
   (a) it does not have any neutron
   (b) it can exist as H^+ and H^-
   (c) it can show +1 and –1 oxidation states
   (d) it can act both as an oxidant and a reductant

2. Which of the following is not a similarity between hydrogen and alkali metals?
   (a) atomicity
   (b) physical state
   (c) reaction with non-metals
   (d) both (a) and (b)

3. Which of the following compounds is possible when hydrogen burns in air?
(a) H_2O (b) H_2O_2
(c) both (a) and (b) (d) no compound is formed

4. Hydrogen from HCl can be prepared using
(a) P (b) Mg
(c) Hg (d) Cu

5. Metal hydride on treatment with water gives
(a) hydrogen (b) acid
(c) H_2O (d) H_2O_2

6. The component present in greater proportion in water gas is
(a) CO (b) H_2
(c) CO_2 (d) CH_4

7. Which of the following metals is not used in the lab preparation of hydrogen?
(a) calcium (b) iron
(c) aluminium (d) sodium

8. The metal that cannot displace hydrogen from dil. HCl is
(a) Fe (b) Cu
(c) Zn (d) Mg

9. Aluminium reacts with conc. HCl and conc. NaOH to liberate the gases ____ respectively.
(a) H_2 and H_2 (b) O_2 and O_2
(c) O_2 and H_2 (d) H_2 and O_2

10. What gas is liberated when alkaline formaldehyde solution is treated with H_2O_2?
(a) CH_4 (b) H_2
(c) CO_2 (d) O_2

Properties of hydrogen

11. Which of the following statements about hydrogen is incorrect?
(a) It is an inflammable gas.
(b) It is the lightest gas.
(c) It is not easily liquefied.
(d) It is a strong oxidising agent.

12. Hydrogen burns in air or oxygen with a/an
(a) invisible pale blue flame
(b) red flame
(c) golden flame
(d) blue flame

13. Ortho and para hydrogen differ
(a) in their molecular weights
(b) in the number of protons
(c) in the nature of spin of electrons
(d) in the nature of spin of protons

14. Which metal adsorbs hydrogen?
(a) Al (b) Fe
(c) Pd (d) K

15. The composition of the nucleus of deuterium is
(a) $1e^-$ and $1p$ (b) $1e^-$ and $1n$
(c) $1n$ and $1p$ (d) $2p$ and $1e^-$

16. Helium is preferred to hydrogen for filling balloons because it is
(a) lighter than air
(b) almost as light as hydrogen
(c) non-combustible
(d) inflammable

17. A metal oxide that is reduced by hydrogen is
(a) Al_2O_3 (b) CuO
(c) CaO (d) Na_2O

18. For the reaction $PbO + H_2 \rightarrow Pb + H_2O$, which of the following statements is wrong?
(a) H_2 is the reducing agent
(b) PbO is the oxidising agent
(c) PbO is oxidised to Pb
(d) H_2 is oxidised to H_2O

19. Hydrogen will not reduce
(a) heated cupric oxide
(b) heated ferric oxide
(c) heated stannic oxide
(d) heated aluminium oxide

20. Which of the following could act as a propellant for rockets?
(a) liquid hydrogen + liquid nitrogen
(b) liquid nitrogen + liquid oxygen
(c) liquid hydrogen + liquid oxygen
(d) liquid oxygen + liquid argon

21. Which of the following hydrides is an electron precise hydride?
(a) B_2H_6 (b) NH_3
(c) H_2O (d) CH_4

22. Only one element of ____ forms the hydride.
(a) Group 6 (b) Group 7
(c) Group 8 (d) Group 9

Miscellaneous

1. Hydrogen resembles the halogens in many respects for which several factors are responsible. Of the following factors, which one is most important in this respect?
(a) Its tendency to lose an electron to form a cation.
(b) Its tendency to gain a single electron in its valence shell to attain a stable electronic configuration.
(c) Its low negative electron gain enthalpy value.
(d) Its small size.

2. Which metal gives hydrogen with all of the following—water, acids, alkalis?
(a) Fe (b) Zn
(c) Mg (d) Pb

3. Why does the H^+ ion always get associated with other atoms or molecules?
(a) Ionisation enthalpy of hydrogen resembles that of alkali metals.
(b) Its reactivity is similar to that of halogens.
(c) It resembles both alkali metals and halogens.
(d) Loss of an electron from the hydrogen atom results in a nucleus that is very small in size when compared to other atoms or ions. Due to its small size, it cannot exist free.

4. Which of the properties of interstitial hydrides is correct?
(a) they give rise to metals fit for fabrication
(b) they generally form non-stoichiometric species
(c) they can be used as hydrogenation catalysts
(d) the hydrogen dissolved in titanium improves its mechanical properties

5. In the following compound, H is covalently bonded in the case of
(a) CaH_2 (b) NaH
(c) SiH_4 (d) BaH_2

6. Tritium is a radioactive isotope of hydrogen. It emits
(a) α-particles (b) β-particles
(c) γ-rays (d) neutrons

7. Hydrogen is evolved by the action of cold dil. HNO_3 on
(a) Fe (b) Cu
(c) Mg (d) Zn

8. Elements which show a unique property in the preparation of hydrogen are
(a) Na, K, Li (b) Mg, Ca, Ba
(c) Al, Zn, Pb (d) Fe, Cu, Ag

9. Which of the following reactions increases the production of dihydrogen from synthesis gas?
(a) $CH_4(g) + H_2O(g) \xrightarrow[\text{Ni}]{1270\text{ K}} CO(g) + 3H_2(g)$
(b) $C(s) + 3H_2O(g) \xrightarrow{1270\text{ K}} CO(g) + H_2(g)$
(c) $CO(g) + H_2O(g) \xrightarrow[\text{Catalyst}]{673\text{ K}} CO_2(g) + H_2(g)$
(d) $C_2H_6 + 2H_2O \xrightarrow[\text{Ni}]{1270\text{ K}} 2CO + 5H_2$

10. Which of the following is a use of hydrogen?
(a) in the preparation of artificial petrol from coal
(b) in the extraction of metals like copper
(c) in oxygen–hydrogen torch
(d) all of these

11. The adsorption of hydrogen is possible on the surface of which metal?
(a) Pt (b) Pd
(c) Ni (d) all of these

12. Which of the following metal oxides can be reduced by hydrogen?
(a) CuO (b) PbO
(c) Fe_2O_3 (d) all of these

13. Which of the following react most violently with dil. acids to give hydrogen?
(a) Na (b) K
(c) Ca (d) Mg

14. In which of the following pairs are both metals below hydrogen in the activity series?
(a) Ca, Cu (b) Mg, Ag
(c) Cu, Ag (d) Al, Ag

Advanced and Olympiads

Single Choice

1. Which of the following reactions is an example of the use of water gas in the synthesis of other compounds?
(a) $CH_4(g) + H_2O(g) \xrightarrow[\text{Ni}]{1270\text{ K}} CO(g) + H_2(g)$
(b) $CO(g) + H_2O(g) \xrightarrow[\text{catalyst}]{673\text{ K}} CO_2(g) + H_2(g)$
(c) $C_nH_{2n+2} + nH_2O(g) \xrightarrow[\text{Ni}]{1270\text{ K}} nCO + (2n+1)H_2$
(d) $CO(g) + 2H_2(g) \xrightarrow[\text{catalyst}]{\text{Cobalt}} CH_3OH(l)$

2. 2 g of aluminium is treated separately with excess of dilute H_2SO_4 and excess of NaOH. The ratio of the volumes of hydrogen evolved is
(a) 1:1 (b) 1:2
(c) 2:1 (d) 2:3

3. Consider the following statements:
I. Atomic hydrogen is obtained by passing hydrogen through an electric arc.
II. Hydrogen gas will not reduce heated aluminium oxide.
III. Finely divided palladium absorbs a large volume of hydrogen gas.
IV. Pure nascent hydrogen is best obtained by reacting Na with C_2H_5OH.
Which of the above statements is/are correct?
(a) I alone (b) II alone
(c) I, II, and III (d) II, III and IV

4. Match the following

Column I	Column II
1. Hydrogenation of oils	(i) Fe filings
2. Haber's process	(ii) Ni
3. Nuclear fusion	(iii) hydrogen
4. Most abundant in the universe	(iv) solar energy

The correct matches are

	1	2	3	4
(a)	(iv)	(iii)	(ii)	(i)
(b)	(iv)	(iii)	(i)	(ii)
(c)	(ii)	(i)	(iv)	(iii)
(d)	(iii)	(i)	(iv)	(ii)

Multiple Choice

5. Select the correct statements regarding hydrogen.
   (a) It is the most abundant element of the universe.
   (b) Hydrogen forms the maximum number of compounds in total.
   (c) Hydrogen has less electronegativity like alkali metals.
   (d) Like halogens, hydrogen has a high value of ionisation energy.

6. Metals which liberate $H_2(g)$ from acids are
   (a) Fe (b) Zn
   (c) Mn (d) Cu

7. Select the incorrect statements.
   (a) Hydrogen is separated from CO by passing the mixture through caustic potash solution.
   (b) All metals react with acids to give hydrogen.
   (c) Hydrogen is dried by passing it through conc. H_2SO_4.
   (d) Very dilute nitric acid reacts with iron to produce hydrogen.

8. Which of the following statements about hydrogen are not true?
   (a) It exists as a diatomic molecule.
   (b) It has one electron in the outermost shell.
   (c) It can lose an electron to form a cation which can exist freely.
   (d) It forms a large number of ionic compounds by losing an electron.

9. Dihydrogen can be prepared on a commercial scale by different methods. In its preparation by the action of steam on hydrocarbons, a mixture of CO and H_2 gas is formed. This is known as ___.
   (a) water gas (b) syngas
   (c) producer gas (d) industrial gas

Matrix Matching

10. Match the following

Column I	Column II
(a) H	(p) Used in the name of perhydrol
(b) H_2	(q) Can be reduced to dihydrogen by NaH
(c) H_2O	(r) Can be used in hydroformylation of olefins
(d) H_2O_2	(s) Can be used for cutting and welding

Integer Type

11. If the most common oxidation states of hydrogen in its compounds are +X and –Y, the sum of X + Y is?

12. How many of these metals can replace hydrogen from dilute acids?
    K, Na, Mg, Zn, Fe, Cu, Ag, Hg, Au

13. How many of these metals can replace hydrogen from dilute acids with violent reactions?
    K, Na, Mg, Zn, Fe, Ca, Ag, Hg, Pb

14. Which of the following metals cannot be used in the laboratory method of hydrogen preparation?
    Zn, Fe, Pb, Al, Ca, Mg, Na, K

15. Which of the following metal oxides cannot be reduced to their corresponding metals?
    Na_2O, K_2O, MgO, CaO, Al_2O_3, CuO, Fe_2O_3

Answer Keys

Fill in the Blanks

1. Cavendish	2. di	3. high
4. sodium zincate	5. sparingly	6. high
7. magnesium	8. iron	9. hydrogen
10. sodium	11. ammonia	

True or False

1. T	2. F	3. F	4. T	5. T
6. F	7. F	8. T	9. F	10. F
11. T				

Very Short Answer Type Questions

1. The name hydrogen (hydro = water, genate = forming) originated from its ability to form water.

2. Hydrogen is found in the earth's crust and in the atmosphere in a free state. It is found in organic compounds in the combined state.

3. Hydrogen is a highly combustible gas; hence, it is called inflammable air.

4. Hydrogen has atomic number 1, and it exists in the first group and 1st period of the periodic table. Hence it is the first element of the periodic table.

5. Hydrogen resembles the alkali metals of group 1 and the halogens of group 17. Hence, it shows dual nature.

6. Hydrogen burns with a pop sound, hence it should be tested with a burning flame. If the burning produces a pop sound, it confirms the presence of hydrogen gas.

7. (i) $Ca + 2H_2O \rightleftharpoons Ca(OH)_2 + H_2$
(ii) $3Fe + 4H_2O \rightleftharpoons Fe_3O_4 + 4H_2$

8. $Zn + H_2O \rightarrow ZnO + H_2$

9. $Zn + 2NaOH \rightarrow Na_2ZnO_2 + H_2$

10. Purified and dry hydrogen is collected over mercury. It will not have any impact.

11. The end of the thistle funnel should be dipped under the acid to prevent the escape of gas from the thistle funnel.

12. Iron < zinc < magnesium < calcium < sodium.

13. (a) Iron oxide is formed with the evolution of hydrogen gas.
(b) Heated magnetic oxide of iron gets reduced by hydrogen.

14. In airships and balloons, a mixture of 15% hydrogen and 85% helium is used as hydrogen is highly inflammable.

Short Answer Type Questions

1. (a) Hydrogen has 1 valence electron in its orbit.
(b) Hydrogen burns with oxygen to form oxides of hydrogen with a pop sound.
$$2H_2 + O_2 \rightarrow 2H_2O$$
(c) Hydrogen acts as a reducing agent.
$$CuO + H_2 \rightarrow Cu + H_2O$$

2. (a) Both alkali metals and hydrogen burn in air to form oxides.
(b) Hydrogen and alkali metals can form oxides on reaction with oxygen.
(c) Hydrogen and alkali metals can form a cation by the loss of an electron.
(d) Both hydrogen atoms and alkali metals are reducing agents.

3. (a) Both the halogens and hydrogen are gases.
(b) Both of them form anions as they are short of 1 electron to attain a stable configuration.
(c) Both hydrogen and the halogens have a valency of 1.
(d) Hydrogen reacts with oxygen to form neutral oxides; halogens react with oxygen to form acidic oxides.

4. (a) For the preparation of hydrogen from water, reactive metals like potassium, sodium and calcium are preferred.
(b) For the preparation of hydrogen from acid, magnesium, aluminium, zinc and iron are preferred.

5. (a) $3Fe + 4H_2O \rightleftharpoons Fe_3O_4 + 4H_2$
(b) This reaction is reversible. If the hydrogen formed is not removed the iron oxide formed is reduced back to iron. Thus, the reaction can proceed continuously.

6. (a) $C + H_2O \xrightarrow{1000°} (CO + H_2) - \Delta$
$$(CO + H_2) + H_2O \xrightarrow[450°C]{Fe_2O_3} CO + 2H_2 + \Delta$$
(b) We can obtain hydrogen from a mixture of hydrogen and carbon monoxide by passing the mixture through an ammoniacal cuprous chloride solution in order to dissolve any uncombined carbon monoxide.
$$CuCl + CO + 2H_2O \rightarrow CuCl.CO.2H_2O$$

7. (a) It is hydrogen gas and it can be tested using litmus paper. It turns blue from red.
(b) $Ca + 2H_2O \rightarrow Ca(OH)_2 + H_2$
(c) The solution turns blue.

8. (a) Sodium (Na)
(b) Magnesium (Mg)
(c) Calcium (Ca)
(d) Zinc (Zn)

9. (a) Hydrogen gas on passing through molten sulfur reacts to give hydrogen sulfide.
$$H_2 + S \rightarrow H_2S$$
(b) Hydrogen burns in the presence of an electric spark with a 'pop' sound in oxygen and with a blue flame, forming water.
$$2H_2 + O_2 \rightarrow 2H_2O$$

10. A mixture of H_2 and O_2 when burnt in a specially designed apparatus is known as an oxygen–hydrogen torch. It produces a very high temperature of nearly 2500°C. This flame is used for cutting and welding metals, melting platinum and quartz, and also for fusing alumina to get rubies and sapphires (gems).

Long Answer Type Questions

1. (a) Zinc and aluminium are considered to have a unique nature for the following reasons.
(i) They can react with acids and alkali to form hydrogen and soluble salts.
$$Zn + 2NaOH \rightarrow Na_2ZnO_2 + H_2$$
$$2Al + 2NaOH + 2H_2O \rightarrow 2NaAlO_2 + 3H_2$$
(ii) They react with both bases and acids to give their salts and water.
$$ZnO + 2HCl \rightarrow ZnCl_2 + H_2O$$
$$ZnO + 2NaOH \rightarrow Na_2ZnO_2 + H_2O$$
(b) (i) $Fe + 2HCl \rightarrow FeCl_2 + H_2$
(ii) $2Al + 6NaOH \rightarrow 2Na_3AlO_3 + 3H_2$
(iii) $Pb + 2KOH \rightarrow K_2PbO_2 + H_2$
(iv) $Zn + 2NaOH \rightarrow Na_2ZnO_2 + H_2$

2. (a) Granulated zinc, dilute HCl or dil. H_2SO_4
(b) $Zn + 2HCl \rightarrow ZnCl_2 + H_2$
(c) It is collected by the downward displacement of water.
(d) For the diagram, see the text part.

3. (a) Hydrogen is collected by the downward displacement of water because it is insoluble in water and it forms an explosive mixture with air.

(b) Hydrogen is a combustible gas but it does not support the combustion process. When a candle is brought near the mouth of a jar containing hydrogen, it burns with the help of the oxygen in the atmosphere. When the candle is pushed inside the jar, oxygen supply gets cut off and hence the fire gets extinguished.

(c) Apparatus for the laboratory preparation of hydrogen should be airtight and away from a naked flame because hydrogen forms an explosive mixture with air.

(d) Potassium and sodium are not used for reaction with dilute hydrochloric acid or dilute sulfuric acid in the laboratory preparation of hydrogen as both potassium and sodium react violently with acids.

4. See the text part.

5. (a) Here P is CuO and Q is Cu

(b) Blue and red litmus paper, when dipped in the colourless liquid, do not change colour. This confirms that the liquid formed is neutral and is water. It changes white anhydrous copper sulfate to a blue salt.

(c) Black copper oxide P on heating with hydrogen reduces copper oxide to reddish-brown copper, and itself gets oxidised to water. Hydrogen is a strong reducing agent and removes oxygen from less active metals. That is, it removes oxygen from heated metal oxides when passed over them and itself gets oxidised to water.

(d) $CuO + H_2 \rightarrow \Delta\ Cu + H_2O$

(e) $Cu + HCl \rightarrow$ No reaction

As copper is less reactive than hydrogen, it cannot displace it from HCl.

Competition Window (Objective Type)

Topic-wise MCQs

1. (d)	2. (d)	3. (c)	4. (b)	5. (a)
6. (b)	7. (a)	8. (b)	9. (a)	10. (b)
11. (d)	12. (a)	13. (d)	14. (c)	15. (c)
16. (c)	17. (b)	18. (c)	19. (d)	20. (c)
21. (d)	22. (a)			

Miscellaneous

1. (b)	2. (b)	3. (d)	4. (b)	5. (c)
6. (b)	7. (c)	8. (c)	9. (c)	10. (d)
11. (d)	12. (d)	13. (b)	14. (c)	

Advanced and Olympiads

Single Choice

1. (d) 2. (a) 3. (c) 4. (c)

Multiple Choice

5. (abd) 6. (abc) 7. (bcd) 8. (cd) 9. (ab)

Matrix Matching

10. (a)—(s); (b)—(r); (c)—(q); (d)—(p)

Integer Type

11. 0 12. 5 13. 3 14. 7 15. 5

Hints and Solutions

Competition Window (Objective Type)

Topic-wise MCQs

1. It can only be a reductant and not an oxidant.

2. Hydrogen is diatomic and a gas while alkali metals are monoatomic solid metals.

5. $MH_2 + 2H_2O \rightarrow M(OH)_2 + 2H_2$

6. Water gas contains about 50 volumes of H_2, 40 volumes of CO and 5 volumes of CO_2 and N_2.

7. The answer is (a) calcium because calcium is expensive.

8. Cu is below hydrogen in the electrochemical series and unable to displace hydrogen from HCl.

12. Hydrogen on burning in air gives a pale blue flame.

13. *Ortho* and *para* hydrogen have different nuclear spins.

19. Hydrogen will not reduce heated aluminium oxide because the Al–O bond is very strong.

20. A mixture of liquid hydrogen and liquid oxygen is used as a propellant for rockets.

Miscellaneous

14. Both Cu and Ag are below hydrogen in the electrochemical series.

Advanced and Olympiads

Single Choice

2. $2Al + 3H_2SO_4 \rightarrow Al_2(SO_4)_3 + 3H_2$

$2Al + 2NaOH + 2H_2O \rightarrow 2NaAlO_2 + 3H_2$

So, the ratio of volumes of hydrogen evolved is 1 : 1.

Multiple Choice

5. Hydrogen has more electronegativity (2.1) while the alkali metals have low electronegativity values.

6. Cu being less electropositive than H cannot liberate hydrogen from acids.

7. The metals which are above hydrogen in the activity series give hydrogen when they react with acids.
Hydrogen is dried by passing through calcium chloride, caustic potash and phosphorous pentoxide.
Very dilute nitric acid reacts with magnesium and manganese to produce hydrogen and not with iron.

Integer Type

11. Hydrogen has +1 and –1 as the common oxidation states; X = +1, Y = –1
Sum of X + Y = (+1) + (–1) = 0

12. K, Na, Mg, Zn, Fe can replace hydrogen from dil. acids.

13. K, Na and Ca can replace hydrogen from dil. acids with a violent reaction.

14. Fe, Pb, Al, Ca, Mg, Na, K cannot be used in the laboratory method of hydrogen preparation.

15. Na_2O, K_2O, MgO, CaO, Al_2O_3 cannot be reduced into their corresponding metals by hydrogen as these metals have a very strong affinity towards oxygen.

CHAPTER 7

Water

LEARNING OBJECTIVES

After studying this unit, you will be able to understand:

- Water, its structure, physiochemical properties
- Solutions, types and concentration units
- Solubility and the factors affecting it
- Crystals and crystallisation
- Hydrated and anhydrous substances
- Efflorescence and deliquescence
- Drying and dehydrating agents
- Soft and hard water

Introduction

Water is the most abundant and most important natural resource in the biosphere. More than 70% of the area of the earth is occupied by water. However, only 2.5% of this water is fresh water including the frozen water in the polar ice caps and glaciers. Water is an oxide of hydrogen with a molecular formula of H_2O having one part of hydrogen and eight parts of oxygen by weight.

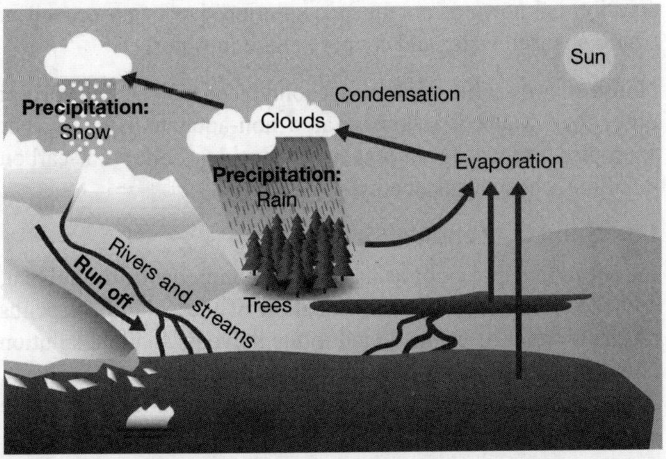

Water (H_2O)

Its chemical name is dihydrogen oxide and its molecular formula is 18 amu. It is a major constituent of all living beings and of the environment in which we live. Almost 70% of our body weight is due to water. It can exist in all the three physical states—solid (ice), liquid (water) and gas (steam). It is present in both the free and the combined states.

Sources of Water

Water in its liquid state is found in oceans and seas. Apart from this, it is also found in rivers, lakes, streams and other water bodies. Water available in the above sources is known as *surface water*. Water present in oceans and seas has appreciable proportions of dissolved salts, predominantly sodium chloride. The presence of these salts imparts a salty taste to water and this water is hence known as saline water. In addition to surface water, water is also available under the earth's surface and is known as *underground water*. This water accumulates due to the seepage of rain water through the soil. This underground water may come out in the form of springs. This can be drawn out artificially by digging wells or with the help of tube wells.

Water as a Universal Solvent

Water is a universal solvent due to its high dielectric constant, high liquid range and ability to dissolve most of the compounds. As water has a unique ability to dissolve a wide range of substances, it is called a universal solvent. The ability of water to form solutions is responsible for all the life activities occurring in nature and also for the sustenance of life on earth.

Water is a good solvent for ionic compounds but bad for covalent compounds as the force of attraction between ions is destroyed or greatly reduced and ions get dissolved in water.

Salts Dissolved in Water

Substances that are apparently insoluble in water actually dissolve in water in traces. Even when we put water in a glass vessel, an extremely small amount of glass dissolves in it. Due to this reason, when distilled water is kept in a sealed bottle for a long time, it leaves etchings on the inside surface of glass.

Tap water, river water and well water contain dissolved solids but rainwater and distilled water do not contain dissolved solids and so concentric rings are not formed in their case.

Importance of Dissolved Salts in Water

- Salts and minerals are essential for the growth and development of plants.
- They provide taste to water.
- They supply the essential minerals needed by our bodies.

Illustration

1. Discuss the properties of water that are responsible for controlling the temperature of our body.

Solution: Our body is made of almost 70% water. Water has a high specific heat. Due to high specific heat capacity, the presence of a large amount of water is able to modify and control the temperature of our body which is warm in winter and cooler in summer.

Solutions and Their Types

Solutions

A solution is a homogeneous mixture of two or more pure substances. For example, lemonade and soda water are examples of solutions. In a solution there is homogeneity at the particle level. For example, lemonade tastes the same throughout. This shows that particles of sugar or salt are evenly distributed in the solution. A solution has a solvent and a solute as its two components (binary solution). The component of the solution that dissolves the other component (usually the component present in larger amount) is known as the solvent while the component of the solution that is dissolved in the solvent (usually present in lesser quantity) is known as the solute. For example, a solution of sugar in water is a solid in liquid solution. Sugar is the solute and water is the solvent. A solution of iodine in alcohol known as 'tincture of iodine', has iodine (solid) as the solute and alcohol (liquid) as the solvent.

Different Types of Solutions

We can divide solutions into different categories as follows.

Based on Amount of Solute in the Solution

Depending upon the dissolution of the solute in the solvent, solutions can be categorised into unsaturated, saturated and supersaturated solutions.

Unsaturated solution: An unsaturated solution is a solution in which a solvent is capable of dissolving some more solute at a given temperature.

Saturated solution: A saturated solution can be defined as a solution in which a solvent is not capable of dissolving any more solute at a given temperature.

Supersaturated solution: A supersaturated solution contains more than the maximum amount of solute that is capable of being dissolved at a particular temperature. When the solvent is reduced, the extra solute will crystallise quickly.

Based on Nature of Solvent

Solutions are of two types, depending on whether the solvent is water or not.

Aqueous solution: When a solute is dissolved in water, the solution is known as an aqueous solution. For example, salt in water, sugar in water and copper sulfate in water.

Non-aqueous solution: When a solute is dissolved in a solvent other than water, it is known as a non-aqueous solution. For example, iodine in carbon tetrachloride, sulfur in carbon disulfide, phosphorus in ethyl alcohol.

Dilute and Concentrated Solutions

Solutions are spoken of as having two components, the solvent, and the solute. Another classification of the solution depends on the amount of solute added to the solvent. A dilute solution contains a small amount of solute in a large amount of solvent. A concentrated solution contains a large amount of solute dissolved in a small amount of solvent.

Based Upon Solute Particle Size

Solutions are of three types—true solutions, suspensions and colloids.

True solution: A solution in which the particles of the solute can be broken down to such a fine state that they cannot be seen even under a powerful microscope is known as a true solution.

Properties of true solutions

- A solution is always homogeneous in nature.
- The properties of the solute are retained in the true solution. For example, a sugar solution is sweet in taste and a solution of salt in water is saline in taste.

- The particles of a solution are smaller than 1 nm (10^{-9} m) in diameter. So, they cannot be seen by the naked eye.
- A true solution does not show the Tyndall effect because of very small particle size. A true solution does not scatter a beam of light passing through the solution so the path of light is not visible. That is, solutions are transparent to light.
- The solute particles in a solution can easily pass through a filter paper; so, a true solution passes through a filter paper. Hence the solute particles cannot be separated from the mixture by the process of filtration.
- The solute particles do not settle down when left undisturbed, that is, a solution is stable.

Types of true solutions: Usually we think of a solution as a liquid that contains either a solid, liquid or a gas dissolved in it. However, we can also have solid solutions (alloys) and gaseous solutions (air).

- **Solution of solid in a solid:** Metal alloys are the solutions of a solid in a solid. For example, brass is a solution of zinc in copper. It is prepared by mixing molten zinc with molten copper and cooling the mixture.
- **Solution of liquid in a liquid:** Vinegar is a solution of acetic acid in water.
- **Solution of gas in a liquid:** Aerated drinks like soda water are gas in liquid solutions. These contain carbon dioxide (gas) as the solute and water (liquid) as the solvent.
- **Solution of gas in a gas:** Air is a mixture of a gas in gas and it is a homogeneous mixture. Its two main constituents are oxygen (21%) and nitrogen (78%) and the other gases (CO_2, Ar) are present in very small quantities.

Suspensions: A suspension is a heterogeneous mixture in which the small particles of a solid are spread throughout a liquid without dissolving in it. In a suspension, the solute particles do not dissolve but remain suspended throughout the bulk of the medium. For example, chalk–water mixture, muddy water, milk of magnesia, sand particles suspended in water, and flour in water.

Properties of suspensions

- A suspension is a heterogeneous mixture.
- The size of the particles in a suspension is greater than 100 nm in diameter.
- The particles of a suspension can be seen with the naked eye or with the help of a simple microscope.
- As the particles of a suspension do not pass through a filter paper, it is possible to separate them by ordinary filtration.
- As the particles of a suspension settle down when a suspension is left undisturbed, a suspension is unstable.
- A suspension is not transparent to light as the particles of a suspension scatter a beam of light passing through it and make its path visible.

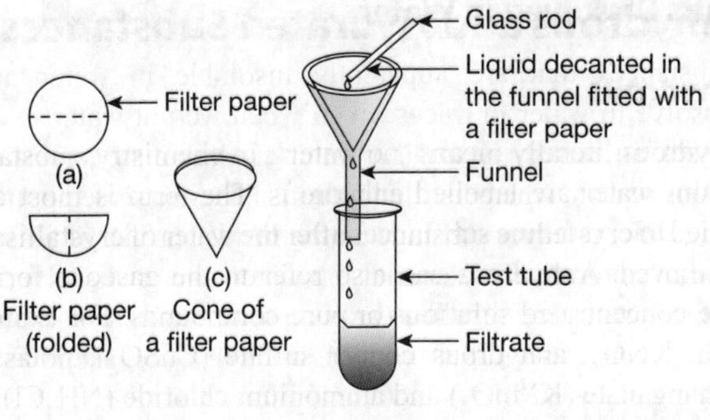

Colloids: A colloid is a type of solution in which the size of the solute particles is intermediate between those in true solutions and those in suspensions. The size of the solute particles in a colloid is larger than those of a true solution but smaller than those of a suspension. The particles of a colloid are uniformly spread throughout the solution. Due to the relatively smaller size of particles, as compared to that of a suspension, the mixture appears to be homogeneous but actually, a colloidal solution is a heterogeneous mixture, for example, milk. A colloid contains dispersed particles and the dispersion medium.

Dispersed particles: The solute particles are known as *dispersed particles*.

Dispersion medium: Solvents are known as the dispersion medium.

Classification of colloids: Colloids are classified according to the physical state of the dispersed phase (solute) and the dispersion medium (solvent). Most of the colloids can be classified into the following eight groups.

Dispersed phase	Dispersion medium	Colloidal system	Examples
Gas	Liquid	Foam or froth	Soap sols, lemonade froth, whipped cream
Gas	Solid	Solid foam	Pumice stone, styrene foam, foam rubber, dried sea foam
Liquid	Gas	Aerosols of liquids	Fog, clouds, mist, fine insecticide sprays
Liquid	Liquid	Emulsions	Milk, cold cream, tonics
Liquid	Solid	Gels	Cheese, butter, boot polish, table jellies
Solid	Gas	Aerosols of solids	Smoke, dust, haze
Solid	Liquid	Sols	Paints, starch dispersed in water, gold sol, muddy water, inks
Solid	Solid	Solid sols	Ruby glass, minerals, gem stones

Anhydrous and Hydrated Substances

Anhydrous

Anhydrous literally means 'no water'. In chemistry, substances without water are labelled anhydrous. The term is most often applied to crystalline substances after the water of crystallisation is removed. Anhydrous can also refer to the gaseous form of some concentrated solutions or pure compounds. For example, NaCl, KNO_3, anhydrous copper sulfate ($CuSO_4$), potassium permanganate ($KMnO_4$) and ammonium chloride (NH_4Cl).

Hydrated

A hydrate is a substance that contains water or its constituent elements. The chemical state of the water varies widely between different classes of hydrates, some of which were so labelled before their chemical structure was understood. This water of crystallisation gives the crystals shape and also colour. For example,

- Blue vitriol (hydrated copper sulfate) is $CuSO_4 \cdot 5H_2O$ and blue in colour.
- Washing soda (sodium carbonate decahydrate) is $Na_2CO_3 \cdot 10H_2O$ and white in colour.
- Epsom salt (magnesium sulfate heptahydrate) is $MgSO_4 \cdot 7H_2O$ and white in colour.
- Glauber's salt (sodium sulfate decahydrate) is $Na_2SO_4 \cdot 10H_2O$ and white in colour.
- Green vitriol [Iron(III) sulfate heptahydrate] is $FeSO_4 \cdot 7H_2O$ and bluish green in colour.
- Plaster of Paris (calcium sulfate semihydrate) is $CaSO_4 \cdot \frac{1}{2} H_2O$ and white in colour.
- Gypsum (hydrated calcium sulfate) is $CaSO_4 \cdot 2H_2O$ and white in colour.

Water of Crystallisation

We have seen in hydrated salts that a definite quantity of water is associated with these salts. This fixed amount of water associated with such compounds is known as water of crystallisation or water of hydration. For example, in $CuSO_4 \cdot 5H_2O$, 5 molecules of water are associated. These water molecules are loosely associated with the salt and can be easily removed by heating above 100°C.

Conversion of Hydrated Salt into Anhydrous Salt

When the water of crystallisation is removed from a hydrated salt, it changes into an anhydrous salt. It is possible by direct heating, heating in dry and hot air, heating under vacuum and by using dehydrating agents such as concentrated H_2SO_4 and P_2O_5.

$$\underset{\text{Hydrated blue salt}}{CuSO_4 \cdot 5H_2O} \underset{\text{Cool}}{\overset{\text{Heat}}{\rightleftharpoons}} \underset{\text{Anhydrous white salt}}{CuSO_4 + 5H_2O}$$

Hygroscopy

Hygroscopy is the phenomenon of attracting and holding water molecules by a substance via either *absorption* or *adsorption* from the surrounding environment, which is usually at normal or room temperature. Such a substance is called a *hygroscopic substance* and it has a tendency to take up moisture from the air. It can be in the solid or liquid state. For example, quick lime (CaO), concentrated sulfuric acid (H_2SO_4), phosphorus pentoxide (P_2O_5) and silica gel (SiO_2).

Illustrations

1. Give the names and formulate of two substances in each case.
   (a) Hydrated substance
   (b) Anhydrous crystalline substance

Solution: The names and formulae of two substances in each case are as follows:

(a) Hydrated substance: Blue vitriol ($CuSO_4 \cdot 5H_2O$), washing soda ($Na_2CO_3 \cdot 10H_2O$).

(b) Anhydrous crystalline substance: Ammonium chloride (NH_4Cl), potassium permanganate ($KMnO_4$).

2. What happens when the given substances are exposed to air?

(i) Sodium chloride (ii) iron (iii) conc. sulfuric acid (iv) table salt (v) sodium carbonate crystals.

In which of these substances will there be
(a) increase in mass
(b decrease in mass
(c) no change in mass

Solution:

(a) Increase in mass: Iron and conc. sulfuric acid (due to formation of hydrated salts).

(b) Decrease in mass: Sodium carbonate crystals (due to loss of water).

(c) No change in mass: Sodium chloride (pure NaCl does not lose or gain water).

Soft and Hard Water

Soft Water

Soft water gives lather easily with soap and it is water containing sodium salts. It is suitable for daily use. For example, rain water, distilled water.

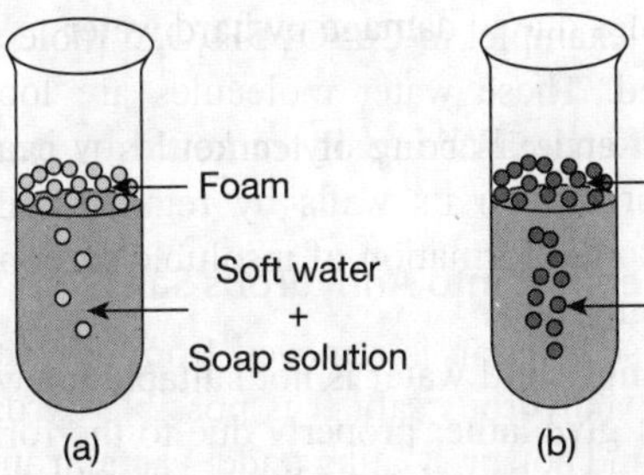

Hard Water

Hard water does not give lather with soap due to the presence of bicarbonates, chlorides and sulfates of magnesium and calcium. For example, sea water, water of rivers, lakes.

Reason for Hardness of Water

Water of some rivers and springs contains dissolved minerals. Due to the presence of insoluble salts of Ca^{+2}, Mg^{+2}, Fe^{+2}, water does not produce lather with soap easily because soaps are sodium or potassium salts of higher fatty acids like sodium palmitate ($C_{15}H_{31}COONa$) and sodium stearate ($C_{17}H_{35}COONa$). When hard water having Ca^{2+}, Mg^{2+} ions is treated with soap solution, a precipitate of calcium or magnesium salts of the higher fatty acids is obtained which is insoluble in water.

$$\underset{\substack{\text{Sodium palmitate}\\\text{(soap) soluble in water}}}{Ca^{2+} + 2C_{15}H_{31}COONa} \rightarrow \underset{\substack{\text{Calcium palmitate}\\\text{(insoluble in water)}}}{(C_{15}H_{31}COO)_2Ca \downarrow} + 2Na^{+}$$

$$\underset{\substack{\text{Sodium stearate}\\\text{(soap, soluble in water)}}}{Mg^{2+} + 2C_{17}H_{35}COONa} \rightarrow \underset{\substack{\text{Magnesium stearate}\\\text{(insoluble in water)}}}{(C_{17}H_{35}COO)_2\ Mg \downarrow} + 2Na^{+}$$

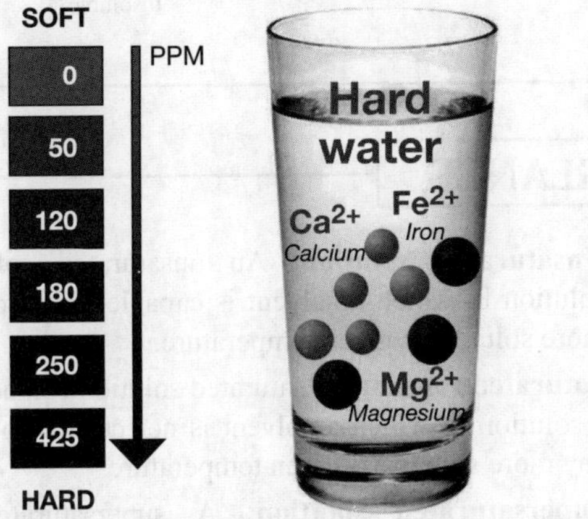

Hence no lather can be produced till all the calcium or magnesium ions get precipitated.

Types of Hardness

Water has two types of hardness—simple or temporary hardness and permanent hardness.

(i) Simple or temporary hardness: It is due to the bicarbonates of Mg^{+2}, Ca^{+2} and the removal of hardness is called *softening of water*. Softening can be done as follows.

(a) By boiling: When water is boiled for nearly 15 minutes the soluble bicarbonates decompose into their insoluble carbonates and carbon dioxides which can be removed by filtration or decantation. As a result, water becomes soft.

$$\underset{\substack{\text{Calcium}\\\text{bicarbonate}\\\text{(soluble)}}}{Ca(HCO_3)_2} \xrightarrow{\Delta} \underset{\substack{\text{Calcium carbonate}\\\text{(Insoluble)}}}{CaCO_3 \downarrow} + CO_2 + H_2O$$

$$\underset{\substack{\text{Magnesium}\\\text{bicarbonate}\\\text{(soluble)}}}{Mg(HCO_3)_2} \xrightarrow{\Delta} \underset{\substack{\text{Magnesium carbonate}\\\text{(Insoluble)}}}{MgCO_3 \downarrow} + CO_2 + H_2O$$

(b) By Clark's method: A calculated quantity of lime [$Ca(OH)_2$] is added to remove temporary hardness in water. Insoluble carbonates are formed which get precipitated and can be easily removed by filtration.

$$Ca(HCO_3)_2 + Ca(OH)_2 \leftrightarrow 2CaCO_3 \downarrow + 2H_2O$$

$$Mg(HCO_3) + 2Ca(OH)_2 \leftrightarrow 2CaCO_3 \downarrow + \underset{\text{Insoluble}}{Mg(OH)_2 \downarrow} + 2H_2O$$

(ii) Permanent hardness: It is due to the presence of soluble chlorides and sulfates of Ca^{+2} and Mg^{+2}. Permanent hardness is removed by the following method.

> **KNOWLEDGE BOOSTER**
>
> Zeolites such as hydrated sodium aluminium silicate ($Na_2Al_2Si_2O_8 \cdot xH_2O$), calgon or sodium hexametaphosphate $(NaPO_3)_6$, EDTA (ethylene diamine tetraacetate) are used to remove the permanent hardness of water. Zeolites have a three dimensional structure and they act as cation exchangers.

Soda lime method: Permanent hardness of calcium ions can be removed by washing soda but not that of magnesium salts. When a calculated amount of washing soda is added to hard water, insoluble carbonates are formed which get precipitated and can be removed by filtration.

$$CaSO_4 + Na_2CO_3 \leftrightarrow \underset{\text{Insoluble}}{CaCO_3 \downarrow} + Na_2SO_4$$

Advantages and Disadvantages of Hard Water

Hard water has a few advantages along with a few disadvantages.

Advantages: Due to the presence of dissolved salts, it has a few advantages.

- The presence of dissolved salts makes water tasty. That's why hard water is used in the manufacture of beverages and wines.
- Hard water has calcium and magnesium salts which are essential for the growth of our bones and teeth.
- Hard water avoids and prevents lead poisoning of water in lead pipes, as calcium sulfate present in hard water forms an insoluble lead sulfate layer inside these pipes.

Disadvantages: Due to the presence of dissolved salts, it has a few disadvantages.

- **Not suitable for producing steam:** Boilers having narrow copper tubes surrounded by fire are used to produce steam. When cold water enters these tubes, it is quickly changes into steam but the dissolved solids in water cannot be vaporised and they get deposited on the inner walls of the tube. Slowly the bore of the tube gets narrower and narrower as a result formation of steam is retarded and the boiler may even burst.

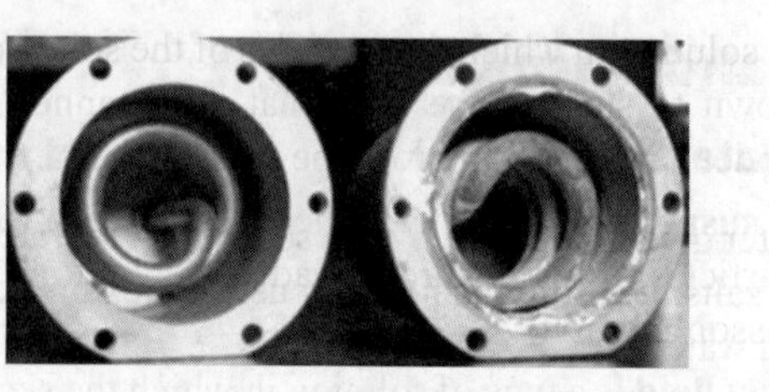

Boiler scales due to damage by hard water

- **Furring of tea kettle:** Furring of tea kettles is caused by the sediment formed on its walls by regular boiling of hard water due to the formation of insoluble carbonates of calcium and magnesium.
- **Unfit for washing:** Hard water is not suitable for washing as soap does not give lather properly due to the formation of soap curd or scum. Soap is a sodium salt of higher fatty acids like steric acid with a formula $C_{17}H_{35}COONa$.

$$\underset{\text{Soap}}{2C_{17}H_{35}COONa} + Ca(HCO_3)_2 \rightarrow \underset{\text{Soap curd}}{Ca(C_{17}H_{35}COO)_2 \downarrow} + 2NaHCO_3$$

- It is better to use synthetic detergents like surf, ariel in place of soap as they are more soluble in water and are almost unaffected by hard water and do not form scum.

Illustration

1. Write the equations to show what happens when temporary hard water is

   (a) boiled (b) treated with slaked lime.

   Solution: The reactions are as follows:

   (a) On boiling temporary hard water, the hardness is removed as follows.

$$\underset{\substack{\text{Calcium}\\\text{bicarbonate}\\\text{(Soluble)}}}{Ca(HCO_3)_2} \xrightarrow{\Delta} \underset{\substack{\text{Calcium carbonate}\\\text{(Insoluble)}}}{CaCO_3 \downarrow} + CO_2 + H_2O$$

$$\underset{\substack{\text{Magnesium}\\\text{bicarbonate}\\\text{(Soluble)}}}{Mg(HCO_3)_2} \xrightarrow{\Delta} \underset{\substack{\text{Magnesium carbonate}\\\text{(Insoluble)}}}{MgCO_3 \downarrow} + CO_2 + H_2O$$

   (b) On treating with slaked lime, the hardness of water is removed and the reactions are as follows.

$$Ca(HCO_3)_2 + Ca(OH)_2 \leftrightarrow 2CaCO_3 \downarrow + 2H_2O$$

$$Mg(HCO_3) + 2Ca(OH)_2 \leftrightarrow 2CaCO_3 \downarrow + \underset{\text{Insoluble}}{Mg(OH)_2 \downarrow} + 2H_2O$$

CHAPTER AT A GLANCE

- **Water** is the most abundant and most important natural resource in the biosphere.
- The chemical name of water is dihydrogen oxide and its molecular formula is H_2O.
- Almost **70%** of our body weight is due to water.
- **Water** can exist in all the three physical states—solid (ice), liquid (water) and gas (steam).
- Water is a **universal solvent** due to its high dielectric constant, high liquid range and ability to dissolve most of the compounds.
- A **solution** is a homogeneous mixture of two or more pure substances.
- **Unsaturated solution:** An unsaturated solution is a solution in which a solvent is capable of dissolving some more solute at a given temperature.
- **Saturated solution:** A saturated solution can be defined as a solution in which a solvent is not capable of dissolving any more solute at a given temperature.
- **Supersaturated solution:** A supersaturated solution contains more than the maximum amount of solute that is capable of being dissolved at a particular temperature. When the solvent is reduced, the extra solute will crystallise quickly.

- A solution in which the particles of the solute can be broken down to such a fine state that they cannot be seen even under a powerful microscope is known as a **true solution**.
- A **suspension** is a heterogeneous mixture in which the small particles of a solid are spread throughout a liquid without dissolving in it.
- A **colloid** is a type of solution in which the size of the solute particles is intermediate between those in true solutions and those in suspensions.
- The **size of the solute particles** in colloids is larger than those of a true solution but smaller than those of a suspension.
- The solute particles are known as **dispersed particles** and the solvent is known as the **dispersion medium.**
- **Anhydrous** means 'no water'. Substances without water are labelled anhydrous.
- A **hydrate** is a substance that contains water or its constituent elements. The fixed amount of water associated with hydrated compounds is known as **water of crystallisation** or water of hydration.
- When **water of crystallisation** is removed from a hydrated salt it changes into an anhydrous salt.
- **Hygroscopy** is the phenomenon of attracting and holding water molecules by a substance via either absorption or adsorption from the surrounding environment, which is usually at normal or room temperature and such a substance is called a **hygroscopic substance.**
- **Soft water** gives lather easily with soap. It is water containing sodium salts.
- **Hard water** does not give lather with soap due to the presence of bicarbonates, chlorides, sulfates, magnesium and calcium.
- **Simple** or **temporary hardness** is due to bicarbonates of Mg^{+2}, Ca^{+2} and the removal of hardness is called softening of water. It is possible by boiling water or by Clark's method using $Ca(OH)_2$.
- **Permanent hardness** is due to the presence of soluble chlorides and sulfates of Ca^{+2} and Mg^{+2} and it is removed by the soda lime method, by using zeolites, calgon and resin.

PRACTICE QUESTIONS

Analyse Your Concepts (School Exam Based)

Fill in the Blanks

Instructions: Complete the following statements with an appropriate word/term to be filled in the blank spaces.

1. Water can exist in _________ physical state.
2. Water occupies nearly _________ of the earth's area.
3. The specific heat capacity of 1 g of water is _______joules.
4. Salts with water of crystallisation are known as _________.
5. The composition of O_2 in dissolved water is _________ than in air.
6. Anhydrous calcium chloride is a _________ agent.
7. Solubility of a gas in water is _________ proportional to the temperature.
8. Potassium nitrate has _________ crystal shape.
9. The conical pillar which grows downwards from the roof is called _________.
10. Only moisture is removed in the process known as _________.
11. For exothermic substances, solubility is _________ proportional to the temperature.
12. The adding of a crystal of a pure substance into its saturated solution for crystallisation is known as _________.

True or False

Instructions: Read the following statements and write your answer as true or false.

1. Water is considered to be a universal solvent.
2. Water has low specific heat.
3. Rain water and distilled water do not contain dissolved salts.
4. Temporary hardness of water can be removed by boiling.
5. Water has maximum density at 40°C.
6. Calcium oxide retains its physical state on exposure to air.
7. Distilled water and boiled water have no taste.
8. A solution is heterogenous in nature.
9. Tincture of iodine is a solution of iodine in alcohol.
10. Solubility of a solid or liquid generally does not depend upon pressure.
11. Silica gel is a hygroscopic substance.
12. Potassium nitrate can have water of crystallisation also.

Very Short Answer Type Questions

1. What is the composition ratio of H : O in water by volume?
2. Write the importance of the suitability of CO_2 and O_2 in water.

3. Write any three factors which affect the solubility of a solid solute in a solvent.
4. What causes the violence associated with torrential rain?
5. Define Henry's law.
6. Solubility of NaCl at 40°C is 36.5 g. What is meant by this statement?
7. What are drying or desiccating agents? Give examples.
8. Write the methods by which a hydrated salt can be made anhydrous.
9. What is the use of solubility of oxygen and carbon dioxide in the water?
10. A hot saturated solution of sodium nitrate forms crystals as it cools. Why?
11. Name the substances which cause (a) temporary hardness, (b) permanent hardness in water.
12. Explain why water is an excellent liquid to be used in cooling systems.

Short Answer Type Questions

1. Why is water considered a compound?
2. How is air dissolved in water different from ordinary air?
3. How is aquatic life benefited by the fact that water has maximum density at 4°C?
4. Define crystallisation and state the different methods of crystallisation.
5. In what way are the properties of water different from the properties of the elements from which it is formed? Discuss.
6. Explain what observations you can understand from the given diagram.

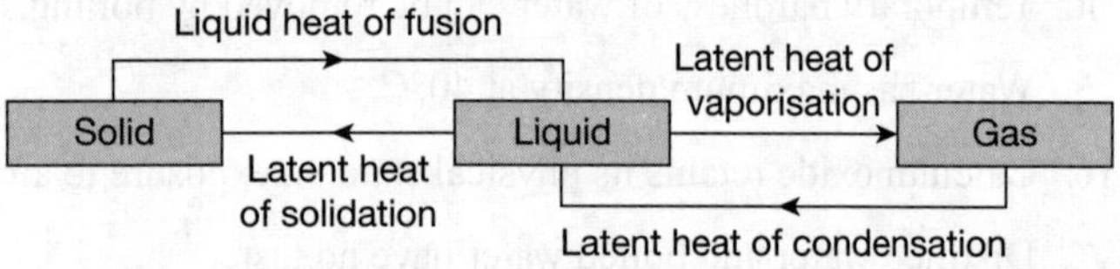

7. Why does table salt become sticky on exposure during the rainy season?
8. Explain the effect of pressure on solubility of gases in water with an example.
9. Write the importance of dissolved impurities in water.
10. Write any two ways by which a saturated solution can be changed to an unsaturated solution.
11. What is the permutit method, how can it be used for softening hard water?
12. What are stalagmites and stalactites? How are they formed?
13. Which test will you carry out to find out if a given solution is saturated, unsaturated or supersaturated?

Long Answer Type Questions

1. Explain why:
   (a) Burns caused by steam are more severe than burns caused by boiling water.
   (b) Rivers and lakes do not freeze easily.
   (c) When distilled water is kept in a sealed bottle for a long time, it leaves etchings on the surface of the glass.
2. (a) Which property of water enables it to modify the climate?
   (b) Density of water varies with temperature. What are its consequences?
   (c) What is the effect of impurities present in the water on the melting point and boiling point of water?
3. What would you observe when crystals of copper(II) sulfate and iron(II) sulfate are separately heated in two test tubes?
4. (a) Define the following terms.
   (i) Soft water, (ii) hard water, (iii) temporary hard water, (d) permanent hard water.
   (b) What are the advantages of soft water and hard water?
5. (a) If you are given some copper sulfate crystals, how would you proceed to prepare its saturated solution at room temperature?
   (b) How can you show that your solution is really saturated?

COMPETITION WINDOW (OBJECTIVE TYPE)

[For NEET, JEE (Main and Advanced), NTSE, KVPY and Olympiads]

Topic-wise MCQs

Water

1. Select the incorrect statement regarding water.
   (a) it is the most important natural resource
   (b) it can exist all three physical states
   (c) it has maximum density at 25°C
   (d) its specific heat value is 80 cal/g
2. Select the correct statement regarding water.
   (a) its chemical name is dihydrogen oxide
   (b) almost 70% of our body weight is due to water
   (c) water is a universal solvent
   (d) all of these

3. The boiling point of water is exceptionally high because
(a) there is a covalent bond between H and O
(b) water molecule is not linear
(c) water molecule is linear
(d) water molecules associate due to hydrogen bonding

4. The H–O–H angle in the water molecule is about
(a) 90° (b) 105°
(c) 135° (d) 180°

5. When two ice cubes are pressed over each other, they unite to form one cube. Which of the following forces is responsible to hold them together?
(a) van der Waals forces
(b) covalent attraction
(c) ionic interaction
(d) hydrogen bond formation

6. Water is oxidised to oxygen by
(a) H_2O_2 (b) $KMnO_4$
(c) ClO_2 (d) fluorine

7. Reaction of potassium with water is
(a) hydrolysis (b) absorption
(c) exothermic (d) endothermic

8. Which of the following is not an advantage of air dissolved in water?
(a) it is good for marine life
(b) it helps in photosynthesis
(c) it provides taste to water
(d) in making shells of snails

Different types of solutions

9. Which of the following is correct about a solution?
(a) it is a homogenous mixture
(b) solute is in less amount than solvent in a binary solution
(c) solutions having water as solvent are aqueous solutions
(d) all are correct

10. Brass is which type of solution?
(a) solid in liquid (b) solid in gas
(c) solid in solid (d) liquid in solid

11. Which of the following is incorrectly matched?

	Type of solution		Examples
(a)	Liquid in liquid	-	Water and alcohol
(b)	Solid in liquid	-	Sugar in water
(c)	Solid in solid	-	Alloys
(d)	Liquid in liquid	-	Soda water

12. The maximum amount of solute is present in
(a) unsaturated solution
(b) saturated solution
(c) supersaturated solution
(d) same in all these

13. In the colloidal state, the particle size ranges from
(a) 10 to 1000 Å (b) 1 to 280 Å
(c) 1 to 10 Å (d) 20 to 50 Å

14. Substances whose solutions can readily diffuse through a parchment membrane are
(a) electrolytes (b) non-electrolytes
(c) colloids (d) crystalloids

15. Which of the following statements is true about a colloidal system?
(a) it can be made out of two gases
(b) it is electrically neutral
(c) it carries a net electric charge
(d) it consists of one phase only

16. In which of the following is the dispersed phase a liquid?
(a) fog (b) paint
(c) smoke (d) volcanic dust

17. In which of the following is the dispersion medium a gas?
(a) smoke (b) cloud
(c) fog (d) all of these

18. Which of the following combinations of colloids is not possible?
(a) gas in gas (b) gas in liquid
(c) solid in liquid (d) liquid in solid

19. Smoke is an example of
(a) a solid dispersed in gas
(b) a solid dispersed in solid
(c) a gas dispersed in liquid
(d) a gas dispersed in solid

20. When the dispersed phase is a liquid and the dispersion medium is a gas, then the colloidal system is called
(a) jelly (b) cloud
(c) emulsion (d) smoke

21. Which of the following forms a colloidal solution in water?
(a) starch (b) barium nitrate
(c) NaCl (d) glucose

22. Tyndall effect will be observed in
(a) vapour (b) solution
(c) sol (d) precipitate

Hydrated and anhydrous substances, hygroscopy

23. The shape of ferrous sulfate is
(a) cubic (b) octahedral
(c) prismatic (d) rhombohedral

24. The number of water molecules present in blue vitriol and green vitriol are respectively
(a) 2, 5 (b) 5, 5
(c) 5, 7 (d) 7, 5

25. Which of the following is not an anhydrous substance?
(a) sucrose (b) potassium permanganate
(c) gypsum (d) ammonium chloride

26. Impure table salt turns moist in air due to impurities of
(a) KNO_3 (b) $MgCl_2$
(c) $CaCl_2$ (d) Both (b) and (c)

27. Ammonia can be dried by using
(a) CaO (b) H_2SO_4
(c) SiO_2 (d) P_2O_5

28. Which of the following is a dehydrating agent?
(a) conc. H_2SO_4 (b) CaO
(c) Al_2O_3 (d) both (a) and (c)

Hard water

29. A type of water which contains soluble salts of Ca and Mg is known as
(a) soft water (b) heavy water
(c) conductivity water (d) hard water

30. Which of the following ions will cause hardness in a water sample?
(a) Ca^{2+} (b) Na^+
(c) Cl^- (d) K^+

31. The process used for the removal of hardness of water is
(a) Baeyer (b) Hoope
(c) Calgon (d) Serpeck

32. Sodium hexa-metaphosphate is known as
(a) permutit (b) calgon
(c) nitrolim (d) natalite

33. Water softening by Clarke's process uses
(a) potash alum (b) calcium bicarbonate
(c) calcium hydroxide (d) sodium bicarbonate

34. Which one of the following is used for reviving the exhausted permutit?
(a) 10% NaCl solution (b) 10% $MgCl_2$ solution
(c) 10% $CaCl_2$ solution (d) HCl solution

35. By which of the following processes can the permanent hardness of water be removed?
(a) washing soda (b) soda lime
(c) sodium chloride (d) sodium bicarbonate

Miscellaneous

1. Reacting with water, an active metal produces
(a) oxygen (b) nitric acid
(c) a base (d) none of these

2. Anhydrous cobalt chloride can be used to remove moisture from the surroundings. Which of the following renders the sample reusable?
(a) exposure to dry air
(b) keeping in contact with another efflorescent substance
(c) exposure to high humid air
(d) keeping in air-tight container

3. When NaCl is added to water, the columbic force of attractions between
(a) Na^+ and partially charged oxygen of water becomes very weak
(b) Cl^- and partially charged hydrogen of water increase
(c) partially charged hydrogen and oxygen of water increase
(d) radicals present in NaCl increase

4. Which one of the following processes will produce hard water?
(a) saturation of water with $CaCO_3$
(b) addition of Na_2SO_4 to water
(c) saturation of water with $MgCO_3$
(d) saturation of water with $CaSO_4$

5. Which of the following does not show much variation of solubility with the change of temperature?
(a) potassium chloride (b) potassium sulfate
(c) sodium chloride (d) sodium sulfate

6. Which of the following changes take place due to the addition of aluminium sulfate to muddy water?
(a) adsorption of mud on aluminium sulfate crystals results in isolation of mud from water
(b) a part of the mud becomes soluble in water
(c) muddy water becomes free from germs and bacteria
(d) coagulation of mud particles takes place

7. When a small crystal of a soluble salt is added to a supersaturated solution,
(a) a saturated solution is formed
(b) an unsaturated solution is formed
(c) some amount of solute is precipitated
(d) both (a) and (c)

8. Which among the following has water of crystallisation?
(a) common salt (b) calcium hydroxide
(c) washing soda (d) baking soda

9. Cane sugar on heating with concentrated H_2SO_4 gives sugar, charcoal and water. It is a case of
(a) drying (b) dehydration
(c) deliquescence (d) efflorescence

10. Cloud or fog is an example of a colloidal system of a
(a) solid dispersed in a gas
(b) solid dispersed in a liquid
(c) liquid dispersed in a gas
(d) gas dispersed in a gas

11. When a colloidal solution is observed under an ultramicroscope, we can see the
(a) shape of the particles
(b) relative size
(c) light scattered by colloidal particles
(d) size of the particles

12. Small liquid droplets dispersed in another liquid is called a/an
(a) emulsion (b) true solution
(c) gel (d) suspension

13. Butter is a colloid formed when
(a) fat is dispersed in water
(b) milk is dispersed in water
(c) water is dispersed in fat
(d) milk is dispersed in fat

14. Which of the following is a colloid?
(a) silicic acid (b) NaCl solution
(c) sugar solution (d) urea solution

15. The chemical formula of zeolite is .....
(a) $Na_2Al_2Si_2O_8.xH_2O$ (b) $Na_2(Na_4(PO_3)_6$
(c) $Ca_2Al_2Si_2O_8$ (d) $K_2Al_2Si_2O_8.xH_2O$

16. What is the formula of Calgon?
(a) $MgSO_4$ (b) Na_3PO_4
(c) $Mg_3(PO_4)_2$ (d) $(NaPO_3)_6$

Advanced and Olympiads

Single Choice

1. Which of the following sets represents only correct statements?
(i) Boiling point of water increases in the presence of impurities
(ii) Latent heat of vaporisation of water is 540 cal/g
(iii) Water has a low dielectric constant
(iv) Rain water and distilled water do not have dissolved solids so no concentric rings are formed
(a) (i), (ii) (b) (ii), (iii)
(c) (i), (ii), (iii) (d) (i), (ii), (iv)

2. Polyphosphates are used as water softening agents because they
(a) form soluble complexes with cationic species
(b) precipitate cationic species
(c) precipitate anionic species
(d) form soluble complexes with anionic species

3. Which of the following sets represents only correct statements?
(i) In a true solution, the size of solute particles is nearly 1 Å
(ii) True solutions can scatter light
(iii) In solids and liquids, solubility is not much dependent on pressure
(iv) Glauber salt shows anomalous solubility
(a) (i), (ii) (b) (ii), (iii)
(c) (i), (ii), (iii) (d) (i), (iii), (iv)

4. The reagent commonly used to determine hardness of water titrimetrically is
(a) disodium salt of EDTA
(b) sodium thiosulfate
(c) sodium citrate
(d) oxalic acid

5. Which of the following sets represent only correct statements?
(i) Gypsum on heating can give plaster of Paris
(ii) Gaseous HCl is anhydrous
(iii) Pure table salt is highly hydrated
(iv) In dehydration new substances are also formed
(a) (i), (ii), (iv) (b) (ii), (iii)
(c) (i), (ii), (iii) (d) (i), (iii), (iv)

6. Among the colloids cheese (C), milk (M) and smoke (S), the correct combination of the dispersed phase and dispersion medium, respectively is
(a) C: solid in liquid; M: solid in liquid; S: solid in gas
(b) C: solid in liquid; M: liquid in liquid; S: gas in solid
(c) C: liquid in solid; M: liquid in solid; S: solid in gas
(d) C: liquid in solid; M: liquid in liquid; S: solid in gas

Multiple Choice

7. Some of the properties of water are described below. Which of them is/are not correct?
(a) water is known to be a universal solvent
(b) hydrogen bonding is present to a large extent in liquid water
(c) there is no hydrogen bonding in the frozen state of water
(d) frozen water is heavier than liquid water

8. Which of the following is/are correct?
(a) temporary hardness of water is due to the presence of bicarbonates of calcium and magnesium
(b) permutit is artificial zeolite
(c) H_2O_2 acts as an oxidising agent in the following reaction:
$$Cl_2 + H_2O_2 \rightarrow O_2 + 2HCl$$
(d) H_2O_2 is used as a bleaching agent for delicate textiles

9. The reaction of H_2O with X liberates gaseous products. Which of the following is/are X.
(a) PbO_2 (b) $KMnO_4/H^+$
(c) PbS (d) Cl_2

10. Hardness of water may be temporary or permanent. Permanent hardness is due to the presence of
(a) chlorides of Ca and Mg in water
(b) sulfates of Ca and Mg in water
(c) hydrogen carbonates of Ca and Mg in water
(d) carbonates of alkali metals in water

11. Which one of the following statements about zeolites is/are true?
(a) They have open structure which enables them to take up small molecules.
(b) They are used as cation exchangers.
(c) Zeolites are aluminosilicates having a three dimensional network.
(d) They cannot remove permanent hardness.

Comprehension Type

Comprehension I: A colloidal solution is a heterogeneous solution which contains particles of intermediate size with a diameter between 1 and 100 nm. A colloid is not a substance but it is a state of a substance which depends upon the molecular size. Colloidal solutions are intermediate between true solutions and suspensions.

12. The size of the colloidal particles lies in the range
(a) 10^{-5}–10^{-7} cm (b) 10–1000 nm
(c) 1–1000 nm (d) 10–1000 mμ

13. The colloidal solution of a solid as the dispersed phase and a gas as the dispersion medium is called a/an
(a) aerosol (b) solid foam
(c) sol (d) gel

14. The colloidal particles can pass through
(a) filter paper but not through animal membrane
(b) animal membrane but not through filter paper
(c) filter paper as well as animal membrane
(d) neither filter paper nor animal membrane

Matrix Matching

15. Match the following

Column I	Column II
(a) solution of a solid in a solid	(p) aerated water
(b) solution of a liquid in a liquid	(q) air
(c) solution of a gas in a liquid	(r) brass
(d) solution of a gas in a gas	(s) vinegar
	(t) bronze

16. Match the following

Column I	Column II
(a) potable water	(p) drinking water
(b) temporary hardness	(q) $CaSO_4$
(c) permanent hardness	(r) $Mg(HCO_3)_2$
(d) heavy water	(s) D_2O

Integer Type

17. In blue vitriol, $CuSO_4 \cdot XH_2O$, what is the value of X?

18. How many of the following compounds have water of crystallisation of more than two?
Plaster of Paris, gypsum, epsom salt, green vitriol, washing soda, glauber salt, white vitriol.

19. How many of these compounds are hygroscopic substances?
Conc. H_2SO_4, P_2O_5, CaO, SiO_2, NH_4Cl, $Na_2CO_3 \cdot 10H_2O$, epsom salt.

20. How many of the following are anhydrous substances?
Potassium nitrate, gypsum, common salt, sugar, potassium permanganate, ammonium chloride, potash alum, plaster of Paris.

Answer Keys

Fill in the Blanks

1. three	2. 70%	3. 4.2
4. hydrated salts	5. more	6. dehydrating
7. inversely	8. prismatic	9. stalactite
10. drying	11. inversely	12. seeding

True or False

1. T	2. F	3. T	4. T	5. F
6. T	7. T	8. F	9. T	10. T
11. T	12. F			

Very Short Answer Type Questions

1. In water, the composition volume ratio of hydrogen and oxygen is H : O that is 2 : 1.
2. CO_2 and O_2 add taste to water for drinking purposes.
3. The three factors on which the solubility of a solid depends are temperature, nature of the solid and nature of solvent.
4. The sudden release of the latent heat of condensation causes the violence associated with torrential rain.
5. According to Henry's law, 'at any given temperature, the mass of a gas dissolved in a fixed volume of a liquid or solution is directly proportional to the pressure on the surface of a liquid'.
6. The solubility of NaCl at 40°C is 36.5 g. That is, 36.5 g of NaCl dissolves in 100 g of water at a temperature of 40°C.
7. These are the substances which can readily absorb moisture from other substances without chemically reacting with them. For example, phosphorus pentoxide (P_2O_5), quick lime (CaO).
8. A hydrated salt can be made anhydrous by heating or by exposing it to dry air.
9. Oxygen dissolved in water is used by aquatic animals for respiration and survival. CO_2 dissolved in water is used by aquatic plants to prepare their food by photosynthesis.
10. A hot saturated solution of sodium nitrate forms crystals as it cools, as its solubility decreases with a decrease in temperature.
11. (a) Hydrogen carbonates of calcium and magnesium make water temporarily hard.
(b) Sulfates and chlorides of magnesium and calcium make water permanently hard.
12. Water is an excellent liquid to be used in cooling systems due to its ability to absorb large quantities of heat, that is, its specific heat = 4.2 J/g°C, so it can be used in cooling systems as a cooling agent.

Short Answer Type Questions

1. Water is considered as a compound because it is made up of two elements, hydrogen and oxygen, combined in a ratio of 1 : 8 by mass.
Mass ratio of elements H_2O
$H : O, 2 \times 1 : 16 \times 1 = 1 : 8$
(Atomic mass of H = 1, O = 16)
Also, the components of water cannot be separated by physical methods but can be separated by electrolysis of water.

2. Oxygen is more soluble in water than nitrogen. Air dissolved in water contains a higher percentage of oxygen (30%–35%) and in ordinary air it is only 21%. In this way, air dissolved in water is different from ordinary air.

3. The property of anomalous expansion of water enables aquatic life to exist because the water freezes on the surface of the water body, but it is still liquid below the ice layer.

4. Crystallisation: It is the process by which crystals of a substance separate out on cooling its hot saturated solution. In the laboratory, crystals may be obtained by the following methods:
   - By cooling a hot saturated solution gently.
   - By cooling a fused mass.
   - By sublimation.
   - By evaporating a saturated solution slowly.

5. The properties of water are different from the properties of elements from which it is formed

Property	Water	Elements (oxygen and hydrogen)
Nature	It is a clear, colourless, odourless, tasteless and transparent liquid.	These are colourless, odourless, tasteless and non-poisonous gases.
Solubility	It can dissolve many things and is known as a universal solvent.	Oxygen and hydrogen are slightly soluble in water.
Density	Pure water has maximum density at 4°C.	Oxygen is heavier than air while hydrogen is the lightest gas.

6. From this diagram, the following observations can be made.
   (i) When solid changes to the liquid, it absorbs heat equal to the latent heat of fusion and when a liquid changes into the solid, it loses heat equal to the latent heat of solidification.
   (ii) When a liquid changes into the gas, it absorbs heat equal to the latent heat of vaporisation and when a gas condenses into the liquid, it loses heat equal to the latent heat of condensation.

7. Table salt becomes sticky on exposure during the rainy season as table salt generally contains a small percentage of magnesium chloride, as an impurity. As these impurities absorb moisture from the air due to their deliquescent nature, it gets wet in the rainy season and becomes sticky.

8. With an increase in pressure, the solubility of a gas in water increases. For example, the solubility of carbon dioxide in water under normal atmospheric pressure is low, but when the water surface is subjected to higher pressure, a lot more of CO_2 gas gets dissolved. Similarly, in the case of soda water, on opening the bottle, the dissolved gas rapidly bubbles out, since the pressure on the surface of the water suddenly decreases.

9. The importance of dissolved impurities in water are:
   - dissolved minerals and salts are essential for the growth of plants,
   - dissolved salts add taste to water, and
   - dissolved salts and minerals in water provide essential minerals required for our body.

10. The two methods are:
    (i) A saturated solution can be changed to an unsaturated solution by heating.
    (ii) A saturated solution can be changed to an unsaturated solution by adding more solvent.

11. See the text part.

12. See the text part.

13. (i) A solution in which more of solute can be dissolved at a given temperature is an unsaturated solution.
    (ii) A solution in which no more solute can be dissolved at a given temperature is a saturated solution at that temperature.
    (iii) A solution in which some solute separates on cooling slightly is a super saturated solution.

Long Answer Type Questions

1. (a) Burns caused by steam are more severe than burns caused by boiling water as 1 g of steam contains 2268 J more energy than 1 g of boiled water.
   (b) Rain water does not leave concentric rings when boiled because rain water does not contain dissolved solids.
   (c) When distilled water is kept in a sealed bottle for a long time, it leaves etching on the surface of glass as the substances which are insoluble in water, actually dissolve in minute traces in water. Even when we drink water from a glass, an extremely small amount of glass dissolves in water, so we can see the etching on the surface of glass when a bottle of distilled water (sealed for a long time) is poured into the glass.

2. (a) Specific heat.
   (b) Water has an unusual physical property. When cooled, it first contracts in volume, as do other liquids, but at 4°C (maximum density), it starts expanding, and continues to do so till the temperature reaches 0°C, the point at which it freezes into ice.

The property of anomalous expansion of water enables marine life to exist in the colder regions of the world, because even when the water freezes on the top, it is still liquid below the ice layer.

(c) Due to the presence of impurities in water, the melting point of water decreases while the boiling point of water increases.

3. When copper(II) sulfate ($CuSO_4 \cdot 5H_2O$) crystals are heated in a hard glass test tube, the following observations are made.

(i) The crystals are converted into a powdery substance.
(ii) The crystals lose their blue colouration on further heating.
(iii) Steaming vapours are produced inside the tube which condense near the mouth of the tube to form a colourless liquid.
(iv) On further heating, steam escapes from the mouth of the tube and water gets collected in a beaker placed under the mouth of tube.
(v) On further heating, the residue changes to a white powder and steam stops coming out.

$$\underset{\text{Blue}}{CuSO_4 \cdot 5H_2O} \xrightarrow{\Delta} \underset{\text{White}}{CuSO_4} + 5H_2O$$

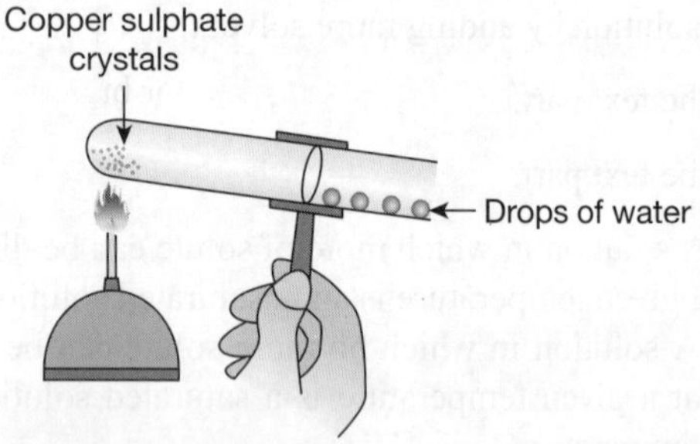

Action of heat on iron(II) sulfate ($FeSO_4 \cdot 7H_2O$): When iron(II) sulfate is heated in a test tube, the following observations are made.

(i) The crystals crumble to white powder and a large amount of steam and gas are given out.
(ii) On strong heating, a brown residue of ferric oxide (Fe_2O_3) is produced and a mixture of SO_2 and SO_3 is given off.

$$FeSO_4 \cdot 7H_2O \xrightarrow{\Delta} FeSO_4 + 7H_2O$$

$$2FeSO_4 \xrightarrow[\text{(Brown residue)}]{\text{Ferric oxide}} Fe_2O_3 + SO_3 + SO_3$$

4. (i) (a) Water is said to be soft water if it readily gives foam with soap.
(b) Water is said to be hard water if it does not readily give foam with soap.
(c) Water that contains only hydrogen carbonates of calcium and magnesium is called temporary hard water.
(d) Water that contains sulfates and chlorides of magnesium and calcium is known as permanent hard water.

Advantages of soft water

(i) With soft water, less soaps and cleansing agents are consumed, which will save money.
(ii) Soft water does not leave deposits of minerals on pipes which will make plumbing work easy.
(iii) Clothes washed using soft water last long and remain bright.

Advantages of hard water

(i) Presence of salts and ions in hard water makes it tasty.
(ii) Calcium and magnesium ions present in hard water are essential for the growth of our bones.
(iii) Hard water checks poisoning of water by lead pipes.

5. Take 100 g of distilled water in a beaker. To this add one gram of copper sulfate crystals. Stir this mixture with the help of a glass rod and dissolve copper sulfate crystals. Similarly, go on dissolving more of copper sulfate, (1 g at a time) with constant and vigorous stirring. A stage is reached when no more copper sulfate dissolves. It is called a saturated solution at this temperature.

Take some of this saturated solution of copper sulfate in a test tube and add some copper sulfate crystals. The crystals do not dissolve but settle down. This indicates that the solution is really saturated.

Competition Window (Objective Type)

Topic-wise MCQs

1. (c)	**2.** (d)	**3.** (d)	**4.** (b)	**5.** (d)
6. (d)	**7.** (c)	**8.** (c)	**9.** (d)	**10.** (c)
11. (d)	**12.** (c)	**13.** (a)	**14.** (d)	**15.** (b)
16. (a)	**17.** (d)	**18.** (a)	**19** (a)	**20.** (b)
21. (a)	**22.** (c)	**23.** (b)	**24.** (c)	**25.** (c)
26. (d)	**27.** (a)	**28.** (d)	**29.** (d)	**30.** (a)
31. (c)	**32.** (b)	**33.** (c)	**34.** (a)	**35.** (b)

Miscellaneous

1. (c)	**2.** (a)	**3.** (b)	**4.** (d)	**5.** (c)
6. (d)	**7.** (d)	**8.** (c)	**9.** (b)	**10.** (c)
11. (c)	**12.** (a)	**13.** (c)	**14.** (a)	**15.** (a)
16. (d)				

Advanced and Olympiads

Single Choice

1. (d)	**2.** (a)	**3.** (d)	**4.** (a)	**5.** (a)
6. (d)				

Multiple Choice

7. (cd) **8.** (abd) **9.** (abd) **10.** (ab) **11.** (abc)

Comprehension Type

12. (c) **13.** (a) **14.** (a)

Matrix Matching

15. (a)—(rt); (b)—(s); (c)—(p); (d)—(q)

16. (a)—(p); (b)—(r); (c)—(q); (d)—(s)

Integer Type

17. 5 **18.** 5 **19.** 4 **20.** 5

Hints and Solutions

Competition Window (Objective Type)

Topic-wise MCQs

1. Water has the maximum density at 4°C and not at 25°C.
3. Association of water molecules due to hydrogen bonding makes its boiling point exceptionally high.
5. Hydrogen bonds are formed between water molecules in ice.
6. Fluorine oxidises water to O_2
   $2F_2 + 2H_2O \rightarrow O_2 + 4HF$
7. Potassium reacts with H_2O liberating H_2 and heat.
8. The taste of water is due to the impurities of dissolved salts in water.
10. Brass is a solid solution containing 70% copper (solvent) and 30% zinc (solute).
11. Soda water is a solution of gas in liquid type and not liquid in liquid type.
13. The size of particles in the colloidal state ranges from 10 to 1000 Å.
    Size of particles .... 10 Å .... 1000Å ....
    True solution — Particles in colloidal state — Suspension
14. Crystalloids can easily be obtained in the crystalline form and their solution can diffuse rapidly through a vegetable or animal membrane (parchment membrane); for example, sugar, urea, salts, acids
16. Fog: liquid in gas phase.
18. Gas in gas will make a homogeneous mixture.
19. Smoke consists of carbon particles (solid) dispersed in air (gas).
20. Clouds consist of fine droplets of water suspended in air.
21. Starch molecules have colloidal dimensions whereas NaCl, glucose and $Ba(NO_3)_2$ are crystalloids and are soluble in water.
22. A sol is a colloidal solution and so, it will show Tyndall effect.
23. The shape of ferrous sulphate is octahedral.
24. Blue vitriol is $CuSO_4{\cdot}5H_2O$ and green vitriol is $FeSO_4{\cdot}7H_2O$. Hence, the number of water molecules are 5, 7 respectively.
25. Gypsum ($CaSO_4{\cdot}2H_2O$) is a hydrated substance.
26. Pure table salt turns moist in air due to impurities of $MgCl_2$ and $CaCl_2$.
27. Ammonia can be dried by using basic drying agents like CaO.
28. Both concentrated H_2SO_4 and Al_2O_3 are dehydrating agents while CaO is a drying agent.
33. In Clarke's process quick lime is added to hard water which produces $Ca(OH)_2$ which further converts bicarbonates into insoluble carbonates.

Miscellaneous

4. $MgCO_3$ and $CaCO_3$ are water insoluble. $CaSO_4$ dissolves in water adding Ca^{2+} ions which are responsible for producing hardness in water.
8. Washing soda has water of crystallisation.
10. Fog is a colloidal system consisting of water droplets dispersed in air.
11. Scattering of light from the surface of colloidal particles – Tyndall effect – is observed, when a colloidal solution is observed under an ultramicroscope.
12. When the dispersed phase and dispersion medium are both liquid, the colloidal system is called as an emulsion. Examples, milk, vanishing cream.
13. Butter is obtained from milk which itself is an emulsion of fat dispersed in water, reverse is true for butter.
14. The substances which cannot pass through a membrane are termed as colloids. For example, silicic acid. Sugar solution, urea and NaCl can pass through the membrane.

Advanced and Olympiads

Single Choice

1. Only statement (iii) is wrong as water has a high dielectric constant.
2. Polyphosphates are used as water softeners because these form soluble complexes with cationic species (Ca^{+2} and Mg^{+2}) present in hard water. The complex calcium and magnesium ions do not form any precipitate with soap and hence water readily produces lather with soap solution.
   $2Ca^{+2} + Na_2[Na_4(PO_3)_6] \rightarrow Na_2[Ca_2(PO_3)_6] + 4Na^+$
   $2Mg^{+2} + Na_2[Na_4(PO_3)_6] \rightarrow Na_2[Mg_2(PO_3)_6] + 4Na^+]$
   Soluble complex
3. A true solution does not scatter light due to the small size of particles.
5. Pure table salt (NaCl) is not hydrated.
6.

Colloid	Dispersed phase	Dispersion medium
Cheese (C)	Liquid	Solid
Milk (M)	Liquid	Liquid
Smoke (S)	Solid	Gas

Multiple Choice

7. In frozen water or ice, there is hydrogen bonding and it is lighter than water.

8. H_2O_2 acts like reducing agent and reduces Cl_2 into HCl.

9. X is an oxidant and it can be PbO_2, $KMnO_4/H^+$ and Cl_2.

11. A zeolite can remove the permanent hardness of water.

Integer Type

17. Blue vitriol is $CuSO_4 \cdot 5H_2O$.

18. Epsom salt, green vitriol, washing soda, glauber salt, white vitriol have more than two water molecules as water of crystallisation (see text part).

19. Conc. H_2SO_4, P_2O_5, CaO, SiO_2 are hygroscopic substances while the rest are efflorescent substances.

20. Potassium nitrate, common salt, sugar, potassium permanganate, ammonium chloride are anhydrous while the rest are hydrated.

CHAPTER 8 Carbon and its Compounds

LEARNING OBJECTIVES

After studying this unit, you will be able to understand:

- The properties of carbon—catenation, tetravalency
- The allotropes of carbon
- The preparation and properties of CO and CO_2

Introduction

Carbon is an element and its symbol is C. It is a non-metal. The name carbon is derived from the Latin word 'carbo' which means 'coal' as carbon is the main constituent of coal. The amount of carbon present in the earth's crust and atmosphere is very small. For example, the earth's crust contains only 0.02% carbon in the form of minerals (like carbonates, coal and petroleum) and the atmosphere has only 0.03% of carbon dioxide gas. In spite of this small amount of carbon available in nature, carbon has great importance in every sphere of life. The importance of carbon can be gauged from the fact that we are ourselves made of carbon compounds called organic compounds. In fact, all the living things like plants and animals, are made up of carbon-based compounds which are known as organic compounds. Hence, carbon is present in all living things.

A large number of things that we use in our daily lives are made of carbon compounds. For example, our food materials like grains, pulses, sugar, tea, coffee, fruits and vegetables are carbon compounds. The materials like cotton, silk, wool, nylon and polyester which are used for making clothes are carbon compounds. The fuels like wood, coal, kerosene, LPG (Liquefied Petroleum Gas), natural gas, CNG (Compressed Natural Gas), petrol and diesel, are carbon compounds. Paper, rubber, plastics, leather, drugs and dyes, are also made of carbon compounds. This means, carbon plays a very important role in our daily life.

Carbon is Tetravalent

A carbon atom has 4 electrons in its valence shell, so it requires 4 more electrons to achieve the stable, 8-electrons inert gas electron arrangement, which it gets by sharing. Hence, the valency of carbon is 4, that is, carbon is tetravalent (tetra = four, valent = valency). The four valencies of carbon are usually represented by putting four short lines on the symbol of carbon

$$-\overset{|}{\underset{|}{C}}-.$$

KNOWLEDGE BOOSTER

This characteristic of the carbon atom by virtue of which it forms four covalent bonds is generally referred to as tetra-covalency of carbon.

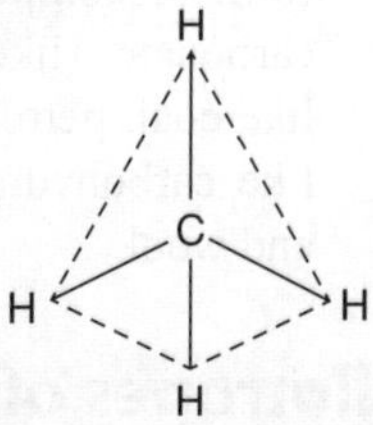

Catenation

The most unique property of carbon is its ability to combine with itself, atom to atom, to form long chains. For example, octane (C_8H_{18}), one of the constituents of petrol, has a chain of 8 carbon atoms, and some of the organic compounds like starch and cellulose contain chains of hundreds of carbon atoms. The property of self-combination of carbon atoms to form long chains is quite useful as it gives rise to an extremely large number of carbon compounds (or organic compounds). This is because a long chain of carbon atoms acts as a backbone to which other atoms can attach in a number of ways to form a very large number of organic compounds.

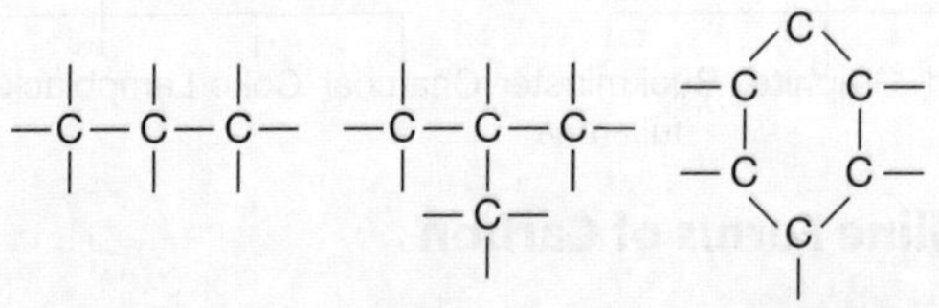

Straight chain   Branched chain   Cyclic chain

The covalent bonds formed between the various carbon atoms are very strong and cannot be not broken easily. The reason for the formation of strong bonds by the carbon atom is its small size. Due to the small size of the carbon atom, their nuclei hold the shared pairs of electrons between atoms strongly, leading to the formation of strong covalent bonds. The carbon atoms can also form strong covalent bonds with the atoms of other elements such as hydrogen, oxygen, nitrogen, sulfur, chlorine, phosphorus and many others. The formation of strong bonds by carbon atoms among themselves and with other elements makes the carbon compounds exceptionally stable.

Occurrence of Carbon

Carbon occurs in nature in the free state (as element) as well as in the combined state (in the form of compounds with other elements). In occurs in the free state in the form of coal, diamond and graphite and in the combined state in the form of CO_2, carbonates and hydrocarbons. It exists in two allotropic forms—crystalline and amorphous.

(i) In the **free state**, carbon occurs in nature mainly in two forms—diamond and graphite. Another naturally occurring form of carbon, known as buckminsterfullerene, has been discovered recently. Only a small amount of carbon occurs as the free element in the earth's crust.

(ii) In the **combined state**, carbon occurs in nature in the form of compounds such as carbon dioxide gas in air, carbonates (like limestone, marble and chalk), fossil fuels like coal, petroleum and natural gas, organic compounds like carbohydrates, fats and proteins, and wood, cotton and wool.

Allotropes of Carbon

Allotropy is the phenomenon in which an element exhibits different physical forms, having almost similar chemical properties. Elements like carbon, phosphorus, sulfur and iron exhibit allotropy. The different physical forms exhibited by the element are known as allotropes. Carbon is an element that can exhibit various allotropic forms which have a wide variety of applications. The allotropic forms exhibited by carbon are mainly of two types—crystalline and non-crystalline (amorphous) forms.

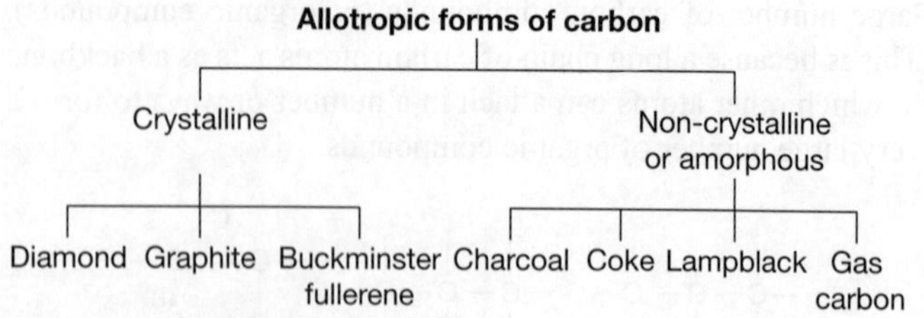

Crystalline Forms of Carbon

Carbon shows three crystalline forms having well defined regular geometrical arrangement of carbon atoms—diamond, graphite and fullerenes.

Diamond

It is a colourless, transparent, quite heavy substance having amazing brilliance (lustre). It is the hardest known natural substance. It does not conduct electricity. It is the purest form of carbon with high density (3.5 g/mL), melting point (7350°C) and refractive index (2.45). On strong heating, diamond gives only carbon dioxide and nothing is left behind. It proves that diamond is composed of only carbon and that is why it is represented by C only. Diamond dies are often used in making cutting tools. It is used as an abrasive.

Structure: A diamond crystal is a rigid and a giant molecule having carbon atoms only. Each carbon atom in the diamond crystal is linked to four other carbon atoms by strong covalent bonds. The four surrounding carbon atoms are at the four vertices (four corners) of a regular tetrahedron. In diamond, the carbon atom is sp^3 hybridised with a tetrahedral structure. Each carbon atom is attached to four other carbon atoms by covalent bonds and this gives diamond a three dimensional polymeric structure. The C–C bond length is 1.54 Å and the bond angle is 109°28'.

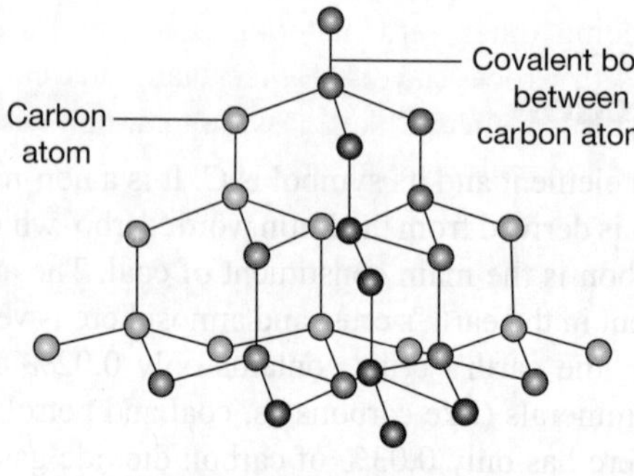

Structure of diamond

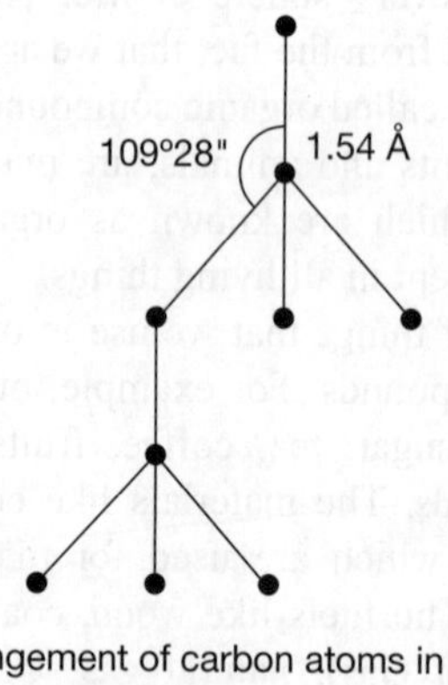

Arrangement of carbon atoms in diamond

Properties

- The high value of boiling point, melting point and density are due to its three dimensional structure which needs high energy for bond breaking. Light rays, X-rays (highest) can pass through it.
- It is a transparent solid with a specific gravity of 3.52.
- Diamond is very hard with very high density due to compact and rigid 3-dimensional arrangement of carbon atoms.
- Due to high refractive index, on proper cutting it produces maximum total internal reflection.
- It is a bad conductor due to the absence of free electrons as each carbon atom in diamond has shared all the four valence electrons.
- It is insoluble in all solvents.

- Pure diamonds are colourless while impure diamond may be coloured. The Kohinoor diamond is 186 carats while the Pit diamond is 136 carats.
- Diamond on heating in vacuum at 1800–2000°C gets converted into graphite.
- Being chemically inert, it does not react with acids and bases.
- When it is heated in air up to 1173 K, CO_2 is formed. On heating with fluorine up to 973 K it gives CF_4.

$$C + O_2 \rightarrow CO_2$$

$$C + 2F_2 \rightarrow CF_4$$

Uses

- As diamond is very hard, it is the right material for cutting and grinding other hard materials, and for drilling holes in the earth's rocky layers. Hence diamonds can be used in cutting instruments like glass cutters, saws for cutting marble and in rock drilling equipment.
- Diamonds are used for making jewellery due to their extraordinary brilliance, which is due to their great ability to reflect and refract light.
- Diamond 'dies' are used for drawing thin wires like the tungsten filament of an electric bulb.
- Sharp-edged diamonds are used by eye-surgeons as a tool to remove cataract from eyes with great precision.

Graphite

It is a grey–black coloured, opaque substance with metallic lustre, soft and slippery in nature. It is more stable and reactive than diamond. It can conduct electricity. On strong heating it gives only CO_2 leaving no other residue so it is composed of carbon only; hence its symbol is C.

Structure: Graphite differs a lot in its structure from diamond and is made up of layers or sheets of carbon atoms. It has a two dimensional sheet-like hexagonal structure in which each carbon atom is sp^2 hybridised and is attached to three other carbon atoms in a hexagonal planar structure (Fig. 8.1). Each carbon atom in a graphite layer is joined to other three other carbon atoms by strong covalent bonds to form flat hexagonal rings. The various layers of carbon atoms in graphite are quite far apart so that no covalent bonds can exist between them; however, they are held together by weak van der Waals forces, so they can slide over one another. Due to its sheet-like structure, graphite is a soft substance. The C–C bond length is 1.42 Å and the distance between two successive layers is 3.35 Å. As these layers are held together by weak van der Waals forces and are slippery, graphite is soft and has low density. The C–C bond length is shorter than that of diamond because a pi bond is formed by the fourth valence electron present in each carbon atom which is free. Due to the presence of these pi electrons which are able to move in these slippery layers, graphite can conduct electricity.

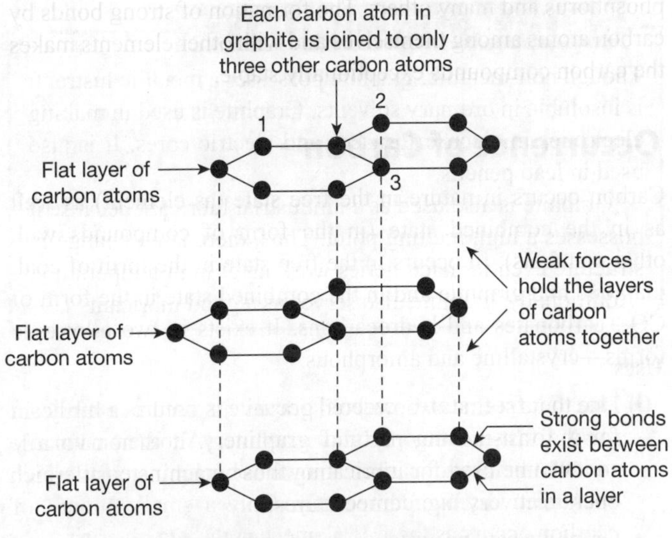

Fig. 8.1 Structure of graphite

Preparation by Acheson method

A mixture of powdered coke, sand, ferric oxide and saw dust is filled in an electric furnace having an inner lining of fire clay bricks. Carbon rods act as cathode and anode. The mixture of sand and coke on the top prevent the supply of air into the furnace. When electric current is passed through the electrodes and the temperature is nearly 3500°C, the following reactions occur to form graphite.

$$SiO_2 + 3C \xrightarrow{\Delta} SiC + 2CO$$

$$SiC \xrightarrow{\Delta} C + Si$$

Properties

- It is a soft, greyish, greasy crystalline solid with a density of 2.5 g/mL.
- Graphite has the lowest energy among the allotropes of carbon.
- It is a conductor of electricity due to the presence of free electrons in its slippery hexagonal sheets. Its conductivity increases with an increase in temperature.
- It is called plumbago or black lead as it leaves a black mark on paper.
- It is thermodynamically more stable than diamond and it has 1.9 kJ less free energy of formation than diamond.
- On heating to 1873 K and at an extremely high pressure (50,000–60,000 atm) it changes into diamond.
- It is a more reactive form than diamond; however, it does not react with dilute acids or alkalies.
- Graphite on heating with conc. HNO_3 gives graphitic acid ($C_{11}H_4O_5$).
- Graphite on oxidation with alkaline $KMnO_4$ gives mellitic acid or benzene hexacarboxylic acid [$C_6(COOH)_6$] and oxalic acid.
- On heating graphite with vapours of some metals such as K, Rb, Cs at nearly 600 K, a paramagnetic compound C_8M is obtained (M = K, Rb, Cs).

KNOWLEDGE BOOSTER

Though non-metallic, graphite possesses a metallic lustre. It is insoluble in ordinary solvents. Graphite is used in making electrodes in electric furnaces and electric cores. It is also used in lead pencils.

Graphite is also used as an industrial lubricant because it possesses a high melting point. Thus, there is no change in structure even in machineries working at high temperature. Additionally, graphite also serves as a good lubricant.

Uses

- Due to its softness, powdered graphite is used as a lubricant for the fast-moving parts of machinery. It is non-volatile so it can be used for lubricating those machine parts which operate at very high temperatures.
- Graphite can be used as a dry lubricant in the form of graphite powder or mixed with petroleum jelly as graphite grease.
- As graphite is a good conductor of electricity, it can be used for making carbon electrodes or graphite electrodes in dry cells and electric arcs. For example, the black coloured 'anode' of a dry cell is made of graphite.
- The carbon brushes of electric motors are also made of graphite.
- As graphite is black in colour and quite soft, it marks black lines on paper. Due to this property, graphite (mixed with clay) is used for making the black cores of our pencils and are called pencil leads.
- It is also used in electroplating, electrotyping, painting, stoves and in making refractory crucibles.
- In nuclear reactors, it is used as a moderator.

Fullerenes

These are the recently discovered crystalline pure allotropes of carbon obtained by evaporation of graphite by laser. C_{60} type of fullerene was discovered by H W Kroto, Richard Smalley and Robert Curl (Nobel prize winners) by applying a laser beam on a sample of graphite.

These are the only pure forms of carbon. Unlike diamond and graphite, these do not have dangling edges or surface bonds which are instrumental in attracting other atoms. These are large spheroidal cage-like molecules of carbon having a molecular formula of C_{60}, C_{70} and so on. They are called buckyballs (Fig. 8.2).

C_{60} is called buckminster fullerene in honour of Robert Buckminster Fuller and its structure is similar to a soccer ball. It is a dark solid at room temperature which is neither very hard nor very soft. It has 12 five membered rings and 20 six membered rings. The five membered rings are connected only to six membered rings while six membered rings are connected both the type of rings. The C–C and C=C bond lengths are 145.3 and 138.3 pm respectively.

C_{70} has also been discovered. It can be easily separated from the soot by extraction into toluene followed by chromatographic separation over alumina.

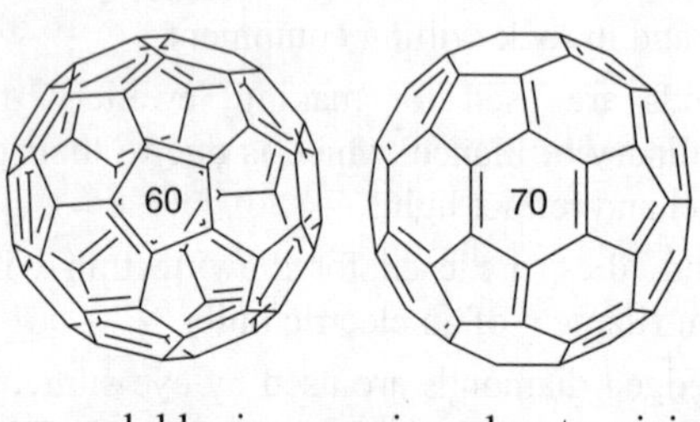

Fullerenes are soluble in organic solvents giving coloured solutions. C_{60} gives a purple colour in toluene while C_{70} gives an orange colour.

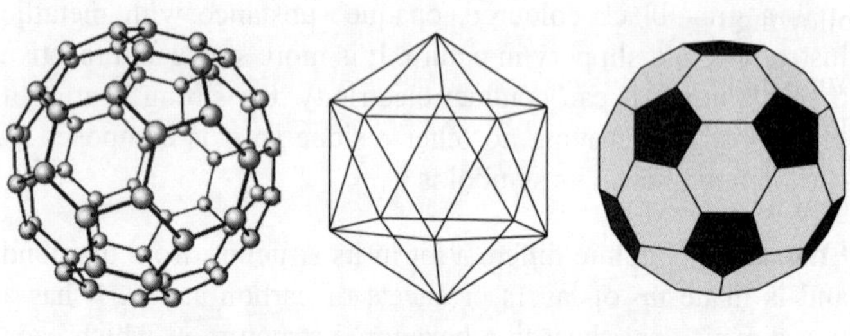

Fig. 8.2 Structure of buckminster fullerene

Illustrations

1. Explain why diamond has a high melting point.

   Solution: The high value of boiling point, melting point and density of diamond are due to its three dimensional structure which needs high energy for bond breaking.

2. If a boy sharpens a pencil at both the ends and then uses its back ends to complete an electric circuit, will the current flow through the electric circuit? Explain it and name the black substance of the pencil.

   Solution: The current will flow through the electric circuit as the black substance of a pencil is graphite and being a good conductor of electricity, the graphite core of pencil allows the electric current to flow through it.

3. A piece of black electrode is used in dry cell which on strong heating in air gives a colourless gas which turns lime water milky. What was the material of the electrode?

   Solution: Graphite is used for making electrodes. So, the piece of black electrode used in the dry cell is made of graphite. This is confirmed by the fact that the piece of electrode, on strong heating in air, gave a colourless gas, carbon dioxide, which turned limewater milky. Hence, the material of the electrode is graphite.

4. How many hexagonal and pentagonal rings are present in C_{60} molecules?

   Solution: It contains 20 hexagonal and 12 pentagonal rings fused together.

Amorphous Forms of Carbon

Amorphous allotropes of carbon are charcoal, coke, lampblack and gas carbon. In these allotropes, the carbon atoms are not arranged in an orderly manner. From X-ray studies, it is clear that these forms are tiny graphite crystals cemented together by impurities.

Coal

It is a crude form of carbon which is formed in nature by the slow decomposition of vegetable matter under the influence of air, heat and pressure. The main types of coal are as follows:

Type	Carbon content	Calorific value (BTU)
Peat	60%	9990
Lignite	70%	11700
Bituminous	80%	14900
Anthracite	90%	16200

The commonly available variety here is bituminous, which is hard like stone and burns with a smoky flame.

Coal can be used as a fuel and in the manufacture of synthetic petrol, coal gas, and so on.

Charcoal

It is the most active form of carbon which can react with many oxidising agents, acids, and so on. It can be obtained by burning wood and cellulose in a limited supply of air. It is of the following types.

Wood charcoal: It is porous and has a very large surface area per unit weight so it is a very effective adsorbent. It floats over water (fills air in pores). It is used as deodorant in toilets, decolouring agent in sugar mills and in cigarette filters. Gas masks are made of wood charcoal. Gun powder has wood charcoal, KNO_3, sulfur.

$$HNO_3 + \text{Charcoal} \rightarrow \text{Artificial tannin}$$

Laboratory preparation: In the laboratory, wood charcoal is prepared by heating wood shavings in a dry hard glass tube in the absence of air. When wood shavings are heated, they get charred and fumes are evolved. These fumes are collected in cold water (Fig. 8.3). Here, the products of destructive distillation of wood are wood charcoal, wood tar, pyroligneous acid and wood gas.

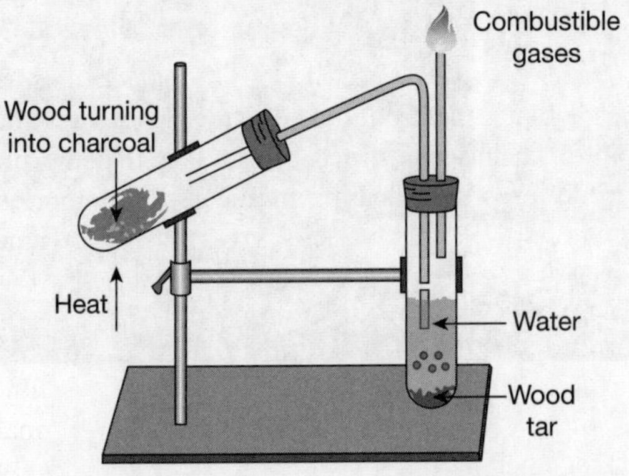

Fig. 8.3 Preparation of wood charcoal

Animal or bone charcoal: Animal bones contain organic matter and calcium phosphate. When the crushed fine powder of the animal bones is processed for destructive distillation, the residue obtained is known as bone charcoal or ivory black. It has maximum adsorption property and is also used as a bleaching agent in removing colour from sugarcane juice. It is also used in the extraction of yellow phosphorus.

Sugar charcoal: It is the purest form of amorphous carbon and it is obtained by destructive distillation of sugar. Sugar crystals lose water and get converted into pure carbon—sugar charcoal. Sucrose or sugar on reaction with concentrated sulfuric acid also gives sugar charcoal. It is used as an adsorbent material in place of activated charcoal and it is used for extracting metals from their oxides.

$$C_{12}H_{22}O_{11} \xrightarrow{H_2SO_4} 12C + 11H_2O$$

Physical Properties

Charcoal is a porous black solid having a huge surface area due to its porosity. It has the unique ability to adsorb specific substances like coloured impurities and poisonous substances.

Chemical Properties

(i) **Reaction with non-metals:** Charcoal can react with non-metals like oxygen, sulfur and hydrogen.

(a) **With oxygen:** Oxides of carbon are formed as follows.

$$C + O_2 \rightarrow CO_2$$

$$C + O_2 \xrightarrow{\text{Excess}} 2CO$$

(b) **With sulfur:** On passing sulfur vapour through red-hot charcoal, a vapour of carbon disulfide is formed.

$$C + 2S \rightarrow CS_2$$

(c) **With hydrogen:** On reaction with hydrogen gas, charcoal gives methane.

$$C + 2H_2 \rightarrow CH_4$$

(ii) **Reducing property:** Both wood charcoal and sugar charcoal behave as good reducing agents when they react with steam or oxides of metals or with oxy acids.

$$C + H_2O(g) \rightarrow CO + H_2$$

$$ZnO + C \xrightarrow{\text{Steam}} Zn + CO$$

$$C + 2H_2SO_4 \xrightarrow{\text{boiling}} CO_2 + 2H_2O + 2SO_2$$

As bone charcoal contains maximum percentage of calcium phosphate, it cannot act as a good reducing agent.

Table 8.1 Other amorphous forms of carbon

	Preparation	Properties	Uses
Coke	It can be prepared by heating coal in the absence of air up to 1300°C in huge iron retorts. The products obtained by the decomposition of coal are coal gas, carbon, coal tar, liquor ammonia and coal gas. Coal – Volatile separation → Coke	It is a greyish black porous solid.	(i) It is used as a household fuel. (ii) It is used in the manufacture of graphite, calcium carbide, water gas and producer gas. (iii) Being an excellent reducing agent, it is used extensively in the extraction of metals like iron copper and lead from their oxide and sulfide ores.
Lampblack	It can be prepared by burning mustard oil, turpentine oil and petroleum in the absence of oxygen. The preparation of lamp black is possible at home by placing a clean dry glass slide over the flame of a mustard oil lamp. After some time, a deep black powdery substance is coated on the slide, which is called lampblack.	It is a light, powdery black substance, with an oily velvet touch due to the presence of the vapours of some amount of oil.	(i) It is used for making black shoe polish and in the manufacture of black carbon paper and carbon ribbons for typewriters. (ii) It is used as a black pigment in inks and paints. (iii) It is used as a stabilising filler for rubber in making tyres and plastics.
Gas carbon	It can be obtained by destructive distillation of coal. The thermal vaporisation of small amounts of carbon from coal on condensation gives a grey solid that is gas carbon.	It is a dull grey solid and a good conductor of electricity.	It is used for making electrodes for dry cells.

Soot: Soot is an airborne contaminant in the environment with several different sources, all of which are a result of some form of pyrolysis (heating at high temperature in the absence of air). This includes soot from coal burning, internal combustion engines, power plant boilers, ship boilers, central steam heat boilers, local field burning, house fires, forest fires, fireplaces and furnaces. These exterior sources also contribute to the indoor environment sources such as smoking of plant matter, cooking, oil lamps, candles, quartz/halogen bulbs with settled dust, fireplaces, exhaust emissions from vehicles and defective furnaces. Soot in very low concentrations is capable of darkening surfaces or making particle agglomerates like those from ventilation systems, appear black. Soot is the primary cause of 'ghosting', the discoloration of walls and ceilings or walls and flooring where they meet. It is generally responsible for the discoloration of the walls above baseboard electric heating units.

Fig. 8.4 Pollution by vehicles

Harmful effects

- Soot, particularly diesel exhaust pollution, accounts for over one quarter of the total hazardous air pollution (Fig. 8.4).
- Long-term exposure to urban air pollution containing soot increases the risk of coronary artery disease.
- Among diesel emission components, particulate matter has been a serious concern for human health due to its direct and broad impact on respiratory organs.
- Soot also tends to form in chimneys of houses possessing one or more fireplaces. If a large deposit collects in one, it can ignite and create a chimney fire.

Uses

- It can be used as a pigment for inks and dyes (in fact both lampblack and carbon black are dyes made of soot).
- Soot is used in the vulcanisation process to treat rubber.
- It is used in toners for laser printers and copiers.

Table 8.2 Uses of carbon

Forms of carbon	Uses
Diamond	Gemstone, cutting, drilling, grinding, polishing, industry
Graphite	Steel manufacture (reducing agent), refractories, pencils, high temperature crucibles, electrodes in electrolytic extraction of elements, neutron moderator in nuclear reactors, high strength composite materials
Coke	Steel manufacture, fuel
Carbon black	Rubber industry, pigments in ink, paints and plastics
Activated charcoal	Decolourising agent in sugar industry, purification of chemicals and gases by adsorption, catalyst
Wood charcoal	Fuel

Illustrations

1. Answer the following

(a) Name one amorphous form of carbon.

(b) Name the most inferior form of coal.

Solution:

(a) Coal is an amorphous form of carbon.

(b) Peat is the most inferior form of coal.

2. Answer the following

(a) What is formed when wood is burnt in a limited supply of air?

(b) Which charcoal is a bad conductor of heat and electricity?

(c) Write any one use of lampblack.

Solution:

(a) Charcoal is formed when wood is burnt in a limited supply of air.

(b) Wood charcoal is a bad conductor of heat and electricity.

(c) Lampblack is used in making black shoe polish.

3. (a) Which charcoal is formed by treating sucrose or sugar with sulfuric acid?

(b) Write any one equation showing the reducing nature sugar charcoal.

Solution:

(a) Sugar charcoal is formed by treating sucrose or sugar with sulfuric acid.

$$C_{12}H_{22}O_{11} \xrightarrow{H_2SO_4} 12C + 11H_2O$$

(b) Sugar charcoal can reduce zinc oxide into zinc.

$$ZnO + C \xrightarrow{\text{Steam}} Zn + CO$$

Compounds of Carbon

Carbon has a great affinity to react with many non-metals such as hydrogen and oxygen, and metals such as Ca and Mg to form a number of inorganic compounds like oxides (CO, CO_2), carbides [SiC (carborundum), CaC_2 (calcium carbide)], metal carbonates (Na_2CO_3, $CaCO_3$) and metal bicarbonates ($NaHCO_3$).

Carbon forms a very large number of organic compounds – hydrocarbons – and their derivatives. For example, CH_4 (methane), C_2H_6 (ethane), HCOOH (formic acid), CH_3COOH (acetic acid), C_2H_5OH (ethanol), and so on.

Carbon Monoxide (: C ≡ O :)

It is not a natural component of atmospheric air but gets added to the atmosphere due to the incomplete combustion of fossil fuels such as coal, wood and petrol (exhaust gases of internal combustion engines and chimney gases) and degradation of organic matter in swamps. It is present in some quantity in volcanic gases.

Preparation

It can be prepared as follows.

(i) It is obtained by the incomplete combustion of carbon.

$$2C + O_2 \xrightarrow[\text{Less air}]{} 2CO$$

(ii) Laboratory method: In the laboratory, it can be prepared by several methods.

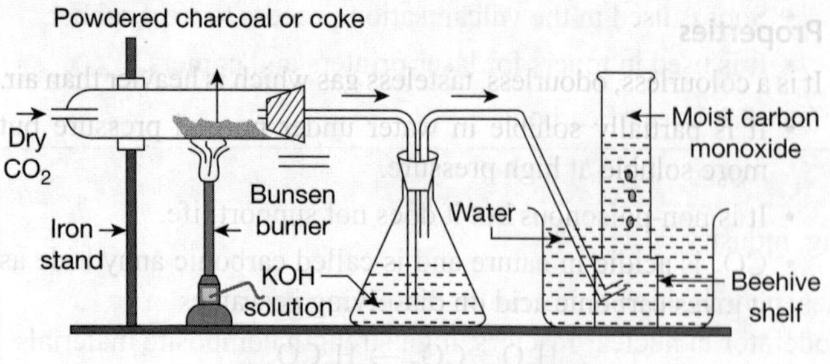

Fig. 8.5 Laboratory method

- **From red-hot charcoal:** when CO_2 is passed over red-hot charcoal, it gets oxidised into carbon monoxide. Powdered charcoal is heated in a glass tube and CO_2 is passed into it, to get carbon monoxide (Fig. 8.5).

$$C + CO_2 \rightarrow 2CO + Heat$$

- **From HCOOH:** When formic acid or sodium formate is heated with concentrated H_2SO_4, pure carbon monoxide is formed.

$$HCOOH \xrightarrow{\text{Conc. } H_2SO_4} H_2O + CO \uparrow$$

$$2HCOONa + H_2SO_4 \rightarrow Na_2SO_4 + 2H_2O + 2CO$$

- **From oxalic acid:** When oxalic acid is treated with concentrated sulfuric acid, carbon monoxide is formed.

$$\begin{array}{c} COOH \\ | \\ COOH \end{array} \xrightarrow{\text{Conc. } H_2SO_4} CO \uparrow + CO_2 \uparrow + H_2O$$

Physical Properties

- It is a neutral (no response to litmus test), highly poisonous, colourless, odourless gas and burns with a blue flame.

$$2CO + O_2 \rightarrow 2CO_2 + heat$$

- It is almost insoluble in water.
- Its density is nearly equal to that of air.
- It is used as a reducing agent in Mond's process (Ni).
- It is a linear molecule in which the carbon atom is *sp* hybridised.
- Carbon monoxide combines with hemoglobin to form carboxyhemoglobin at any or all of the oxygen-binding sites of hemoglobin, and also acts to increase the stability of the bond between hemoglobin and oxygen. It thus reduces the ability of the hemoglobin molecule to release the oxygen bound to other oxygen-binding sites. That is, it destroys the oxygen carrying capacity of blood as it can absorb oxygen and death may occur due to suffocation. It is known as asphyxia (a condition arising when the body is deprived of oxygen, causing unconsciousness or death; suffocation).

$$Hemoglobin + CO \rightarrow \underset{(Stable)}{Carboxyhemoglobin}$$

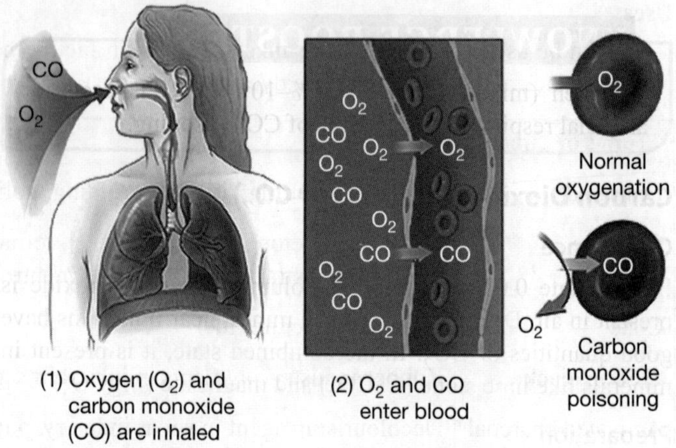

Chemical Properties

(i) Combustion: Carbon monoxide burns in air with a blue flame in an exothermic and explosive reaction. However, it does not support combustion.

$$2CO + O_2 \rightarrow 2CO_2$$

(ii) Reducing properties: It is a strong reducing agent so it is widely used for the extraction of metals from their oxides. For example,

$$Fe_2O_3 + 3CO \xrightarrow{600-900°C} 2Fe + 3CO_2$$

$$ZnO + CO \rightarrow Zn + CO_2$$

(iii) Reaction with chlorine: It combines with chlorine in the presence of sunlight to form phosgene.

$$CO + Cl_2 \rightarrow \underset{Phosgene}{COCl_2}$$

(iv) Reaction with metals: Carbon monoxide reacts with metals to form metal carbonyls.

$$M + xCO \rightarrow M[CO]_X$$

Here, M may be Ni, Cr, Mo, Fe, Co, and so on. For example,

$$Ni + 4CO \xrightarrow{325-345K} \underset{\text{Nickel carbonyl}}{Ni(CO)_4}$$

(v) Reaction with dihydrogen: When it reacts with dihydrogen at 425–675 K in the presence of a catalyst, methanol is formed.

$$CO + 2H_2 \xrightarrow[ZnO+Cu]{425-675\ K} \underset{Methanol}{CH_3OH}$$

Uses

- It is used as a reducing agent and in the preparation of metal carbonyls.
- It is used in war gas preparation (phosgene).
- It is used in the refining of nickel in Mond's process.
- It is used in the preparation of methanol, phosgene, synthetic petrol.
- It is used in gaseous fuel as water and producer gas.

KNOWLEDGE BOOSTER

Carbogen (mixture of O_2 + 5%–10% CO_2) is used for artificial respiration for victims of CO poisoning.

Carbon Dioxide ($O = C = O$ or CO_2)

Occurrence

In free state 0.03%–0.05 % by volume of carbon dioxide is present in air. Dry wells, caves and mines, near lime kilns have good quantities of CO_2. In the combined state, it is present in minerals like lime stone ($CaCO_3$) and magnesite ($MgCO_3$).

Preparation

It is prepared by the following methods.

(i) **From carbonates and bicarbonates:** Metal carbonates and bicarbonates on thermal decomposition (by heating) give carbon dioxide.

$$M\text{–}CO_3 \xrightarrow{\Delta} CO_2\uparrow + MO$$

For example,

$$CaCO_3 \xrightarrow{\Delta} CaO + CO_2$$

$$M\text{–}HCO_3 \xrightarrow{\Delta} M\text{–}CO_3 + CO_2\uparrow + H_2O$$

For example, $Ca(HCO_3)_2 \xrightarrow{\Delta} CaO + 2CO_2 + H_2O$

(ii) **Laboratory method:** When metal carbonates or bicarbonates are treated with mineral acids, CO_2 is formed. Some marble chips taken in a conical flask are treated with dilute hydrochloric acid (Fig. 8.6) to get CO_2.

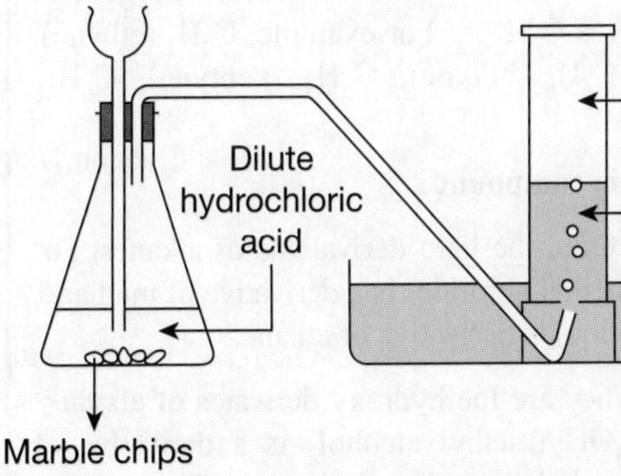

Fig. 8.6 Preparation of CO_2

$$M - CO_3 + HCl \rightarrow CO_2 + M - Cl + H_2O$$

$$MHCO_3 + HCl \rightarrow MCl + H_2O + CO_2$$

For example,

When $CaCO_3$ is treated with dilute HCl, carbon dioxide is formed.

$$CaCO_3 + 2HCl \xrightarrow{\Delta} CaCl_2 + CO_2 + H_2O$$

(iii) **By burning of charcoal or coke:** When charcoal or coke is burnt in adequate supply of air, the major product is carbon dioxide and this reaction is highly exothermic.

$$C + O_2 \rightarrow CO_2 + \text{Heat}$$

Properties

It is a colourless, odourless, tasteless gas which is heavier than air.

- It is partially soluble in water under normal pressure but more soluble at high pressure.
- It is non-poisonous but it does not support life.
- CO_2 is acidic in nature and is called carbonic anhydride as it gives carbonic acid on dissolving in water.

$$H_2O + CO_2 \rightarrow H_2CO_3$$

Property	CO	CO_2
BP (K)	81.5	194.5
C–O bond length (pm)	112.8	116.3
Density (g/L) at 273 K	1.250	1.977
ΔH_f (kJ mol^{-1})	110.5	–393.5
MP (K)	68	216.4 (at 5.2 atm)
Structure	$:C \equiv O:$	$O = C = O$

Chemical Properties

(i) **Non-combustible nature and non-supporter of combustion:** It is neither combustible nor does it support it. However, active metals like Na, K and Mg continue to burn in an atmosphere of CO_2.

$$CO_2 + 4Na \rightarrow 2Na_2O + C$$

$$CO_2 + 2Mg \rightarrow 2MgO + C$$

CO_2 can be used as a fire extinguisher except in the case of active metals like Na, Mg.

(ii) **Reduction:** When carbon dioxide is passed over red-hot coke, it gets reduced to carbon monoxide.

$$CO_2 + \underset{\text{Red-hot coke}}{C} \rightarrow 2CO$$

(iii) **Acidic nature:** It is an acidic oxide as it forms carbonic acid on dissolving in water and can form salts with bases.

$$CO_2 + H_2O \rightarrow H_2CO_3$$

$$2NaOH + CO_2 \rightarrow Na_2CO_3 + H_2O$$

$$Na_2CO_3 + CO_2 + H_2O \rightarrow 2NaHCO_3$$

(iv) **Action on lime water:** CO_2 turns lime water milky insoluble calcium carbonate.

$$Ca(OH)_2 + CO_2 \rightarrow \underset{\text{Milky}}{CaCO_3\downarrow} + H_2O$$

This milkiness disappears when more CO_2 is passed as soluble calcium bicarbonate is formed.

$$CaCO_3 + H_2O + CO_2 \rightarrow \underset{\text{Soluble}}{Ca(HCO_3)_2}$$

$$Ca(HCO_3)_2 \rightarrow CaCO_3 + CO_2 + H_2O$$

(v) **Formation of carbonates:** It reacts with some basic oxides like Na_2O, K_2O to form their carbonates.

$$K_2O + CO_2 \rightarrow K_2CO_3$$

$$Na_2O + CO_2 \rightarrow Na_2CO_3$$

(vi) Photosynthesis: CO_2 is changed into a carbohydrate (glucose) by plants in the presence of sunlight and chlorophyll.

$$6CO_2 + 6H_2O \xrightarrow[\text{Sunlight}]{\text{Chlorophyll}} C_6H_{12}O_6 + 6O_2$$

$$6 \text{ x } CO_2 + 5 \text{ x } H_2O \rightarrow (C_6H_{10}O_5) \text{ x} + 6 \text{ x } O_2$$

KNOWLEDGE BOOSTER

Dry ice: CO_2 gas can be easily liquefied into a mobile liquid which can be further changed into a white solid mass of CO_2. Solid carbon dioxide is known as dry ice since it does not wet the surface on which it melts. It is used as a coolant for preserving fresh articles in the food industry and for making cold baths in the laboratory. Under the name *dry cold*, it is used as a refrigerant. It can provide not only a cold atmosphere but also an inert atmosphere which can help in the killing of unwanted bacteria.

Uses

- It is used in cold drinks and aerated water.
- CO_2 (15%) is used as a fire extinguisher (except for Mg-fires).
- O_2 + 5%–10% CO_2 is carbogen which is used for artificial respiration in the case of CO poisoning and pneumonia patients.
- It is used in the manufacture of sodium carbonate by the Solvay method.
- Dry fire extinguisher is $SiO_2 + NaHCO_3$.
- Foaming fire extinguisher is $NaHCO_3 + Al_2(SO_4)_3$.
- **Carbon** is an element and its symbol is C. It is a non-metal.

KNOWLEDGE BOOSTER

Compounds of carbon and hydrogen (hydrocarbons)

Carbon has a great affinity towards hydrogen and it reacts with hydrogen to form a large number of compounds. Those compounds which contain carbon and hydrogen alone are called hydrocarbons. For example, CH_4 (methane), C_2H_6 (ethane), C_2H_4 (ethene), C_2H_2 (ethyne), and so on. These hydrocarbons and their derivatives (formed by replacing hydrogen atom/s with other atoms or groups such as X, OH) are called organic compounds. The word organic indicates living things as most of the organic compounds have natural sources. For example, formic acid (HCOOH) is obtained from formica (red ants) and acetic acid (CH_3COOH) or vinegar is obtained from mycoderma aceti.

Classification of hydrocarbons

Hydrocarbons are of the following types.

Alkanes: All compounds in which carbon and hydrogen are attached with single bonds are called alkanes. The general formula of alkanes from which all the members of the family can be derived is C_nH_{2n+2} (n = number of carbon atoms). For example, CH_4 (methane), C_2H_6 (ethane), C_3H_8 (propane), C_4H_{10} (butane), C_5H_{12} (pentane), C_6H_{14} (hexane), and so on.

Alkenes: Those unsaturated hydrocarbons which have one or more double bonds are called alkenes. Their general formula is C_nH_{2n}. For example, C_2H_4 (ethene), C_3H_6 (propene), C_4H_8 (butene), C_5H_{10} (pentene), and so on.

Alkynes: Those unsaturated hydrocarbons which contain one or more triple bonds are called alkynes. The general formula of alkynes is C_nH_{2n-2}. For example, C_2H_2 (ethyne), C_3H_4 (propyne), C_4H_6 (butyne), C_5H_8 (pentyne), C_6H_{10} (hexyne).

Some other organic compounds

Halides (RX): they are the halo derivatives of alkanes. For example, CH_3Cl (methyl chloride) is a derivative of methane, C_2H_5Cl (ethyl chloride) is derivative of ethane.

Alcohol (ROH): They are the hydroxy derivates of alkanes. For example, CH_3OH (methyl alcohol) is a derivative of methane, C_2H_5OH (ethyl alcohol) is derivative of ethane.

Amines (RNH_2): they are the amino derivatives of alkanes. For example, CH_3NH_2 (methyl amine) is a derivative of methane, $C_2H_5NH_2$ (ethyl amine) is derivative of ethane.

CHAPTER AT A GLANCE

- The characteristic of carbon by virtue of which it forms four covalent bonds is generally referred to as **tetra covalency** of carbon.
- **Catenation:** The most outstanding (or unique) property of carbon is its ability to combine with itself, atom to atom, to form long chains.
- Allotropy is a phenomenon by which an element exhibits different physical forms, having almost similar chemical properties. Elements like carbon, phosphorus, sulfur and iron show **allotropy**.
- Carbon shows three **crystalline forms** having well defined regular geometrical arrangement of carbon atoms—diamond, graphite, fullerenes.
- **Diamond** is a colourless, transparent, quite heavy substance having amazing brilliance (lustre). It is the hardest natural known substance. It does not conduct electricity.
- **Graphite** is a grey-black opaque substance with a metallic lustre, and is soft and slippery in nature. It is more stable and reactive than diamond. It can conduct electricity.
- The **Acheson method** is used to prepare graphite.
- Diamond has a tetrahedral structural unit while graphite has hexagonal sheet-like structural units.
- **Fullerenes** are crystalline pure allotropes of carbon obtained by evaporation of graphite by laser.
- C_{60} is called **buckminster fullerene** in honour of Robert Buckminsterfuller and its structure is similar to that of a soccer ball. It is a dark solid at room temperature which is neither very hard nor very soft. It has 12 five-membered rings and 20 six-membered rings.
- **Amorphous** allotropes of carbon are charcoal, coke, lamp black and gas carbon. In these allotropes, the carbon atoms are not arranged in an orderly manner.
- **Coal** is a crude form of carbon which is formed in nature by the slow decomposition of vegetable matter under the influence of air, heat and pressure.
- **Charcoal** is the most active form of carbon which can react with many oxidising agent, acids, and so on.
- **Wood charcoal** is porous and has a very large surface area per unit weight so it is a very effective adsorbent. It floats over water (fills air in pores). It is used as a deodorant and disinfectant in toilets.
- When the crushed fine powder of animal bones is processed for destructive distillation, the residue obtained is known as **bone charcoal** or ivory black.
- **Sugar charcoal** is the purest form of amorphous carbon and it is obtained by destructive distillation of sugar. Sugar crystals lose water and get converted into pure carbon (sugar charcoal).
- Both wood charcoal and sugar charcoal behave as good **reducing agents** when they react with steam or oxides of metals or with oxyacids. As **bone charcoal** contains the maximum percentage of calcium phosphate, it cannot act as a good reducing agent.
- **Soot** is an airborne contaminant in the environment having several different sources, all of which are the result of some form of pyrolysis.
- **Carbon monoxide (CO)** can be obtained by incomplete combustion of carbon, from red-hot charcoal by passing CO_2 and by heating formic acid and oxalic acid. It is a neutral (no response to litmus test), highly poisonous, colourless, odourless gas and burns with a blue flame.
- **Carbon monoxide** combines with hemoglobin to form **carboxyhemoglobin** at any or all of the oxygen-binding sites of hemoglobin, and also acts to increase the stability of the bond between hemoglobin and oxygen, reducing the ability of the hemoglobin molecule to release oxygen bound to other oxygen-binding sites. Thus, it destroys the oxygen carrying capacity of blood as it can absorb oxygen.
- Carbogen (mixture of O_2 + 5%–10% CO_2) is used for artificial respiration for victims of CO poisoning.
- **Carbon dioxide (CO_2)** is an acidic oxide which can be obtained by heating metal carbonates and bicarbonates.
- CO_2 turns limewater **milky** due to the formation of insoluble calcium carbonate. This **milkiness** disappears when more CO_2 is passed as soluble calcium bicarbonate is formed.
- **Solid carbon dioxide** is known as **dry ice** since it does not wet the surface on which it melts.
- **CO_2** (15%) is used as a fire extinguisher (except for Mg-fires).

PRACTICE QUESTIONS

Analyse Your Concepts (School Exam Based)

Fill in the Blanks

Instructions: Complete the following statements with an appropriate word/term to be filled in the blank spaces.

1. The name 'carbon' is derived from the Latin word ________.
2. ________ is present in both living and non-living things.
3. The tendency of an element to exist in two or more forms but in the same physical state is called ________.
4. ________ and ________ are the two major crystalline allotropes of carbon.
5. The soft crystalline form of carbon is ________.
6. ________ is used in the manufacture of black carbon paper.
7. ________ is the hardest substance that occurs naturally.
8. Colour is imparted to diamond due to the presence of ________.
9. Crystallisation of carbon to form diamond takes place when ________ metal expands during solidification.
10. The form of carbon which is used as a lubricant at high temperature is ________.
11. Bone charcoal contains maximum percentage of ________.
12. Carbon monoxide is slightly ________ than air.
13. Diamond on heating in vacuum around 2000°C gets converted into ________.
14. Gas carbon is used as ________ in dry cells.

True or False

Instructions: Read the following statements and write your answer as true or false.

1. Carbon constitutes 0.03% of the earth's crust.
2. Carbon and its compounds are some of our major sources of fuels.
3. Fullerenes are the amorphous allotrope of carbon.
4. Graphite is the purest form of carbon.
5. Lamp black is used as a filler in making rubber tyres.
6. Diamond is a good conductor of electricity.
7. Sugar charcoal is a good reducing agent.
8. Coloured diamonds are costlier than colourless and transparent diamonds.
9. Diamond is chemically almost inert.
10. Graphite has layers of hexagonal carbon bonding.
11. Diamond is thermodynamically more stable than graphite.
12. Diamond is insoluble in all solvents.
13. Carbon monoxide is neither combustible nor a supporter of combustion.
14. Sugar on reaction with conc. H_2SO_4 gives sugar charcoal.
15. The form of carbon which is known as black lead is graphite.

Very Short Answer Type Questions

1. Name the element whose allotropic form is graphite.
2. Which of the following molecules is called buckminster fullerene?
   C_{90}, C_{60}, C_{70}, C_{120}
3. What is the number of 5-membered and 6-membered rings in buckminster fullerene?
4. What is meant by carbonisation?
5. Explain how sugar charcoal is prepared.
6. How is bone charcoal prepared?
7. Write the name of the purest form of charcoal.
8. Explain the reducing property of wood charcoal.
9. Mention any two physical properties of carbon dioxide.
10. Give two properties and three uses of lamp black.
11. Define catenation.
12. What is the electron dot structure of carbon dioxide which has the formula CO_2?

Short Answer Type Questions

1. Name the following:
   (a) The hardest naturally occurring substance.
   (b) A greyish-black non-metal that is a good conductor of electricity.
   (c) The third crystalline form of carbon.
2. Define the following terms.
   (a) Allotropy (b) Carat (c) Crystal
3. Why is graphite slippery?

4. Why is graphite used as a lubricant?
5. How is coke formed? Give some uses of coke.
6. What are the two properties of carbon which lead to the huge number of carbon compounds we see around us?
7. Carbon, a group 14 element in the periodic table, is known to form compounds with many elements. Write an example of a compound formed with
   (a) chlorine (group 17 of the periodic table)
   (b) oxygen (group 16 of the periodic table)
8. Explain the reactivity of graphite with oxygen and sodium carbonate with chemical equations.
9. How is gas carbon formed? Write one property and one use of gas carbon.
10. Give two uses of:
    (a) graphite (b) diamond.

Long Answer Type Questions

1. A solid element P has four electrons in the outermost shell of its atom. An allotrope Q of this element is used as a dry lubricant in machinery and also in making pencil leads.
   (a) What is element P?
   (b) Name the allotrope Q.
   (c) State whether allotrope Q is a good conductor or non-conductor of electricity.
   (d) Name one use of allotrope Q (other than lubrication and pencil leads).
   (e) Name two other allotropes of element P.
2. The solid element X exhibits the property of catenation. It is also present in the form of a gas Y in the air which is utilised by plants in photosynthesis. An allotrope Z of this element is used in glass cutters.
   (a) What is element X?
   (b) What is the gas Y?
   (c) Name the allotrope Z.
   (d) State another use of allotrope Z (other than in glass cutters).
   (e) Name another allotrope of element X which exists as spherical molecules.
   (f) Name yet another allotrope of element X which conducts electricity.
3. Carbon, a member of group 14, forms a large number of carbon compounds (estimated to be about three million). Why is this property not exhibited by other elements of this group? Explain.
4. Answer the following questions:
   (a) Why is graphite a good conductor of electricity but not diamond?
   (b) Why is diamond very hard?
   (c) What are fullerenes? Name the most common fullerenes.
   (d) What impurity is present in a black diamond?
   (e) Explain the softness of graphite with reference to its structure.
5. Write differences between graphite and diamond.

COMPETITION WINDOW (OBJECTIVE TYPE)

[For NEET, JEE (Main and Advanced), NTSE, KVPY and Olympiads]

Topic-wise MCQs

Introduction

1. Carbon exists in the atmosphere in the form of
   (a) carbon monoxide only
   (b) carbon monoxide in traces and carbon dioxide
   (c) carbon dioxide only
   (d) coal
2. In the combined state, carbon occurs as
   (a) coal (b) diamond
   (c) graphite (d) petroleum
3. Carbon occurs as the native element in the earth's crust in the form of
   (a) carbon dioxide
   (b) gas carbon
   (c) diamond
   (d) methane
4. The property of self-combination of the atoms of the same element to form long chains is known as
   (a) catenation
   (b) protonation
   (c) coronation
   (d) carbonation
5. Carbon forms innumerable compounds because
   (a) it has four electrons in its outermost shell
   (b) it behaves as metal as well as a non-metal
   (c) carbon atoms can form long chains
   (d) it combines with other elements to form covalent compounds
6. The pair of elements which exhibits the property of catenation is
   (a) silicon and carbon
   (b) carbon and sodium
   (c) chlorine and carbon
   (d) sodium and silicon

7. Carbon forms four covalent bonds by sharing its four valence electrons with four univalent atoms; for example, hydrogen. After the formation of four bonds, carbon attains the electronic configuration of
(a) helium (b) neon
(c) argon (d) krypton

8. Which of the following types of bonds can be formed by a carbon atom?
(a) single bond
(b) double bond
(c) triple bond
(d) all of these

9. Select the incorrect statement.
(a) Carbon exists as the native element in the form of coal
(b) Coal is an impure form of carbon
(c) Carbon is present in animals and plants in the form of fats and proteins
(d) Carbon occurs in the gaseous state in kerosene and LPG

Allotropic forms of carbon

10. One of the following is not an allotrope of carbon. This is
(a) cumene
(b) buckminster fullerene
(c) diamond
(d) graphite

11. Which of the following is the purest form of carbon?
(a) diamond
(b) graphite
(c) charcoal
(d) coal

12. A diamond-toothed saw is usually used for cutting
(a) marble slabs
(b) asbestos sheets
(c) steel girders
(d) logs of wood

13. The pencil leads are made of mainly
(a) lead (b) graphite
(c) lithium (d) charcoal

14. Buckminster fullerene is an allotropic form of the element
(a) fluorine (b) phosphorus
(c) carbon (d) sulfur

15. The number of carbon atoms joined in a spherical molecule of buckminster fullerene is
(a) seventy (b) ninety
(c) fifty (d) sixty

16. Which among the following properties of coke is exploited in the preparation of graphite by the Acheson process?
(a) highly porous nature
(b) low melting point
(c) bad conductor of electricity
(d) good conductor of heat

17. Graphite cannot be used as a lubricant in space because
(a) there is no atmosphere in space and hence there is no adsorbed air and water between layers of graphite
(b) there is no atmosphere in space and hence graphite sublimes in space
(c) absence of external pressure transforms crystalline graphite to the amorphous form
(d) none of these

18. Coke can be used as
(a) printer ink
(b) electrodes
(c) an oxidising agent
(d) a household fuel

19. Which of the following allotropes of carbon is used for the manufacture of fullerenes?
(a) coke (b) charcoal
(c) graphite (d) diamond

20. Graphite on heating with conc. HNO_3 gives
(a) benzoic acid (C_6H_5COOH)
(b) graphitic acid ($C_{11}H_4O_5$)
(c) oxalic acid ($H_2C_2O_4$)
(d) acetic acid (CH_3COOH)

Compounds of carbon

21. Carbon can form which type of compounds?
(a) inorganic compounds
(b) organic compounds
(c) both of these
(d) none of these

22. Which of the following compounds of carbon are possible?
(a) oxides (b) carbides
(c) carbonate (d) all of these

23. Which of the following methods can be used to prepare carbon monoxide in the laboratory?
(a) by heating formic acid with sulfuric acid
(b) by passing CO_2 over red-hot charcoal
(c) by heating oxalic acid with sulfuric acid
(d) all of these

24. Calcium carbonate on reaction with dil. HCl gives
(a) CO (b) CO_2
(c) C_3O_2 (d) both (a) and (b)

25. Which of the following can turn calcium hydroxide into a milky precipitate?
(a) CO
(b) CO_2
(c) both of these
(d) none of these

26. Which of the following statements is incorrect?
(a) CO can reduce oxides of less reactive metals
(b) CO on heating with metals can form metal carbonyls
(c) CO combines with chlorine to give a poisonous gas
(d) CO_2 can also reduce less active metal oxides

Miscellaneous

1. Buckminster fullerene is an allotropic form of
(a) phosphorus (b) sulfur
(c) carbon (d) tin

2. Diamond is used for
(a) making the electrodes of electric furnaces
(b) making crucible for melting metals
(c) cutting and drilling rocks and glass
(d) making carbon brushes for electric motors

3. Identify the endothermic reaction among the following.
(a) burning of water gas
(b) burning of methane
(c) passage of O_2 through red-hot carbon
(d) passage of CO_2 through red-hot carbon

4. Nature of the products obtained by the complete combustion of methane are
(a) basic, neutral
(b) acidic, basic
(c) acidic, neutral
(d) both are neutral

5. Which of the following carbonates does not decompose to give CO_2?
(a) K_2CO_3 (b) $ZnCO_3$
(c) $NaHCO_3$ (d) $CaCO_3$

6. A compound P made of two non-metals, in a reaction with another non-metal Q gives the hardest substance and carbon monoxide. Identify the compound P and non-metal Q.
(a) SiC, O_2 (b) CO, Si
(c) CO_2, Si (d) SiO_2, C

7. Why do black shoe polishes have an oily feel?
(a) vapour of some amount of oil is left over after its preparation
(b) these are made by dissolving charcoal in oil
(c) shoe polish is a mixture of oil and powdered graphite
(d) on exposure to air, these undergo chemical changes and produce oily substances

8. Black shoe polish has an oily feel because
(a) it contains liquefied vapours of hydrocarbons
(b) it produces oily substances when it comes in contact with air
(c) it is a mixture of oil and charcoal powder
(d) it is a solution of gas carbon and oil

9. Coal is a
(a) biofuel (b) nuclear fuel
(c) fossil fuel (d) none of these

10. The formation of soot particles during the burning of a candle is due to
(a) unburnt carbon particles left behind because of insufficient oxygen for burning
(b) burning of hydrocarbons
(c) burning of hydrogen
(d) all of these

11. Graphite on oxidation with alkaline $KMnO_4$ gives
(a) benzene hexacarboxylic acid [$C_6(COOH)_6$]
(b) oxalic acid
(c) graphitic acid
(d) both (a) and (b)

12. Which of the following is not correctly matched?
(a) Wood charcoal — It is used as a deodorant and disinfectant
(b) Bone charcoal — It is used in the extraction of yellow phosphorus
(c) Sugar charcoal — It is used as an excellent household fuel
(d) Coke — It is used in the manufacture of water gas

13. Select the incorrect statement.
(a) Carbon monoxide can be used as a fuel.
(b) Excess of CO in a room may even cause death.
(c) Excess of CO_2 may even cause death.
(d) Solid carbon dioxide (dry ice) is a refrigerant.

14. Which of the following compounds of carbon are possible?
(a) silicon carbide (SiC)
(b) calcium carbide (CaC_2)
(c) sodium bicarbonate
(d) all of these

15. Which of the following is correctly matched?
(a) Alkane — C_nH_{2n+2}
(b) Alkene — C_nH_{2n}
(c) Alkyne — C_nH_{2n-2}
(d) All of these

Advanced and Olympiads

Single Choice

1. Substance P is used extensively for small scale production of metals. It is also used as a deodorant and disinfectant and is one of the constituents of gun powder. Identify P.
(a) bone charcoal (b) wood charcoal
(c) coke (d) sugar charcoal

2. Why should we not light a candle in a closed room with people?
(a) methane gas which is poisonous is formed
(b) carbon monoxide gas which reduces the ability of blood to carry oxygen is formed
(c) the CO_2 formed causes breathlessness
(d) carbon particles are formed which are dangerous for the respiratory tract

3. Which of the following statements are usually correct for carbon compounds? These
(i) are good conductors of electricity
(ii) are poor conductors of electricity
(iii) have strong forces of attraction between their molecules
(iv) do not have strong forces of attraction between their molecules

(a) (i) and (iii) (b) (ii) and (iii)
(c) (i) and (iv) (d) (ii) and (iv)

4. Reaction: $2CaO + C \rightarrow CaC_2 + 2CO$
$2CaC_2 + 2H_2O \rightarrow X + Y$
If X is involved in the neutralisation reaction with an acid, then Y may be
(a) $Ca(OH)_2$ (b) CaH_2
(c) CH_4 (d) C_2H_2

5. Select the set having correct statements.
(i) In diamond, carbon is tetrahedral.
(ii) The melting point of diamond is more than that of graphite.
(iii) The density of fullerene is more than that of diamond.
(iv) Fullerene becomes a super conductor when it combines with noble gases.
(a) (i), (ii)
(b) (ii), (iii)
(c) (i), (ii), (iv)
(d) (i), (ii), (iii) and (iv)

Multiple Choice

6. The existence of an element in different allotropic forms is due to the
(a) different amounts of energy associated in the formation of each allotrope
(b) different arrangement of atoms
(c) similar arrangement of atoms
(d) different methods of formation

7. amorphous forms of carbon are
(a) lampblack
(b) gas carbon
(c) graphite
(d) diamond

8. Select the correct statements.
(a) CO is neutral while CO_2 is acidic.
(b) Both CO and CO_2 support combustion.
(c) CO is more poisonous and harmful than CO_2.
(d) Solid CO_2 is known as dry ice.

9. Select the correct statement.
(a) Wood charcoal is one of the constituents of gun powder.
(b) Bone charcoal is used in the extraction of yellow phosphorus.
(c) Gas carbon is used in making shoe polishes.
(d) Lampblack is used to make black pigments and ink.

10. Select the correct statements.
(a) Formic acid on heating with sulfuric acid gives CO.
(b) Oxalic acid on heating with sulfuric acid gives CO and CO_2.
(c) Carbon dioxide forms carboxy haemoglobin with haemoglobin.
(d) In photosynthesis, CO_2 is converted into glucose.

Comprehension Type

Comprehension I: Carbon has the remarkable ability to form compounds in which several carbon atoms remain joined together in chains or rings and it has also the tendency to unite with a number of other elements such H, N, O, halogens and sulfur. This enables carbon to form a large number of different compounds. The chemistry of the compounds formed by carbon is generally known as organic chemistry since many such carbon compounds are the essential constituents of all living bodies such as plants and animals. Examples of organic compounds are hydrocarbons, carbohydrates such as sugars and starch, various organic dyes such as methyl orange, indigo.

11. The chemistry of carbon compounds is known as *organic chemistry* because
(a) carbon compounds are the most important constituents of living organisms—plants, animals and micro-organisms
(b) carbon forms more than ten million compounds
(c) carbon atoms form double and triple bonds
(d) carbon is a tetravalent element

12. Which of the following compounds are organic compounds?
(a) H_2CO_3 (b) C_2H_5OH
(c) C_2H_5Cl (d) both (b) and (c)

13. Ethylene (C_2H_4) is an unsaturated compound because
(a) the valency of carbon is 2
(b) it contains only hydrogen bonded to carbon
(c) because it contains a double bond between the two carbon atoms
(d) because it is an unstable compound

Matrix Matching

14. Match the following

Column I	Column II
(a) diamond	(p) amorphous
(b) graphite	(q) tetrahedral
(c) fullerene	(r) hexagonal
	(s) bucky ball structure

Integer Type

15. Carbogen is a mixture of *x*% O_2 and *y*% CO_2. What is the value of *y*?

16. Out of the substances given below,
lamp black, coke, fullerene, coal, diamond, sugar, charcoal, graphite, wood charcoal, diamond, bone charcoal, soot
What is the total number of crystalline forms of carbon?

17. If the formula of the alkanes is C_nH_{10}, the value of *n* is?

18. In C_{60} buckminster fullerene, if x is the number of 6 membered rings and y is the number of 5 membered rings. What is the difference x – y?

Answer Keys

Fill in the Blanks

1. carbo
2. carbon
3. allotropy
4. crystalline, non-crystalline
5. graphite
6. lampblack
7. diamond
8. impurities
9. iron
10. graphite
11. calcium phosphate
12. lighter
13. graphite
14. electrodes

True or False

1. T	2. T	3. F	4. F	5. T
6. F	7. T	8. F	9. T	10. T
11. F	12. T	13. F	14. T	15. T

Very Short Answer Type Questions

1. Graphite is the allotropic form of carbon.
2. C_{60} is called buckminsterfullerene.
3. The number of 5 membered and 6 membered rings in buckminsterfullerene are 12 and 20 respectively.
4. Heating of a substance in the absence of air is called carbonisation.
5. Destructive distillation of sugar gives charcoal.
6. Destructive distillation of animal bones gives charcoal.
7. Sugar charcoal is the purest form of charcoal.
8. When wood charcoal is heated with oxides of less active metals, the metals get reduced to free metals.
9. Carbon dioxide is highly soluble in water and heavier than air.
10. Lampblack is light, powdery black, having a velvety touch. It is used in the manufacture of black carbon paper, carbon ribbons, black shoe polish.
11. Catenation is a unique property of carbon because of which a large number of organic compounds are formed. It is the ability of carbon atoms to form long chains with other carbon atoms through the sharing of electrons.
12. $\ddot{\underset{\cdot\cdot}{O}}::C::\ddot{\underset{\cdot\cdot}{O}}$ or O=C=O

Short Answer Type Questions

1. (a) Diamond (b) graphite and (c) fullerenes.
2. **(a) Allotropy:** The phenomenon due to which an element exists in two or more forms in the same physical state with identical properties but with different physical properties is known as allotropy.

   (b) Carat: The weight of a diamond is expressed in carats (1 carat = 0.2 g).

   (c) Crystal: A homogeneous solid in which particles such as atoms, molecules or ions are arranged in a definite pattern due to which they have a definite geometrical shape with plane surfaces is called a crystal. For example, sugar and sodium chloride.
3. In graphite, various layers of carbon atoms are quite far apart and the carbon atoms are held together by weak van der Waals forces; so, they can slide over one another.
4. As the structure of graphite contains flat hexagonal parallel layers joined together by weak van der Waals forces, these layers can slide over one another easily, thus making graphite a good lubricant.
5. Destructive distillation of coal gives coke. Coke is used as a fuel, reducing agent, for the manufacture of graphite, CaC_2, water gas and producer gas.
6. Two properties of carbon which lead to the huge number of carbon compounds we see around us are:
   (i) Carbon has six valence electrons leading to a high valency.
   (ii) Covalent bonding happens easily with carbon atoms and numerous others such as oxygen, chlorine, nitrogen, sulfur, hydrogen, and so on.
7. (a) Carbon tetrachloride (CCl_4)
   (b) Carbon dioxide (CO_2)
8. Graphite reacts with oxygen to give CO_2 and CO

   $$3C(s)+2O_2(g)\xrightarrow{700°C}CO_2(g)+2CO(g)+\text{Heat}$$

   On heating the fused mixture of sodium carbonate and graphite, sodium oxide and carbon monoxide gas are formed as follows:

   $$Na_2CO_3(s)+C(s)\xrightarrow{\text{Fuse}}Na_2O(s)+2CO(g)$$
9. Gas carbon is obtained by destructive distillation of coal. It is a dull grey solid and a good conductor of electricity. It is used in making electrodes for dry cells.
10. (a) Two uses of graphite are:
    - For making the electrodes of electric furnaces.
    - For making crucibles for melting metals due to its high melting points.

    (b) Two uses of a diamond are:
    - Diamond is used in jewellery as a gem.
    - It is used for cutting and drilling rocks, glass.

Long Answer Type Questions

1. (a) Carbon (b) graphite (c) good conductor of electricity (d) electrodes (e) diamond and buckminster fullerene.

2. (a) Carbon (b) carbon dioxide (c) diamond (d) for making jewellery (e) buckminsterfullerene (f) graphite.

3. (i) Tetravalency: Carbon can easily form four covalent bonds with other atoms.

(ii) Catenation: Carbon can easily form covalent bonds with other carbon atoms and the carbon–carbon bonds are more stable than the other element to same element bonds formed by other member of group 14.

(iii) Formation of multiple bonds: Carbon can form not only single bonds, it can also form double or triple bonds with the same or different elements. Since these properties are not exhibited by other elements of this group, therefore, the number of carbon compounds is very large as compared to the other elements of group 14.

4. (a) One valence electron of each carbon atom remains free in a graphite molecule, thus making graphite a good conductor of electricity. Whereas in diamond, they have no free mobile electrons. That is why diamond is a bad conductor of electricity.

(b) A diamond is a giant molecule. There are 4 valence electrons in the carbon atom where each carbon atom is linked to four neighbouring carbon atoms thus forming a rigid tetrahedral structure. It is the strong bonding that makes diamond the hardest substance.

(c) Fullerenes: Fullerenes are the third crystalline form of carbon. They have been found to exist in interstellar dust as well, though they were discovered only recently in the geological formations of the earth. Common fullerenes are C-32, C-60, C-70 and C-76.

(d) Copper oxide present in black diamond is an impurity.

(e) Each carbon atom is linked to three neighbouring carbon atoms in a graphite molecule forming a hexagonal arrangement of atoms. These hexagonal grouping of carbon atoms are arranged as layers or sheets piled one on top of the other. The layers are held together by weak forces such that they can slide over one another. That is why graphite is soft.

5. Differences between diamond and graphite.

Diamond	Graphite
1. Pure diamond is colourless and transparent.	1. Graphite is greyish black, opaque and shiny.
2. It is the hardest naturally occurring substance.	2. It is soft and greasy to the touch.
3. It has a high density of 3.5 g/cm^3.	3. It has a low density of 2.39 g/cm^3.
4. It is a bad conductor of electricity.	4. It is a good conductor of electricity.
5. It burns in air at 900°C to form carbon dioxide.	5. It burns in air at 700°C to form carbon dioxide.

Competition Window (Objective Type)

Topic-wise MCQs

1. (b)	**2.** (d)	**3.** (c)	**4.** (a)	**5.** (c)
6. (a)	**7.** (b)	**8.** (d)	**9.** (d)	**10.** (a)
11. (a)	**12.** (a)	**13.** (b)	**14.** (c)	**15.** (d)
16. (c)	**17.** (a)	**18.** (d)	**19.** (c)	**20.** (b)
21. (c)	**22.** (d)	**23.** (d)	**24.** (b)	**25.** (b)
26. (d)				

Miscellaneous

1. (c)	**2.** (c)	**3.** (d)	**4.** (c)	**5.** (a)
6. (d)	**7.** (a)	**8.** (a)	**9.** (c)	**10.** (a)
11. (a)	**12.** (c)	**13.** (c)	**14.** (d)	**15.** (d)

Advanced and Olympiads

Single Choice

1. (b) **2.** (b) **3.** (d) **4.** (d) **5.** (c)

Multiple Choice

6. (abd) **7.** (ab) **8.** (acd) **9.** (abd) **10.** (abd)

Comprehension Type

11. (a) **12.** (d) **13.** (c)

Matrix Matching

14. (a)—(q), (b)—(r), (c)—(s), (d)—(p)

Integer Type

15. 5 **16.** 3 **17.** 4 **18.** 8

Hints and Solutions

Competition Window (Objective Type)

Topic-wise MCQs

9. Carbon occurs in the liquid state in kerosene, LPG, and so on.

11. Diamond is the purest form of carbon.

20. Graphite on heating with conc. HNO_3 gives graphitic acid $(C_{11}H_4O_5)$.

21. Carbon can form both organic compounds (CH_4) and inorganic compounds (CO_2).

22. Carbon can form all these types of compounds—CO, SiC, $CaCO_3$

26. CO_2 can never act as a reducing agent.

Miscellaneous

8. Shoe polish contains lampblack which has an oily feel because of liquefied vapours of hydrocarbons left behind during its preparation.

11. Graphite on oxidation with alkaline $KMnO_4$ gives mellitic acid or benzene hexacarboxylic acid $[C_6(COOH)_6]$ and oxalic acid.

12. Sugar charcoal is not used as a household fuel.

Advanced and Olympiads

Single Choice

4. $2CaO + C \rightarrow CaC_2 + 2CO$

$$2CaC_2 + 2H_2O \rightarrow \underset{X}{Ca(OH)_2} + \underset{Y}{C_2H_2}$$

5. Fullerene has less density than both graphite and diamond.

Multiple Choice

8. Both CO and CO_2 do not support combustion.

9. Lampblack is used in making shoe polish and not gas carbon.

10. Only carbon monoxide forms carboxyhemoglobin and not carbon dioxide.

Integer Type

15. Carbogen is a mixture of 95% O_2 and 5% CO_2.

16. Diamond, fullerenes and graphite are the crystalline forms of carbon.

17. The general formula of an alkane is C_nH_{2n+2} so, in C_nH_{10}, n is 4 and the alkane is C_4H_{10}.

18. In C_{60} buckminsterfullerene, the number of 6 membered rings or x = 20; the number of 5 membered rings or y = 12. Hence, x – y = 20 – 12 = 8.

CHAPTER

9 Synthetic Fibres and Plastics

LEARNING OBJECTIVES

After studying this unit, you will be able to understand:
- Polymers and their types
- Natural and synthetic fibres
- Plastics and their types

Introduction

We all are aware that carry bags are made of polythene, toothbrushes are made of nylon and clothes are made of rayon, as these things are parts of our daily life. All these materials are polymers. Polymers may be natural (cotton, wool, silk), that is, obtained from living things like plants and animals, they may also be synthetic (artificial, such as nylon, rayon). Rayon and nylon are synthetic fibres with a lot of uses. We are also aware of plastics from which most of the household things are made. In this unit we will study all about fibres and plastics.

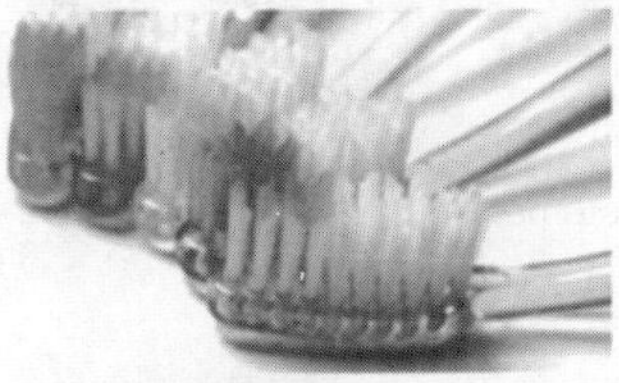

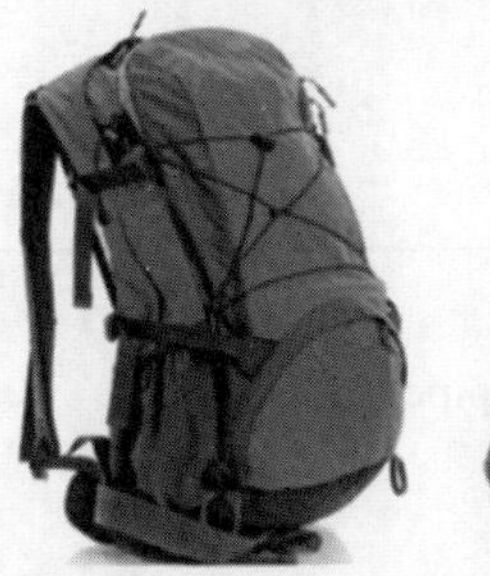

Polymers

A polymer is a single large chain-like molecule in which the repeating units derived from small molecules (called monomers) are joined together. The process by which monomers are transformed into a polymer is called *polymerisation*. The number of monomers which are joined together in a polymer is called the *degree of polymerisation*. There are many natural polymers, often called biopolymers, such as carbohydrates and proteins. Most synthetic polymers are long chain organic molecules and contain thousands of monomer units; such molecules have very high molecular mass and are often called macromolecules. If polymers are made of a single monomer, they are called homopolymers; for example, polythene, PVC. If two or more types of monomers form a polymer, it is called a copolymer. For example, Buna-S is a polymer of buta-1,3-diene and styrene.

Classification of Polymers

Natural Polymers or Biopolymers

Natural polymers include starch, cellulose and natural rubber. Starch is a polymer of glucose. Cellulose is also a polymer of glucose. It is made by plants from glucose produced during photosynthesis. Protein is obtained as a result of polymerisation of α-amino acids. Natural rubber is a polymer consisting of repeating units of the hydrocarbon 2-methyl-1,3-butadiene (isoprene).

$$n\underset{\text{Isoprene}}{CH_2=\underset{\displaystyle CH_3}{\overset{|}{C}}-CH=CH_2} \xrightarrow{\text{Polymerisation}} \underset{\text{Polyisoprene}}{(CH_2-\underset{\displaystyle CH_3}{\overset{|}{C}}=CH-CH_2)_n}$$

Synthetic Polymers

Synthetic polymers are artificial polymers. For example, polythene, polystyrene, PVC, bakelite, nylon and dacron.

Classification Based on Synthesis

(i) Addition or chain growth polymers: It involves the repeated addition of monomers to the polymer chain. The monomers are unsaturated compounds. These are chain growth polymers. For example,

	Monomer	**Polymer**
(a)	Ethylene	Polythene
(b)	Propylene	Polypropylene
(c)	Butadiene	Polybutadiene
(d)	Tetraflouoroethylene	Polytetrafluoroethelene (Teflon or PTFE)
(e)	Vinyl chloride	Polyvinyl chloride (PVC)

(ii) Condensation or step growth polymers: It involves a series of condensation reactions involving two monomers. Each monomer normally contains two functional groups. For example,

Monomer	Polymer
(a) Adipic acid and hexamethylene diamine	Nylon-66
(b) Phenol and formaldehyde	Bakelite
(c) Terephthalic acid and ethylene	Glycol polyester (Terylene)

Classification Based on Structure

Linear polymers: Such polymers have a long straight chain of monomers and have a close packed structure. They have high tensile strength, high density and high melting point. For example, polythene, nylon.

Branched chain polymers: These polymers have branches along the main chain of the monomers. Such polymers have low melting point, low tensile strength and low density due to the lack of close packing. For example, amylopectin, glycogen.

Three dimensional network or crossed-linked polymers: Such polymers have three dimensional cross-linked structures. They are hard, brittle and rigid. For example, bakelite, melmac.

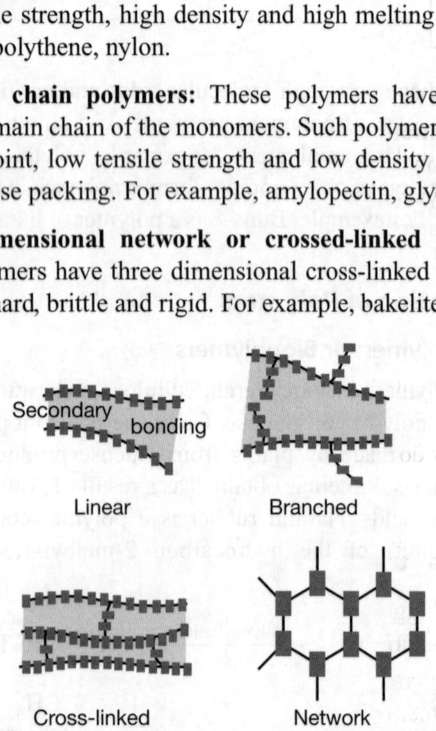

Classification Based on Molecular Forces

Elastomers: In elastomers, the polymer chains are held together by the weakest intermolecular forces or van der Waals forces. For example, vulcanised rubber.

Fibres: Such polymers are used for making fibres and possess high tensile strength and high modulus. This can be attributed to the stronger intermolecular forces like hydrogen bonding. For example, polyamides (nylon-66).

Thermoplastics: The intermolecular forces of attraction in thermoplastic polymers are intermediary to those of elastomers and fibres. As a result, these can be easily moulded by heating. In thermoplastic polymers, there is no cross-linking between chains. For example, polythene, polystyrene. Thermoplastic polymers are formed by addition and condensation reactions and have a linear structure and soften on heating.

Thermosetting polymers: These polymers are generally made from relatively low molecular mass semi-fluid polymers which when heated in a mould, become infusible and form an insoluble hard mass. This happens due to extensive cross-linking between different polymer chains. For example, bakelite, urea–formaldehyde resins. Thermosetting polymers are formed by condensation reactions and have a cross-linked structure and do not soften on heating.

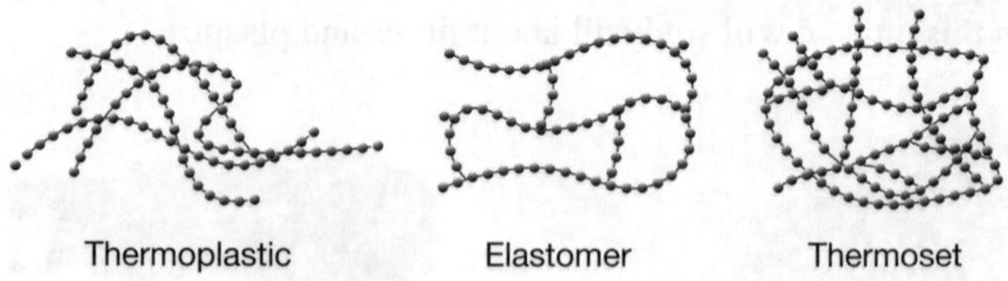

Table 9.1 Some commercially important polymers and their uses

Name of polymer	Monomer	Structure	Uses
Polypropene	Propene	$\left(CH_2-CH(CH_3) \right)_n$	Manufacture of ropes, toys, pipes, fibres.
Polystyrene	Styrene	$\left(CH_2-CH(C_6H_5) \right)_n$	As insulator, wrapping material, manufacture of toys, radio and television cabinets.
Polyvinyl chloride (PVC)	Vinyl chloride	$\left(CH_2-CH(Cl) \right)_n$	Manufacture of rain coats, hand bags, vinyl flooring, water pipes.
Urea–formalclehyle resin	(a) Urea (b) Formaldehyde	$\left(NH-CO-NH-CH_2 \right)_n$	For making unbreakable cups and laminated sheets.

(Continued)

Name of polymer	Monomer	Structure	Uses
Glyptal	(a) Ethylene glycol (b) Phthalic acid	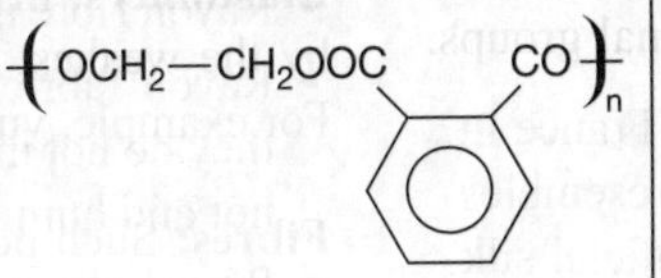	Manufacture of paints and lacquers.
Bakelite	(a) Phenol (b) Formaldehyde	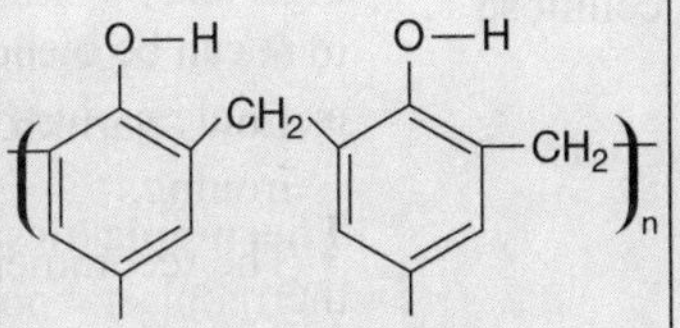	For making combs, electrical switches, handles of utensils and computer discs.

Illustrations

1. Give three advantages of polythene over natural materials.

Solution: Three advantages of polythene over natural materials are:

- it is strong but flexible,
- it can be rolled into sheets, and
- it is water resistant.

2. Give one use each of bakelite, nylon and acrylic.

Solution: Bakelite is used for making plugs and switches. Nylon is used in the textile industry to produce clothes. Acrylic is used as a substitute of natural wool for knitting sweaters and blankets.

Synthetic Fibres

Those fibres which are manufactured artificially from chemicals are known as synthetic fibres. The most commonly used synthetic fibres are rayon, nylon, polyester (terylene) and acrylic.

Characteristics of Synthetic Fibres

Synthetic fibres have these important properties:

(i) High strength: They can withstand a lot of force without breaking in comparison to natural fibres. That is, they have high tensile strength. Clothes made from synthetic fibres are stronger, do not wrinkle easily.

(ii) Elasticity: They are extremely elastic and can be stretched to some extent depending on the nature of the fibre. On releasing the pressure, they can regain their original shape. That is why nylon is used to make socks.

(iii) Water absorption: They can absorb only a little water, so the clothes made of such fibres can dry very fast.

Advantages of Synthetic Fibres

- They are strong, tough and durable.
- They are cheaper and need little care.
- They are easy to maintain and they are resistant to moths and insects.
- They are water-resistant and easy to dry.
- They can be dyed in different colours and they do not bleed colour when washed.
- They do not wrinkle easily.
- Unlike natural fibres, they do not shrink after washing.

Disadvantages of Synthetic Fibres

- The biggest disadvantage of synthetic fibres is that they melt on heating. If the clothes catch fire, it can be disastrous.
- They are heavy chemicals which may harm our skin. Another disadvantage is that the clothes made of synthetic fibres are not suitable to wear during the hot summer weather.
- As they have less air spaces in them compared to natural fibres and do not 'breathe' so well, they cannot absorb sweat also. These properties make them unsuitable for hot and humid weather.
- These are not comfortable in summers as they cannot absorb sweat.
- They trap the heat in our body.
- They are not very comfortable and may cause skin problems like rashes or allergies.
- They are not biodegradable as they are not decomposed by natural processes.
- They are highly inflammable so they catch fire easily.

Some Common Synthetic Fibres

Rayon

Rayon is the first artificial fibre and was developed in France in 1884 as a cheaper alternative to silk textiles. Rayon resembles silk in its shiny appearance so it is also known as artificial silk (polysaccharide). It can be produced from natural cellulose (obtained from wood pulp) by chemical treatment.

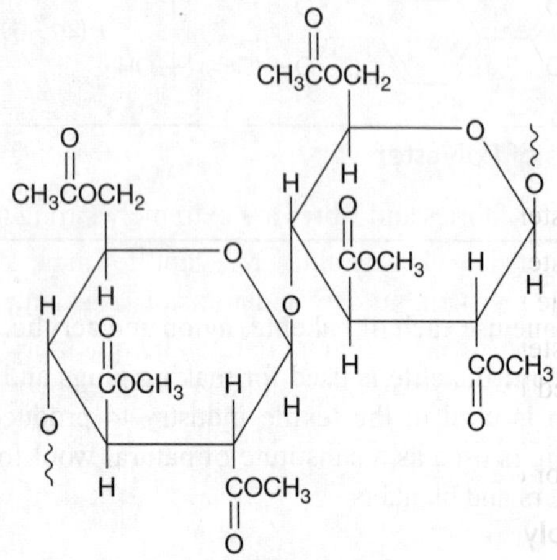

Cellulose acetate
(Acetate rayon)

Wood pulp is first dissolved in an alkaline solution and the thick liquid produced is then passed through tiny holes of a metal cylinder which is called a spinneret to make fibres. The fibres are hardened by passing them into a bath of sulfuric acid. These fibres are then spun into yarn and woven into cloth.

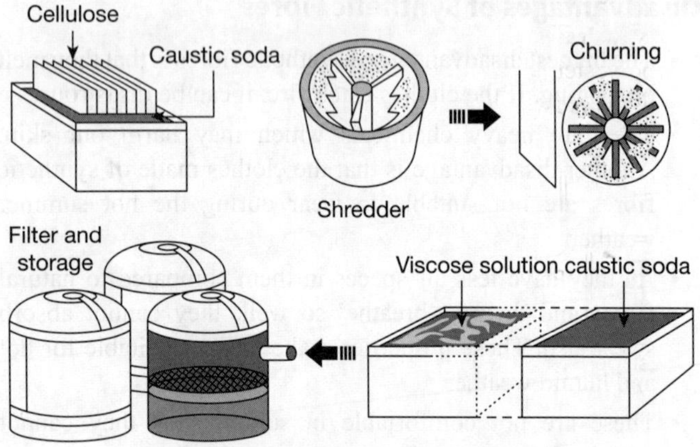

Properties of Rayon

- Rayon fibre is almost as comfortable as natural fibres (silk).
- Rayon fabrics are soft, smooth, cool and comfortable but they do not insulate body heat, making them ideal for use in hot and humid climates.
- Rayon can easily be dyed in a variety of colours.
- It can be blended with cotton or wool.
- Low warmth, good to drape, creases easily and needs ironing.
- The recommended care for rayon fibres is dry cleaning.

Uses of Rayon

(i) It is the most commonly used semi-synthetic fibre and is used in:
- textile industry for making textile (fabrics) and dress materials,
- the manufacture of tyre cords,
- the medical field for making bandages and surgical dressing.

(ii) Rayon's thin fibres have the ability to allow clothing to breathe more than other fabrics, which makes it a good choice for the design of sports wear and sundresses that are expected to provide coolness and comfort.

(iii) Rayon when mixed with wool is used to make durable carpets.

Nylon

It is the first completely synthetic fibre. It was first introduced by Wallace Carotters in 1935. It was made simultaneously in New York (NY) and London (LON) so it has been given the name NYLON. The chemical compounds used in making nylon were coal, water and air but now it can be obtained from petroleum products (petrochemicals). Actually, nylon is a thermoplastic polymer made up of repeating units of a chemical known as an amide. That is, nylon is a polyamide.

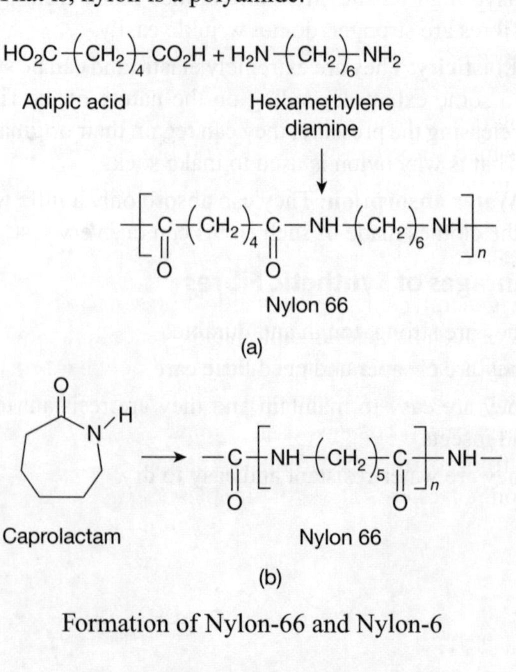

Formation of Nylon-66 and Nylon-6

Properties of Nylon

- Nylon is a very strong but light weight material having very good strength. Its strength is not lost with age.
- As it possesses good elasticity, it stretches but also regains its original shape easily.
- It is moth and alkali resistant.
- In the wet condition also, it is stable and strong.
- It is wrinkle resistant.
- It is inert to many chemicals and solvents.
- Nylon fabrics absorb very little water, so nylon clothes are very easy to wash and dry.
- Nylon fabrics have low resistance to sunlight.

Uses of Nylon

Nylon is used for making the following.

- Ropes used for mountaineering, tyre cords, parachute fibres.
- Seat belts, fishing nets, car seats, strings for sports rackets.
- Machine parts in the form of plastics.
- Carpets, textile fibres, bristles for brushes, track suits, combat uniforms, swim wear and stockings.
- Material in dress socks, swim wear, shorts, track pants, active wear, wind breakers, draperies and bedspreads.
- A substitute for metals in ball bearings and gears.

KNOWLEDGE BOOSTER

At high temperatures, nylon melts. Therefore, when ironing items made of nylon, we should set our iron at a low setting and turn the item over onto the wrong side.

Polyesters

It is a general name for synthetic fibres having many ester groups. Alcohols and organic acids react together to give compounds known as esters which get polymerised to form polyesters. Terylene or dacron is the most common example. The chemical compounds used in making polyester fibres are made from petroleum products (petrochemicals) and most common polyester is also a thermoplastic polymer like nylon.

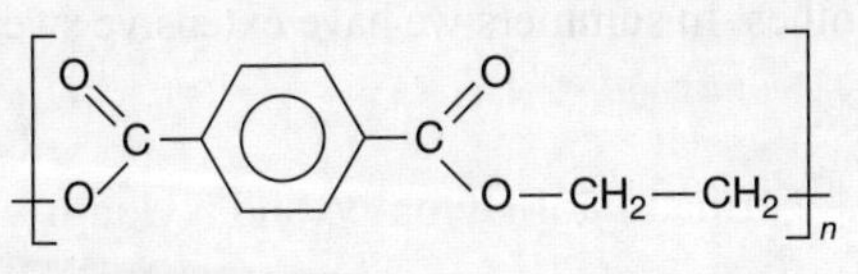

Polyester blended with cotton and wool are known as polycot and polywool respectively. Polycot contains 55% polyester and 45% cotton.

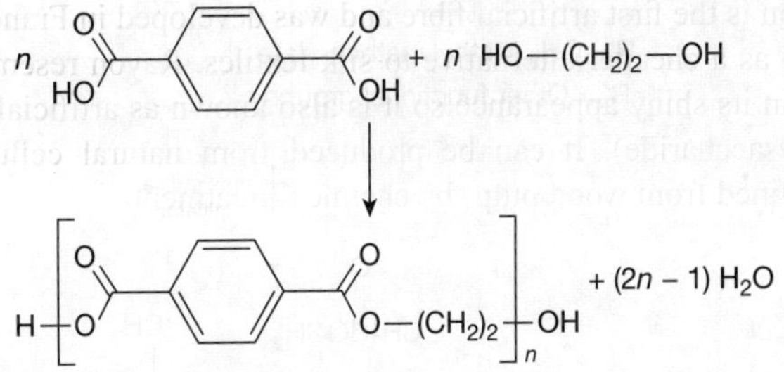

Properties of Polyester

- Polyester fabrics and fibres are extremely strong and tough.
- Polyester is very durable, resistant to most chemicals, wrinkle resistant, mildew and abrasion resistant.
- Polyester is hydrophobic in nature and quick drying. It can be used for insulation by manufacturing hollow fibres.
- Polyester retains its shape and hence is good for making outdoor clothing for harsh climates.

Uses of Polyester

- It is used for making sarees, shirts, trousers and other dress materials.
- It is used for making fire-hoses and conveyer belts.
- It is used for making sails for sailing boats.
- Due to its high strength and tenacity, polyester is used to make ropes in industries.
- Polyester is also used for making nets, thread, raincoats, jackets, clothing and medical textiles.
- Now PET bottles are one of the most popular uses of polyester. They are used in making soft drink bottles and utensils for packaging.

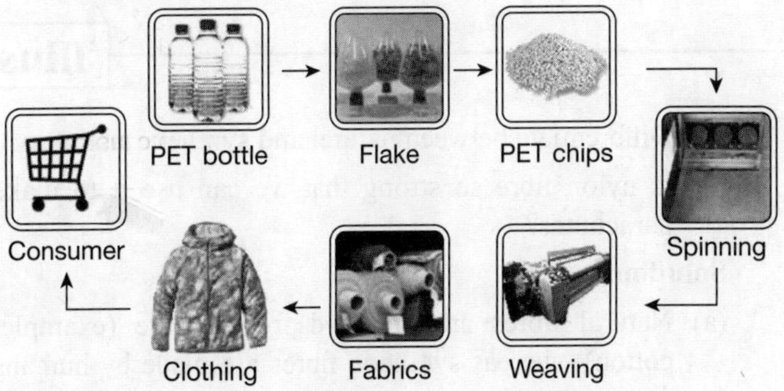

Acrylic

Acrylic is an example of a synthetic fibre that resembles natural wool and is much cheaper and durable than wool itself. It is made from a chemical known as acrylonitrile (vinyl cyanide) through polymerisation. It is light weight, soft and warm, having a wool-like feel, can be dyed easily, does not shrink or wrinkle and is also insect resistant. Acrylic is a commonly used clothing material. It is easy to maintain and is often used as an alternative to expensive woolen fabrics such as cashmere. However, it tends to form fuzz.

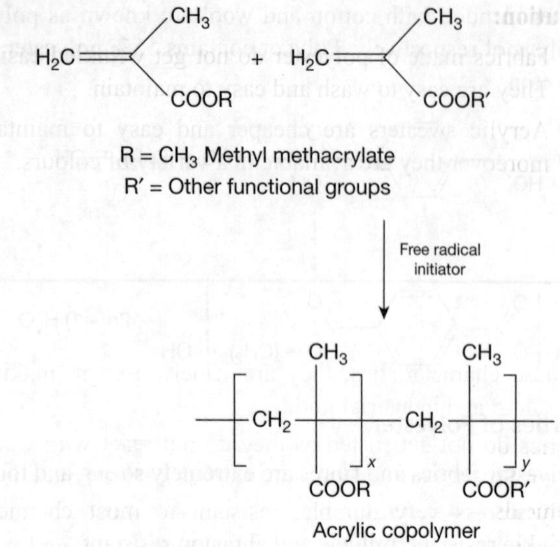

Properties of Acrylic

- It is light weight, soft and warm.
- It is more flexible and shock resistant than glass.
- It can be dyed easily and it does not shrink or wrinkle.
- It is abrasion resistant and resistant to UV and chemical damage.
- Acrylic can transmit or filter ultraviolet light and is easily cleaned.
- It can be easily cut, is corrosion resistant and a good insulator.
- It is resistant to moths and insects.

Uses

The main applications of acrylic include:

- In the manufacture of windows, aircraft canopies, sunscreens, automobile tail lights, hobby crafts, lighting fixtures, furniture, sign boards, table tops, windshields, decorating panels, aquariums, camera lenses, toys, incubators, appliances and security shields.
- It is used for making blankets, shawls and sweaters.
- It can be blended with wool to make material for suits and trousers.

KNOWLEDGE BOOSTER

Mineral fibres: The most common mineral fibres are glass and metal fibres. Fibre glass, made from a specific glass formula and optical fibres made from purified natural quartz are also fibres that come from natural raw material. Metallic fibres are drawn from ductile metals like copper, silver, gold. Carbon fibres are made up of pure carbon.

Illustrations

1. (a) Differentiate between natural and synthetic fibres.

   (b) Is nylon fibre so strong that we can use it to make parachutes?

   Solution:

   (a) Natural fibres are obtained from nature (example, cotton) whereas synthetic fibres are made by humans (example, nylon).

   (b) Yes, nylon is very strong and it can be used for making parachutes.

2. Why is it not advisable to burn plastic and synthetic fabrics?

   Solution: Burning of plastics and synthetic fabrics produces poisonous gases which cause severe air pollution. So, it is not advisable to burn plastic and synthetic fibres.

3. (a) Shashwat took with him some nylon ropes, when he was going for rock climbing. Can you tell why he selected nylon ropes instead of ropes made up of cotton or jute?

   (b) Shanvi wants to buy shirts for summer. Should she buy cotton shirts or shirts made from synthetic material? Advise Shanvi, giving your reason.

   Solution:

   (a) Nylon ropes are strong, plastic and lighter as compared to cotton and jute ropes. A nylon thread is actually stronger than a steel wire of similar thickness.

   (b) She should buy cotton shirts. This is because cotton has more capacity to hold moisture than synthetic clothes. In summers we have extensive sweating which is easily soaked by cotton shirts and hence, cotton clothes are much better than the clothes made from synthetic material.

4. (a) What are the advantages of using fabrics made of polyester?

(b) Why do people prefer to wear acrylic sweaters than pure wool sweaters?

Solution:

(a) Fabrics made of polyester do not get wrinkled easily. They are easy to wash and easy to maintain.

(b) Acrylic sweaters are cheaper and easy to maintain moreover they are available in a variety of colours.

Plastics

The first plastic was made by Alexander Parks in 1962. Plastics are polymers just like synthetic fibres. However, they can be moulded or shaped, usually by heat and pressure. Plastic is an organic material having high molecular weight and specific properties. A few plastics are linear and a few are cross-linked. Plastics of different types are the most widely used artificial materials. They are made from chemicals with petroleum origin (petrochemicals). However, a few may be partially natural. Some of the most commonly used plastics are polythene, polyvinyl chloride, bakelite, polystyrene and teflon.

Properties

- They are strong, tough, withstand high pressure and are stretchable.
- They are light and easy to handle. Due to their light weight and high tensile strength, they have a longer life. Because of these characteristics, they are widely used in making household and industrial goods.
- Plastics do not get rusted as they do not react with water and air. So, they are used to store various kinds of materials, chemicals and water.
- They can be moulded into any shape on heating, and thus can be reused.
- They are non-conductors of heat and electricity. As they are insulators, electric wires are covered with plastics. Handles of screw drivers and frying pans are made of plastics.
- They may be transparent, translucent or opaque.

Classification of Plastics

Plastics can be classified into these two types—thermoplastics and thermosetting plastics (Table 9.2). The applications of plastics are listed in Table 9.3.

Table 9.2 Differences between thermoplastics and thermosetting plastics

Thermoplastic	Thermosetting plastic
A plastic material that can be melted repeatedly by heating and can be moulded again and again into different shapes is known as a thermoplastic material. A thermoplastic material becomes soft on heating and hard on cooling. Such plastics can be moulded into toys, beads, buckets, telephone and television cases. For example, polythene, polyvinyl chloride (PVC) and polystyrene. 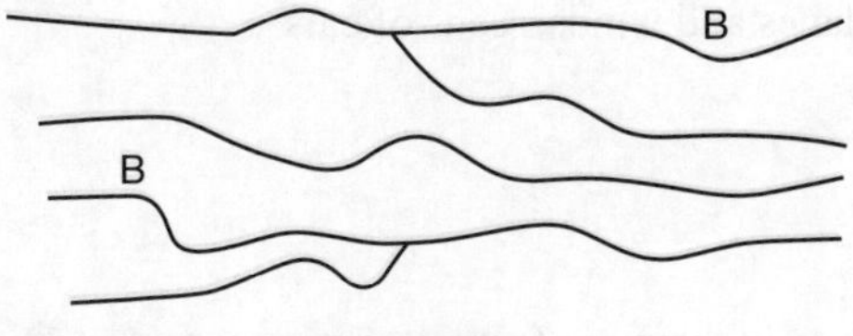 Thermoplastic B: Branch	A plastic substance, which once set, does not soften on heating is known as a thermosetting plastic. A thermosetting plastic can be used only once. It is used in a situation where resistance to heat is important. For example, bakelite, melamine and urea–formaldehyde resin. 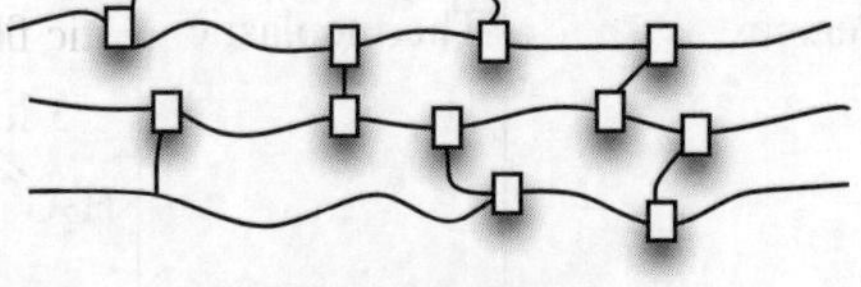 Thermosettings □: Crosslinking point
Effect of heat: Soften at 70–90°C.	Do not soften easily.
Strength: Comparatively weaker.	Comparatively stronger.
Reusability: Can be reused after melting.	Cannot be remelted or reused again.
Flexibility: They are flexible.	They are had and rigid.

(Continued)

Thermoplastic	Thermosetting plastic
Articles made of thermoplastics 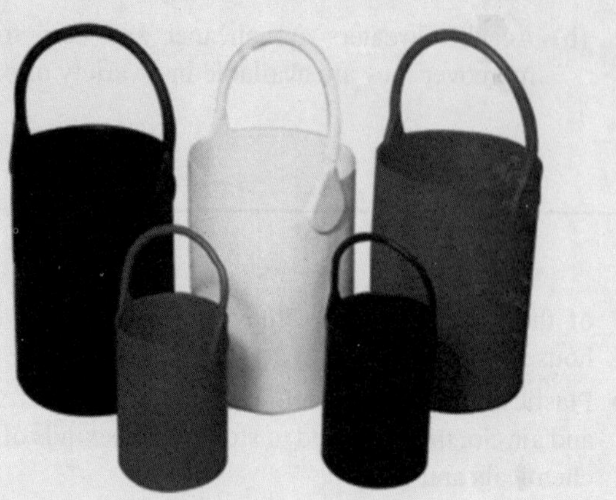	Articles made of thermosetting plastic 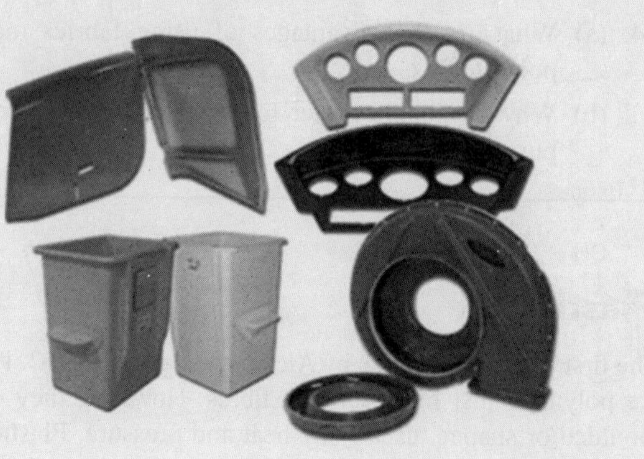

Table 9.3 Applications of some routine thermoplastic and thermosetting materials

S.No.	Name of plastic	Type of plastic	Monomer	Uses of plastic
1.	(i) LDPE (low density polyethene)	Thermoplastic	Ethene (CH_2=CH_2)	It is chemically inert, tough, flexible and a poor conductor of electricity. It is used in insulation of electricity carrying wires and manufacture of squeeze bottles, toys and flexible pipes.
	(ii) HDPE (high density polyethene)	Thermoplastic	Ethene (CH_2=CH_2)	It is comparatively more chemically inert, tougher and harder. It is used for making buckets, dustbins, bottles and pipes.
2.	Polytetrafluoro ethene (teflon)	Thermoplastic	Tetrafluoroethene (CF_2=CF_2)	It is used in making oil seals and gaskets and also used for non-stick surface coated utensils as it is chemically inert and resistant to attack by corrosive reagents.
3.	Polyacrylonitrile (PAN) or orlon	Thermoplastic	Vinyl cyanide (CH_2=CHCN)	It is used as a substitute for wool in making commercial fibres as orlon or acrilan.
4.	Polypropene	Thermoplastic	Propene (CH_3—CH=CH_2)	It is used in the manufacture of ropes, toys and pipes.
5.	Polystyrene	Thermoplastic	Vinyl benzene (C_6H_5—CH=CH_2)	It is used as an insulator, wrapping material in the manufacture of toys and television cabinets.
6.	Perspex	Thermoplastic	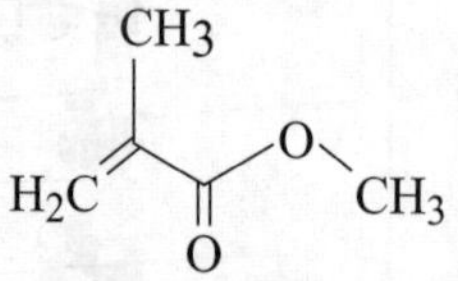 Methyl methacrylate	It is used as a substitute for glass for making windows in aeroplanes and windscreens of cars.
7.	Polyvinyl chloride (PVC)	Thermoplastic	Vinyl chloride (CH_2=CHCl)	It is used in the manufacture of rain coats, hand bags, vinyl flooring and water pipes.
8.	Bakelite	Thermosetting	Phenol (C_6H_5OH) + Formaldehyde (HCHO)	It is used in making combs, phonograph records, electrical switches and handles of utensils.
9.	Melamine	Thermosetting	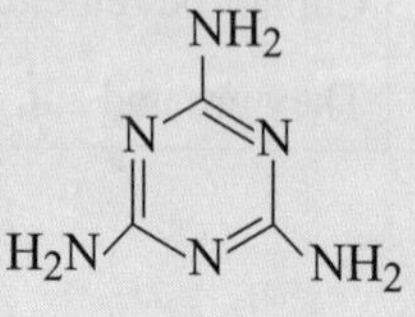 Melamine	It is used in the manufacture of unbreakable crockery.

KNOWLEDGE BOOSTER

Formation of bakelite

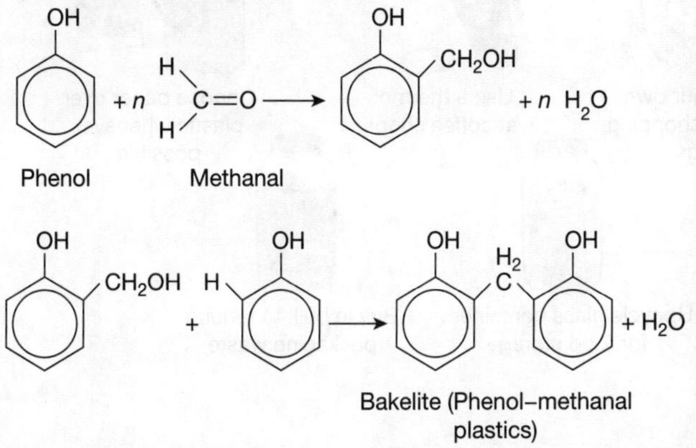

Plasticisers

Some plastics do not soften very much on heating. These can be easily softened by the addition of some organic compounds which are known as plasticisers. For example, polyvinyl chloride (PVC) is very stiff and hard but it is made soft by adding di-*n*-butyl phthalate (a plasticiser). Some common plasticisers are dialkyl phthalates, cresyl phthalate.

Uses of Plastics

The main uses of plastics are:

- For making water bottles and waterproof containers.
- For making plastic wrap materials, which are used to keep food fresh in refrigerators.
- As packing materials.
- For making table cloths, bathroom curtains, raincoats and waterproof shoes.
- In making of glass-reinforced plastic (GRP) or glass fibre-reinforced plastic (GFRP). It is a fibre reinforced polymer made of a plastic matrix reinforced by fine fibres made of glass.

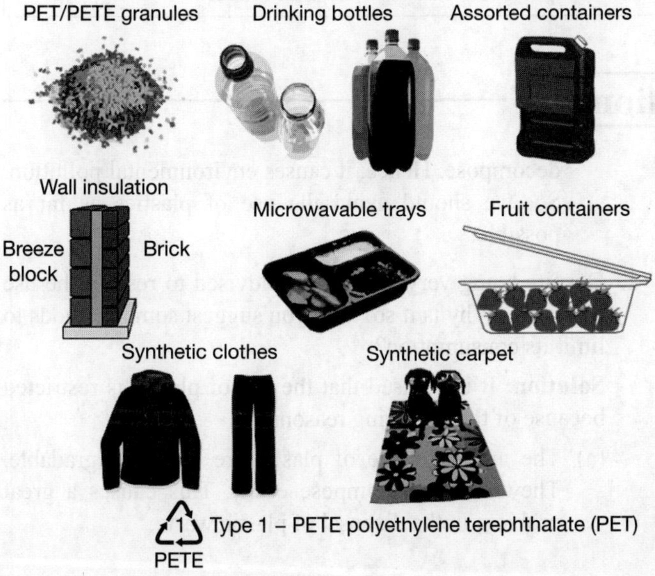

Disadvantages of Plastics

The main disadvantages of plastics are the following.

- A large part of our household waste is made of plastics. We throw away plastic bags in which we buy foodstuff, vegetables and other materials. The plastic bags used for packing are not biodegradable and take thousands of years to disintegrate.
- On burning, plastic gives off poisonous fumes. For example, PVC on burning gives fumes of hydrogen chloride.
- The poisonous gaseous products produced by the decomposition of plastic can cause cancer.
- They harm the environment by choking the drains.
- Plastic bags are a great threat to aquatic life.
- They are non-renewable resources.
- Recycling plastic is expensive.

Plastics and the Environment

The large-scale use of plastic is now a global threat for our environment. It is one of the major toxic pollutants. Plastic wastes are a major environmental problem as it is the cause of many health problems. Some side effects of using plastics are as follows.

- Plastic is a non-biodegradable substance composed of toxic chemicals. It pollutes the soil, air and water.
- Articles like polythene bags when mixed with garbage, choke the respiratory system of animals such as cows.
- Polythene bags cause blockage of drains and gutters.
- Plastic bags are a big threat to aquatic life as they are harmful for aquatic animals.
- On burning, plastics evolve poisonous fumes and the smoke is full of carbon particles.
- Plastic bags filled with rain water become breeding grounds for mosquitoes (cause of malaria, dengue).
- Dumping of plastic wastes on land makes land infertile.
- Plastic buried under the soil prevents the percolation rain water and the run-away water into the soil.

Measures to Reduce Plastic Pollution

- We should try to reduce plastic pollution by following the 4R principle.
- As a responsible citizen remember the 4R principle—Reduce, Reuse, Recycle and Recover.

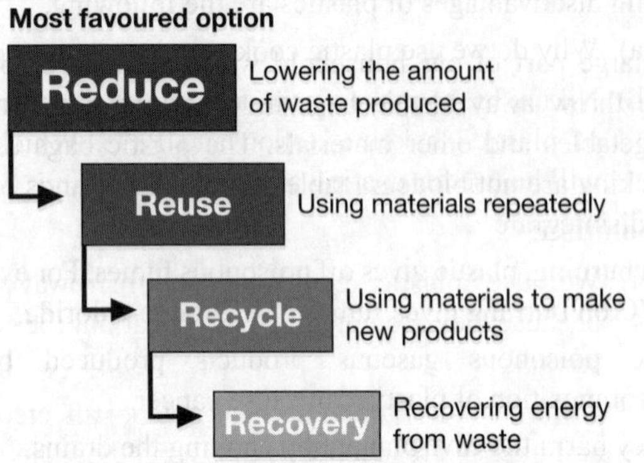

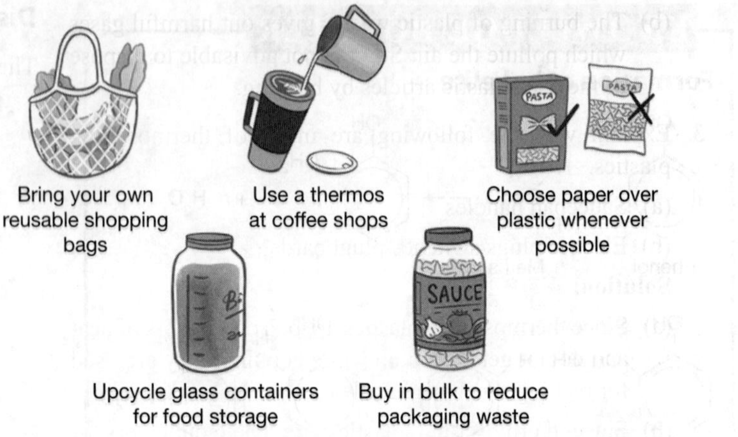

Biodegradable and Non-biodegradable Substances

Materials around us can be divided into two types based on the time taken by them for their decomposition or degeneration (Table 9.4).

Table 9.4 Differences between biodegradable and non-biodegradable substances

Biodegradable	Non-biodegradable
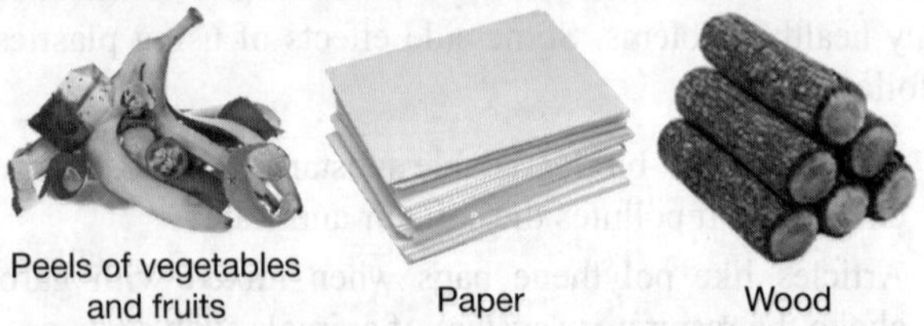 Peels of vegetables and fruits, Paper, Wood	 Plastic bag, Glass bottle, Can
A material which gets decomposed through natural processes such as action by microorganisms is known as a biodegradable material.	A material which is not easily decomposed by natural processes is known as a non-biodegradable material. It may take several years to get decomposed.
• The term biodegradable is used to name any material that is capable of undergoing biodegradation. • Biodegradable materials easily and quickly undergo biodegradation. • Rate of biodegradation is very high. • Examples include paper, clothes, fruit. • Natural rubber, nylon (from castor oil).	• The term non-biodegradable is used to name any material that cannot be degraded by biological means. • Non-biodegradable materials do not undergo biodegradation. Or, they take a very long time for the biodegradation to occur. • Rate of biodegradation is very low. • Examples include plastics, polyethylene and many other synthetic polymers like bakelite, melamine.

Illustrations

1. Explain why: (a) Plastic articles are available in all possible shapes and sizes.

(b) Plastic is used for making a large variety of articles of daily use and these articles are very attractive. But it is advised to avoid the use of plastic as far as possible.

Solution:

(a) Plastics can be easily moulded, so they are used to make a large variety of articles having different shapes and sizes.

(b) The disposal of plastic waste is a major problem as plastic is non-biodegradable. It takes several years to decompose. Hence, it causes environmental pollution. So, we should avoid the use of plastics as far as possible.

2. Despite being very useful, it is advised to restrict the use of plastic. Why is it so? Can you suggest some methods to limit its consumption?

Solution: It is advised that the use of plastic is restricted because of the following reasons:

(a) The articles made of plastic are non-biodegradable. They do not decompose easily. This causes a great problem in the disposal of plastic waste.

(b) The burning of plastic wastes gives out harmful gases which pollute the air. So, it is not advisable to dispose of the used plastic articles by burning.

3. Explain why the following are made of thermosetting plastics.
(a) Saucepan handles
(b) Electric plugs/switches/plugboards

Solution:

(a) Since thermosetting plastics are bad conductors of heat and do not get heated up while cooking, they are used for making saucepan handles.

(b) Since thermosetting plastics are bad conductors of electricity and electric current does not pass through such plastics, they are used for making electric plugs/switches/plugboards.

4. (a) Why do we use plastic cookware in microwave ovens?
(b) Nikunj has to make a non-stick pan. He has three types of plastic—bakelite, teflon and PVC. Which plastic will he use for coating and why?

Solution:

(a) We use plastic cookware in microwave ovens, because they are not affected by heat but the food gets cooked.

(b) Nikunj will use teflon because it is not affected by heat and does not react chemically with other substances.

CHAPTER AT A GLANCE

- A **polymer** is a large but single chain-like molecule in which the repeating units derived from small molecules (called monomers) are held together. The process by which monomers are transformed into a polymer is called **polymerisation**.
- There are many natural polymers, often called **biopolymers**, such as carbohydrates and proteins.
- If polymers are made of a single monomer, they are called **homopolymers**.
- If two or more types of monomers form a polymer, the polymer is called a **copolymer**.
- **Natural rubber** is a polymer consisting of repeated units of the hydrocarbon 2-methyl-1,3-butadiene (isoprene).
- **Starches** are polymers of glucose. **Cellulose** is also a polymer of glucose.
- **Addition** or **chain growth polymers** involves the repeated addition of monomers to the polymer chain.
- **Condensation** or **step growth polymers** involves a series of condensation reactions involving two monomers.
- **Linear polymers** have long straight chains of monomers. They have high tensile strength, high density, high melting point.
- **Branch chain polymers** have branches along the main chain of monomers. They have low melting points, low tensile strength, low density due to lack of close packing.
- **Three dimensional network** or **cross-linked polymers** have three dimensional cross-linked structures. They are hard, brittle and rigid.
- In **elastomers** the polymer chains are held together by the weakest intermolecular forces or van der Waals forces.
- **Fibres** are used for making polymers that possess high tensile strength and high modulus. This can be attributed to the strong intermolecular forces like hydrogen bonding.
- The intermolecular forces of attraction in **thermoplastic** polymers are intermediary to those of elastomers and fibres. As a result, these can be easily moulded by heating. In thermoplastic polymers there is no cross-linking between chains.
- **Thermosetting** polymers are made from relatively low molecular mass semi-fluid polymers which when heated in a mould become infusible and form an insoluble hard mass.
- Those fibres which are man-made and manufactured artificially from chemicals are known as **synthetic fibres**. The most commonly used synthetic fibres are rayon, nylon, polyester (terylene) and acrylic.
- **Synthetic fibres** have high strength, high elasticity and water absorption property.
- **Rayon** resembles silk in its shiny appearance, so it is also known as artificial silk. It can be produced from natural cellulose by chemical treatment and the cellulose is obtained from wood pulp.
- **Nylon** is a thermoplastic polymer made up of repeating units of a chemical known as an amide (nylon is a polyamide).
- Alcohols and organic acids react together to give compounds known as esters which get polymerised to form **polyesters**.
- **Polyester** blended with cotton and wool are known as polycot and polywool respectively. Polycot contains 55% polyesters and 45% wool.
- **Acrylic** is an example of a synthetic fibre that resembles natural wool and is much cheaper and durable than wool itself. It is made from a chemical known as acrylonitrile through polymerisation.
- **Plastic** is an organic material having high molecular weight and specific properties. Few plastics are linear and few are cross-linked.
- Some plastics do not soften very much on heating. These can be easily softened by the addition of some organic compounds which are known as **plasticisers**.
- We should try to reduce **plastic pollution** by following the **4R** principle, that is, Reduce, Recycle, Reuse and Recover.

- A material which gets decomposed through natural processes like action by microorganisms is known as a **biodegradable material**.
- A material which is not easily decomposed by natural processes is known as a **non-biodegradable material**. It may take several years to get decomposed.

PRACTICE QUESTIONS

Analyse Your Concepts (School Exam Based)

Fill in the Blanks

Instructions: Complete the following statements with an appropriate word/term to be filled in the blank spaces.

1. A polymer is a chain of many small units joined together which are known as _______.
2. Polymers having the same type of monomers are called _______.
3. Bakelite is an example of _______ polymer.
4. Synthetic fibres are also called _______ fibres.
5. _______ bags cause clogging of drains.
6. The first fully synthetic fibre is _______.
7. A fibre similar to wool is _______.
8. Rubber is an example of _______ type of polymer.
9. A plastic used for making crockery is _______.
10. Synthetic fibres are also called _______ or _______ fibres.
11. Synthetic fibres are synthesised from a raw material called _______.
12. Like synthetic fibres, plastic is also a _______.

True or False

Instructions: Read the following statements and write your answer as true or false.

1. The basic unit of a polymer is known as a monomer.
2. Nylon and polyester are examples of homopolymers.
3. The fabric terry wool is obtained by mixing terylene and wool.
4. Clothes of synthetic fibres are very comfortable in summer also.
5. Synthetic fibre does not melt on heating.
6. Thermoplastics become soft on heating, can be melted and reshaped.
7. It is risky to wear synthetic clothes while working in the kitchen.
8. Perspex is an example of a thermosetting polymer.
9. Most of the plastics are biodegradable.
10. The 4R principle can be used to reduce plastic pollution.
11. Di-*n*-butyl phthalate is an example of a plasticiser.
12. Glass and metal fibres are examples of mineral fibres.

Very Short Answer Type Questions

1. Write any two examples of natural polymers.
2. Write any two examples of copolymers.
3. Which type of structure is possible in bakelite?
4. Define elastomers.
5. Among elastomers, fibres, thermoplastics and thermosetting plastics, which has the weakest and strongest force?
6. Explain why some fibres are called synthetic.
7. Give examples which indicate that nylon fibres are very strong.
8. Name the plastic whose sheets are used for packing liquids.
9. Classify the following into thermosetting plastics and thermoplastics—combs, bakelite, melamine, toys.
10. Cotton is a natural polymer. What is its chemical name?
11. A synthetic fibre which looks like silk is obtained by chemical treatment of wood pulp. It is therefore known as artificial silk. What is its common name?
12. Terrycot is made by mixing two types of fibres. Write the names of the fibres.

Short Answer Type Questions

1. Differentiate between a polymer and a macromolecule with an example.
2. Mention some uses of plastic in healthcare and cookware.
3. What type of polymer is rubber? What is vulcanisation?
4. A bucket made of plastic does not rust like a bucket made of iron. Why?
5. Give examples which indicate that nylon fibres are very strong.
6. Explain why plastic containers are favoured for storing food.

7. Differentiate between biodegradable and non-biodegradable materials.
8. What are the disadvantages of wearing synthetic fabrics?
9. Explain the difference between thermoplastics and thermosetting plastics.
10. Should the handle and bristles of a toothbrush be made of the same material? Explain your answer.

Long Answer Type Questions

1. Define polymerisation and classify polymers on the basis of their structures.
2. (a) How is plastic useful in healthcare industry?
   (b) 'Avoid plastics as far as possible'. Comment on this advice.
3. (a) 'Even though plastics are very useful, they are not environment friendly.' Justify the statement.
   (b) Give some methods by which pollution due to plastics can be solved.
4. Write some advantages and disadvantages of synthetic fibres.
5. Write the monomers and uses of following polymers.
   (a) PVC (b) perspex (c) teflon (d) bakelite (e) nylon-66.
6. Describe an activity to show that thermoplastics are poor conductors of electricity.
7. Take two pieces of cloth of the same size, about half a metre square each. One of these should be from a natural fibre and the other could be a synthetic fibre. You can take the help of your friends in selecting these pieces. Soak the pieces in different mugs each containing the same amount of water. Take the pieces out of the containers after five minutes and spread them in the sun for a few minutes. Compare the volume of the water remaining in each container.

NCERT Corner

Textbook Exercises

1. Mark the correct answer.
   Rayon is different from synthetic fibres because
   - It has a silk-like appearance.
   - It is obtained from wood pulp.
   - Its fibres can also be woven like those of natural fibres.
2. Why is teflon used as tape for sealing purposes?
3. Name the first fully synthetic fibre.
4. Why is nylon used for making parachutes?
5. Name the material used for making ropes for rock climbing.
6. What is polyester?
7. Name the plastic used for making polythene bags.
8. What name is given to plastics which can be reset a number of times?
9. Why are the chemicals in the laboratory stored in plastic containers?
10. Can you store pickles in plastic containers? Why?
11. Why are the electric wires covered with plastic?
12. Tin takes about 100 years to degenerate. Is it biodegradable or non-biodegradable?
13. Is it advisable to use recycled plastic containers for storing food?
14. As a responsible citizen, what are the 4Rs we should remember?
15. Explain the difference between thermoplastics and thermosetting plastics.
16. How is rayon made? Give one advantage of using rayon.
17. Give some uses of PET.
18. Give the composition of polycot and polywool.
19. What properties of plastics make it useful?
20. Why is melamine used for making kitchenware?
21. Buckets made of plastics are better. Why?
22. Should the handle and bristles of a toothbrush be made of the same material? Explain your answer.
23. Explain why the following are made of thermosetting plastics:
    (a) saucepan handles
    (b) electric plugs/switches/plug boards.
24. Give examples to show that plastics are non-corrosive in nature.
25. Give three advantages of rayon.
26. What are the advantages of nylon?
27. Give some uses of nylon.
28. Give the characteristics of synthetic fibres.

Exemplar Problems

Short Answer Type Questions

1. Categorise the materials of the following products into 'can be recycled' and 'cannot be recycled'.
Telephone instruments, plastic toys, cooker handles, carry bags, ballpoint pens, plastic bowls, plastic covering on electrical wires, plastic chairs, electrical switches.

2. Give examples to show that plastics are non-corrosive in nature.

3. 'Avoid plastics as far as possible'. Comment on this advice.

4. 'Manufacturing synthetic fibres is actually helping the conservation of forests.' Comment.

5. Select the articles from the following list which are biodegradable.
(a) Paper (b) Woollen clothes
(c) Wood (d) Aluminium can
(e) Plastic bag (f) Peels of vegetables

6. A lady went to the market to buy a blanket. The shopkeeper showed her blankets made of acrylic fibres as well as those made of wool. She preferred to buy an acrylic blanket. Can you guess why?

7. PVC (polyvinyl chloride) is a thermoplastic and is used for making toys, chappals, and so on. Bakelite is a thermosetting plastic and is used for making electrical switches, handles of various utensils, and so on. Can you write the major difference between these two types of plastics?

8. Match items in Column A with the terms in Column B.

Column A	Column B
(a) Nylon	(i) Non-stick coating
(b) PET	(ii) Electric switches
(c) Rayon	(iii) Parachutes
(d) Thermosetting plastics	(iv) Polyester
(e) Teflon	(v) Artificial wool

9. Unscramble the jumbled words (of materials) given below, related to synthetic fibres.
(a) Anory (b) Lopmery
(c) Relyteen (d) Laspict
(e) Yespolter (f) Felton

Long Answer Type Questions

1. Write the importance of synthetic polymers in our life.

2. Give an activity which shows that synthetic fibres are stronger than cotton fibres.

3. Crossword puzzle: Complete the crossword given below with the help of the clues given.

Across

(1) Substance used as synthetic wool. (7)
(2) A plastic used for making containers and carry bags. (9)
(3) Substance made up of large number of smaller molecules. (7)
(4) Another name for this compound is artificial silk. (5)

Down

(5) A type of fibre obtained naturally from cocoon. (4)
(6) A synthetic fibre classified as polyester. (8)
(7) A polymer used for making ropes. (5)

COMPETITION WINDOW (OBJECTIVE TYPE)

[For NEET, JEE (Main and Advanced), NTSE, KVPY and Olympiads]

Topic-wise MCQs

Polymers

1. A naturally occurring polymer is
(a) cellulose (b) polyester
(c) nylon (d) PVC

2. Which is not a polymer?
(a) sucrose (b) enzyme
(c) starch (d) teflon

3. On the basis of mode of formation, polymers can be classified as
(a) addition polymers only
(b) condensation polymers only
(c) copolymers
(d) both addition and condensation polymers

4. Which of the following is not an example of an addition polymer?
(a) polystyrene (b) nylon
(c) PVC (d) propylene

5. Which of the following is a copolymer?
(a) nylon-66 (b) polyethylene
(c) polyvinyl chloride (d) natural rubber

6. Rubber is a polymer of
(a) pyrene (b) ethylene
(c) isoprene (d) urea

7. Which pair of species given below produce bakelite?
(a) phenol, methanol (b) phenol, NaOH
(c) phenol, urea (d) phenol, formaldehyde

8. Which of the following polymers do not involve cross-linkages?
(a) melmac (b) bakelite
(c) polythene (d) vulcanised rubber

9. Which of the following is a biodegradable polymer?
(a) cellulose (b) nylon-6
(c) polythene (d) polyvinyl chloride

10. Which of the following is a fully fluorinated polymer?
(a) neoprene (b) teflon
(c) thiokol (d) PVC

Synthetic fibres

11. Synthetic fibres are obtained from
(a) petroleum (b) plants
(c) animals (d) all of these

12. The first artificial fibre is
(a) rayon (b) cotton
(c) nylon (d) polyester

13. Pick the synthetic fibre out of the following.
(a) cotton (b) nylon
(c) jute (d) wool

14. Which of the following is a source of rayon?
(a) wool (b) PET
(c) wood pulp (d) silk

15. Which of the following synthetic fibres is a good absorbent?
(a) polyester (b) acrylic
(c) nylon (d) rayon

16. Which of the following are synthetic fibres?
(a) wool, silk and cotton
(b) jute, ramie and hemp
(c) orlon, polypropylene and nylon
(d) all of these

17. A synthetic fibre obtained from renewable resources is
(a) rayon (b) nylon
(c) polyester (d) acrylic

18. The major raw materials for rayon are obtained from
(a) petroleum (b) wood
(c) coal (d) both coal and petroleum

19. Terrywool is obtained by mixing
(a) polyester and cotton (b) cotton and silk
(c) cotton and wool (d) polyester and wool

20. Which of these fabrics will you prefer on a hot and humid day?
(a) wool (b) nylon
(c) silk (d) cotton

21. Synthetic fibres are
(a) nylon, polyester and jute
(b) rayon, nylon and polyester
(c) rayon and cotton
(d) none of these

22. Polycot is obtained by mixing
(a) nylon and wool (b) polyester and wool
(c) nylon and cotton (d) polyester and cotton

23. The material similar to silk in appearance is
(a) nylon (b) rayon
(c) polyester (d) terylene

24. Clothes made of which fabric are best suited for a hot climate?
(a) cotton (b) nylon
(c) acrylic (d) polycot

25. Synthetic fibres are a result of extensive research on
(a) semi-synthetic fibres (b) optical fibres
(c) natural fibres (d) glass fibres

26. Which of the following is a semi-synthetic fibre?
(a) cellulose acetate (b) acrylic
(c) nylon (d) polyester

27. Which of the following synthetic fibres has most applications?
(a) acrylic (b) polyolefin
(c) nylon (d) polyester

Plastic

28. Plastics are
(a) polymers (b) solvents
(c) salts (d) acids

29. Plastics
(a) can be coloured and moulded into any desired shape
(b) can be spun into fibres to make cloth and carpets
(c) are light
(d) all of these

30. Which of the following is a thermosetting plastic?
(a) polypropylene (b) bakelite
(c) polyvinyl chloride (d) polyester

31. The plastics which do not remould again on heating are called
(a) polyester plastics
(b) polymer plastics
(c) thermosetting plastics
(d) thermoplastics

32. Which is a thermosetting plastic?
(a) melamine (b) polythene
(c) PVC (d) nylon

33. The most suitable material for the preparation of handles of cooking utensils is
(a) polythene (b) PVC
(c) nylon (d) bakelite

34. The material which is commonly used for making kitchen containers is
(a) PVC (b) acrylic
(c) teflon (d) PET

35. Electrical switches are made of
(a) nylon (b) bakelite
(c) polythene (d) melamine

36. Plastic used for coating non-stick pans is
(a) PVC (b) ester
(c) bakelite (d) teflon

37. Pickles are kept in plastic containers because plastic containers are
(a) non-corrosive (b) light
(c) colourful (d) cheap

38. Which of the following is a thermoplastic?
(a) bakelite (b) melamine
(c) polythene (d) jute

39. Which of the following can be recycled?
(a) plastic bowls
(b) ballpoint pens
(c) telephone instruments
(d) electrical switches

40. Which material is best suited for covering electric wires?
(a) plastic bowls (b) PVC
(c) polystyrene (d) nylon

Miscellaneous

1. Why is polyester not suitable for summer wear?
(a) it is very soft and thin
(b) it does not dry easily
(c) it crumples easily
(d) it does not absorb sweat

2. Which of the following is correct about blended fabrics?
(a) they are made by blending synthetic fibres with natural fibres
(b) they have combined properties of each fibre
(c) they are less expensive
(d) all of the above

3. The polymer used in making non-stick cookware is
(a) teflon (b) melamine
(c) rayon (d) nylon

4. Styrofoam cups are used to serve hot tea or coffee because
(a) they are very flexible
(b) they are easy to clean
(c) they are good conductors of heat
(d) they are poor conductors of heat

5. Toothpaste tubes are made of plastic because
(a) it is flexible and light
(b) it can be coloured in different colours
(c) it can hold more amount of toothpaste
(d) it is biodegradable

6. Which of the following substances can be used to make the insulation of wires?
(a) wood (b) glass
(c) plastic (d) rubber

7. The ability of a material to blend without breaking and return to its original shape is
(a) flexibility (b) conductivity
(c) tensile strength (d) hardness

8. The strength of a material to withstand a weight without breaking is
(a) hardness (b) metallic strength
(c) tensile strength (d) flexibility

9. Mark the correct statement
(a) A flexible object will never change its shape
(b) Strength of an object refers to its ability to support a load without breaking
(c) Metals do not conduct electricity
(d) Bakelite is good conductor of heat

10. The plastic which cannot be recycled is
(a) petrochemicals (b) elastomer
(c) jute (d) bakelite

11. What are the physical properties associated with plastic materials?
(a) do not corrode, low density and shiny
(b) poor conductors of heat, easily moulded and strong
(c) strong, transparent, high melting point
(d) brittle, flexible and elastic

12. Cotton on burning gives the smell of burning paper because
(a) paper is also obtained from plants
(b) cotton is obtained from plants
(c) paper is obtained from cotton plant only
(d) both cotton and paper are obtained from plants

13. When we burn wool why do we get the smell of burning of hair?
(a) wool is obtained from jute plants
(b) wool is obtained from hair of sheep
(c) wool is obtained from nylon
(d) wool is obtained from cotton plants

14. Four 'R's in the 4R principle are reduce, reuse,
(a) regain and retain (b) reform and regain
(c) recycle and recover (d) recycle and retain

15. What are the qualities of synthetic fibres which make them more popular than natural fibres?
(a) they are more durable than natural fibres
(b) they are cheaper than natural fibres
(c) both (a) and (b)
(d) none of these

16. Which is a biodegradable substance?
(a) PVC pipes
(b) thermoplastics
(c) thermosetting plastics
(d) cotton cloth

17. Non-stick teflon coated cookware are generally recommended to heart patients because
(a) non-stick cookware does not absorb oil
(b) non-stick cookware does not absorb water
(c) non-stick cookware is biodegradable
(d) non-stick cookware enables faster cooking

18. To which of the following does the 4R principle apply the most?
(a) natural fibres (b) plastics
(c) synthetic fibres (d) wool

19. Which is the strongest fibre amongst the following?
(a) acrylic (b) cotton
(c) rayon (d) nylon

20. To make PVC, a flexible plastic, the additive used is called
(a) antioxidant (b) plasticiser
(c) filler (d) stabiliser

Advanced and Olympiads

Single Choice

1. Which of the following groups contains all synthetic substances?
(a) nylon, terylene, wool
(b) cotton, polycot, rayon
(c) PVC, polythene, bakelite
(d) acrylic, silk, wool

2. Plastics are bad conductors of heat and electricity, which of the following prove it?
(i) Electrical wires have plastic covering.
(ii) Refrigerators and freezers have a plastic core.
(iii) Cooking pans have plastic handles.
(iv) Plastics are used to make parachutes.
(a) (i) and (iii) (b) (i), (ii) and (iv)
(c) (i) and (ii) (d) (i), (ii) and (iii)

3. Select the set having correct statements.
(i) Plastics can be moulded into different shapes and sizes and can be reused.
(ii) Metals are cheaper than plastics.
(iii) Plastics are used in industries and for making household articles as they are very light, strong and durable.
(iv) Electric wires are covered with plastics as they are good conductors of heat and electricity.
(v) Plastics are used to store many chemicals as they cannot be corroded easily.
(a) (i), (iii) and (v) (b) (i), (ii), (iii), (iv) and (v)
(c) (i), (ii) and (iv) (d) (iii), (iv) and (v)

4. Why is it not advisable to wear synthetic clothes while bursting fire crackers?
(i) All synthetic fibres are prepared by using raw materials of petroleum origin.
(ii) Synthetic fibres catch fire easily.
(iii) On heating, synthetic fibres melt and stick to the body of the person wearing it.
(a) (i) and (iii) (b) (ii) only
(c) (i) only (d) (ii) and (iii)

5. Which of the following represents the correct match for items in Column I with those in Column II.

Column I	Column II
(a) Nylon	(i) Thermoplastic
(b) PVC	(ii) Thermosetting plastic
(c) Bakelite	(iii) Fibre

(a) (a)—(ii), (b)—(iii), (c)—(i)
(b) (a)—(iii), (b)—(i), (c)—(ii)
(c) (a)—(ii), (b)—(i), (c)—(iii)
(d) (a)—(iii), (b)—(ii), (c)—(i)

6. Choices for the correct combination of elements from Column I and Column II are given as options (a), (b), (c) and (d) out of which one is correct.

Column I	Column II
(a) Bakelite	(p) Used to make water tanks
(b) Nylon	(q) Used to make ropes, mats
(c) Coir	(r) Used to make radio, telephone set
(d) PVC	(s) Used to make fishing net

(a) (a)—(r), (b)—(s), (c)—(q), (d)—(p)
(b) (a)—(s), (b)—(r), (c)—(q), (d)—(p)
(c) (a)—(q), (b)—(s), (c)—(p), (d)—(r)
(d) (a)—(q), (b)—(r), (c)—(s), (d)—(p)

7. Which of the following properties of nylon make it most suitable for making toothbrush bristles?
(i) It is strong when wet.
(ii) It absorbs little water.
(iii) It is clear, transparent and heavy.
(iv) It is not acted upon by bacteria and fungus.
(a) (i) and (iii) (b) (i) and (iv)
(c) (i), (ii) and (iv) (d) (i), (ii) and (iii)

8. Which one of the following properties should we consider when choosing a material for food containers?
(i) can be converted into various shapes
(ii) able to resist high temperature also
(iii) a good insulator of heat
(iv) non-corrosive
(a) (i), (ii), (iii) and (iv) (b) (i), (ii) and (iv)
(c) (ii) and (iii) (d) (i) and (iv)

Multiple Choice

9. Which of the following are animal fibres?
(a) jute (b) silk
(c) wool (d) cotton

10. Which of the following are incorrectly matched?

Material	Natural source	Artificial source
(a) Thermosetting plastics	Yes	No
(b) Jute	Yes	No
(c) Cellulose	No	Yes
(d) Polystyrene	Yes	No

11. Out of the following, which is/are natural fibres?
(a) cotton (b) silk
(c) jute (d) rayon

12. Select the correct statement about plastics.
(a) Plastics are polymers having long chains
(b) Plastics are made by the process of polymerisation
(c) Plastics are organic materials of high molecular weight
(d) Plastic is cheaper

13. Which of the following are thermoplastic polymers?
(a) melamine (b) polythene
(c) nylon (d) PVC

14. Which of the following is/are common property(properties) of plastics?
(a) non-reactive
(b) light in weight
(c) durable
(d) good conductor of electricity

15. Out of the following, which is/are biodegradable?
(a) vegetable peels
(b) plastic bags
(c) cotton
(d) jute

16. For which polymers are the monomers are correctly given?

Polymer		Monomer
(a) Nylon-6	—	Caprolactam
(b) Bakelite	—	Phenol, formaldehyde
(c) Nylon-66	—	Adipic acid, butan-1,4-di-amine
(d) Teflon	—	Tetrafluoro ethene

Comprehension Type

Comprehension I: A polymer is a large molecule which is formed by addition or condensation reaction between two or more small molecules (monomers). Polymers may be natural, semi-synthetic or synthetic. They may be linear, branched and cross-linked, based on their structures. They can be further classified as elastomers, fibres, thermoplastics and thermosetting polymers on the basis of nature of forces present between the monomers.

17. Which of the following is correctly matched?
(a) Silk, wool, cotton — natural polymers
(b) Nylon-66, polyester, rayon — Synthetic polymers
(c) Nylon-66, polyester, rayon — Fibres
(d) All are correctly matched

18. In which polymer is the monomer not correctly matched?
(a) Nylon-66 — Adipic acid, hexamethylene diamine
(b) Teflon — $CF_2{=}CF_2$
(c) Orlon — $CH_2{=}CHCl$
(d) Polystyrene — $C_6H_5{-}CH{=}CH_2$

19. Which set has only addition polymers?
(a) polythene, polyester (b) nylon-66, polyester
(c) polystyrene, polythene (d) nylon-6, polystyrene

Comprehension II: Tensile strength of a fibre determines its strength. It is the maximum amount of stress that it can be subjected to before the material ruptures or tears away. Materials from fibres are made in various strengths and degrees of durability. The tensile strength of a polymer depends not only on the arrangement of molecules that make up the polymer but also on the orientation of the polymer. Polymers arrange themselves as long chains and when pulled lengthwise, these chains can be stretched to a great length before breaking.

20. Which of the following fibres has minimum tensile strength?
(a) wool (b) silk
(c) cotton (d) polyester

21. Parachutes and the ropes for rock climbing are made up of nylon and not of steel as
(a) steel is more flexible than nylon
(b) nylon is stronger than steel
(c) nylon is cheaper than steel
(d) nylon is biodegradable while steel is not

22. Why is nylon the strongest fibre?
(a) Nylon fibres are very thin and light-weight; so are strong and elastic.
(b) Nylon is wrinkle free; so, the fibre is very strong.
(c) Nylon fibres resemble silk in structure; so, they are very strong and elastic.
(d) The molecules in each filament fall into parallel lines, are cross-linked by hydrogen bonds, giving the nylon fibre both strength and elasticity.

Matrix Matching

23. Match the following

Column I	Column II
(a) A fibre obtained by chemical treatment of wood pulp	(p) Melamine
(b) A fibre stronger than steel wire	(q) Acrylic
(c) A fibre known as artificial wool	(r) Rayon
(d) A fire-proof plastic	(s) Nylon

24. Match the following

Column I	Column II
(a) Polyester	(p) Prepared by using wood pulp
(b) Teflon	(q) Used for making parachutes and stockings
(c) Rayon	(r) Used to make non-stick cookware
(d) Nylon	(s) Fabrics do not wrinkle easily

Integer Type

25. The number of thermoplastics out of the following is: Polythene, polypropylene, bakelite, melamine, teflon, PVC, orion

26. How many species are biodegradable materials in the list? Leaves, fruits, nylon, polythene, cotton clothes, coir, jute, polycot, paper, PVC

27. How many of these are condensation polymers? Bakelite, nylon-66, polyester, rayon, polythene, polystyrene, melamine

28. In nylon-66, both the monomers have chains of how many carbon atoms in them?

29. In nylon-6 the monomer has a ring structure of how many atoms?

30. What is the number of nitrogen atoms present in melamine?

Answer Keys

Fill in the Blanks

1. monomer	2. homopolymer
3. thermosetting	4. man-made
5. polythene	6. nylon
7. acrylic	8. elastomer
9. melamine	10. man-made, artificial fibres
11. petrochemicals	12. polymers

True or False

1. T	2. F	3. T	4. F	5. F
6. T	7. T	8. F	9. F	10. T
11. T	12. T			

Very Short Answer Type Questions

1. Silk and cellulose are two examples of natural polymers.
2. Nylon and polyester are examples of copolymers.
3. Bakelite has a cross-linked three dimensional network structure.
4. The polymer having elastic character is known as an elastomer. For example, rubber.
5. Elastomers have the weakest force while thermosetting plastics have the strongest force of attraction.
6. Some fibres are called synthetic because they are made by human beings.
7. Parachutes and nylon ropes are used for rock climbing.
8. Polythene sheets are used for packing liquids.
9. Thermosetting plastics: bakelite, melamine
Thermoplastics: combs, toys
10. Chemical name of natural polymer of cotton is cellulose.
11. Its common name is rayon. Rayon is often regarded as artificial silk.
12. Terrycot is made by mixing terylene and cotton.

Short Answer Type Questions

1. A polymer is made of many small units (monomers) to give a large molecule. For example, polythene is made of a large number of ethene molecules.
A macromolecule is a large single molecule. For example, hemoglobin.
2. Healthcare: Plastic is used in packing of tablets and medicines, making of threads used in stitching wounds, making gloves.
Cookware: Plastic is used in making microwave oven cookware, non-stick cookware and handles of sauce pans.
3. Rubber is a type of elastomer polymer. For example, natural rubber is a polymer of 2-methyl-buta-1,3-diene. When natural rubber is heated with 3%–5% of sulfur, it is called vulcanisation. It makes rubber more elastic and non-sticky. It also gives it more tensile strength.
4. Plastic is a non-reactive material. When left exposed to moisture and air, it does not react with air and water (which are essential for corrosion), so it does not rust.
5. The following examples indicate that nylon fibres are very strong.
   - They are used for making parachutes and ropes for rock climbing.
   - They are used in making seat-belts, fishing nets, tyre cords, string for sports rackets and musical instruments.
6. Food can be stored in plastic containers because
   - plastics are non-reactive,
   - plastics are light and strong, and
   - plastics are durable.

7.

Biodegradable materials	Non-biodegradable materials
(i) A material which gets decomposed through natural processes, such as the action of bacteria. (ii) Examples, cotton, jute.	(i) A material which does not get decomposed by natural processes. (ii) Example, plastics.

8. Synthetic fabric catches fire very easily. It melts and sticks to the body of the person wearing it. During summers, synthetic fibres do not absorb sweat and a person wearing it feels uncomfortable.

9.

Thermoplastics	Thermosetting plastics
(i) These plastics soften on heating and can be bent easily.	(i) These plastics when moulded once, can't be softened again.
(ii) They do not lose their plasticity.	(ii) They lose their plasticity.
(iii) Examples are polyethene, PVC.	(iii) Examples are bakelite, melamine.

10. No, the handle and bristles of a toothbrush should not be made of the same material. This is because our gums are soft and the bristles should be made of soft material (nylon) so that it does not harm the gums. On the other hand, the handles should be made of hard material (plastic) so that it can give a firm grip.

Long Answer Type Questions

1. See text part.
2. (a) In healthcare, plastics are used
   - for packaging of tablets,
   - to make threads for stitching wounds,
   - in syringes,
   - surgical gloves, and
   - for making a number of medical instruments.

   (b) Plastics are very useful, but they cause serious environmental and health concerns.
   - Plastics are non-biodegradable.
   - Careless disposal of plastic bags, chokes drains and blocks the soil.
   - If eaten by cows, it can kill them.
   - Plastic bags can also contaminate foodstuffs due to poisonous dyes getting absorbed into food
3. (a) Plastics are very useful, but their disposal is a big problem. Plastics are non-biodegradable and take many years to decompose. They release a lot of poisonous fumes when burnt, so they cause environmental pollution.

   (b) Pollution due to plastics can be solved by the following methods:
   - avoid the use of plastics as far as possible,
   - use bags made of jute or cloth,
   - biodegradable and non-biodegradable wastes should be separated and disposed of separately, and
   - plastics should be recycled and reused.
4. See text part.
5. See text part.
6. In order to show that thermoplastics are poor conductors of electricity, you have to make a circuit using a bulb, wires, cell, clips, and so on. Now arrange a circuit as shown in the given figure. Leave a gap between two ends of the wire. Place a thermoplastic in the gap. Observe the bulb.
   It is observed that on switching on the current, the bulb does not glow. This shows that thermoplastic is a poor conductor of electricity.

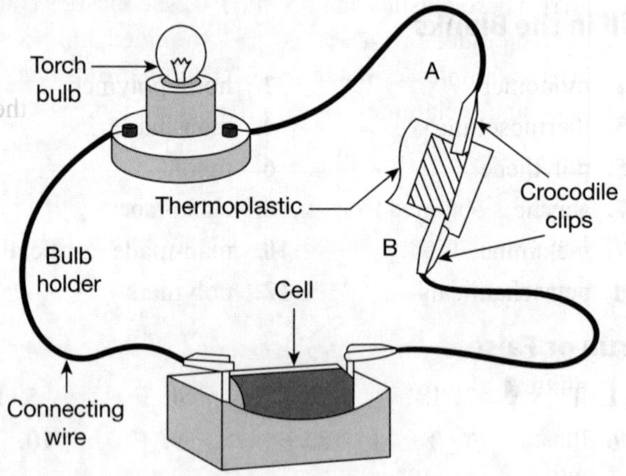

An electric circuit to test materials for their electric conductance

7. It is observed that the volume of water of the container in which natural fibre is soaked contains less water compared to the container in which synthetic fibre is soaked. Thus, a natural fibre absorbs more water as compared to a synthetic fibre. When both the fibres were spread out in the sun, it was observed that the synthetic fibre took less time to dry than the natural fibre.

NCERT Corner

Textbook Exercises

1. It is obtained from wood pulp.
2. Because of its toughness, teflon is used as a tape for sealing purposes.
3. Nylon.
4. Nylon is used for making parachutes, because it is very strong, elastic and light.
5. Nylon.

6. Polyester is made up of the repeating units of an ester.
7. Polythene bags are made of plastic which is a polymer of ethene.
8. Thermoplastics.
9. Plastics are non-reactive, so they can be used for storing chemicals.
10. Plastics are non-reactive, so pickles can be stored in them.
11. Plastic is a poor conductor of electricity; therefore, the wires are covered with it.
12. Tin is non-biodegradable.
13. No, we should not use recycled plastic containers for storing food because they contain colouring agents.
14. Reduce, Reuse, Recycle and Recover.
15.

Thermosetting plastics	Thermoplastics
(i) These plastics can be moulded. (ii) Examples, bakelite and melamine.	(i) These plastics can be moulded and reset a number of times. (ii) Examples, polythene and PVC.

16. Rayon is obtained from wood pulp by treating it chemically. It is cheaper than silk, but can be woven like silk fibre.
17. PET is used for making bottles, utensils, fibres and wires.
18. Polycot is a mixture of polyester and cotton; polywool is a mixture of polyester and wool.
19. Plastics can be recycled, reused, coloured, melted, rolled into sheets or made into wires, so they are very useful.
20. Melamine resists fire and can tolerate heat better than other plastics. So, it is used for making kitchenware.
21. Buckets made of plastic are better because they are light, strong and durable.
22. No, the handle and bristles of toothbrush cannot be made of the same plastic. The handle is hard but the bristles are soft and flexible.
23. (i) Saucepan handles are made of thermosetting plastics because they are bad conductors of heat.
    (ii) Electric plugs/switches/plug boards are made of thermosetting plastics because they are bad conductors of electricity.
24. (i) Buckets made of iron get rusted but those made of plastic do not rust.
    (ii) Food can be stored in plastic containers, as they are resistant to chemicals in the food.
    (iii) Teflon coated pans are used instead of metallic pans as teflon does not react with chemicals in the food.
25. Advantages of rayon are:
    (i) it is cheap,
    (ii) it can be dyed in many colours, and
    (iii) it can be woven like silk.
26. Nylon fibre is strong, elastic and light. It is lustrous and easy to wash.
27. Nylon can be used for making socks, ropes, tents, toothbrushes, sleeping bags, curtains, parachutes and ropes for rock climbing.
28. Synthetic fibres dry up soon, are durable, less expensive, readily available and easy to maintain.

Exemplar Problems

Short Answer Type Questions

1.

Can be recycled	Cannot be recycled
Plastic toys, carry bags, plastic bowls, plastic covering on electrical wires, plastic chairs	Telephone instruments, cooker handles, ballpoint pens, electrical switches

2. The literal meaning of non-corrosive is, resistant to destruction by chemical action. The following are the examples that show that plastics are non-corrosive in nature.
   - Plastic containers do not react with items stored in it.
   - They do not get rusted when exposed to moisture and air.
   - They do not decompose when left in the open for a long period.
3. Plastics must be avoided as far as possible. The materials made of plastics are non-biodegradable. The use of plastics has a bad effect on the environment. When plastics are burnt, it releases a lot of poisonous fumes into the atmosphere causing air pollution. These plastic materials when eaten by animals (such as cows) can cause their death. The waste plastic thrown carelessly gets into water drains and sewers, and blocks them. In a nutshell, plastics can be considered a threat to our environment.
4. In the manufacturing of synthetic fibres, we use only chemical substances and no natural materials, thus, in turn, we conserve forests.
5. The biodegradable articles are:
   (a) paper, (b) woollen clothes, (c) wood, (f) peels of vegetables.
6. Acrylic blankets are cheap, light in weight, more durable and are available in a variety of colours and designs. They can be easily washed at home.
7. Thermoplastics get deformed easily on heating and can be bent easily on heating. On the other hand, thermosetting

plastics when moulded once cannot be softened on heating.
Thermoplastics → Polythene and PVC
Thermosetting plastics → Bakelite and melamine

8. The correct matching is as given:
(a)—(iii), (b)—(iv), (c)—(v), (d)—(ii), (e)—(i)

9. (a) Rayon (b) Polymer (c) Terylene (d) Plastic (e) Polyester (f) Teflon

Long Answer Type Questions

1. The uses of synthetic polymers in our life are given below.
   - Nylon: It is used for making ropes for rock climbing, fishing nets, raincoats, parachutes and tyre cords.
   - Acrylic: It is used for sweaters, track suits, linings for boots and gloves and in furnishing fabrics and carpets.
   - Terylene: It is used in the textile industry to make clothes like sarees, tapestry and dress material. It is also mixed with natural fibre like cotton and wool to make more variety of clothes.
   - Plastic: It is used to store food items, water, milk, pickles, dry food, and so on. Plastic containers seem most convenient because they are light weight, low priced, have good strength and easy handling as compared to metals. Plastics are used in cars, aircrafts and spacecrafts.

2. Activity: Take threads of cotton and nylon (synthetic fibres) of about same length and thickness. Hang the cotton thread from a stand. Attach a pan at the other end of the thread. Add weights, one by one, to the pan until the thread breaks (see figure below) and note the total weight required to break the cotton thread. This gives a measure of the tensile strength of the thread. Repeat the same with the nylon thread.

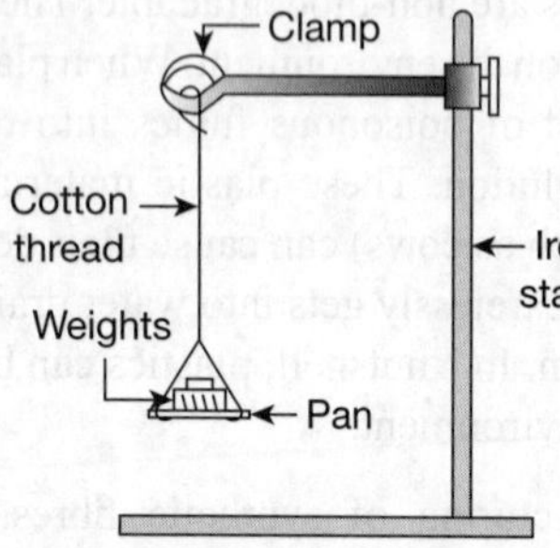

Arrangement to determine the relative strengths of natural fibres and synthetic fibres

Conclusion: Much more weight is required to break the nylon thread showing that it has a greater strength that the cotton thread.

3. **Across**

1. Acrylic	2. Polythene
3. Polymer	4. Rayon

Down

5. Silk	6. Terylene
7. Nylon	

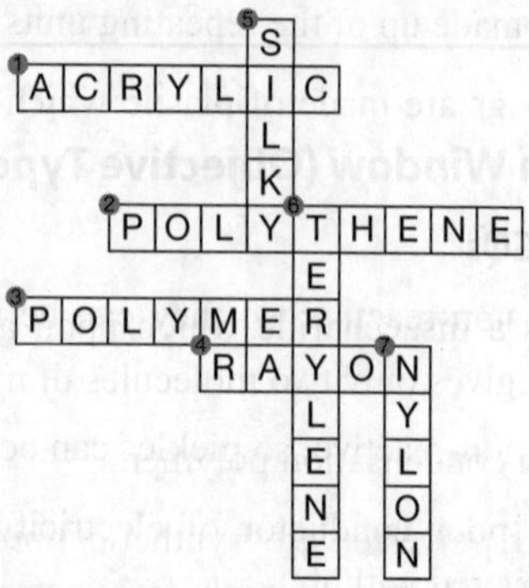

Competition Window (Objective Type)

Topic-wise MCQs

1. (a)	**2.** (a)	**3.** (d)	**4.** (b)	**5.** (a)
6. (c)	**7.** (d)	**8.** (c)	**9.** (a)	**10.** (b)
11. (a)	**12.** (a)	**13.** (b)	**14.** (c)	**15.** (d)
16. (c)	**17.** (a)	**18.** (b)	**19.** (d)	**20.** (d)
21. (b)	**22.** (d)	**23.** (b)	**24.** (a)	**25.** (c)
26. (a)	**27.** (d)	**28.** (a)	**29.** (d)	**30.** (b)
31. (c)	**32.** (a)	**33.** (d)	**34.** (d)	**35.** (b)
36. (d)	**37.** (a)	**38.** (c)	**39.** (a)	**40.** (b)

Miscellaneous

1. (d)	**2.** (d)	**3.** (a)	**4.** (d)	**5.** (a)
6. (c)	**7.** (a)	**8.** (c)	**9.** (b)	**10.** (d)
11. (b)	**12.** (d)	**13.** (b)	**14.** (c)	**15.** (c)
16. (d)	**17.** (a)	**18.** (b)	**19.** (d)	**20.** (b)

Advanced and Olympiads

Single Choice

1. (c)	**2.** (d)	**3.** (a)	**4.** (d)	**5.** (c)
6. (a)	**7.** (c)	**8.** (a)		

Multiple Choice

9. (bc)	**10.** (acd)	**11.** (abc)	**12.** (abcd)	**13.** (bcd)
14. (abc)	**15.** (acd)	**16.** (abd)		

Comprehension Type

17. (d)	**18.** (c)	**19.** (c)	**20.** (c)	**21.** (b)
22. (d)				

Matrix Matching

23. (a)—(r); (b)—(s); (c)—(q); (d)—(p)

24. (a)—(s), (b)—(r), (c)—(p), (d)—(q)

Integer Type

25. 5	**26.** 6	**27.** 5	**28.** 6	**29.** 7
30. 6				

Hints and Solutions

Competition Window (Objective Type)

Topic-wise MCQs

2. Sucrose is a disaccharide which upon acid or enzymatic hydrolysis gives only two molecules of monosaccharides.
4. Nylon is a condensation polymer.
11. Nylon and polyester are synthetic fibres obtained from plastic material we get mainly from petroleum.
13. Nylon is a synthetic fibre. It is made without using any natural raw material. Cotton, jute and wool are natural fibres.
14. Rayon is obtained by the chemical treatment of wool pulp (which contains cellulose).
15. Rayon has appreciable absorbent properties.
16. Orlon, polypropylene and nylon are prepared synthetically by chemical methods thus are synthetic fibres.
17. Rayon, as it is obtained from cellulose.
18. Wood is the major raw material for rayon.
19. Terrywool is obtained by mixing polyester (synthetic fibre) and wool (natural fibre).
20. Wool retains air between the fibres. That is why it is fluffy. This enables it to retain body heat. Woolen clothes are therefore good for wearing in cold weather.
21. Rayon, nylon and polyester are synthetic fibres.
22. Polycot is obtained by mixing polyester and cotton. Polycot = polyester + cotton
23. Rayon resembles silk in appearance; therefore, rayon is also called artificial silk.
27. Polyester alone accounts for around 60% use of synthetic fibre.
28. Plastics are polymers.
29. All are correct options.
30. Bakelite is a thermosetting plastic.
31. Once set, thermosetting plastics cannot be remoulded because they do not soften on heating.
32. Melamine is a thermosetting plastic which when moulded once, cannot be softened by heating.
33. Bakelite is used for making the handles of various cooking utensils because it is a poor conductor of heat and does not become soft on getting heated.
34. PET is polyethylene terephthalate. As a plastic, it is very light weight and is commonly used for making kitchen containers.

Miscellaneous

1. Due to the low water absorption capacity of polyester, it does not absorb sweat and hence, becomes uncomfortable in summers.
2. Teflon is a polymer of tetrafluoroethene. Due to its inert nature and high melting point, it is used in coating non-stick cookware.
4. A cup made of styrofoam containing hot tea or coffee can be held in the hands since it does not conduct heat.
5. Due to its flexible nature, it is easy to squeeze out toothpaste.
6. Since plastic is a bad conductor of electricity, it is used for covering the wires to insulate them.
7. Flexibility or elasticity is the property of regaining the original shape after being pulled and releasing the pressure.
9. It is also called tensile strength.
11. Plastics are strong, poor conductors of heat and easy to mould.
12. Cotton gives the smell of burning paper because cotton and paper are obtained from plants.
13. Wool is made from the hair of sheep; so, it gives the smell of burning hair.
14. The four 'R's in the 4R principle are reduce, reuse, recycle and recover.
16. Cotton cloth is the only natural substance; hence, it is biodegradable.
17. Due to the teflon coating, they absorb very less oil. Hence, food can be cooked in very little oil without food sticking to the pan.
18. The 4R principle is mostly applicable to plastics which can be recycled and reused.
19. Nylon is a fibre that has H-bonds as intermolecular forces.
20. A plasticiser is a substance added to plastics to make them flexible and soft.

Advanced and Olympiads

Single Choice

1. PVC, polythene and bakelite are synthetic substances.
2. Nylon is used to make parachutes.
3. Statements (ii) and (iv) are incorrect. Electric wires are covered with plastics since they are poor conductors of heat and electricity. Plastics are cheaper than metals.
4. Synthetic fibres catch fire easily and they also melt on heating and stick to the body of the person wearing it so, it is not advisable to wear synthetic clothes while working in a kitchen or bursting fire crackers.

5. Nylon → Fibre, PVC → Thermoplastic, bakelite → Thermosetting plastic

8. Food grade material should have all these properties.

Multiple Choice

9. Wool and silk are animal fibres.

10. The fibres which we get from natural resources are called natural fibres—cellulose, jute, cotton. Thermosetting and polystyrene have artificial sources and are not natural.

13. Melamine, nylon and PVC are thermoplastic polymers.

14. Plastics are non-reactive, light in weight and durable but they do not conduct electricity.

15. Nylon-66 is a polymer of adipic acid and hexamethylene diamine.

Comprehension Type

18. Orlon is made of CH_2=CHCN (vinyl cyanide).

19. Polythene, polystyrene are addition polymers while nylon-6, nylon-66 and polyester are condensation polymers.

22. Monomers are linked with each other in straight lines to form parallel fibres cross-linked by bonds called hydrogen bonds.

Integer Type

25. Except bakelite and melamine, all the rest are thermoplastics.

26. Leaves, fruits, cotton, coir, jute and paper are biodegradable.

27. Bakelite, nylon-66, polyester, rayon and melamine are condensation polymers.

28. In nylon-66 both the monomers, adipic acid and hexamethylene diamine, have a chain of 6-carbon atoms.

29. The monomer of nylon-6 is caprolactam which has a 7-membered ring structure.

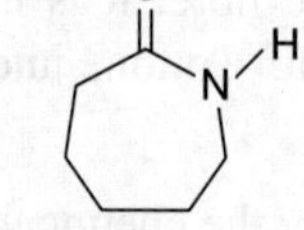

Caprolactam

Melamine has 6 N atoms.

30.

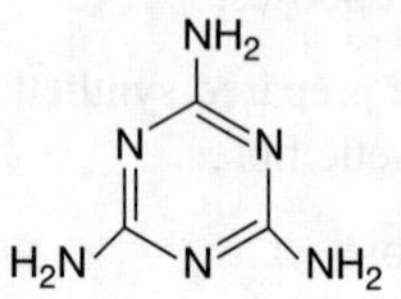

CHAPTER 10 Materials: Metals and Non-metals

LEARNING OBJECTIVES

After studying this unit, you will be able to understand:

- Metals and non-metals
- Various properties of metals and non-metals
- The extraction of metals
- Alloys

Introduction

Element

An element is a substance that cannot be split up into two or more simpler substances by the usual chemical methods of applying heat, light or electric energy because an element is made of only one type of atoms. For example, oxygen is an element as it cannot be split up into two or more simpler substances by applying heat, light or electricity.

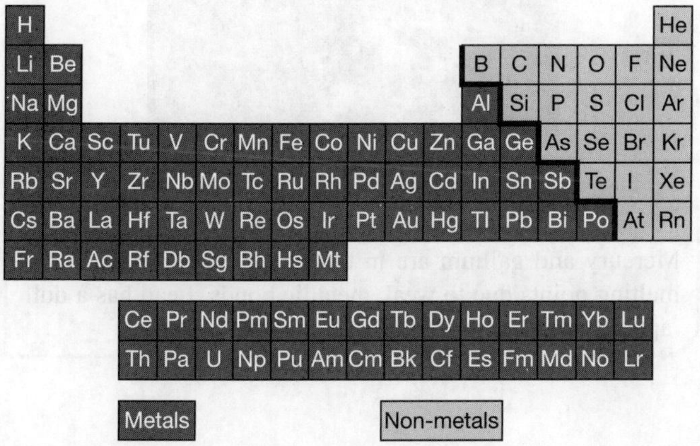

Types of Elements

In order to conveniently study elements, they can be roughly divided into non-metals, metalloids and metals.

Comparative Study of Metals and Non-metals

Metals

The majority of elements are metals and they occupy the largest portion of periodic table. Those elements which have 1,2,3 electrons in their valence cells, are lustrous, malleable, ductile (can be drawn into sheets or pulled into wires) and have high densities and tensile strength are metals. For example, except hydrogen, the rest of the *s*-block elements are metals. All elements of the *d* and *f*-block are metals. The lower members of the groups of elements of the *p*-block are also metals such as Pb, Bi, Po and so on.

Position of Metals in Periodic Table

The metals can be arranged on the basis of their electronic configuration as follows.

Group	Metals
1	The alkali metals
2	The alkaline earth metals
13–18	The transition metals
3	The coinage metals (called so because they were used to make coins earlier)
4	The zinc group metals
Lower part of 13-16	Metals like Al, Sn, Pb

Illustrations

1. Write the correct position of iron and aluminium in the periodic table.

   Solution: Aluminium is present in the *p*-block, III period and group13.

   Iron is present in the *d*-block, IV period and in group 10.

2. Name any two metals which belong to the alkali metals.

   Solution: Sodium and potassium belong to the alkali metals.

Physical Properties

Metals are characterised by some specific physical properties which differentiate metals from non-metals.

(i) Solid nature: Metals exist in the solid state as strong metallic bonds are present in the metal lattices.

(ii) Metallic lustre: Metals are shiny with a bright metallic lustre because of the to and fro oscillation of mobile (free) electrons. They can be polished to give a highly reflective surface.

(iii) Hardness, density, melting point and boiling point: Metals are hard and have high values of density, melting point and boiling point due to strong metallic bonds.

Density, MP, BP $\propto$ Strength of metallic bond $\propto$ Number of valence electrons

- Alkali metals (Na, K) are soft, cuttable with a knife and have low density, low melting point due to weak metallic bonds (have only 1 valence electron).

(iv) Malleable and ductile nature: Metals (except Zn, Hg and Ga) are highly malleable (converted into sheets) and highly ductile (converted to wires) as the metallic bond is omni directional. For example, silver wires may be of 0.0008 mm diameter and gold foil can be of less than 0.0001 mm thickness.

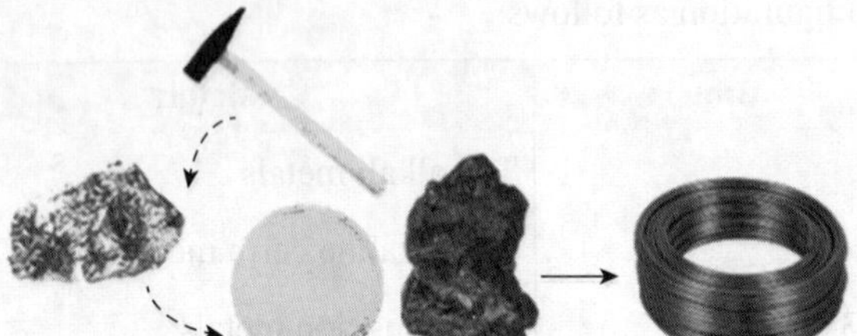

(v) Thermal and electrical conductivity: Metals (except W) are good conductors of heat, electricity and sound due to the presence of free electrons in the metal lattices.

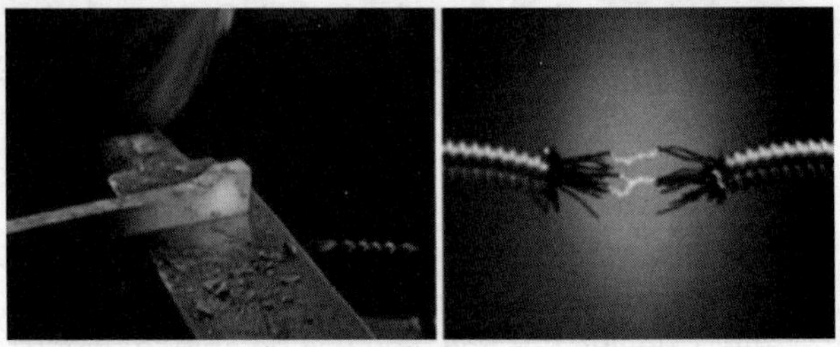

(vi) Tensile strength: Metals have very high tensile strength.

Tensile strength $\propto$ Metallic bond strength

(vii) Magnetic properties: Most of the metals show magnetic properties such as paramagnetism and ferromagnetism due to the presence of unpaired electrons.

Magnetic moment (μ) is given as

$$\mu = \sqrt{n(n+2)}\ \text{BM}$$

Here n = number of unpaired electrons.

KNOWLEDGE BOOSTER

Mercury and gallium are in the liquid state and have low melting points due to weak metallic bonds. Lead has a dull appearance.

Illustrations

1. Why are metals used in making cutting tools?

Solution: As metals are hard and quite strong, they are used for making cutting tools.

2. Name any two metals which lack in malleability and ductility.

Solution: Zinc and mercury lack in malleability and ductility.

Chemical Properties

Most of the metals show these chemical properties:

(i) Reaction with air or oxygen: Most of the metals form metal oxides on burning in air.

1, 2,	Al, Zn, Fe, Pb, Cu,	Ag, Au
Highly reactive bright flame	Reacts on heating	No reaction

Metal + Oxygen $\rightarrow$ Metal oxide

For example, $4Na + O_2 \rightarrow 2Na_2O$

$2Ca + O_2 \rightarrow 2CaO$

All metals do not react equally with air as the tendency of metals to lose electrons differs. For example, highly reactive metals (group 1) rapidly tarnish in air so they are placed under kerosene to avoid this. This reaction is vigorous and explosive also.

Many other metals (group 2, Zn, Al) tarnish in air after some time as they react slowly with O_2 because they are covered with a thin layer of oxide (protective oxide layer) which prevents further oxidation.

Iron reacts with moist air to give red-brown coloured rust, which is hydrated ferric oxide.

$$4Fe + 3O_2 + 2xH_2O \rightarrow 2Fe_2O_3.xH_2O$$

- Iron does not burn on heating; however, iron filings burn vigorously when sprinkled in the flame of the burner.

(ii) Reaction with water: Metals react with water to form metal oxides and hydrogen gas. Metal oxides are soluble in water and easily form metal hydroxides.

Metal + Water → Metal oxide + Hydrogen

Metal oxide + Water → Metal hydroxide

For example, $2Na(s) + 2H_2O(l) \rightarrow 2NaOH(aq) + H_2(g) +$ heat energy

$$Ca(s) + 2H_2O(l) \rightarrow Ca(OH)_2(aq) + H_2(g)$$

Alkali metals react quickly and vigorously and the reactivity increases down the group. Alkaline earth metals have a lower reactivity compared to the alkali metals. For example, magnesium cannot react with cold water but reacts with hot water. However, calcium reacts even with cold water.

Metals like Mg, Al, Fe and Zn do not react with cold water so steam is used for the reaction.

Metal + steam → Metal oxide + Hydrogen

$$Mg(s) + H_2O(s) \rightarrow MgO(s) + H_2(s)$$

$$2Al(s) + 3H_2O(g) \rightarrow Al_2O_3(s) + 3H_2(g)$$

$$3Fe(s) + 4H_2O(g) \rightarrow Fe_3O_4(s) + 4H_2(g)$$

KNOWLEDGE BOOSTER

The reactivity of different metals with water in decreasing order:

K > Na > Ca > Mg > Al > Zn > Fe > Pb > Cu > Ag > Au

A few metals like Cu, Ag, Au and Pb do not react with cold or hot water.

(iii) Reaction with acids: Metals react with acids to form a metal salt along with evolution of hydrogen.

Metal + Acid → Salt + Hydrogen

For example, $2Na + 2HCl \rightarrow 2NaCl + H_2$

$$Ca + H_2SO_4 \rightarrow CaSO_4 + H_2O$$

Alkali metals and alkaline earth metals react readily with dilute HCl or H_2SO_4. However, the reaction of Zn is slow.

$$Zn + H_2SO_4 \rightarrow ZnSO_4 + H_2$$

When a little copper sulfate is taken with Zn, the reaction gets fast as Zn can displace Cu (catalyst).

When metals react with dil. HCl, metal chloride and bubbles of hydrogen are formed. The reactivity of some metals for this reaction is as follows:

Mg > Al > Zn > Fe

Cu cannot react with dil. HCl.

KNOWLEDGE BOOSTER

When metals react with HNO_3, H_2 is not evolved as HNO_3 being a strong oxidising agent oxidises H_2 into H_2O and get reduced into its oxides such as N_2O, NO, NO_2. However, only with Mg and Mn, hydrogen is evolved.

$$Mg + 2HNO_3 \rightarrow Mg(NO_3)_2 + H_2$$

(iv) Reaction with dil. nitric acid: When dil. nitric acid is added to a mixture containing Cu, Zn and Al, only Cu reacts with HNO_3 as follows:

$$Cu + 2HNO_3 + 2H^+ \rightarrow Cu^{2+} + 2NO_2 + 2H_2O$$

Al, Fe, Zn do not react with conc. HNO_3 due to the formation of a stable oxide layer on these metals.

(v) Reaction with metal salt solution (displacement reaction): A more reactive metal can displace a less reactive metal from its salt solution. It is a case of displacement reaction.

For example, $Fe(s) + \underset{\text{Blue}}{CuSO_4} \rightarrow \underset{\text{Pale green}}{FeSO_4} + Cu(s)$

- During the reaction, blue colour disappears and iron metal acquires a pink brown colour due to deposition of copper.

$$Sn + PbCl_2 \rightarrow SnCl_2 + Pb$$

When the added metal is not more reactive than the metal to be displaced from the metal salt, the reaction does not take place.

For example, $Fe + MgSO_4 \rightarrow$ No reaction

This is because, Fe has less reactivity than Mg.

$Cu + Pb(NO_3)_2 \rightarrow$ No reaction

As Cu has less reactivity than Pb.

Illustrations

1. Between Mg and Zn, which burns with a bright flame to form the oxide?

Solution: Magnesium reacts quickly and vigorously with air with a bright flame to form an oxide while Zn reacts slowly with air.

$$2Mg + O_2 \rightarrow 2MgO$$

$$2Zn + O_2 \rightarrow 2ZnO$$

2. Explain why lead on reaction with dil. HCl does not displace hydrogen.

Solution: Lead on reaction with dil. HCl does not displace hydrogen because a thin film of $PbCl_2$ is formed which protects the metal from the action of the acid.

3. (a) Name any two metals which can react with cold water.
 (b) Name any two metals which are stored in kerosene.

Solution:
(a) Potassium and calcium can react with cold water easily.
(b) Sodium and potassium are stored in kerosene.

Non-metals

Those elements which in general are non-lustrous, electronegative, bad conductors of heat and electricity, are non-metals. These elements have a tendency to accept electrons to form anions due to their electronegative nature. For example, the following are non-metals: hydrogen (H), carbon (C), nitrogen (N), oxygen (O), phosphorus (P), sulfur (S), selenium (Se), halogens (F, Cl, Br, I) and noble gases (He, Ne, Ar, Kr, Xe, Rn).

Position in Periodic Table

Except hydrogen, the rest of the non-metals are placed in the *p*-block in groups 13 to 18. They are 24 in number and are as follows:

Solids (12)	B, C, Si, Ge, P, As, Sb, S, Se, Te, I, At
Liquid (1)	Br
Gases (11)	H, N, O, F, Cl, He, Ne, Ar, Kr, Xe, Rn

Physical Properties

- **Physical state:** They exist mainly in the gaseous state, bromine is a liquid and C, P, S, Se and I_2 are solids.
- **Lustre:** They are dull except for iodine and graphite.
- **Softness:** They are soft and brittle. They break easily into a powdery mass on tapping with a hammer. For example, coal and sulfur. Diamond is the hardest natural substance and quite brittle also.
- **Density:** They have low values of densities. However, the densities of iodine and diamond are high.
- **Non-sonorous:** As they are non-sonorous, they do not create a ringing sound when they are hit.
- **Conductivity:** They are very poor conductors of heat and electricity. In fact, some of them are used as insulators. However, carbon (graphite) can conduct electricity due to the presence of free electrons in its hexagonal slippery structure.
- **Non-ductile and non-malleable:** They are neither ductile nor malleable except for carbon fibres.
- **Alloys:** They do not form alloys and amalgams except for carbon and phosphorus which can be used in alloys.

Illustrations

1. Name any two non-metals which are lustrous.
 Solution: Iodine and graphite are quite lustrous.

2. Name any two non-metals which are mono-atomic.
 Solution: Helium and neon are mono-atomic.

Chemical Properties

(i) **Action of air or oxygen:** Most of the non-metals can react with air on ignition. However, phosphorus can react with air even at room temperature. For example,

$$2H_2 + O_2 \xrightarrow[\text{Popping sound}]{\text{Ignition}} 2H_2O$$

$$C(s) + O_2(s) \xrightarrow{\text{Ignition}} CO_2(g)$$

$$P_4 + 5O_2 \xrightarrow{\text{Ignition}} P_4O_{10}(s)$$

These non-metallic oxides radially dissolve in water to give oxyacids. For example,

$$CO_2 + H_2O(l) \rightarrow \underset{\text{Carbonic acid}}{H_2CO_3(aq)}$$

$$P_4O_{10} + 6H_2O(l) \rightarrow \underset{\text{Phosphoric acid}}{4H_3PO_4(aq)}$$

(ii) **Action of water:** Most of the non-metals do not react with water but chlorine dissolves in water:

$$2Cl_2 + 2H_2O \xrightarrow[\text{or heat}]{\text{Sunlight}} 2HCl + \underset{\text{(Hypochlorous acid)}}{2HOCl}$$

When steam is passed over red hot coke, water gas is formed:

$$\underset{\text{(red hot)}}{C} + \underset{\text{(steam)}}{H_2O} \rightarrow \underset{\text{(water gas)}}{CO + H_2}$$

(iii) **Reaction with hydrogen:** Non-metals on heating with hydrogen form hydrides:

$$N_2(g) + 3H_2(g) \xrightleftharpoons{500°C} \underset{\text{ammonia}}{2NH_3(g)}$$

$$H_2 + Cl_2 \rightarrow 2HCl$$

(iv) **Reaction with chlorine:** Non-metals on reaction with chlorine form covalent chlorides. For example,

$$2P + 3Cl_2 \rightarrow \underset{\text{Phosphorus trichloride}}{2PCl_3}$$

$$2P + 5Cl_2 \rightarrow \underset{\text{Phosphorus pentachloride}}{2PCl_5}$$

(v) **Reaction with acids:** Most of the non-metals do not react with acids. However, on heating with concentrated sulfuric acid or concentrated nitric acid, some non-metals react. For example,

$$S(s)+2H_2SO_4(\text{conc.})\xrightarrow{\text{heat}}3SO_2(g)+2H_2O(l)$$

$$S(l)+6HNO_3(\text{conc.})\xrightarrow{\text{heat}}H_2SO_4(aq)+6NO_2(g)+2H_2O(l)$$

Mostly non-metals do not react with water or acids as they cannot displace hydrogen as metals do.

(vi) Displacement reaction: A more reactive non-metal can displace a less reactive non-metal from its salt. For example,

$$Cl_2 + 2NaBr \rightarrow 2NaCl + Br_2$$

$$Cl_2 + 2NaI \rightarrow 2NaCl + I_2$$

Illustrations

1. Name any two oxides other than water which are neutral in nature.

Solution: CO and NO are neutral in nature.

2. Which type of compounds are formed between non-metals?

Solution: Compounds formed by non-metals are covalent in nature.

For example, CO_2, PCl_5.

Uses of Non-metals

- Sulfur is used in the manufacture of sulfuric acid and it is used to make antiseptic creams. Sulfur is also used in the vulcanisation of rubber.
- Hydrogen is used in welding metals. Liquid hydrogen is used as a rocket fuel.
- Oxygen is essential for all living beings for breathing.
- Phosphorus is used in safety materials and in fire crackers.
- Nitrogen is used in the manufacture of ammonia and fertilisers. It is also used as a preservative due to its inert nature.
- Diamond, another form of carbon, is the hardest substance known. Hence, it is used to cut glass.
- Graphite, a form of carbon, is used in pencil leads and as electrodes.
- Chlorine is used in the treatment of water to kill germs.
- Silicon is used to make computer chips; silica gel is used to absorb moisture.

Difference between Metals and Non-metals

Table 10.1 Differences in physical properties of metals and non-metals

Property	Metals	Non-metals
Physical state	They are mostly solids and are non-volatile	They may be gas, solid or liquid and are volatile
Density	They have high densities except for sodium and potassium	They have low densities
Metallic lustre	They have metallic lustre and can be polished	They do not have metallic lustre so cannot be polished (exceptions are graphite and iodine)
Malleability and ductility	They are malleable and ductile except for Bi	They are neither malleable nor ductile
Thermal and electrical conductivity	Metals are good conductors of heat and electricity except for lead. Ag and Cu are the best conductors	They are poor conductors of heat and electricity except for graphite
Alloy formation	Due to similar sizes metals can readily form alloys	They do not form alloys except for phosphorus, carbon and silicon

Table 10.2 Differences in chemical properties of metals and non-metals

Property	Metals	Non-metals
Number of valence electrons	They have 1,2,3 valence electrons	They have 4 to 7 valence electrons except H and He which have 1 and 2 valence electrons respectively
Electrochemical nature	They are electropositive, form cations and during electrolysis they are liberated at the cathode	They are electronegative, form anions and during electrolysis they are liberated at the anode

(Continued)

Property	Metals	Non-metals
Solubility	Mostly they dissolve by chemical reactions	Most of the non-metal dissolve without any chemical reactions
Nature of oxides	Their oxides are mainly basic in nature $Na_2O + H_2O \rightarrow 2NaOH$	Their oxides are mainly acidic in nature $2NO_2 + H_2O \rightarrow HNO_2 + HNO_3$
Action of acids	They dissolve in mineral acids to form the salt and release H_2 $Zn + 2HCl \rightarrow ZnCl_2 + H_2$	Most of the non-metal do not dissolve in acids; however, S, P, I_2 can react with concentrated HNO_3 $I_2 + 10HNO_3 \rightarrow 2HIO_3 + 10NO_2 + 4H_2O$
Hydrides	Only a few metals can form stable hydrides LiH, CaH_2	Most of the non-metals form stable hydrides NH_3, CH_4

KNOWLEDGE BOOSTER

Metalloids

Certain elements which show properties of both metals and non-metals are called metalloids. They are, boron (B), silicon (Si), germanium (Ge), arsenic (As), antimony (Sb), tellurium (Te) and polonium (Po).

The Activity Series of Metals

The activity series or reactivity series of metals represents a list of metals arranged in the order of their decreasing chemical activity. Thus, the most reactive metal is potassium and is placed at the top of the list and the lest active metal is gold which is placed at the bottom of the list.

Activity (reactivity) of elements

Metals			Non metals		
Potassium	K	Most active metal			
Sodium	Na				
Calcium	Ca		Fluorine	F	Most active
Magnesium	Mg		Chlorine	Cl	
Aluminium	Al		Bromine	Br	
Zinc	Zn		Iodine	I	Least active
Iron	Fe				
Lead	Pb				
[Hydrogen]	[H]				
Copper	Cu				
Mercury	Hg				
Silver	Ag				
Gold	Au				
Platinum	Pt	Least active metal			

Reactivity in decreases order

KNOWLEDGE BOOSTER

A more reactive element displaces the less reactive element from its salt solution.

For example, $2KBr + Cl_2 \rightarrow 2KCl + Br_2$

Main Points of the Activity Series

- Any metal placed higher up in the series displaces any metal below it from its salt solution.
- The larger the differences in the positions of the metals in the series, more rapidly does the displacement occur.
- Metals above hydrogen can displace hydrogen from water and from dilute acid but the metals below hydrogen cannot do it.
- Oxides of the metals K, Na, Ca and Mg cannot be reduced by H_2, CO or C.
- Oxides and nitrates of less reactive metals (Hg, Ag, Au) decompose to give the metals on being strongly heated.
- Metals below copper do not rust easily.
- Hydrogen, though a non-metal, has been included in this series. It occupies the position based on the ease of formation of the positive ions.

Occurrence and Extraction of Metals

Occurrence

Elements can occur as follows.

Elements in the Atmosphere

The atmosphere mainly contains nitrogen (78.09%), oxygen (20.95%) and other gases (about 1%).

Elements in the Sea

Sea is the major source of elements like Br, I, Ni, Cu, Zn, Sn and Au.

Elements in the Earth's Crust (Lithosphere)

Elements occur in two states in the earth's crust.

Free or native state: Less reactive metals or noble metals with least electropositive nature are present in the free or native state. Examples are copper, silver, gold, platinum.

KNOWLEDGE BOOSTER

Gangue or matrix

(i) interferes with the process of extraction,
(ii) makes the metal impure, and
(iii) increases the cost of production of the pure metals.

Hence, all efforts are made to remove the gangue from an ore.

Most abundant elements in the lithosphere

	O	Si	Al	Fe	Ca	Na
%	48.60	26.30	7.73	4.75	3.45	2.74

Combined state (minerals): Reactive metals occur in the combined state and are known as *minerals*. Those minerals from which metals can be profitably extracted are called *ores*. These metals are generally associated with rocky materials, sand, clay, which is known as *gangue* or *matrix*.

Illustrations

1. Name the two most abundant metals in the earth's crust.

   Solution: Al and Fe are the two most abundant metals in the earth's crust.

2. Which of these metals can have both sulfide and oxide ores?

   Ag, Zn, Fe, Al, Cu

   Solution: Here Zn, Cu, Fe can have both sulfide and oxide ores.

General Metallurgical Processes

Metals and their compounds exist as natural elements called *minerals*. Ore is that form of a mineral from which a metal can be easily and cheaply extracted. The process of extraction of pure metal from its ore is called *metallurgy*. Some important types of ores are given in Table 10.1.

Table 10.1 Some important types of ores

Oxides	Carbonates	Halides	Sulfides	Sulfates
Bauxite ($Al_2O_3.2H_2O$)	Marble or limestone ($CaCO_3$)	Rock salt ($NaCl$)	Zinc blende (ZnS)	Gypsum ($CaSO_4.2H_2O$)
Hematite ($Fe_2O_3.xH_2O$) Magnetite (Fe_3O_4)	Magnesite ($MgCO_3$)	Cryolite (Na_3AlF_6)	Iron pyrites (FeS_2)	Epsom salt ($MgSO_4.7H_2O$)
Zincite (ZnO)	Calamine ($ZnCO_3$)	Horn silver ($AgCl$)	Cinnabar (HgS)	Anglesite ($PbSO_4$)
Cuprite (Cu_2O)	Siderite ($FeCO_3$)	Fluorspar (CaF_2)	Galena (PbS)	Baryte ($BaSO_4$)

Steps Involved in Extraction

Extraction of a metal involves the following processes.

1. **Crushing (pulverisation) of ore:** Big lumps of ore obtained from the earth's crust are crushed into smaller pieces with the help of jaw crushers and grinders. This process is known as crushing of ore.
2. **Concentration of the ore:** Removal of impurities from the crushed ore is known as *concentration* or *benefaction* or *dressing of ore*. It can be done by the following methods.

   (i) Hand picking: Selected pieces of ores are picked up.

   (ii) Levigation (hydraulic washing)

   Principle: The difference in the densities of the ore and the gangue is the main criterion.

   The crushed ore is washed in a stream of water. The lighter impurities are swept away while the heavier ore particles settle down. Iron ores and tin ores are concentrated by this method (mainly oxides and carbonates). It is also known as *gravity separation* or *hydraulic washing*.

   (iii) Magnetic separation

   Principle: Magnetic properties of the ores.

   This process is meant for the separation of magnetic ore from impurities. In this method, the powdered ore is placed over a leather belt which moves over two rollers, one of which is magnetic.

   When the crushed ore is passed over the magnetic roller, magnetic ore particles are attracted by it and fall below it while impurities fall away from the magnetic roller.

   Chromite [$Fe(CrO_2)$] from siliceous gangue, rutile (TiO_2) from chlorapatite and wolframite ($FeWO_4$) from cassiterite are separated by this method.

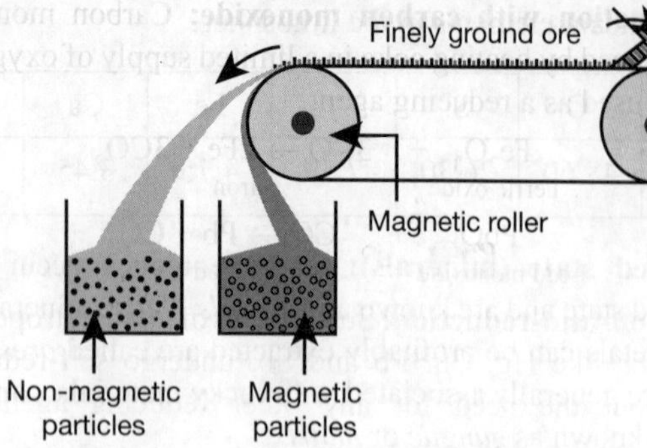

Fig. 10.1 Magnetic separation (schematic)

(iv) Froth floatation process: This process is commonly used for sulfide ores and is based upon different wetting characteristics of the ore and gangue particles.

Finely powdered ore is mixed with water, pine oil (frother) and ethylxanthate or potassium ethylxanthate (collector) in a big tank. The whole mixture is agitated with air. The ore particles wetted with oil come into the froth and are taken off by skimming, while the impurities wetted with water settle at the bottom.

The foaming agent is pine oil and froth stabilisers are cresol, anisole, and so on. The collectors are ethylxanthate and potassium ethylxanthate.

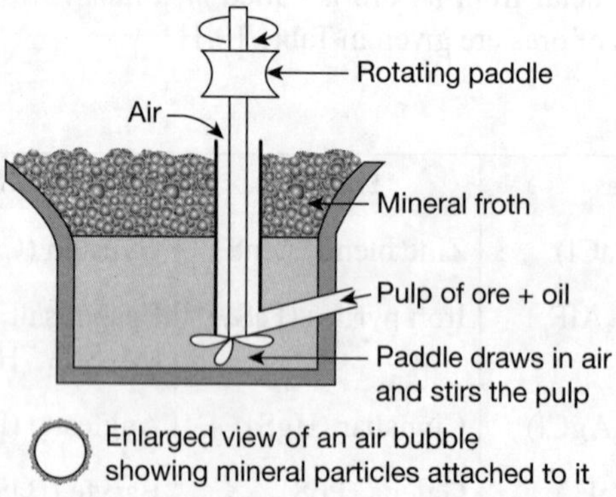

Fig. 10.2 Froth floatation process

(v) Chemical method (leaching): In this method, the powdered ore is treated with a suitable chemical reagent which dissolves the ore while impurities remain insoluble in that reagent. Bauxite is separated from Fe_2O_3, SiO_2, TiO_2 with the help of NaOH. Al_2O_3 gets dissolved in NaOH while the rest are insoluble.

$$Al_2O_3 + 2NaOH \rightarrow 2NaAlO_2 + H_2O$$

$$NaSiO_2 + 2H_2O \rightarrow Al(OH)_3 \downarrow + 2NaOH$$

$$2Al(OH)_3 \rightarrow Al_2O_3 + 3H_2O$$

3. Calcination: Calcination is the process of heating the ores below their melting points in the absence of air to remove volatile impurities such as water, CO_2 and organic matter.

$$\underset{\text{Limestone}}{CaCO_3} \rightarrow \underset{\text{Calcium oxide}}{CaO} + CO_2 \uparrow$$

$$\underset{\text{Bauxite}}{Al_2O_3.2H_2O} \rightarrow \underset{\text{Alumina}}{Al_2O_3} + 2H_2O \uparrow$$

During calcination, the ore becomes porous, volatile impurities are removed and carbonate ore decomposes into oxides.

4. Roasting: Roasting is the process of heating the ore in excess of air in order to convert metals into their oxides and water insoluble sulfides into water soluble sulfates.

It is also called de-electronation of ores. For example,

$$\underset{\text{Pyrites}}{2FeS} + 3O_2 \xrightarrow{\Delta} 2FeO + 2SO_2$$

$$\underset{\text{Zinc sulfide}}{ZnS} + 2O_2 \xrightarrow{\Delta} ZnSO_4$$

During roasting, volatile impurities are removed. S, As, Sb are removed as SO_2, As_2O_3, Sb_2O_3 respectively and sulfide ores are converted into their oxides. Both calcination and roasting occur mainly in the reverberatory furnace.

Illustrations

1. Which type of ores cannot be concentrated by the gravity separation method?

Solution: The gravity separation method cannot be used for sulfide ores.

2. What is the requirement for the magnetic separation method to remove impurities from an ore?

Solution: In this method either the ore must be magnetic and impurities non-magnetic or vice versa since this method is used to the separate magnetic species from the non-magnetic.

3. Differentiate between calcination and roasting.

Solution:

Calcination	Roasting
1. Ore is heated in the absence of air.	1. Ore is heated in excess of air.
2. Carbonate and hydrated ores are calcinated.	2. Mostly sulfide ores are heated.
3. Moisture and CO_2 are removed.	3. Volatile impurities of phosphorus (P) and sulfur (S) are removed as P_2O_5, SO_2.

Reduction of Metal Oxides into Free Metals

Reduction is carried out using several methods, as described below.

(i) Reduction with carbon (smelting): Reduction of oxides of less electropositive metals like Pb, Fe, Zn, Sb and Cu is carried out by heating them with coal or coke in a blast furnace in the presence of limited air. For example,

$$\underset{\text{Zinc oxide}}{ZnO} + \underset{\text{Coke}}{C} \xrightarrow{\Delta} \underset{\text{Zinc}}{Zn} + CO$$

$$Fe_2O_3 + 3C \rightarrow 2Fe + 3CO$$

(ii) Reduction with carbon monoxide: Carbon monoxide produced by heating coke in a limited supply of oxygen, is also used as a reducing agent.

$$\underset{\text{Ferric oxide}}{Fe_2O_3} + 3CO \rightarrow \underset{\text{Iron}}{2Fe} + 3CO_2$$

$$\underset{\text{Lead monoxide}}{PbO} + CO \rightarrow \underset{\text{Lead}}{Pb} + CO_2$$

(iii) Self or Auto-reduction: Sulfide ores of less electropositive metals like Hg, Cu, Pb and Sb, undergo self-reduction without the need for any other reducing agent. For example,

$$2PbO + \underset{\text{Lead sulfide}}{PbS} \rightarrow \underset{\text{Lead}}{2Pb} + SO_2$$

Illustrations

1. Why can't the oxides of sodium, calcium and aluminium be reduced by C or CO or H_2?

Solution: As the oxides of these metals have great affinity towards oxygen, they cannot be reduced by C or CO or H_2.

2. Differentiate between pyrometallurgy and hydrometallurgy.

Solution: Pyrometallurgy is the cheapest and the most common method of reducing metal on a large scale, mainly by the carbon reduction process.

For example, $ZnO + C \xrightarrow{\Delta} Zn + CO$

In hydrometallurgy, precious metals (Ag, Au) are obtained by their displacement from aqueous salts solutions by using more active metals (Zn).

For example,

$$\underset{(M=Ag,Au)}{2Na[M(CN)_2]} + Zn \rightarrow Na_2[Zn(CN)_4] + \underset{(M=Ag,Au)}{2M}$$

Refining or Purification of Metals

The metals obtained after reduction may still contain some objectionable impurities which are removed by refining using following methods.

Physical Methods

Liquation, distillation and zone refining (fractional crystallisation).

Liquation: This method is meant for refining of readily fusible metals like Sn, Pb and Bi, which have less fusible impurities.

Distillation: Volatile metals such as zinc and mercury are refined by distillation. The crude metal is heated in a retort. As a result, the pure metal gets distilled and impurities are left behind in the retort.

Zone refining (fractional crystallisation): It is based on the fact that the metal and the impurities have a difference in their solubilities. Highly pure metal is obtained by this method. It can be used for Ge, Si, Ga, and so on.

Chemical Methods

These include cupellation, poling, van Arkel, electrorefining, and so on.

Cupellation: By this method, lead (the impurity) is removed from silver.

Poling: It is used for the metals having their oxides as impurities. For example, Cu_2O is removed from blister copper, SnO_2 from tin.

van Arkel method: This method is used mainly for the purification of Ti and Zr (space technology metals) using iodine. It is called *vapour phase refining*.

$$\underset{\substack{\text{Impure}\\\text{metal}}}{Ti + 2I_2} \xrightarrow{500\text{ K}} TiI_4 \xrightarrow{1700\text{ K}} \underset{\substack{\text{Pure}\\\text{metal}}}{Ti} + 2I_2$$

Electrorefining: Highly electropositive metals like Cu, Ag, Au, Al, Zn, Sn, Pb are purified by this method. The pure metal gets deposited at the cathode while the impurities get deposited at the anode after electrolysis.

Alloys

An alloy is a homogeneous mixture of two or more metals or of one or more metals with certain non-metallic elements. The properties of alloys are mostly different from those of the components.

For example,

(i) Pure gold is too soft to be used alone, so a small percentage of copper is added to it.

(ii) Steel having up to 10% of elements such as Cr, Mo and Ni with a low percentage of carbon is known as alloy steel. These added elements improve the hardness and toughness.

Purpose of Making Alloys

The purpose of an alloy is to improve the specific use of the primary component and not to adulterate or degrade it. Alloys are made to change the property of their major constituents to achieve a specific objective.

Table 10.3 Some common alloys

Principal metal	Alloy's name	Composition	Properties	Used for making
Aluminium	1. Magnalium	90%–95% Al 10%–5% Mg		1. Scientific instruments 2. Aircraft 3. Metal mirrors 4. Beams of balance 5. Light tools 6. Household appliances
	2. Duralumin	95% Al 4% Cu 0.5% Mg 0.5% Mn	1. Highly ductile 2. Light but as strong as steel 3. Hard and resistant to corrosion	1. Pressure cooker 2. Bodies of aircraft, buses and tube trains 3. Light tools
Zinc and copper	Bronze	80% Cu 18% Sn 2% Zn	1. Can take up polish 2. Hard and easily cast	1. Coins 2. Bearings 3. Utensils 4. Statues 5. Medals
	Brass	40%–30% Zn 60%–70% Cu	1. Can be easily cast 2. Malleable and ductile 3. Yellow in colour 4. Resists corrosion	1. Electrical goods 2. Screws and handles 3. Musical instruments 4. Parts of watches 5. Cartridges, containers 6. Decorative hardware, utensils
	German silver	50% Cu 30% Zn 20% Ni	1. White and light like silver 2. Malleable and ductile 3. High electrical resistance	1. Decorative articles 2. Electric heaters 3. Rheostats 4. Resistors
	Gun metal	88% Cu, 8% Sn, 1% Zn, 1% Pb	1. Hard and brittle 2. Easily cut	1. Barrels of cannons 2. Bearings, gears
Iron	1. Stainless steel	73% Fe, 18% Cr, 8% Ni, 1% C	1. Resists corrosion 2. Lustrous, hard 3. Resistant to acids and alkalis	1. Utensils 2. Cutlery 3. Ornamental pieces 4. Surgical instruments
	2. Manganese steel	85% Fe, 1% C, 14% Mn	Durable, tough and hard	1. Safe 2. Rock drills and 3. Armour plates
	3. Tungsten steel	84% Fe, 5% W 1% C	Very hard	1. Cutting tools for high speed lathes

CHAPTER AT A GLANCE

- An **element** is a substance that cannot be split up into two or more simpler substances by the usual chemical methods of applying heat, light and electric energy.
- **Metals** occupy the largest portion of the periodic table. Those elements which have 1,2,3 electrons in their valence cells, are lustrous, malleable, ductile (can be drawn into sheets or pulled into wires) and have high densities, tensile strength are metals.
- Density, MP, BP ∝ Strength of metallic bond ∝ Number of valence electrons.
- Metals (except Zn, Hg and Ga) are highly **malleable** (beaten into sheets) and highly **ductile** (drawn into wires) as the metallic bond is omni-directional.
- **Metals** have a high tensile strength, show thermal and electrical conductance and magnetic properties.
- Most of the metals form metal oxides on burning in air. Metals react with water to form metal oxides and hydrogen gas. Metal oxides are soluble in water and easily form metal hydroxides.

- The **reactivity** of different metals with water in **decreasing order**:
  K > Na > Ca > Mg > Al > Zn > Fe > Pb > Cu > Ag > Au
- A more reactive metal can displace a less reactive metal from its salt solution. It is a case of displacement reaction.
- Those elements which in general are non-lustrous, electronegative, bad conductors of heat and electricity are **non-metals**. These elements have a tendency to accept electrons to form anions due to their electronegative nature. **Non-metals** are soft with low density and non-sonorous nature.
- Most of the non-metals can react with air on ignition. However, phosphorus can react with air even at room temperature.
- Most of the non-metals do not react with water but chlorine dissolves in water to give hypochlorous acid.
- Non-metals on heating with hydrogen form hydrides such as HX, NH_3, H_2S and so on.
- A more reactive non-metal can displace a less reactive non-metal from its salts.
- Certain elements which show properties of both metals and non-metals are called **metalloids**. They are, boron (B), silicon (Si), germanium (Ge), arsenic (As), antimony (Sb) and tellurium (Te).
- Reactive metals occur in the combined state and are known as **minerals**. Those minerals from which metals can be profitably extracted are called **ores**.
- These metals are generally associated with rocky materials, sand, clay and so on, which is known as **gangue** or **matrix**.
- The process of extraction of a pure metal from its ore is called **metallurgy**.
- Removal of impurities from the crushed ore is called **concentration** or benefaction or dressing of ore.
- In **levigation**, the crushed ore is washed in a stream of water. The lighter impurities are swept away while the heavier ore particles settle down.
- **Magnetic separation** process is meant for the separation of magnetic ores from the impurities.
- **Froth floatation** process is commonly used for sulfide ores and is based upon different wetting characteristics of the ore and gangue particles.
- In **leaching**, the powdered ore is treated with a suitable chemical reagent which dissolves the ore while impurities remain insoluble in that reagent.
- **Calcination** is the process of heating the ores below their melting points in the absence of air to remove volatile impurities like water, CO_2 and organic matter.
- **Roasting** is the process of heating the ore in excess of air in order to convert metals into their oxides and water-insoluble sulfides into water soluble sulfates.
- **Smelting** is the reduction of oxides of less electropositive metals like Pb, Fe, Zn, Sb and Cu. It is carried out by heating them with coal or coke in a blast furnace in the presence of limited air.
- Sulfide ores of less electropositive metals such as Hg, Cu, Pb and Sb undergo **self-reduction** without using any reducing agent.
- **Electrorefining:** Highly electropositive metals like Cu, Ag, Au, Al, Zn, Sn, Pb are purified by this method. The pure metal gets deposited at the cathode while impurities get deposited at the anode after electrolysis.
- An **alloy** is a homogeneous mixture of two or more metals or of one or more metals with certain non-metallic elements. The properties of alloys are different from those of the components.

PRACTICE QUESTIONS

Analyse Your Concepts (School Exam Based)

Fill in the Blanks

Instructions: Complete the following statements with an appropriate word/term to be filled in the blank spaces.

1. The number of metals is much ____________ than non-metals.
2. Phosphorus is a very ____________ non-metal.
3. Metals are ____________ conductors of heat and ____________.
4. Mercury and gallium exist in the ____________ state.
5. Examples of metals are ____________, ____________ and ____________.
6. Examples of non-metals are ____________, ____________ and ____________.
7. Iron is ____________ reactive than copper.
8. Metals react with acids to produce ____________ gas.
9. The strong electrostatic force of attraction present between metal kernels is called ____________.

10. The gravity separation method is not suitable for ________ ores.
11. The concentration of aluminium ore (boxide) by leaching is done by using ________.
12. ZnO is reduced into zinc by using ________.
13. The constituent metals of brass are ________.
14. The percentage of silver in German silver is ________.
15. ________ is the member of the family of noble metals (platinum/iron).

True or False

Instructions: Read the following statements and write your answer as true or false.

1. Most of the elements in the periodic table are metals.
2. Both metals and non-metals are sonorous.
3. Generally, non-metals react with acids.
4. Sodium is a very reactive metal.
5. Alkali metals and halogens (non-metals) are highly reactive.
6. On heating in air, Au and Pt form oxides.
7. Non-metals have high tensile strength.
8. In froth flotation, the ore is present in form of froth on the top layer.
9. German silver contains silver metal.
10. During calcination and roasting, volatile impurities are removed.
11. Copper is extracted mainly from copper pyrites.
12. Copper displaces zinc from zinc sulfate solution.
13. Metals can be extracted from minerals cheaply and easily.
14. Every ore is a mineral but every mineral cannot be an ore.
15. Both carbon and carbon monoxide are used as reductants in metallurgy.

Match the Following

Instructions: Each question contains statements given in two columns which have to be matched. Statements (a, b, c, d) in column I have to be matched with statements (p, q, r, s) in column II.

1.

Column I (Ores)	Column II (Metal)
(a) Carnalite	(p) Zinc
(b) Calamine	(q) Titanium
(c) Ilmentie	(r) Magnesium
(d) Chalcopyrite	(s) Copper

2.

Column I (Alloy)	Column II (Constituent)
(a) Stainless steel	(p) Fe, Cr, Ni
(b) German silver	(q) Cu, Ni, Zn
(c) Brass	(r) Cu, Sn
(d) Bronze	(s) Cu, Zn

Very Short Answer Type Questions

1. Name two metals which are both ductile as well as malleable.
2. Name one non-metal and one metal which are in the liquid state at room temperature.
3. Name any two metals which can displace hydrogen from dilute HCl.
4. Why are metals highly electropositive in nature?
5. What kind of oxides are formed when non-metals react with oxygen?
6. Why is sodium kept immersed in kerosene?
7. Which one is more reactive, Cu or Pt?
8. Between Zn and Ca, which burns with bright flame to form an oxide?
9. Why are cutting tools usually made of metals?
10. Name two metals which are found in nature in the free state.
11. Which gas is produced when dilute hydrochloric acid is added to a reactive metal?
12. Write the chemical reaction when iron reacts with dilute H_2SO_4.
13. Which metals do not corrode easily?
14. A non-metal X exists in two different forms Y and Z. Y is the hardest natural substance, whereas Z is a good conductor of electricity. Identify X, Y and Z.
15. An element forms an oxide A_2O_3 which is acidic in nature. Identify A as a metal or non-metal.
16. A solution of $CuSO_4$ was kept in an iron pot. After a few days the iron pot was found to have a number of holes in it. Explain the reason in terms of reactivity. Write the equation of the reaction involved.

Short Answer Type Questions

1. Write one example of each of:
   (a) a metal which is so soft that it can be cut with knife and a non-metal which is the hardest substance.
   (b) a metal and a non-metal which exist as liquid at room temperature.
2. State three reasons for the following facts:
   (a) Sulfur is a non-metal.
   (b) Magnesium is a metal.

One of these reasons must be supported with a chemical equation.

3. Explain the meanings of malleable and ductile.

4. Write one example of each of the following:
   (a) Most malleable metal and most ductile metal.
   (b) The best conductor of heat and the poorest conductor of heat.
   (c) A metal with the highest melting point and a metal with the lowest melting point.

5. Name a metal/non-metal:
   (a) Which makes iron hard and strong.
   (b) Which is alloyed with any other metal to make an amalgam.
   (c) Which is used to galvanise iron articles.
   (d) Whose articles when exposed to air form a black coating.

6. You are given a hammer, a battery, a bulb, wires and a switch.
   (a) How could you use them to distinguish between samples of metals and non-metals?
   (b) Assess the usefulness of these tests in distinguishing between metals and non-metals.

7. Which of the following listed metals can displace zinc from its salt solution? Give reasons for your answer along with the chemical equation. Copper, lead, magnesium, silver.

8. Mention the names of the metals for the following:
   (a) Two metals which are alloyed with iron to make stainless steel.
   (b) Two metals which are used to make jewellery.

9. Explain the following statements:
   (a) Most metal oxides are insoluble in water but some of these dissolve in water. What are these oxides and their solutions in water called?
   (b) At ordinary temperature, the surface of metals such as magnesium, aluminium, zinc, and so on, is covered with a thin layer. What is the composition of this layer? State its importance.
   (c) Some alkali metals can be cut with a knife.

10. What type of oxides are formed when non-metals combine with oxygen?

11. What are alloys? Give any three examples of alloys.

12. Shanvi treated a lustrous, divalent element M with sodium hydroxide. She observed the formation of bubbles in the reaction mixture. She made the same observations when this element was treated with hydrochloric acid. Suggest how can she identify the produced gas. Write chemical equations for both the reactions.

13. A non-metal A is an important constituent of our food and forms two oxides B and C. Oxide B is toxic whereas C causes global warming.
   (a) Identify A, B and C.
   (b) To which group of the periodic table does A belong?

14. Give two examples each of the metals that are good conductors and poor conductors of heat respectively.

15. Name one metal and one non-metal that exist in the liquid state at room temperature. Also name two metals having a melting point less than 310 K (37°C).

Long Answer Type Questions

1. Compare in tabular form the reactivities of the following metals with cold and hot water:
   (a) Sodium
   (b) Calcium
   (c) Magnesium

2. (a) Write two properties of gold which make it the most suitable metal for ornaments.
   (b) Name two metals which are the best conductors of heat.
   (c) Name two metals which melt when you keep them on your palm.

3. (a) Of the three metals X, Y and Z, X reacts with cold water, Y with hot water and Z with steam only. Identify X, Y and Z and also arrange them in the order of increasing reactivity.
   (b) An element P burns with a golden flame in air. It reacts with another element Q, atomic number 17 to give a product R. An aqueous solution of product R on electrolysis gives a compound S and liberates hydrogen. Identify P, Q, R and S. Also write down the equations for the reactions involved.

4. Two ores P and Q were taken. On heating ore P gives CO_2 whereas, ore Q gives SO_2. What steps will you take to convert them into metals?

5. Samples of four metals A, B, C and D were taken and added to the following solution one by one. The results obtained have been tabulated as follows.

Metal	Iron (IIII) sulfate	Copper (IIII) sulfate	Zinc sulfate	Silver nitrate
A	No reaction	Displacement		
B	Displacement		No reaction	
C	No reaction	No reaction	No reaction	Displacement
D	No reaction	No reaction	No reaction	No reaction

Use the table above to answer the following questions about the metals A, B, C and D.

(a) Which is the most reactive metal?

(b) What would you observe if B is added to a solution of copper (II) sulphate?

(c) Arrange the metals A, B, C and D in the order of decreasing reactivity.

6. Complete the crossword given in figure with the help of the clues.

Across

1. Which is generally hard, ductile, malleable and sonorous.
2. A metal is called so, since it can be drawn into wires.
3. Metal bells are used because of this property.

Down

4. A metal generally used for making jewelry.
5. A metal which is liquid at room temperature.
6. A metal which reacts with acid as well as base to form hydrogen gas.
7. Substances used to enhance the growth of plants.
8. Property by virtue of which metals can be beaten into thin sheets.

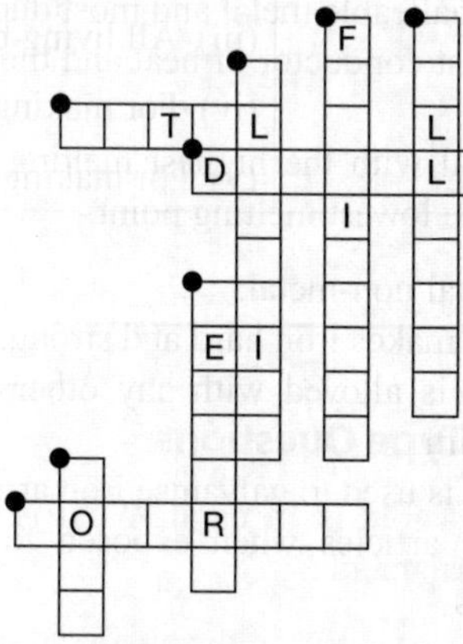

NCERT Corner

In-Text Questions

1. Non-metals cannot be drawn into wires. Why?

2. Explain the term 'metallurgy'.

3. State the general steps involved in the metallurgy of a metal.

4. Why is aluminium used in making aeroplanes?

5. What type of oxides are formed by metals?

6. How does phosphorus occur in nature?

7. Give the different forms of silica in nature.

8. Name the non-metal used in vulcanisation.

9. Name one metal which is not malleable.

10. Name the element used for making containers of dry cells.

11. Which metal is used for making radiators of cars?

12. Name the metal whose salt is used for making photographic films.

13. Which non-metal is essential for our life and all living beings inhale it during breathing?

14. Name two major non-metals which are present in fertilisers and enhance the growth of plants.

15. Which non-metal is used to disinfect water?

16. A purple-coloured non-metal forms a brown solution in alcohol which is applied on wounds as an antiseptic. Name the non-metal.

17. Zinc sulfate forms a colourless solution in water. Will you observe any colour on adding copper turnings into it?

18. Why are bells made of metals?

19. Which liquid metal is used for making thermometers?

20. A doctor prescribed a tablet to a patient suffering from iron deficiency. The tablet does not look like iron. Explain.

Textbook Exercises

1. Paheli bought a statue made of copper. To her surprise, it acquired a dull green coating after a couple of months. Explain the reason.

2. In the given figure, you find that the bulb glows when an iron nail is placed between two ends of wire. Complete the following sentences on the basis of this fact.

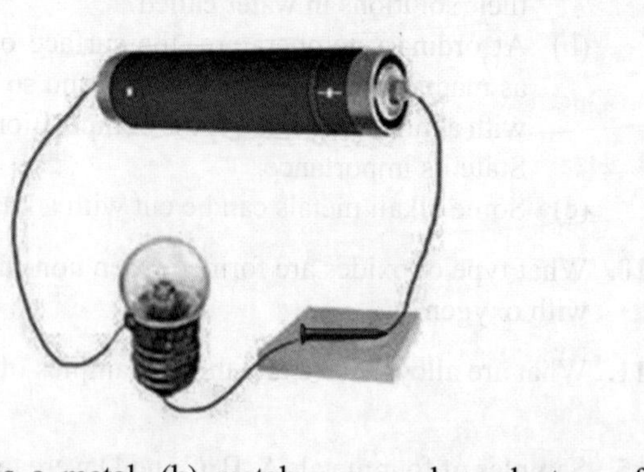

(a) Iron is a metal, (b) metals are good conductors of electricity.

3. If in the figure given above, the iron nail is replaced by a wooden stick, will the bulb glow or not? Justify your answer.

4. Paheli prepared a blue coloured solution of copper sulfate in beaker A and placed an iron nail in it. Boojho prepared a yellowish green solution of ferrous sulfate in beaker B and placed a copper wire in it. What changes will they observe in the two beakers after an hour?

5. Match the Column A with Column B.

Column A (Substance)	Column II (Application)
(a) Oxygen	(i) For making crackers
(b) Copper	(ii) For disinfecting water
(c) Sulphur	(iii) All living beings inhale during breathing
(d) Iron	(iv) For making electric wires
(e) Chlorine	(v) For making rails

Exemplar Problems

Short Answer Type Questions

1. The purity of gold is 15 carat. What is the percentage of gold in the ornaments?
2. Give two uses of sulfur in the chemical industry.
3. How is sulfur useful in agriculture? How is sulfur useful in medicine?
4. Give two uses of platinum.
5. Gaurav knows that wires can be made from copper and aluminium. He tries to make wires from sulfur and carbon. Will he succeed? Give the reason.

Long Answer Type Questions

1. Some properties are listed in the following table. Distinguish between metals and non-metals on the basis of these properties.

Properties	Metals	Non-metals
1. Appearance		
2. Hardness		
3. Malleability		
4. Ductility		
5. Heat conduction		
6. Conduction of electricity		

2. Can you store the lemon pickle in an aluminium utensil? Explain.
3. Match the substances given in column A with their uses given in column B.

Column A	Column B
(i) Gold	(a) Thermometers
(ii) Iron	(b) Electric wire
(iii) Aluminium	(c) Wrapping food
(iv) Carbon	(d) Jewellery
(v) Copper	(e) Machinery
(vi) Mercury	(f) Fuel

4. One day Reeta went to a jeweller's shop with her mother. Her mother gave an old gold jewellery piece to the goldsmith to polish. Next day when they brought the jewellery back, they found that there was a slight loss in its weight. Can you suggest a reason for the loss in weight?
5. Give reasons, why:
   (a) Silver is used in jewellery.
   (b) Copper is used in electrical wiring.
   (c) Sodium is stored in kerosene.
6. Taking the examples of magnesium and sulfur explain how metals and non-metals produce oxides with different characteristics.
7. Compare the following chemical properties of metals and non-metals.
   (a) Formation of ions
   (b) Action with dilute acids
   (c) Action with hydrogen
8. What happens when
   (a) Hydrochloric acid is poured on aluminium foil?
   (b) Sodium is placed in water?
   (c) Sulfur dioxide is dissolved in water?
   (Write the chemical equation of the reaction involved)
9. (a) A copper spoon had fallen into a container containing dil. HCl. What would happen to it in three days?
   (b) Give reasons for the following.
   - Metals are used for making bells.
   - We can't use pure gold to make jewellery.

   (c) A metal ribbon burns in air with a bright white light and forms a white powder.
   - Which metal is this?
   - Give the equation of the reaction taking place.
   - Will the metallic oxide formed be acidic or basic in nature?

COMPETITION WINDOW (OBJECTIVE TYPE)

[For NEET, JEE (Main and Advanced), NTSE, KVPY and Olympiads]

Topic-wise MCQs

Non-metals and their general features

1. A majority of non-metals are present as
(a) solids (b) liquids
(c) gases (d) both (a) and (c)

2. The most abundant element in the universe is
(a) hydrogen (b) oxygen
(c) helium (d) carbon

3. The most abundant element in the earth's crust is
(a) iron (b) oxygen
(c) silicon (d) aluminium

4. Which of the following statements regarding non-metals is true?
(a) Non-metals are non-lustrous with a dull appearance. Graphite, an allotrope of carbon and iodine have shining lustrous surfaces.
(b) Non-metals replace hydrogen from acids.
(c) Non-metals are of two types only—solids and gases.
(d) Generally, non-metals react with oxygen to form basic oxides.

5. Select the property that is associated with non-metals.
(a) low melting point
(b) poor conductor of electricity
(c) low density
(d) all of the above

6. Which of the following non-metals is a liquid?
(a) carbon (b) bromine
(c) phosphorus (d) sulfur

7. Generally, non-metals are not conductors of electricity. Which of the following is a good conductor of electricity?
(a) diamond (b) graphite
(c) sulfur (d) fullerene

8. Which of the following is a non-metal?
(a) graphite (b) platinum
(c) tungsten (d) mercury

9. Which of the following is not a metal?
(a) copper (b) sulfur
(c) aluminium (d) iron

10. Non-metals do not conduct electricity because
(a) they are electron donors
(b) they are electron acceptors
(c) they have free electrons
(d) they do not have free electrons

11. For the vulcanisation of rubber, which one of the following non-metals do we use?
(a) chlorine (b) carbon
(c) sulfur (d) phosphorus

12. Which of these elements is used as an antiseptic in medicine?
(a) nitrogen (b) iodine
(c) carbon (d) oxygen

13. Which of the following non-metals reacts and catches fire on exposure to air?
(a) phosphorus (b) nitrogen
(c) sulfur (d) hydrogen

Metals and their general features

14. Which of the following metals occurs in the free state in nature?
(a) gold (b) sodium
(c) aluminium (d) calcium

15. The most abundant metal in earth crust is
(a) Cu (b) Al
(c) Fe (d) Zn

16. Metals are generally solid. Which of the following metals is in the liquid state at room temperature?
(a) mercury (b) silver
(c) aluminium (d) sodium

17. The ability of metals to be drawn into thin wires is known as
(a) ductility (b) malleability
(c) sonorosity (d) conductivity

18. Select the metal that is soft
(a) potassium (b) lead
(c) aluminium (d) copper

19. Mercury is used in thermometers because
(a) it is a liquid
(b) it expands on heating
(c) it does not stick to the glass
(d) all of these

20. Tungsten is used in electric bulbs because
(a) it has high tensile strength
(b) it has high density
(c) it is sonorous
(d) it has high melting point

21. Which of the following properties is not responsible for copper to be used as electrical conduction wires?
(a) ductility
(b) colour
(c) good conductor of electricity
(d) it is solid

22. Which of the following can be beaten into thin sheets?
(a) zinc (b) phosphorus
(c) sulfur (d) oxygen

23. Materials which can be drawn into wires are called ductile. Which of the following is not a ductile material?
(a) silver (b) copper
(c) sulfur (d) aluminium

24. Metals are generally hard. Which of the following metals is an exception and can be cut with a knife?
(a) iron (b) sodium
(c) gold (d) magnesium

25. When a substance X is a hit with a hammer, it expands in size but does not break. This is because the substance is
(a) elastic (b) malleable
(c) ductile (d) hard

Reactivity of metals

26. The most characteristic property of metals is their tendency to
(a) lose electrons (b) gain electrons
(c) form basic oxides (d) form hydrides

27. Metals combine with oxygen to form an oxide. The nature of oxide formed is ________
(a) neutral (b) amphoteric
(c) acidic (d) basic

28. Which of the following metals has no reaction even with steam?
(a) iron (b) silver
(c) sodium (d) calcium

29. Metals generally react with dilute acids to produce hydrogen gas. Which one of the following metals does not react with dilute hydrochloric acid?
(a) magnesium (b) aluminium
(c) iron (d) copper

30. Which of the following reacts with cold water vigorously?
(a) carbon (b) sodium
(c) magnesium (d) sulfur

31. The metal which produces hydrogen gas on reaction with dilute hydrochloric acid as well as sodium hydroxide solution is
(a) copper (b) iron
(c) aluminium (d) sodium

32. Which one of the following metals does not react with cold as well as hot water?
(a) Na (b) Ca
(c) Mg (d) Fe

33. Sodium reacts with cold water to form
(a) sodium hydride and oxygen
(b) sodium hydroxide and oxygen
(c) sodium hydroxide and hydrogen
(d) none of these

34. The reactivities of iron, magnesium, sodium and zinc towards water are in the order
(a) Na > Mg > Zn > Fe (b) Mg > Na > Fe > Zn
(c) Fe > Mg > Na > Zn (d) Zn > Na > Mg > Fe

35. Which of the following methods is suitable for preventing an iron frying pan from rusting?
(a) applying grease
(b) applying a coating of zinc
(c) applying paint
(d) all of the above

36. An element A forms an oxide AO which turns red litmus blue. Identify A.
(a) a metalloid (b) a noble gas
(c) a metal (d) a non-metal

37. When steam is passed through zinc,
(a) zinc hydroxide is formed
(b) zinc oxide is formed
(c) water is formed
(d) hydrochloric acid is formed

38. When a metal M is added to dilute HCl solution, there is no evolution of gas. The metal M is
(a) Ag (b) Zn
(c) K (d) Na

Extraction of metals

39. Chalcopyrite is an ore of
(a) calcium (b) copper
(c) magnesium (d) gallium

40. Which of the following is a fluoride mineral of aluminium?
(a) feldspar (b) cryolite
(c) mica (d) porcelain

41. Which of the following is an important carbonate mineral of magnesium?
(a) epsomite (b) carnallite
(c) dolomite (d) olivine

42. Which of the following is a carbonate ore?
(a) limonite (b) carnallite
(c) horn silver (d) siderite

43. The removal of gangue from the ore is called
(a) concentration (b) liquation
(c) hand picking (d) leaching

44. Cassiterite is concentrated by
(a) hand picking
(b) levigation
(c) electromagnetic separation
(d) Wilfley table method

45. When the densities of the ore particles and gangue particles are different, then the ore is concentrated by
(a) froth flotation process
(b) gravity separation method
(c) electromagnetic
(d) hand picking

46. Which of the following ores is concentrated by the froth flotation process?
(a) carbonate (b) chloride
(c) sulfide (d) silicate

47. Bauxite is concentrated by
(a) chemical method (b) roasting
(c) froth flotation process (d) magnetic separation

48. Wolframite ore is separated from tinstone ore by the process of
(a) smelting
(b) calcination
(c) roasting
(d) electromagnetic separation

49. Roasted ore is reduced to metal by
(a) poling (b) smelting
(c) liquation (d) roasting

50. The active metals are extracted from their ores by the process of reduction with
(a) hydrogen (b) carbon
(c) electric current (d) aluminium

51. If the melting points of the metals are lower than those of the impurities, then the crude metals are purified by the process of
(a) cupellation (b) liquation
(c) bessemerisation (d) distillation

52. Which of the following processes is used to remove reducible oxides from the metal?
(a) liquation (b) cupellation
(c) poling (d) electrorefining

53. Which of the following processes is not used for the purification of solid impurities?
(a) sublimation (b) distillation
(c) vapourisation (d) crystallisation

Alloy

54. Two metals are melted and mixed together. The resulting mass is cooled to solidify. What is this solid called?
(a) an oxide (b) a mineral
(c) an alloy (d) a non-metal

55. Alloys are homogeneous mixtures of a metal with a metal or non-metal. Which among the following alloys contains a non-metal as one of its constituents?
(a) amalgam (b) brass
(c) bronze (d) steel

56. The metal which is present in brass, bronze and German silver is
(a) Cu (b) Al
(c) Mg (d) Zn

57. Stainless steel contains iron and
(a) Zn + Pb (b) Fe + Cr + Ni
(c) Cr + Ni (d) Cr + Zn

58. Bell metal is an alloy of
(a) Cu + Zn (b) Cu + Ni
(c) Cu + Pb (d) Cu + Sn

59. Which among the following alloys contains mercury as one of its constituents?
(a) stainless steel (b) alnico
(c) solder (d) zinc amalgam

Miscellaneous

1. Which of the following properties is generally not shown by metals?
(a) electrical conduction (b) sonorous in nature
(c) dullness (d) ductility

2. Which of the following elements produces a basic oxide on reacting with oxygen?
(a) sulfur (b) chlorine
(c) magnesium (d) phosphorus

3. A number of elements are available in the earth's crust but the most abundant elements are
(a) Al and Fe (b) Fe and Cu
(c) Cu and Ag (d) Fe and Cu

4. Boojho has learnt that non-metals on beating with a hammer are generally broken into pieces. Which of the following is a non-metal?
(a) iron nail (b) aluminium wire
(c) copper plate (d) piece of coal

5. Which one of the following four metals would be displaced from the solution of its salts by the other three metals?
(a) Mg (b) Ag
(c) Zn (d) Cu

6. Generally metallic oxides are basic and non-metallic oxides are acidic in nature. Solution of which of the following oxides in water will change the colour of blue litmus to red?
(a) sulfur dioxide (b) iron oxide
(c) magnesium oxide (d) copper oxide

7. While extracting an element from its ore, the ore is ground and leached with dilute KCN solution to form the soluble product, potassium argentocyanide. The element is
(a) silver (b) manganese
(c) lead (d) chromium

8. The substance that will be flattened on beating with a hammer is a
(a) crystal of iodine (b) lump of sulfur
(c) piece of coal (d) zinc granule

9. Stainless steel is a very useful material. In stainless steel, iron is mixed with
(a) Ni and Cr (b) Cu and Cr
(c) Ni and Cu (d) Cu and Au

10. Which of the following reactions is an example of auto reduction?

(a) $Cu_2O + C \rightarrow 2Cu + CO$
(b) $Fe_3O_4 + 4CO \rightarrow 3Fe + 4CO_2$
(c) $Cu_2O + Cu_2S \rightarrow 3Cu + SO_2$
(d) $Cu^{2+}(aq) + Fe(s) \rightarrow Cu(s) + Fe^{2+}(aq)$

11. If copper is kept open in air, it slowly loses its shining brown surface and gains a green coating. It is due to the formation of
(a) $CuSO_4$ (b) $CuCO_3$
(c) $Cu(NO_3)_2$ (d) CuO

12. Which of the following oxide(s) of iron would be obtained on prolonged reaction of iron with steam?
(a) FeO (b) Fe_2O_3
(c) Fe_3O_4 (d) Fe_2O_3 and Fe_3O_4

13. Generally, metals react with acids to give its salt and hydrogen gas. Which of the following acids does not give hydrogen gas on reacting with metals (except Mn and Mg)?
(a) H_2SO_4 (b) HCl
(c) HNO_3 (d) all of these

14. Sulfide ores of metals are usually concentrated by the froth flotation process. Which one of the following sulfide ores is an exception and is concentrated by chemical leaching?
(a) argentite (b) galena
(c) copper pyrites (d) sphalerite

15. Reaction between X and Y forms a compound Z. X loses an electron and Y gains an electron. Which of the following properties is not shown by Z?
(a) has high melting point
(b) has low melting point
(c) conducts electricity in molten state
(d) occurs as solid

16. The electronic configurations of three elements X, Y and Z are
X: 2, 8; Y: 2, 8, 7 and Z: 2, 8, 2. Which of the following is correct?
(a) X is a metal
(b) Y is a metal
(c) Z is a non-metal
(d) Y is a non-metal and Z is a metal

17. Although metals form basic oxides, which of the following metals form an amphoteric oxide?
(a) Na (b) Ca
(c) Al (d) Cu

18. Electrical wires have a coating of an insulting material. The material, generally used is
(a) sulfur (b) graphite
(c) PVC (d) all can be used

19. The process of eating away of metals layer by layer due to the formation of a metal compound on the surface is called
(a) corrosion
(b) vulcanisation
(c) galvanisation
(d) amalgam formation

20. Which of the following methods of purification is generally used when a metal of high degree of purity is required?
(a) van Arkel
(b) liquation
(c) electrorefining
(d) zone refining

Advanced and Olympiads

Single Choice

1. Which of the following metals exist in their native state in nature?
(i) Cu (ii) Au
(iii) Zn (iv) Ag
(a) (i) and (ii) (b) (ii) and (iii)
(c) (ii) and (iv) (d) (iii) and (iv)

2. The reason(s) why aluminium is preferred over copper for making overhead transmission cables is,
(i) aluminium is lighter than copper
(ii) aluminium is harder than copper
(iii) aluminium is less soluble in water than copper
(a) (i) and (ii) (b) (i), (ii) and (iii)
(c) (i) only (d) (iii) only

3. Aluminium is used for making cooking utensils. Which of the following properties of aluminium are responsible for the same?
(i) good thermal conductivity
(ii) good electrical conductivity
(iii) ductility
(iv) high melting point
(a) (i) and (ii) (b) (i) and (iii)
(c) (ii) and (iii) (d) (i) and (iv)

4. What happens when calcium is treated with water?
(i) it does not react with water
(ii) it reacts violently with water
(iii) it reacts less violently with water
(iv) bubbles of hydrogen gas formed stick to the surface of calcium
(a) (i) and (iv) (b) (ii) and (iii)
(c) (i) and (ii) (d) (iii) and (iv)

5. $P + CuSO_4 \rightarrow PSO_4 + Cu$
$Q + CuSO_4 \rightarrow QSO_4 + Cu$, in these reactions P and Q may be ________.
(a) Zn, Ag (b) Ag, Al
(c) Fe, Ag (d) Zn, Fe

6. 2 mL each of concentrated HCl, HNO_3 and a mixture of concentrated HCl and concentrated HNO_3 in the ratio of 3 : 1 were taken in test tubes labelled as A, B and C. A small piece of metal was put in each test tube. No change occurred in test tubes A and B but the metal got dissolved in test tube C respectively. The metal could be
(a) Al (b) Au
(c) Cu (d) Pt

7. An unknown metal A when placed in copper sulfate solution gives a red brown deposit. When placed in magnesium sulfate solution, it gives no reaction. Identify element A.
(a) potassium (b) sodium
(c) iron (d) calcium

8. Which of the following alloys contain Cu and Zn?
(i) bronze (ii) brass
(iii) gun metal (iv) type metal
(a) (ii), (iii), (iv) (b) (i), (ii), (iii)
(c) (i), (ii) (d) (i), (iii), (iv)

Multiple Choice

9. Which of the following statements are true?
(a) Mercury is the only liquid metal.
(b) Bromine is the only liquid non-metal.
(c) Metals are good conductors of heat and electricity.
(d) Gold, silver and zinc are the most malleable metals.

10. Which of the following contain a metallic element?
(a) sodium chloride (b) silicon dioxide
(c) magnesium oxide (d) calcium hydroxide

11. Leaching is used for which metal/s?
(a) Hg (b) Al
(c) Ag (d) Au

12. Which among the following statements is correct for magnesium metal?
(a) It burns in oxygen with a dazzling white flame.
(b) It reacts with cold water to form magnesium oxide and evolves hydrogen gas.
(c) It reacts with hot water to form magnesium hydroxide and evolves hydrogen gas.
(d) It reacts with steam to form magnesium hydroxide and evolves hydrogen gas.

13. Which of the following reactions occur during calcinations?
(a) $2FeS_2 + \frac{11}{2}O_2 \rightarrow Fe_2O_3 + 4SO_2$
(b) $CaCO_3 \rightarrow CaO + CO_2$
(c) $ZnS + \frac{3}{2}O_2 \rightarrow ZnO + SO_2$
(d) $Al_2O_3.xH_2O \rightarrow Al_2O_3 + xH_2O$

14. Which metals react with water?
(a) Na (b) Mg
(c) Cu (d) Fe

15. Which of the following statement is/are correct?
(a) Iron does not form an amalgam.
(b) Dental amalgams are the mixtures of mercury with a silver–tin alloy.
(c) Sodium amalgam is more reactive than sodium metal.
(d) Solder and type metal are alloys of lead.

16. Which of the following is/are a property of aluminium?
(a) it is a good conductor of heat and electricity
(b) it is malleable
(c) it is heavy
(d) it is ductile

17. Which of the following are properties of a non-metal?
(a) forms anions (b) forms acidic oxides
(c) high tensile strength (d) low density

Comprehension Type

Comprehension I: An alloy may be defined as a metallic material consisting of two or more elements. Metals, when melted together, tend to dissolve in one another to form alloys. However, it is known that platinum and iron do not dissolve in mercury and zinc is almost insoluble in molten lead. Metals form alloys not only with one another but also with certain non-metallic elements such as carbon, silicon and phosphorous. Usually, the alloys are solid but they may be liquids also. For example, certain amalgams (alloys of mercury). Alloys may also have a variety of colours different from those of the constituent elements. They are usually harder, corrosion resistant and are generally more useful in industry than the parent metals.

18. An alloy is
(a) an inter-metallic compound
(b) a solid material containing two or more metallic elements
(c) a metallic material containing at least one metallic element
(d) a metallic material containing at least one non-metallic element.

19. Which one of the following pairs of elements might form an alloy?
(a) zinc and lead (b) iron and carbon
(c) iron and mercury (d) mercury and platinum

20. Amalgams are
(a) highly coloured alloys
(b) always solids
(c) alloys which contain mercury as one of the components
(d) alloys which have great resistance to abrasion

Matrix Matching

21. Match the following

Column I	Column II
(a) Sodium	(p) On burning gives an acidic gas
(b) Phosphorus	(q) Reacts neither with acids nor with bases
(c) Copper	(r) It is so soft that it can be cut with a knife
(d) Charcoal	(s) Burns spontaneously on exposure to air
	(t) Acquires a dull green coating on exposure to air

22. Match the following

Column I	Column II
(a) Gun metal	(p) Lead + tin
(b) German silver	(q) Copper + tin + zinc
(c) Brass	(r) Copper + zinc
(d) Solder	(s) Copper + zinc + nickel

Integer Type

23. How many of the following elements are metals?
C, H, Br, B, Na, Mg, Al, Fe, Cu

24. How many of these metals are more reactive than hydrogen?
Na, Mg, Al, Zn, Sn, Pb, Cu, Ag, Au

25. How many of these ores are oxide ores?
Cinnabar, zinc blende, bauxite, hematite, magnetite, zincite, horn silver

26. Find the total number of atoms present in chalcopyrite?

27. How many water molecules are present in the general formula of bauxite ore?

28. How many metals are present German silver?

Answer Keys

Fill in the Blanks

1. more
2. reactive
3. good, electricity
4. liquid
5. iron, sodium and nickel
6. sulfur, chlorine and oxygen
7. more
8. hydrogen
9. metallic bond
10. sulfide
11. NaOH
12. coke or carbon
13. copper and zinc
14. zero
15. platinum

True or False

1. T	2. F	3. F	4. T	5. T
6. F	7. F	8. T	9. F	10. T
11. T	12. F	13. F	14. T	15. T

Match the Following

1. (a)—(r), (b)—(p), (c)—(q), (d)—(s)
2. (a)—(p), (b)—(q), (c)—(s), (d)—(r)

Very Short Answer Type Questions

1. Gold and silver are ductile and malleable.
2. Bromine (non-metal) and mercury (metal) are liquids at room temperature.
3. Zinc and iron can displace H_2 from HCl.
4. Metals are highly electropositive elements as they can lose electrons very easily to form positively charged ions.

For example, $Na \rightarrow Na^+ + e^-$

$Mg \rightarrow Mg^{2+} + 2e^-$

5. Acidic oxides are formed when non-metals react with oxygen.

$$C + O_2 \rightarrow CO_2$$

6. Sodium is a very reactive metal. It reacts vigorously with the oxygen (in air) and even catches fire. So, to protect sodium, it is kept immersed in kerosene.
7. Copper (Cu) is more reactive than platinum (Pt) as copper is above Pt in the reactivity series.
8. Ca reacts and burns with a bright flame on heating in air while Zn reacts slowly.
9. Cutting tools are usually made up of metals as metals are hard and strong.
10. Gold and platinum are two metals that are found in nature in the free state.
11. When a reactive metal reacts with hydrochloric acid, hydrogen gas is produced. A reactive metal displaces the hydrogen from acid and releases hydrogen gas.
12. The chemical reaction when iron reacts with dilute H_2SO_4 is

$$Fe(s) + \text{dil. } H_2SO_4(aq) \rightarrow FeSO_4(aq) + H_2(g)$$

13. Metals which are less reactive and lie at the bottom of reactivity series (silver, gold, platinum) do not react with the atmospheric gases and hence do not corrode easily.
14. X: carbon; Y: diamond and Z: graphite.
15. Since the oxide of the element is acidic, A will be a non-metal.
16. Fe is more reactive than Cu. Therefore, Fe displaces Cu from $CuSO_4$ and forms $FeSO_4$.

$$Fe + CuSO_4 \rightarrow FeSO_4 + Cu$$

Short Answer Type Questions

1. (a) Sodium is a soft metal and carbon (diamond) is the hardest non-metal.
(b) Mercury is a metal (liquid) and bromine is a non-metal (liquid).
Note: Mercury is a metal as it has free electrons as do all metals.

2. (a) Sulfur is a non-metal.
   - It is a poor conductor of heat and electricity.
   - It is neither malleable for ductile.
   - It forms acidic oxides.

$$S + O_2 \rightarrow SO_2$$

$$SO_2 + H_2O \rightarrow H_2SO_3 \text{ (Sulfurous acid)}$$

SO_2 is an acidic oxide.

Note: Sulfur is an ingredient of gunpowder and is used in making matches and as an antiseptic and fungicide.

(b) Magnesium is a metal.
   - It is a good conductor of heat and electricity.
   - It is malleable and ductile.
   - It forms basic oxides.

$$2Mg + O_2 \rightarrow 2MgO$$

$$MgO + H_2O \rightarrow \underset{\text{Magnesium hydroxide}}{Mg(OH)_2}$$

MgO is basic in nature.

3. **Malleable:** A metal is said to be malleable if it can be beaten into thin sheets. For example, silver and gold are highly malleable metals.

Ductile: A metal is said to be ductile when it can be drawn into thin wires. For example, platinum and gold are highly ductile metals.

4. (a) Gold is the most malleable and most ductile metal.
(b) Silver is the best conductor and lead is the poorest conductor.
(c) Tungsten has the highest melting point and mercury has the lowest melting point.

5. (a) Carbon (non-metal) makes iron hard and strong.
(b) Mercury is alloyed to make an amalgam.
(c) Zinc is used to galvanise iron articles.
(d) Silver articles form a black coating when exposed to air.

Note: Pure iron is a very soft and flexible (on heating) metal. When mixed with a small amount of carbon (app. 0.05%), it becomes hard.

6. (a) The sample should be beaten with a hammer. If it can be beaten into thin sheets (that is, it is malleable), then it is a metal, if not, it is a non-metal.
(b) We can use the battery, bulb, wires and a switch to set up a circuit with the sample. If the sample conducts electricity, then it is a metal, if not, it is a non-metal. Since metals have free electrons, they are good conductors of electricity. Non-metals do not have free electrons, they are hence bad conductors of electricity.

The above tests are useful in distinguishing between metals and non-metals as these are based on the physical properties. No chemical reactions are involved in these tests.

7. Metals that are more reactive than zinc can displace it from its salt solution. So, magnesium can displace Zn from its salt solution.

$$Mg(s) + ZnSO_4(aq) \rightarrow MgSO_4(aq) + Zn(s)$$

8. (a) Nickel and chromium are alloyed with iron to make stainless steel.
(b) Gold and platinum are used to make jewellery.

9. (a) Solution of metal oxides in water is called an alkali. For example, NaOH (sodium hydroxide).
(b) This layer is called the oxide layer. It makes the metal less reactive and due to this layer, the metal cannot get corroded and remains intact.
(c) Some alkali metals are very soft so they can be cut with a knife. For example, sodium metal.

Note: Due to presence of weak metallic bonding, alkali metals are very soft.

10. Non-metals combine with oxygen to form neutral and acidic oxides.

$$C + \frac{1}{2}O_2 \rightarrow CO \text{ (neutral oxide)}$$

$$N_2 + O_2 \rightarrow NO \text{ (neutral oxide)}$$

$$C(s) + O_2 \rightarrow CO_2 \text{ (acidic oxide)}$$

$$N_2 + 2O_2 \rightarrow 2NO_2 \text{ (acidic oxide)}$$

$$S(s) + O_2(g) \rightarrow SO_2(g) \text{ (acidic oxide)}$$

11. Alloys are homogeneous mixtures of two or more metals or a metal and a non-metal.

Some examples of alloys are given below.

Brass	Cu + Zn	metal + metal
Steel	Fe + Ni + Cr + C	metals + non-metals
Bronze	Cu + Zn	metal + metal

12. The produced gas can be identified by bringing a burning match stick near the reaction vessel, a pop sound is heard.

$$M + 2NaOH \rightarrow Na_2MO_2 + H_2$$

$$M + 2HCl \rightarrow MCl_2 + H_2$$

The element is a metal.

13. (a) A is carbon, B is carbon monoxide and C is carbon dioxide.
(b) A belongs to Group 14 of the periodic table.

14. (a) Good conductors: Ag and Cu
(b) Poor conductors: Pb and Hg

15. Metal: mercury (Hg); non-metal: bromine (Br)

Two metals with melting points less than 310 K are cesium (Cs) and gallium (Ga).

Long Answer Type Questions

1.

Metal	Reactivity	Reactions
Sodium (Na)	It reacts vigorously with cold water.	$2Na(s) + 2H_2O(l) \rightarrow 2NaOH(aq) + H_2(g)$
Calcium (Ca)	It reacts slowly with cold water and moderately with hot water.	$Ca(s) + 2H_2O(l) \rightarrow Ca(OH)_2 + H_2(g)$ (Hot)
Magnesium (Mg)	Magnesium does not react with cold water. It reacts slowly with hot water and vigorously with steam.	$Mg(s) + 2H_2O(l) \rightarrow Mg(OH)_2 + H_2(g)$ (Hot) $Mg(s) + H_2O(l) \rightarrow MgO + H_2(g)$ (Steam)

2. (a) Gold is an inert metal and it has a shiny surface. It is also malleable and ductile. Therefore, it is used for making ornaments.
   (b) Copper and aluminium.
   (c) Cesium and gallium.

3. (a) X is alkali metal, Na or K
   Y is an alkaline earth metal, Mg or Ca
   Z is Fe
   Increasing reactivity series: Na > Mg> Fe
   (b) P = Na; Q = Cl_2; R = NaCl; S = NaOH

$$2Na + Cl_2 \rightarrow 2NaCl$$

$$2NaCl(aq) + 2H_2O(l) \rightarrow 2NaOH(aq) + Cl_2(g) + H_2(g)$$

4. Since ore P gives CO_2 and ore Q gives SO_2, the ores are MCO_3 and MS.
   A can be obtained

$$MCO_3 \xrightarrow{\text{Calcination}} MO + CO_2$$

$$MO + C \xrightarrow{\text{Reduction}} M + CO$$

B can be obtained

$$2MS + 3O_2 \xrightarrow{\text{Roasting}} 2MO + 2SO_2$$

$$MO + C \rightarrow M + CO$$

5. From the given table we can write the following observations:
   - A is less reactive than Fe but more reactive than Cu.
   - B is more reactive than Fe but less reactive than Zn.
   - C is more reactive than Ag but less reactive than Fe, Cu and Zn.
   - D is less reactive than Fe, Cu, Zn, Ag.

   Hence:
   (a) B is the most reactive.
   (b) The blue colour of the copper sulfate solution disappears, and reddish-brown copper metal is deposited on the metal B.
   (c) B > A > C > D is the order of reactivity.

6.

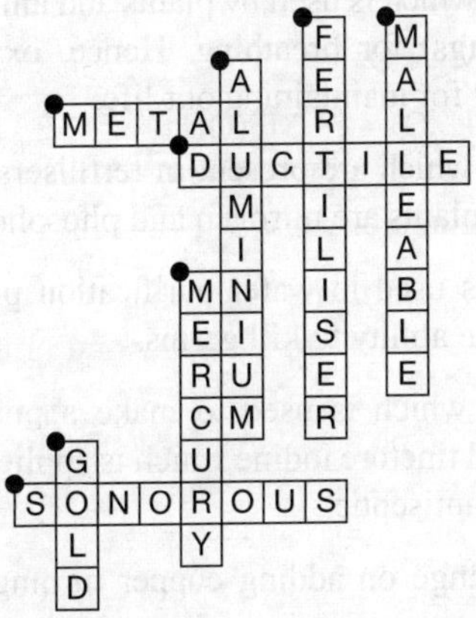

Across
1. Metal
2. Ductile
3. Sonorous

Down
4. Gold
5. Mercury
6. Aluminium
7. Fertiliser
8. Malleable

NCERT Corner

In-Text Questions

1. Non-metals are not ductile, so they cannot be drawn into wires.
2. Metallurgy is the science of extracting metals from their ores and purifying them for various uses.
3. The general steps involved in metallurgy are:
   (i) Concentration of ore.
   (ii) Reduction of the metal compound.
   (iii) Refining of metal.
4. Aluminium is used in making aeroplanes, as it is light and has high resistance to corrosion when exposed to air (which aircrafts require the most).

5. Metals form basic oxides. For example, Na_2O, CaO.
6. Phosphorus occurs in nature in the combined state as it has a strong affinity for oxygen.
7. Silica occurs in nature as ordinary sand, flint, quartz and opal.
8. Sulfur.
9. Zinc or arsenic is not malleable.
10. Zinc is used for making containers of dry cells.
11. Copper is used for making radiators of cars.
12. Silver salt (AgBr) is used for making photographic films.
13. Oxygen is a non-metal which is used by plants and animals (including human beings) for breathing. Hence, oxygen (non-metal) is essential for maintaining our life.
14. Two major non-metals which are present in fertilisers and enhance the growth of plants are nitrogen and phosphorus.
15. Chlorine (non-metal) is used in water purification plants because chlorine has the ability to kill germs.
16. Iodine is a non-metal which is used to make a purple-coloured solution called tincture iodine which is applied on cuts and wounds as an antiseptic.
17. There is no colour change on adding copper turnings to zinc sulfate solution because copper is less reactive than zinc and it cannot displace zinc from its salt solution. So, displacement reaction does not take place.
18. Metals are sonorous. The ringing sound produced by metals is known as sonorosity. So, the bells are made of metals.
19. Mercury metal is used for making thermometers.
20. Iron is a metal but the tablet is not made up of iron metal. It is made up of a salt of iron. So, it does not look like iron.

Textbook Exercises

1. When a copper statue (or copper vessel) is exposed to moist air for long, it acquires a dull green coating. The green material is a mixture of copper hydroxide [$Cu(OH)_2$] and copper carbonate ($CuCO_3$) formed due to the reaction of copper with moist air. The reaction is as follows:

$$\underbrace{2Cu + H_2O + CO_2 + O_2}_{\text{Moist air}} \longrightarrow \underbrace{Cu(OH)_2 + CuCO_3}_{\text{Green coating}}$$

The green coating is basic copper carbonate and the formation of green coating on copper objects show the corrosion of copper.

2. (a) Iron (b) conductors.
3. No, the bulb will not glow as wood is a non-conductor of electricity.
4. In beaker A_1, a reddish-brown layer of copper will deposit on the iron nail and blue coloured solution of copper sulfate will become yellowish green due to the formation of iron sulfate. This is because iron is more reactive than copper and it displaces copper from its salt solution (copper sulfate solution). In beaker B_1, no change is observed because copper being less reactive, does not displace iron (more reactive) from the ferrous sulfate solution.
5. The correct matching is as given:
(a)—(iii), (b)—(iv), (c)—(i), (d)—(v), (e)—(ii)

Exemplar Problems

Short Answer Type Questions

1. 24 carat purity of gold =100
∴ 1 carat purity or gold = 100/24
15 carat purity of gold = (100 × 15)/24 = 62.5%
2. (i) It is used in the manufacture of sulfuric acid.
(ii) It is used in the manufacture of carbon disulfide, which is used as an industrial solvent.
3. Sulfur powder is an excellent insecticide and fungicide. It is used in spraying fruit trees.
(i) Sulfur is the main constituent of skin ointments.
(ii) Metallic sulfides of sulfur are used in the preparation of Ayurvedic medicines.
4. (i) It is used as a catalyst in the manufacture of sulfuric and nitric acids.
(ii) Platinum catalytic converters use platinum as a catalytic agent.
5. No, he will not succeed as sulfur and carbon are non-metals. Non-metals are not ductile, so they cannot be drawn into wires.

Long Answer Type Questions

1.

Properties	Metals	Non-metals
1. Appearance	Have metallic lustre	Dull
2. Hardness	Hard	Soft
3. Malleability	Malleable	Non-malleable
4. Ductility	Ductile	Non-ductile
5. Heat conduction	Good conductors	Bad conductors
6. Conduction of electricity	Good conductors	Bad conductors/ insulators

2. No, we cannot store lemon pickle in an aluminium utensil as aluminium is a metal and metals can readily react with acids to produce hydrogen. When aluminium comes in contact with lemon, which is acidic, will react to give hydrogen and the pickles will get spoilt.
3. (i)—(d), (ii)—(e), (iii)—(c), (iv)—(f), (v)—(b), (vi)—(a).
4. The gold jewellery is dipped into an acidic solution called aqua regia (a mixture of 3 parts hydrochloric acid and 1 part

nitric acid) for polishing. On dipping the gold jewellery in the acid solution, the outer layer of gold dissolves and the inner shiny layer appears. This causes a slight loss in its weight.

5. (a) Silver does not corrode and it is malleable and ductile, so, it can be used in jewellery.
   (b) Copper is used in electrical wiring as it is a good conductor of electricity.
   (c) Sodium has a low ignition temperature. It oxidises quickly and burns when exposed to air. It can only be stored in a liquid hydrocarbon like mineral oil or kerosene.

6. Magnesium burns in oxygen to form magnesium oxide, which dissolves in water to form magnesium hydroxide(an alkali).

$$2Mg + O_2 \rightarrow 2MgO$$
$$MgO + H_2O \rightarrow Mg(OH)_2$$

Magnesium hydroxide changes red litmus into blue.
Sulfur burns in air to form sulfur dioxide, which dissolves in water to form sulfurous acid – an acid – which turns blue litmus into red.

$$S + O_2 \rightarrow SO_2$$
$$SO_2 + H_2O \rightarrow H_2SO_3$$

7. (a) Formation of ions.
   Metals form cations whereas non-metals form anions.
   (b) Action with dilute acids.
   Metals react with dilute mineral acid to liberate hydrogen.
   Non-metals do not react with dilute mineral acids.
   (c) Action with hydrogen.
   Metals do not react with hydrogen but non-metals react with hydrogen to form hydrides.

8. (a) $\underset{\text{Aluminium}}{2Al} + \underset{\text{Hydrochloric acid}}{6HCl} \longrightarrow \underset{\text{Aluminium chloride}}{2AlCl_3} + \underset{\text{Hydrogen}}{3H_2}$
   (b) $\underset{\text{Sodium}}{2Na} + \underset{\text{Water}}{2H_2O} \longrightarrow \underset{\text{Sodium hydroxide}}{2NaOH} + \underset{\text{Hydrogen}}{H_2}$
   (c) $\underset{\text{Sulphur dioxide}}{SO_2} + \underset{\text{Water}}{H_2O} \longrightarrow \underset{\text{Sulphurous acid}}{H_2SO_3}$

9. (a) Nothing will happen as copper does not react with hydrochloric acid.
   (b) • Metals have the property of sonorosity so they are used for making bells.
   • Pure gold cannot be used for making jewellery because it is very soft.
   (c) • Magnesium.
   • $2Mg + O_2 \rightarrow 2MgO$
   • Basic in nature.

Competition Window (Objective Type)

Topic-wise MCQs

1. (d) 2. (a) 3. (b) 4. (a) 5. (d)
6. (b) 7. (b) 8. (a) 9. (b) 10. (d)
11. (c) 12. (b) 13. (a) 14. (a) 15. (b)
16. (a) 17. (a) 18. (a) 19 (d) 20. (d)
21. (b) 22. (a) 23. (c) 24. (b) 25. (b)
26. (a) 27. (d) 28. (b) 29. (d) 30. (b)
31. (c) 32. (d) 33. (c) 34. (a) 35. (b)
36. (c) 37. (b) 38. (a) 39. (b) 40. (b)
41. (c) 42. (d) 43. (a) 44. (c) 45. (b)
46. (c) 47. (a) 48. (d) 49. (b) 50. (c)
51. (b) 52. (c) 53. (b) 54. (c) 55. (d)
56. (a) 57. (c) 58. (d) 59. (d)

Miscellaneous

1. (c) 2. (c) 3. (a) 4. (d) 5. (b)
6. (a) 7. (a) 8. (d) 9. (a) 10. (c)
11. (b) 12. (c) 13. (c) 14. (a) 15. (b)
16. (d) 17. (c) 18. (c) 19. (a) 20. (d)

Advanced and Olympiads

Single Choice

1. (c) 2. (c) 3. (d) 4. (d) 5. (d)
6. (b) 7. (c) 8. (c)

Multiple Choice

9. (abc) 10. (acd) 11. (bcd) 12. (acd) 13. (bd)
14. (abd) 15. (abd) 16. (abd) 17. (abd)

Comprehension Type

18. (b) 19. (a) 20. (c)

Matrix Matching

21. (a)—(r), (b)—(s), (c)—(t), (d)—(p)
22. (a)—(q); (b)—(s); (c)—(r), (d)—(p)

Integer Type

23. 5 24. 6 25. 4 26. 4 27. 2 28. 3

Hints and Solutions

Competition Window (Objective Type)

Topic-wise MCQs

1. Non-metals are mainly available in the form of solids and gases; there is only one non-metal (bromine) that is a liquid at room temperature.

8. Graphite is an allotropic form of carbon.

9. Copper, aluminium and iron are metals but sulfur is not a metal.

10. Electrons in non-metals are not free or mobile.

12. Tincture of iodine is used on wounds as an antiseptic.

13. Phosphorus is a non-metal which reacts and catches fire on exposure to air.

14. Gold occurs in the free state since it is not reactive.

16. Mercury metal is in the liquid state at room temperature.

20. Due to its very high melting point, tungsten can be used as a filament in an electric bulb.

21. The colour of copper is not responsible for copper to be used as electrical conduction wires.

23. Ductility is a characteristic property of metals. Sulfur is a non-metal, so it is not ductile. When stretched, sulfur breaks into pieces and does not form wires.

24. Metals are generally hard. This means that most of the metals cannot be cut easily. But sodium metal is soft and can be easily cut with a knife.

25. Malleability is the property of beating the metal into thin sheets.

27. Metal oxides are basic in nature.

29. The less reactive metals such as copper, silver and gold do not react with dilute hydrochloric acid to produce hydrogen gas.

30. The vigour (or intensity) of reaction of a metal with water depends on its chemical reactivity. Sodium is a very reactive metal, so it reacts violently even with cold water.

31. Aluminium reacts with hydrochloric acid as well as sodium hydroxide solution to produce hydrogen gas.

36. Metallic oxides turn red litmus blue, hence AO is a basic oxide made from metal A.

37. $Zn + \underset{\text{(Steam)}}{H_2O} \rightarrow ZnO + H_2$

38. Silver is present below hydrogen in the reactivity series, hence it does not react with dilute HCl.

54. An alloy is a mixture of two metals or a metal and non-metal made by mixing them in their molten form.

56. Brass is Cu + Zn, Bronze is Cu + Sn, German silver is Cu + Zn + Ni.

Miscellaneous

2. Only metallic oxides are basic in nature.

4. A piece of coal is a non-metal and it breaks into small pieces when hammered.

5. Reactivity series Mg > Zn > Cu > Ag.

6. Sulfur is a non-metal and it forms an acidic oxide. Sulfur dioxide (SO_2) turns blue litmus red.

8. Metals can be beaten into thin sheets (or flattened) with a hammer. Non-metals break into small pieces when hammered. So, zinc granules will get flattened on beating with a hammer.

14. Silver is extracted from argentite by the MacArthur and Forest process (leaching process).

$$Ag_2S + 4NaCN \rightleftharpoons 2NaAg(CN)_2 + Na_2S$$

$$2NaAg(CN)_2 + Zn \rightarrow Na_2Zn(CN)_4 + 2Ag]$$

Advanced and Olympiads

Single Choice

2. Because of its light weight, aluminium is preferred over copper for making overhead transmission cables.

5. As Zn and Fe are more reactive than copper, they can displace copper while Ag and Al cannot do that.

7. Iron is more reactive than copper and thus displaces copper from copper sulfate. It is less reactive than magnesium and therefore does not react with magnesium sulfate.

$$Fe + CuSO_4 \rightarrow FeSO_4 + \underset{\text{(Brown deposit)}}{Cu}$$

$$Fe + MgSO_4 \rightarrow \text{No reaction}$$

8. Brass: Cu = 80% and Zn = 20%; bronze: Cu = 80%, Zn = 10% and Sn = 10%; gun metal: Cu = 90% and Sn = 10%; type metal: Pb = 70%, Sb = 20% and Sn = 10%. So, both brass and bronze contain Cu and Zn.

Multiple Choice

9. Gold and silver are the most malleable metals whereas zinc is non-malleable and brittle.

10. Sodium chloride has sodium, magnesium oxide has magnesium and calcium hydroxide has calcium as metallic elements. Silicon dioxide contains silicon which is a non-metal.

14. Copper does not react with water at ordinary or high temperature.

15. Sodium amalgam is less reactive than sodium metal.

16. Aluminium is a light metal.

17. Only metals have high tensile strength.

Integer Type

23. The metals are Na, Mg, Al, Fe and Cu while the non-metals are C, H, Br and B.

24. Na, Mg, Al, Zn, Sn, Pb are more reactive than hydrogen while Cu, Ag, Au are less reactive than hydrogen.

25. Bauxite, hematite, magnetite, zincite are oxide ores while cinnabar, zinc blende are sulfide ores and horn silver is a halide ore.

26. Chalcopyrite is $CuFeS_2$ and it has 4 atoms.

27. Bauxite is $Al_2O_3.2H_2O$.

28. German silver contains 60% Cu, 20% Zn and 20% Ni.

CHAPTER
11 Coal and Petroleum

LEARNING OBJECTIVES

After studying this unit, you will be able to understand:

- Fossil fuels and coal
- Petrol and petroleum products

Introduction

A fuel is a material containing stored energy. When a fuel is burned, energy is released mainly as heat along with some light. This heat energy can be used for various purposes such as, cooking food, heating water and for running generators in thermal power stations, machines in factories and engines of motor vehicles. Most of the common fuels are either free carbon or carbon compounds. For example, the fuels such as coal, coke and charcoal contain free carbon while the fuels like kerosene, petrol, LPG and natural gas are all carbon compounds.

When carbon in any form (coal, coke, charcoal) is burned in oxygen (air), it forms carbon dioxide gas and releases a large amount of heat and some light as follows:

$$\underset{\substack{\text{Carbon}\\\text{(Coal, coke or charcoal)}}}{C} + \underset{\substack{\text{Oxygen}\\\text{(From air)}}}{O_2} \rightarrow \underset{\text{Carbon dioxide}}{CO_2} + \text{Heat} + \text{Light}$$

Natural Resources

The resources that can be obtained from nature are known as *natural resources*. For example, air, water, soil. On the basis of their availability in nature, natural resources are classified into the following two types.

Inexhaustible Natural Resources

These resources are present in unlimited quantity in nature and are not likely to be exhausted by human activities. For example, sunlight, air, wind.

Exhaustible Natural Resources

The amount of these resources in nature is limited and they can be exhausted by human activities. For example, forests, wildlife, minerals, coal, petroleum and natural gas.

- Renewable natural resources are inexhaustible and can be replaced within a reasonable period of time through natural processes. Non-renewable natural resources are exhaustible and cannot be replaced within a reasonable period of time.

Fossil Fuels

Most of the fuels which are used today are obtained from coal, petroleum and natural gas. Actually, coal, petroleum and natural gas are called fossil because they were formed by the decomposition of the remains of the pre-historic plants and animals (fossils) buried under the earth, long long ago.

How are Fossil Fuels Used?

Fossil fuels are the largest sources of energy for electricity generation. Natural gas was the largest source – about 80% – of India's electricity generation in 2019. Natural gas is used in steam turbines and gas turbines to generate electricity. Fossil fuels are burned in power stations to heat water which generates steam to drive large turbines. In a power station, the turbines drive a generator to produce electricity which is then fed into the grid system (Fig. 11.1).

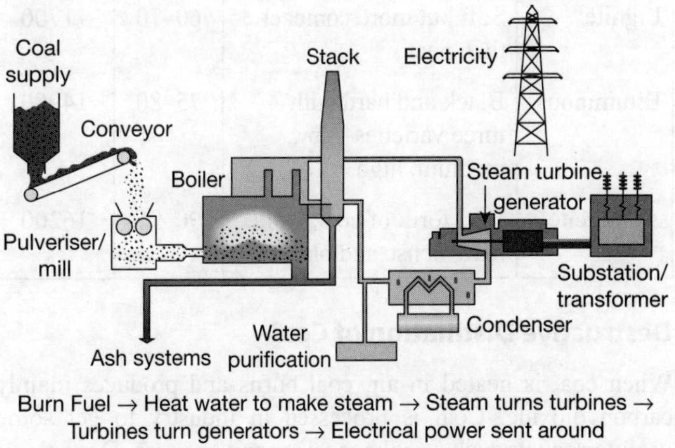

Fig. 11.1 Generation of electricity from fossil fuels

Coal

You may have seen coal, or heard about it. It is as hard as stone and is black in colour. Coal is one of the fuels used to cook food. Earlier, it was used in railway engines to produce steam to run the engine. It is also used in thermal power plants to produce electricity. Coal is also used as a fuel in various industries. Coal is a complex mixture of compounds of carbon, hydrogen and oxygen and some free carbon. Fractional amounts of nitrogen and sulfur compounds are also present in coal.

How was Coal Formed?

Coal was formed by the decomposition of large land plants and trees buried under the earth millions of years ago. It is believed that millions of years ago, due to earthquakes and volcanoes, forests were buried under the surface of the earth and got covered with sand, clay and water. As there is high temperature and high pressure inside the earth, in the absence of air, wood was converted into coal. This phenomenon is called carbonisation. As coal is obtained from the remains of vegetation, it is a fossil fuel.

Table 11.1 Types of coal

Type	Property	Carbon content (%)	Calorific value (BTU)
Peat	It is in the first stage of transformation from wood to coal. It has low calorific value and burns like wood.	50–60	9990
Lignite	Soft but more compact than peat.	60–70	11700
Bituminous	Black and hard with three varieties—low, medium, high.	75–80	14900
Anthracite	Best form of coal. Quite hard, dense and black.	90–95	16200

Destructive Distillation of Coal

When coal is heated in air, coal burns and produces mainly carbon dioxide. Coal is processed in industry to get some useful products such as coke, coal tar and coal gas. Destructive distillation is possible by heating coal strongly in the absence of air around 1000°C (Fig. 11.2). The following products are formed during destructive distillation.

Coke

It is formed as a residual product during the process of destructive distillation. It is a tough, porous and black substance. It is almost a pure form of carbon. Coke is used in the manufacture of steel and in the extraction of many metals like iron. It is one of the constituents needed in the blast furnace.

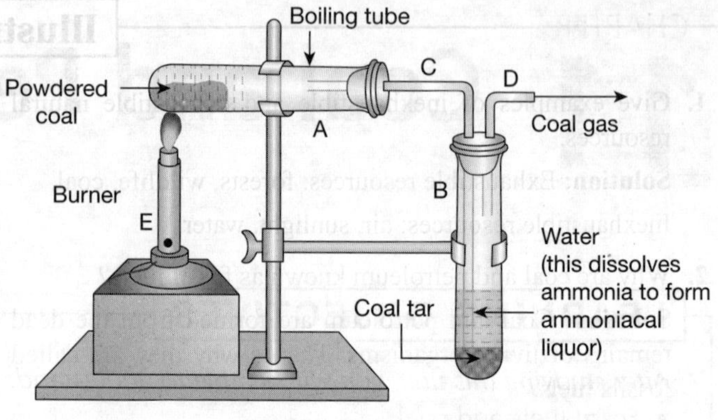

Fig. 11.2 Destructive distillation of coal tar

Coal Tar

It is a black, thick liquid with an unpleasant smell. It is a mixture of nearly 200 substances. Products obtained from coal tar are used as starting materials for manufacturing various substances used in everyday life and in industry, such as, synthetic dyes, drugs, explosives, perfumes, plastics, paints, photographic materials and roofing materials. Naphthalene balls (obtained from coal tar) are used to repel moths and other insects. Nowadays a petroleum product, bitumen, is used in place of coal tar for metalling the roads.

Coal Gas

Coal gas is obtained during the processing of coal to get coke. It has hydrogen, methane, carbon monoxide and few other gases. It is used as an excellent fuel due to its high calorific value, mainly in industries situated near the coal processing plants.

Ammonium Compounds

During distillation of coal, a mixture of compounds is formed which on dissolving in water forms ammoniacal liquor. It can be used in the manufacture of nitrogenous fertilisers such as ammonium sulfate and ammonium super phosphate.

Gaseous Fuels

Coke reacts with steam and air to give two important gaseous fuels known as water gas and producer gas. Both the gases are used as industrial fuels.

$$\text{Water gas: } C + H_2O \xrightarrow{\text{Red hot}} \underbrace{CO + H_2}_{\text{Water gas}}$$

$$\text{Producer gas: } 2C + \underbrace{O_2 + 4N_2}_{\text{Air}} \xrightarrow{\text{Red hot}} \underbrace{2CO + 4N_2}_{\text{Producer gas}}$$

KNOWLEDGE BOOSTER

Amorphous forms of carbon

The amorphous forms of carbon are coke, wood charcoal, animal charcoal, sugar charcoal, lampblack and gas carbon. These can be obtained in small quantities when wood, bones, and so on, are heated in the absence of or a supply of little air.

Illustrations

1. Give examples of inexhaustible and exhaustible natural resources.

   Solution: Exhaustible resources: forests, wildlife, coal.

   Inexhaustible resources: air, sunlight, water.

2. Why are coal and petroleum known as fossil fuels?

   Solution: Coal and petroleum are formed from the dead remains of living organisms. That is why they are called fossils fuels.

3. Define destructive distillation. Name the residue formed by destructive distillation of coal.

   Solution: Destructive distillation is the process of heating a substance in absence of air. Coke is formed by destructive distillation of coal.

4. Describe the characteristics and uses of coke.

   Solution: Characteristics of coke are: pure form of carbon, tough and porous.

   Coke is used for the extraction of metals and in the manufacture of steel.

Petroleum

Petroleum is a dark coloured, viscous and foul-smelling crude oil. The name petroleum means rock oil (petra = rock; oleum = oil). It is called petroleum as it is found under the crust of the earth, trapped in rocks. The crude petroleum is a complex mixture of several solid, liquid and gaseous hydrocarbons mixed with water, salt and earth particles. Hence, the crude petroleum oil is not a single chemical compound, it is a mixture of compounds. The fuels such as petrol, kerosene, diesel and LPG are obtained from petroleum. Petroleum is also known as *black gold* as it is very widely used as a fuel in industries and vehicles.

How was Petroleum Formed?

It is believed that millions of years ago, the microscopic plants and animals which lived in seas, died. Their bodies sank to the bottom of the sea and were soon covered with mud and sand. Petroleum and natural gas were formed by the decomposition of the remains of these plants and animals buried under the sea millions of years ago. The chemical effects of pressure, heat and bacteria, converted the remains of microscopic plants and animals into petroleum and natural gas just as they converted forest trees into coal. This conversion took place in the absence of oxygen or air. The petroleum thus formed got trapped between two layers of impervious rocks (non-porous rocks) forming an oil trap. Natural gas is above this petroleum oil.

The fuels like coal and petroleum have some nitrogen and sulfur in them too. So, when coal and petroleum fuels (petrol and diesel) are burnt, they lead to the formation of oxides of nitrogen and sulfur which go into the air. These oxides of nitrogen and sulfur are the major pollutants in the environment.

Refining of Petroleum

Petroleum is found with natural gases in rocks known as oil wells, from where it can be taken out by drilling. Petroleum is a dark oily liquid with an unpleasant odour. It is a mixture of various constituents such as petroleum gas, petrol, diesel, lubricating oil and paraffin wax. The process of separating the various constituents/fractions of petroleum is known as refining. It is carried out in a petroleum refinery by the process of *fractional distillation*.

Crude oil or petroleum obtained from the oil well is a mixture of many liquids. Each liquid boils and changes into vapour at a different temperature. This temperature is the boiling point of that specific liquid. The crude oil heated to a temperature of 673 K is fed in at the bottom of the fractionating column and heated further. The liquid that has the lowest boiling point changes into vapour first and rises upwards. The next volatile liquid changes into its vapour state at a higher temperature and rises upward. At different parts of the column, as the vapours rise, they cool and condense separately into liquids on a tray and are collected separately. A few gases reach the top of the column without condensing.

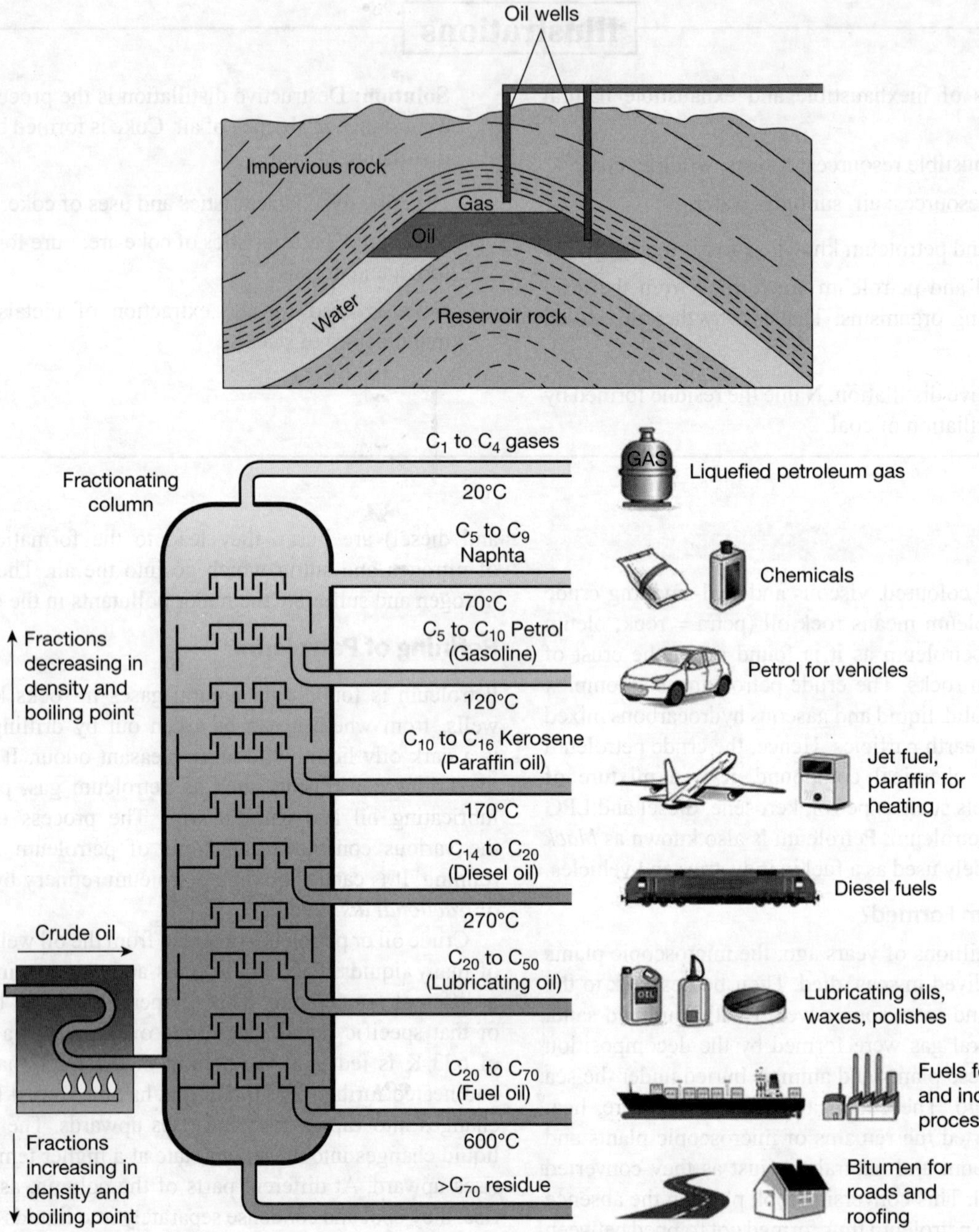

Fig. 11.3 Refining of petrol

Table 11.2 Various constituents of petroleum and their uses

S. No.	Fractions	Boiling range (°C)	Number of carbon atoms per molecule	Uses
1.	Petroleum gas	Below 40	1–4	It is the most commonly used domestic and industrial fuel.
2.	Petrol (gasoline)	40–75	5–10	Petrol is obtained from this fraction. Fuel for cars and other vehicles.
3.	Naphtha	75–150	7–14	Feedstock (raw material) for petrochemical industry (that produces plastic, detergents).
4.	Kerosene	160–250	11–16	It is commonly used as a domestic fuel for cooking using oil stoves. It is also used in jet engines and aircraft engines as fuel.

(Continued)

S. No.	Fractions	Boiling range (°C)	Number of carbon atoms per molecule	Uses
5.	Diesel	250–300	16–20	It is obtained from this fraction which is used in the automobile industry as fuel for diesel engines.
6.	Lubricating oil	300–350	20–35	It is used as lubricant for machines and also for making waxes and polishes.
7.	Bitumen (Asphalt)	Above 350	More than 70	For paving road surfaces as a substitute for coal tar.

Petrochemicals

Those substances which are obtained from petroleum and natural gas are known as petrochemicals. They are used in the manufacture of synthetic fibres like nylon, polyester, polythene, plastics and detergents.

Natural Gas

Natural gas is a very important fossil fuel because it is easy to transport through pipes. It has nearly 95% methane (CH_4). It is found deep inside the earth along with the petroleum deposits. Natural gas is also used as a starting material for the manufacture of a number of chemicals and fertilisers. India has vast reserves of natural gas. In our country, natural gas has been found in Tripura, Rajasthan, Maharashtra and in the Krishna–Godavari delta.

CNG (Compressed Natural Gas): Natural gas is stored under high pressure as compressed natural gas (CNG). CNG is used for power generation. It is now being used as a fuel for transport vehicles because it is less polluting. It has a high calorific value and does not produce any toxic gas or smoke on combustion. The great advantage of CNG is that it can be used directly for burning in homes and factories where it can be supplied through pipes. Such a network of pipelines exists in Vadodara (Gujarat), some parts of Delhi and other places.

LPG (Liquified Petroleum Gas): LPG is commonly called cooking gas. It has mainly propane (C_3H_8) and butane (C_4H_{10}) in the liquefied form. It is a cleaner fuel and does not leave any residue or cause any pollution. As the hydrocarbons present in it do not have any smell, a compound ethyl mercaptan (C_2H_5SH) is added for smell in order to discover any leakage of LPG. When butane (or LPG) burns in air in the burner of a gas stove, then it forms carbon dioxide and water vapour, with the evolution of a lot of heat (and some light). Due to this, butane (or LPG) is an excellent fuel. Its calorific value is 55000 cal/g. It is mainly used as a household fuel.

KNOWLEDGE BOOSTER

Combustion or Burning

The process of burning of a carbon compound in air to give carbon dioxide, water, heat and light, is called combustion. Most of the carbon compounds burn in air to produce a huge amount of heat. For example, alkanes burn in air to produce a huge amount of heat due to which they are excellent fuels. For example, when methane (natural gas) burns in a sufficient supply of air, then carbon dioxide and water vapour are formed and a lot of heat is also produced.

$$\underset{\text{Methane (Natural gas)}}{CH_4} + \underset{\text{Oxygen (From air)}}{2O_2} \xrightarrow{\text{Combustion}} \underset{\text{Carbon dioxide}}{CO_2} + \underset{\text{Water}}{2H_2O} + \text{Heat} + \text{Light}$$

Some Natural Resources are Limited

We have studied in the beginning of this chapter that some natural resources are exhaustible—fossil fuels, forests, minerals. We know that coal and petroleum are fossil fuels. The known reserves of these will last at most a few hundred years. Moreover, burning of these fuels is a major cause of air pollution. Their use is also linked to global warming. It is therefore necessary that we use these fuels only when absolutely necessary. This will result in better environment, smaller risk of global warming and their availability for a longer period of time. In India, the Petroleum Conservation Research Association (PCRA) advises people how to save petrol/diesel while driving. Their tips are:

- drive at a constant and moderate speed as much as possible,
- switch off the engine at traffic lights or at a place where you have to wait for more than 1 minute,
- ensure correct tyre pressure in the vehicles,
- ensure regular maintenance of the vehicles, and
- use these fuels only in the case of necessity.

Few Alternative Sources of Energy

Now few non-conventional sources of energy are also being tapped. For example, solar energy, wind energy, various forms of ocean energy such as tidal energy, ocean wave energy and ocean thermal energy.

Solar Energy

Solar energy can be used directly to heat water. This hot water may be used directly in homes or used to generate steam in order to generate electricity. Solar dishes and parabolic reflectors can be used to concentrate the sun's energy and focus it.

The reflectors are so designed that they can follow the sun from dawn to dusk. Solar energy can also be converted directly into electricity, using photovoltaic or solar cells. The electricity is stored in batteries and used as needed. Solar energy can be used directly for cooking using solar cookers.

Wind Energy

Wind energy is a form of solar energy. Wind energy or wind power describes the process by which wind is used to generate electricity. Wind turbines convert the kinetic energy in the wind into mechanical power. A generator can convert mechanical power into electricity. Modern windmills are used to turn turbines and generate electricity.

Hydroelectric Power

Hydropower plants capture the energy of falling water to generate electricity. A turbine converts the kinetic energy of falling water into mechanical energy. Then a generator converts the mechanical energy from the turbine into electrical energy. Hydropower has the ability to generate electricity without emitting greenhouse gasses. However, it can also cause environmental and social threats, such as damaged wildlife habitat, harmed water quality, obstructed fish migration and diminished recreational benefits of rivers.

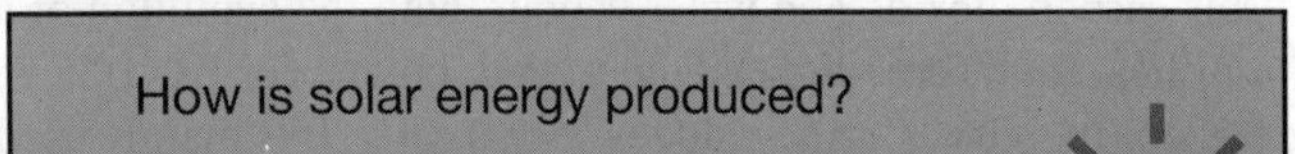

(a)

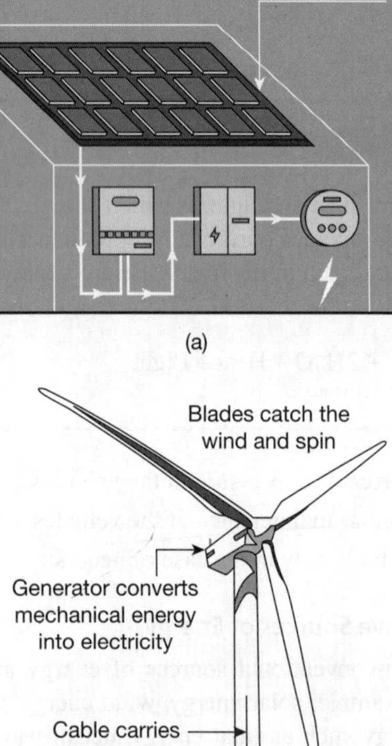

(b)

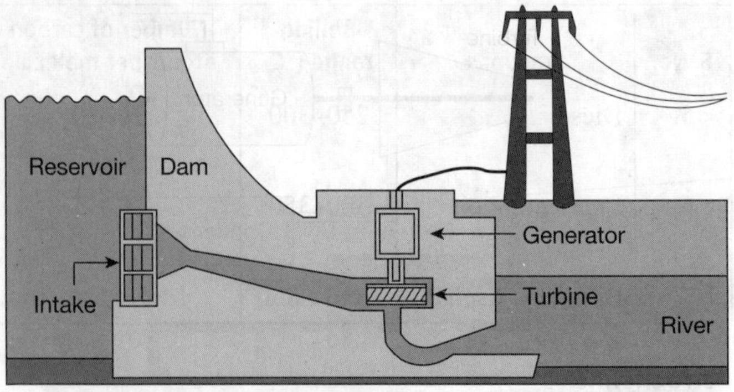

(c)

Fig. 11.4 (a) Solar energy (b) Wind energy (c) Hydroelectric power

Tidal Energy

Tidal energy is created using the movement of the tides, where the intensity of the water from the rise and fall of tides is a form of kinetic energy. Tidal power surrounds gravitational hydropower, which uses the movement of water to push a turbine to generate electricity. Tidal energy is produced through the use of tidal energy generators. These large underwater turbines are placed in areas with high tidal movements; they are designed to capture the kinetic motion of the ebbing and surging of ocean tides in order to produce electricity. The use of tidal energy could also decrease the need for nuclear power, with its associated radiation risks. Changing tidal flows by damming a bay or estuary could, however, result in negative impacts on aquatic and shoreline ecosystems, as well as navigation and recreation.

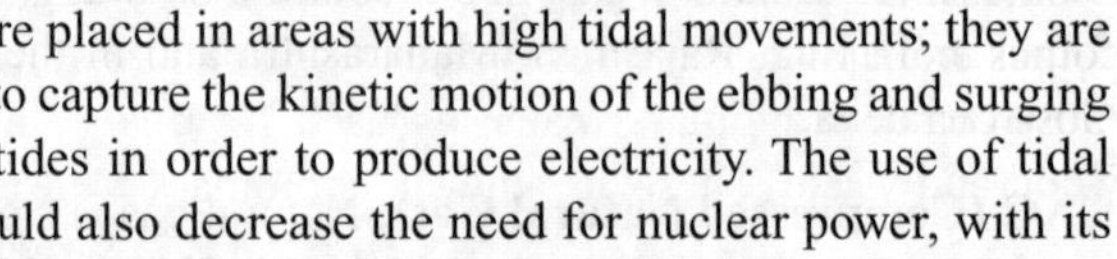

Geothermal Energy

Geothermal energy is heat within the earth (Fig. 11.5). The word geothermal comes from the Greek words *geo* (earth) and *therme* (heat). When water is pumped into hot rocks in the earth crust far below the ground level, the internal heat of the rocks converts the water into steam which can be used to drive turbines to generate electricity. Geothermal energy is a renewable energy source because heat is continuously produced inside the earth. People use geothermal heat for bathing, to heat buildings and to generate electricity. It is a major source of energy in Iceland.

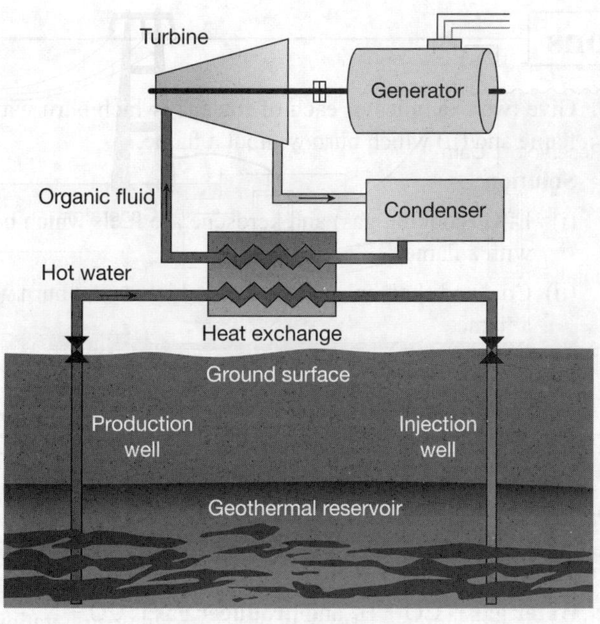

Fig. 11.5 Geothermal energy

Biomass Energy

Biomass is organic, meaning, it is made of material that comes from living organisms such as plants and animals. The most common biomass materials used for energy are plants, wood and waste. These are known as biomass feedstocks. The energy obtained from these species is called *biomass energy* (Fig. 11.6). Biomass energy can also be a non-renewable energy source. Most electricity generated from biomass is produced by direct combustion. Biomass is burned in a boiler to produce high-pressure steam. This steam flows over a series of turbine blades, causing them to rotate. The rotation of the turbine drives a generator, producing electricity. In India, methane (CH_4) generators are used to get bioenergy as methane is generated by the digestion of animal waste as biogas. It can be used for cooking, heating and lighting.

Sugarcane is fermented to make ethyl alcohol (power alcohol) which can be burnt to generate power. The cane can be crushed and the pulp or bagasse can be combusted to make steam to drive turbines.

KNOWLEDGE BOOSTER

Hydrogen as a fuel

Hydrogen is a zero-carbon fuel when burned with oxygen. It can be used in fuel cells or internal combustion engines. It has begun to be used in commercial fuel cell vehicles, such as passenger cars, and has been used in fuel cell buses for many years. It is also used as a fuel for spacecraft propulsion.

$$2H_2(g) + O_2(g) \rightarrow 2H_2O(l) + \text{Heat}$$

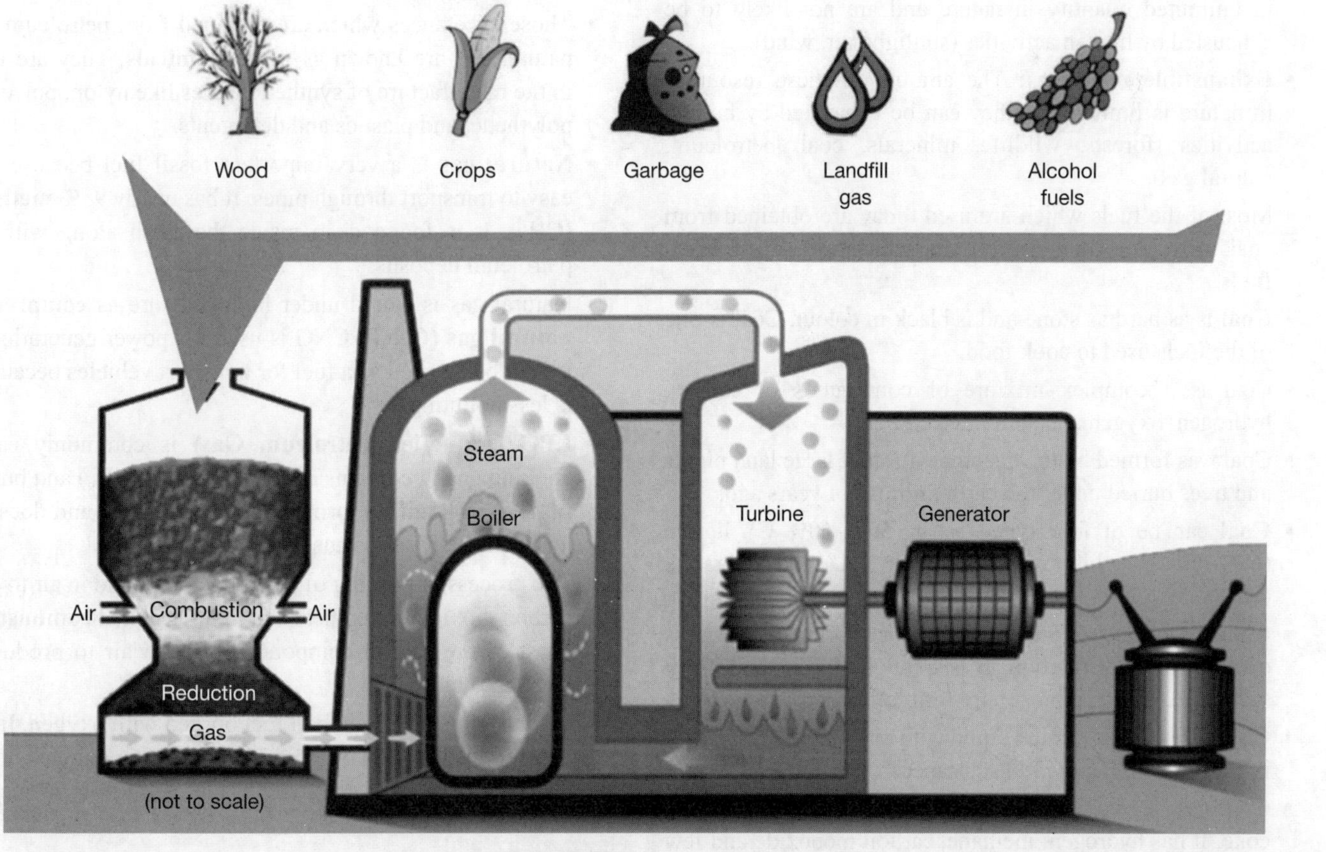

Fig. 11.6 Biomass energy

Illustrations

1. Explain the process of formation of petroleum. Name two places in India where it is found.

Solution: Petroleum is formed by the decomposition of the remains of aquatic plants and animals. In India, petroleum is obtained from oil wells in Assam and Bombay High.

2. Does Compressed Natural Gas (CNG, used these days in automobiles) undergo spontaneous or rapid combustion?

Solution: CNG undergoes rapid combustion in the automobiles' internal combustion engine, it does not ignite by itself.

3. Give two examples of each of fuels (i) which burn with a flame and (ii) which burn without a flame.

Solution:

(i) LPG (cooking gas) and kerosene are fuels which burn with a flame.

(ii) Coal and charcoal are two fuels which do not burn with a flame.

CHAPTER AT A GLANCE

- A **fuel** is a material possessing stored energy. When a fuel is burned, energy is released mainly as heat along with some light.
- Most of the common fuels are either free carbon or carbon compounds. For example, the fuels such as coal, coke and charcoal contain free carbon while the fuels like kerosene, petrol, LPG and natural gas are all carbon compounds.
- The resources that can be obtained from nature are known as **natural resources** (air, water, soil).
- **Inexhaustible resources:** These resources are present in unlimited quantity in nature and are not likely to be exhausted by human activities (sunlight, air, wind).
- **Exhaustible resources:** The amount of these resources in nature is limited and they can be exhausted by human activities (forests, wildlife, minerals, coal, petroleum, natural gas).
- Most of the fuels which are used today are obtained from coal, petroleum and natural gas which are called **fossil fuels**.
- **Coal** is as hard as stone and is black in colour. Coal is one of the fuels used to cook food.
- Coal is a complex mixture of compounds of carbon, hydrogen, oxygen and some free carbon.
- Coal was formed by the decomposition of large land plants and trees buried under the earth millions of years ago.
- **Coal** can be of four types—peat (50%–60% C), lignite (60%–70% C), bituminous (75%–80% C) and anthracite (90%–95% C).
- **Coke** is formed as a residual product during the process of destructive distillation. It is a tough, porous and black substance. It is an almost pure form of carbon.
- **Coal tar** is a black, thick liquid with an unpleasant smell. It is a mixture of nearly 200 substances.
- **Coal gas** is obtained during the processing of coal to get coke. It has hydrogen, methane, carbon monoxide and few other gases.
- **Water gas** is $CO + H_2$ and **producer gas** is $CO + N_2$.
- **Petroleum** is a dark coloured, viscous and foul-smelling crude oil. The name petroleum means rock oil (petra = rock; oleum = oil).
- **Petroleum** oil and natural gas were formed by the decomposition of the remains of extremely small plants and animals buried under the sea millions of years ago.
- The process of separating the various constituents/fractions of petroleum is known as **refining**. It is carried out in a petroleum refinery by the process of **fractional distillation**.
- Those substances which are obtained from petroleum and natural gas are known as **petrochemicals**. They are used in the manufacture of synthetic fibres like nylon, polyester, polythene, and plastics and detergents.
- **Natural gas** is a very important fossil fuel because it is easy to transport through pipes. It has nearly 95% **methane** (CH_4). It is found deep inside the earth along with the petroleum deposits.
- Natural gas is stored under high pressure as **compressed natural gas (CNG)**. CNG is used for power generation. It is now being used as a fuel for transport vehicles because it is less polluting.
- **LPG (Liquefied Petroleum Gas)** is commonly called cooking gas. It contains mainly propane (C_3H_8) and butane (C_4H_{10}) in liquified form. It is a cleaner fuel and does not leave any residue or cause any pollution.
- The process of burning of a carbon compound in air to give carbon dioxide, water, heat and light, is called **combustion**. Most of the carbon compounds burn in air to produce a huge amount of heat.
- **Hydrogen** is a zero-carbon fuel burned with oxygen. It can be used in fuel cells or internal combustion engines. It has begun to be used in commercial fuel cell vehicles, such as passenger cars.

PRACTICE QUESTIONS

Analyse Your Concepts (School Exam Based)

Fill in the Blanks

Instructions: Complete the following statements with an appropriate word/term to be filled in the blank spaces.

1. Forests and coal are _______ natural resources.
2. Coal and petroleum are formed from the dead remains of organisms and are known as _______.
3. Coal is one of the _______ used to cook food.
4. When heated in air, coal burns and produces mainly _______ gas.
5. Coal tar is a black, thick _______ with an _______ smell.
6. The process of separation of different constituents from petroleum is called _______.
7. The least polluting fuel for vehicles is _______.
8. The slow process of conversion of dead vegetation into coal is called _______.
9. During the processing of coal to get coke, coal tar and _______ are also obtained.
10. Excessive burning of fossil fuels is a major cause of _______ pollution.

True or False

Instructions: Read the following statements and write your answer as true or false.

1. Fossil fuels can be made in the laboratory.
2. Resources which are present in unlimited quantity in nature are called exhaustible natural resources.
3. Wildlife is an exhaustible natural resource.
4. Oxygen in air is an exhaustible natural resource.
5. Under high temperature and pressure, dead plants get slowly converted to coal.
6. Coal tar is a mixture of various substances.
7. CNG is a less polluting fuel than petrol and diesel.
8. Coke is an almost pure form of carbon.
9. Kerosene is not a fossil fuel.
10. Methane is the main component of natural gas.
11. LPG contains ethene and ethane mainly.

Very Short Answer Type Questions

1. What are natural resources?
2. Classify the following as exhaustible and inexhaustible natural resources—air, sunlight, water, forests, wildlife, coal.
3. Define carbonisation.
4. Explain why fossil fuels are exhaustible natural resources.
5. What name is given to the process of heating a substance in the absence of air?
6. How is coke obtained?
7. How is coal tar formed?
8. Why is petroleum called a fossil fuel?
9. Write down any four amorphous forms of carbon.
10. What is meant by refining of petroleum and where is it done?
11. How can bitumen be used?
12. Why is petroleum also known as black gold?
13. Can we make coal and petroleum in the laboratory from dead organisms?
14. Give the full form of (a) LPG and (b) CNG.
15. How is petroleum gas obtained?
16. Name the petroleum product used as fuel for stoves, lamps and jet aircrafts.

Short Answer Type Questions

1. Sunlight and air are inexhaustible natural resources. Comment on it.
2. (a) Give two disadvantages of using fossil fuels.
   (b) Why should the fossil fuels be used with care?
3. What is meant by fractional distillation? What is the principle on which it is based?
4. How is natural gas formed? Name some places in India where natural gas has been found.
5. Name the products obtained when coal is heated in the absence of air and write any two uses of its products.
6. What are petrochemicals? Give the uses of petrochemicals obtained from petroleum.

7. Name the agency in India which advises people how to save petrol/diesel. What tips are given by them?

8. State one use of each of the following:
   (a) charcoal (b) bone charcoal
   (c) coke

9. Give two advantages of the judicious use of energy.

10. (a) Give two uses of diesel.
    (b) How can petrol be used?

11. The underlined words in the following sentences have been jumbled up. Write them in their correct form.
    (a) Loca is obtained from mines.
    (b) Umpetlore is a fossil fuel.
    (c) Renrfvie is a place where various fractions of petroleum are separated.
    (d) Keenrose is a fuel used in jet crafts.
    (e) Nutsoilh is an example of inexhaustible natural resources.

Long Answer Type Questions

1. (a) How was petroleum formed in nature?
   (b) From which natural substance are liquid fuels formed?
   (c) Which gas is the main constituent of LPG?
   (d) What is CNG? Give one use of CNG.
   (e) Why do green leaves not catch fire easily?

2. (a) How have fossil fuels been formed?
   (b) Why are fossil fuels non-renewable sources of energy?
   (c) Give one reason why LPG is a better fuel than coal.
   (d) Why is a smelling agent added to LPG?
   (e) Give one advantage of the modem chullah over the traditional chullah.

3. Draw a labelled diagram showing the fractional distillation of petroleum. Give the uses of any three fractions.

4. We read in newspapers that burning of fuels is a major cause of global warming. Explain why.

5. Why is petrol an exhaustible natural resource whereas sunlight is not? Explain.

Exemplar Problems

Short Answer Type Questions

1. Explain why fossil fuels are exhaustible natural resources.

2. Describe how coal is formed from dead vegetation. What is this process called?

3. What are the advantages of using CNG and LPG as fuels?

4. Describe the characteristics and uses of coke.

5. Explain the process of the formation of petroleum.

6. You are provided with a mixture of petroleum and water. Can you suggest a method to separate the two?

7. What does CNG stand for and why is it considered to be a better fuel than petrol?

8. Write the characteristics and some important uses of coal.

9. Fill in the blanks and complete the story.
   About 300 million years ago, the earth had dense _______ in low lying wetland areas. Due to natural processes like _______ these forests got buried under the _______. As more _______ deposited over them, they were compressed. The _______ also rose as they sank deeper and deeper. Under high _______ and high _______, dead plants got slowly converted into coal.

Long Answer Type Questions

1. Name the products obtained and their uses when coal is processed in industry.

2. We say fossil fuels will last only for a few hundred years. Comment.

3. While driving, what are the tips we must follow to save petrol/diesel/natural gas?

4. Imagine that all the exhaustible natural resources are exhausted by human activities. Do you think survival of living beings would be possible? If yes, why? If not, why not?

5. Write some important uses of the various constituents of petroleum.

6. Coal reserves are said to be enough to last for another hundred years. Do you think we need to worry in such a case? Why or why not?

7. What steps would you suggest for the judicious use of fuels?

8. Complete the crossword in figure with the help of the clues:

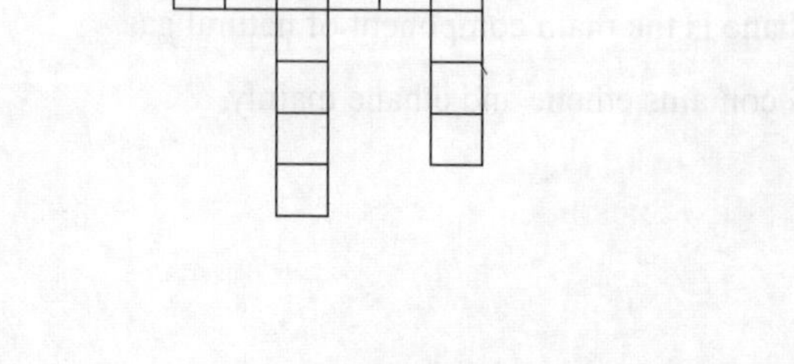

Across

1. Fuels obtained from dead remains of living organisms. (6, 5)
2. A process by which the various constituents of petroleum are separated. (8)
3. A porous black substance obtained from coal. (4)
4. Another name for motor fuel. (6)

Down

5. The substance obtained by carbonisation. (4)
6. Fuel for heavy motor vehicles. (6)
7. A petroleum product used for road surfacing. (7)
8. Dead remains of sea animals got converted into ______. (9)

COMPETITION WINDOW (OBJECTIVE TYPE)

[For NEET, JEE (Main and Advanced), NTSE, KVPY and Olympiads]

Topic-wise MCQs

Fossil fuels

1. Exhaustible natural resources are
(a) unlimited in quantity
(b) not dependent on nature
(c) limited in quantity
(d) not exhausted by human activities

2. Fossil fuels are obtained from
(a) remains of non-living materials
(b) dead remains of birds only
(c) dead remains of insects only
(d) dead remains of living organisms

3. Air is a natural resource and cannot be exhausted by human activities. It is known as an inexhaustible natural resource. Which of the following is another inexhaustible natural resource?
(a) coal (b) petroleum
(c) sunlight (d) minerals

4. A brownish-black sedimentary rock is known as
(a) charcoal (b) coke
(c) coal (d) coal tar

5. Which of the following fuels is expected to last longest?
(a) natural gas (b) petroleum
(c) coal (d) wood

6. Peat is a type of
(a) charcoal (b) coke
(c) coal (d) none of these

7. The most pure form of a carbon fuel is
(a) coal (b) coke
(c) charcoal (d) coal gas

8. The conversion of wood into coal occurs by a biochemical process that takes over millions of years and is known as ______.
(a) pyrolysis (b) destructive distillation
(c) catenation (d) carbonisation

9. Coal tar is a
(a) dark, oily liquid
(b) black, thick liquid
(c) gas
(d) tough, porous and black substance

10. Nowadays coal gas is used as a source of
(a) electricity (b) steam
(c) light (d) heat

11. Which type of coal has the highest percentage of carbon?
(a) peat (b) lignite
(c) anthracite (d) bituminous

12. The non-combustible element of producer gas is
(a) H_2 (b) H_2O
(c) CO (d) N_2

13. What percentage of carbon does lignite coal contain?
(a) 64% (b) 88%
(c) 60% (d) 67%

14. The gas which is present in the coal mines and is the cause of explosion in coal mines is
(a) hydrogen
(b) carbon dioxide
(c) methane
(d) carbon monoxide

15. Coke is not used in the preparation of
(a) iron (b) diamond
(c) $CO + H_2$ (d) $CO + N_2$

16. Coke, coal gas and coal tar are products of
(a) destructive distillation of coal
(b) destructive distillation of wood
(c) burning wood in air
(d) dissolving the coal in water

17. Black soot which is deposited on kerosene lamps after burning is called
(a) gas carbon (b) charcoal
(c) coke (d) lampblack

18. Coal is used in generating electricity in a
(a) hydro power station
(b) geothermal power station
(c) thermal power station
(d) nuclear power station

Petroleum products

19. The fossil fuel found below the sea is
(a) petrol (b) petroleum
(c) kerosene (d) diesel

20. The natural petroleum contains
(a) compounds of sulfur
(b) saturated hydrocarbons
(c) cyclic saturated hydrocarbons
(d) all of these

21. The origin of petroleum is indicated by the fact that
(a) petroleum was formed from living sea organisms which after death got buried in the sea bed for a long period of time
(b) its constituents can be separated by fractional distillation
(c) oil fields are located with the help of a seismograph
(d) petroleum contains traces of chlorophyll

22. Crude petroleum is also called
(a) brown gold (b) white gold
(c) black gold (d) yellow gold

23. Gasoline is obtained from crude petroleum oil by
(a) steam distillation (b) pyrolysis
(c) fractional distillation (d) vacuum distillation

24. Kerosene is used as a fuel
(a) for electric generators
(b) in home and industry
(c) for stoves, lamps and jet aircraft
(d) for heavy motor vehicles

25. The white semi-solid fraction of petroleum used for making vaseline is
(a) asphalt (b) lubricating oil
(c) paraffin wax (d) fuel oil

26. Petroleum refining is
(a) cracking of petroleum to get gaseous hydrocarbons
(b) purification of petroleum
(c) distillation of petroleum to get different fractions
(d) obtaining aromatic compounds from aliphatic compounds present in petroleum

27. Gasoline is the name of the
(a) mixture of uncondensed gases produced in the distillation of crude oil
(b) mixture of residue and gas oil obtained in the distillation of crude oil
(c) crude oil
(d) fraction of petroleum condenses within the range of 343–393 K and has composition of C_7–C_9

28. Out of the following fractions of petroleum, the one having the lowest boiling point is
(a) gasoline (b) heavy oil
(c) kerosene (d) diesel oil

29. Which of the following are the constituents of petroleum?
(a) kerosene, coal gas (b) coal tar, bitumen
(c) CNG, coal tar (d) diesel, paraffin

30. Synthetic polymers, paints and plastics are obtained from
(a) volatile chemicals (b) condensed chemicals
(c) petrochemicals (d) none of these

Natural gas and sources of energy

31. Wind, sun and hydropower are
(a) renewable (b) non-renewable
(c) synthetic sources (d) none of these

32. The unit of calorific value of combustion of fuels is
(a) kilojoule (b) joule
(c) kilojoule/kilogram (d) kilogram

33. Natural gas is a very important fossil fuel because it is
(a) a tough porous and black substance
(b) used as a fuel in many industries
(c) a mixture of various constituents
(d) easy to transport through pipes and emits no pollutants

34. Natural gas mainly contains
(a) methane (b) propane
(c) butane (d) ethane

35. LPG is a mixture of
(a) methane and butane (b) methane and ethane
(c) methane and propane (d) propane and butane

36. Which of the following forms of energy is environment friendly?
(a) wind (b) peat
(c) coal (d) petroleum

37. Besides the risk of pollution, fossil fuels also pose a risk of
(a) leakage (b) explosion
(c) global warming (d) water pollution

38. Out of the following, which fuel is best used in the homes?
(a) wood (b) CNG
(c) LPG (d) kerosene

Miscellaneous

1. Which of the following is a pair of exhaustible natural resources?
(a) coal and soil (b) air and sunlight
(c) water and petroleum (d) wildlife and minerals

2. The process of heating coal in the absence of air is called
(a) destructive distillation
(b) fractional distillation
(c) distillation
(d) none of these

3. Which fraction of petroleum is used in making skin ointments and vaseline?
(a) bitumen (b) fuel oil
(c) lubricating oil (d) paraffin wax

4. Ethyl mercaptan is added to LPG
(a) to give smell to it (b) to make it liquid
(c) to give colour to it (d) to give volume to it

5. $C_{12}H_{22}O_{11} \xrightarrow[H_2SO_4]{Conc.} 12C + 11H_2O$
Which of the following is obtained in the above reaction?
(a) coke (b) wood charcoal
(c) animal charcoal (d) sugar charcoal

6. Butane gas is used for filling cylinders to be used as LPG because it is
(a) stored in gaseous state only in the cylinder
(b) the cheapest gas available
(c) easily available
(d) easily compressed into a liquid and stored in cylinders

7. Various materials which are obtained from nature are called natural resources. Which of the following is not a natural resource?
(a) minerals (b) water
(c) soil (d) plastic

8. Coal is processed in industries to get some useful products. Which of the following is not obtained from coal?
(a) coke (b) coal tar
(c) coal gas (d) CNG

9. Coal is formed from the remains of
(a) vegetation only
(b) animals only
(c) both vegetation and animals
(d) neither vegetation nor animals

10. Which substance is formed by the carbonisation of dead vegetation?
(a) coal (b) coke
(c) coal gas (d) coal tar

11. Naphthalene balls are obtained from coal tar and are used as
(a) mosquito repellent (b) honeybee repellent
(c) moth repellent (d) snake repellent

12. Which of the following is not a constituent of petroleum?
(a) paraffin wax (b) lubricating oil
(c) petrol (d) coke

13. Petroleum was formed from organisms
(a) living on the land (b) living on the plants
(c) living in the sea (d) living on the rocks

14. Coal gas is a mixture of
(a) $C_4H_{10} + H_2$ (b) $CH_4 + H_2 + CO$
(c) $C_2H_6 + H_2 + O_2$ (d) $C_4H_{10} + H_2O$

15. We can use coke
(a) in printer ink
(b) as electrode
(c) as an oxidising agent
(d) as a reducing agent

16. Different varieties of coal differ in their
(a) carbon content
(b) number of hydrogen atoms
(c) moisture
(d) volatile nature

17. Coal is mainly carbon; it also has some other elements like
(a) sulfur, phosphorus, iodine and oxygen
(b) bromine, nitrogen, phosphorus and hydrogen
(c) oxygen, hydrogen, nitrogen and sulfur
(d) chlorine, nitrogen, sulfur and helium

18. Choose the correct statement from the following.
(a) It is difficult to transport natural gas through pipes.
(b) The disadvantage of natural gas is that it cannot be used directly for burning in homes.
(c) Natural gas is stored under high pressure as compressed natural gas.
(d) Natural gas cannot be used for power generation.

Advanced and Olympiads

Single Choice

1. Which of the following sources of energy can be a good substitute to coal in a power station?
(i) energy from water (ii) geothermal energy
(iii) energy from plants (iv) energy from petrol
(a) (iv) only (b) (ii) and (iii) only
(c) (i) and (ii) only (d) (i), (ii) and (iii) only

2. Out the following statements. Find the correct ones from the given set.
(i) Coal, petroleum and natural gas are known as fossil fuels.
(ii) Coal and natural gas are two exhaustible substances.
(iii) Coke is used in the manufacture of steel.
(iv) Fossil fuels are present in limited quantities.
(a) (i), (ii) and (iii) (b) (i), (ii), (iii) and (iv)
(c) (i) and (ii) (d) (i) and (iv)

3. Study the following statements and select the set showing correct ones.
(i) Fossil fuels are present in limited quantities.
(ii) Coal, petroleum and natural gas are known as fossil fuels.
(iii) Coke is used in the manufacture of steel.
(iv) Coal and natural gas are examples of exhaustible substances.
(a) (i) and (ii) (b) (iii) and (iv)
(c) (ii) and (iii) (d) (i), (ii), (iii) and (iv)

4. Arrange the following liquids in the decreasing order of their boiling points:
petrol, coal tar, lubricating oil, kerosene.
(a) kerosene > petrol > coal tar > lubricating oil
(b) petrol > coal tar > lubricating oil > kerosene
(c) coal tar > lubricating oil > kerosene > petrol
(d) lubricating oil > kerosene > petrol > coal tar

5. Coal is a fossil fuel. However, it cannot be prepared in a lab or industry as the formation of coal
   (i) is a very slow process
   (ii) needs very high pressure and high temperature
   (iii) needs very low pressure and low temperature
   (iv) causes air pollution
   Select the correct alternative
   (a) (i) and (iii)
   (b) (iii) and (iv)
   (c) (i) and (ii)
   (d) (ii) and (iv)

6. B is formed when A is heated in the absence of air. B is a tough, porous and black substance. Both A and B are carbon-rich materials. What could A and B be?
   (a) A = petroleum, B = diesel
   (b) A = coal, B = coal tar
   (c) A = petroleum, B = petrol
   (d) A = coal, B = coke

Multiple Choice

7. Select the one that is derived from a fossil fuel.
   (a) kerosene (b) diesel
   (c) LPG (d) biogas

8. Which of the following is/are not true?
   (a) coal as a fuel will last forever
   (b) fossil fuels do not create any pollution
   (c) the demand for energy has decreased over the years
   (d) we must use fossil fuels sparingly

9. Select the correct statements.
   (a) Petroleum products are called petrochemicals.
   (b) Petroleum is refined in a petroleum refinery.
   (c) CNG and LPG are clean fuels.
   (d) Coal does not cause any pollution when combusted.

10. Select the correct statements.
    (a) Air, wind and sunlight are examples of inexhaustible natural resources.
    (b) Peat has the highest percentage of carbon.
    (c) Now asphalt is used as a substitute for coal tar for making roads.
    (d) Hydrogen will be the fuel of the future.

11. Select the correct statements.
    (a) When coal burns in insufficient amount of oxygen, carbon monoxide is formed.
    (b) Burning of coal in a sufficient amount of oxygen produces carbon dioxide.
    (c) LPG is considered to be a good fuel for domestic use.
    (d) Charcoal is a better fuel than kerosene to be used as a fuel for cooking at home.

12. Which of the following is/are petroleum products?
    (a) petrol
    (b) kerosene
    (c) bees' wax
    (d) paraffin wax

13. Which of the following products is/are obtained from fossil fuels?
    (a) wood (b) coke
    (c) coal tar (d) coal

14. Select the incorrect statement.
    (a) Coal tar is mixture of various substances.
    (b) Petroleum is a mixture of various oxides of carbon.
    (c) LPG has a lower calorific value than wood.
    (d) The innermost zone of the flame is yellow.

15. Which amongst the following is/are the free state of carbon?
    (a) diamond (b) charcoal
    (c) petrol (d) coke

Comprehension Type

Comprehension I: During the process of oil refining, LPG (Liquified Petroleum Gas) can be obtained. LPG is compressed under high pressure and supplied in cylinders in the liquid form for household use. LPG undergoes pyrolysis to give olefins like ethylene and propylene. These olefins undergo polymerisation to form polymers and plastics like polythene and polypropylene. CNG can be obtained along with different fuels which are extracted from oil wells. CNG can also be compressed into liquid form and can be used as a fuel in vehicles.

16. LPG mainly has
    (a) methane, ethane and propane
    (b) methane, propane and butane
    (c) methane, ethane and hexane
    (d) ethane, propane and hexane

17. The main products formed during the pyrolysis of LPG are
    (a) ethene and butene
    (b) ethene and propylene
    (c) propylene and butylene
    (d) ethylene and benzene

18. Select the correct statements
    (a) CNG is called condensed natural gas
    (b) CNG is called Compressed Natural gas
    (c) Both LPG and CNG are cleaner fuels causing very less pollution
    (d) Both (b) and (c) are correct

Matrix Matching

19. Match the following

Column I	Column II
(a) Used for road surfacing	(p) Black gold
(b) Natural gas	(q) Vaseline and candles
(c) Petroleum	(r) Bitumen
(d) Paraffin wax	(s) CNG

20. Match the following

Column I	Column II
(a) Water gas	(p) Compressed Natural Gas
(b) Producer gas	(q) Condensed Natural Gas
(c) CNG	(r) Liquified Petroleum Gas
(d) LPG	(s) $CO + N_2$
	(t) $CO + H_2$

21. Match the following

Column I (Type of Coal)	Column II (% of carbon)
(a) Anthracite	(p) 60
(b) Bituminous	(q) 70
(c) Lignite	(r) 80
(d) Peat	(s) 90

Integer Type

22. How many of the following are examples of exhaustible natural resources:
sunlight, air, wind, petroleum, forest, coal, natural gas, minerals

23. If in gasoline the minimum number of carbon atom is x and maximum is 10, find the value of x.

24. 1 g of LPG produces nearly $n \times 10$ kJ of heat. What is the value of n?

25. In the given reaction $Fe_2O_3 + nC \rightarrow 2Fe + nCO$. What is the value of n?

26. If x is the number of amorphous forms of carbon, find the value of x.

Answer Keys

Fill in the Blanks

1. exhaustible
2. fossil fuels
3. fuels
4. carbon dioxide
5. liquid, unpleasant
6. refining
7. CNG
8. carbonisation
9. coal gas
10. air

True or False

1. F 2. F 3. T 4. F 5. T
6. T 7. T 8. T 9. F 10. T
11. F

Very Short Answer Type Questions

1. Material obtained from nature are known as natural resources.

2. Exhaustible resources: forests, wildlife, coal.
Inexhaustible resources: air, sunlight, water.

3. The slow process of conversion of dead vegetation into coal is known as carbonisation.

4. Fossil fuels are exhaustible natural resources as they will be exhausted by human activities since they are limited in nature.

5. The process of heating a substance in the absence of air is known as destructive distillation.

6. Coke is obtained by heating coal in the absence of air.

7. When destructive distillation of coal is done, the vapours which condense in the water form coal tar.

8. Petroleum is called a fossil fuel as it is obtained from the bodies of dead organisms.

9. Amorphous forms of carbon: coal, charcoal, coke and lampblack.

10. The process of separating the different fractions of petroleum is known as refining and it is carried out in a refinery.

11. Bitumen can be used for making paints and for surfacing the roads.

12. Due to its great commercial importance, petroleum is known as black gold.

13. No, their formation is a very slow process and conditions for their formation cannot be reached in the laboratory.

14. (a) LPG: Liquefied Petroleum Gas.
(b) CNG: Compressed Natural Gas.

15. Petroleum gas is obtained during fractional distillation of petroleum.

16. Kerosene is a petroleum product which is used as fuel for stoves, lamps and jet aircrafts.

Short Answer Type Questions

1. These resources are present in an unlimited quantity in nature and are not likely to be exhausted by human activities. There is a never ending supply of these resources in nature. These resources can be used again and again. They last forever.

2. (a) Fossil fuels cause air pollution and global warming.
(b) If we use fossil fuels carefully, it will result in a better environment, less risk of global warming and they will be available for a longer period.

3. Petroleum is mixture of various hydrocarbons having different boiling points. As the number of carbon atoms

increase, the boiling point also increases. This property is used to separate the different components of petroleum and is called fractional distillation.

4. Natural gas is formed below the earth's surface. It is formed above the petroleum. In India, natural gas is found in Tripura, Jaisalmer, in the offshore areas of Mumbai and in the Krishna–Godavari delta.

5. Coke is formed when coal is heated in the absence of air and coke is used as a fuel and as a reducing agent in the extraction of metals.

6. Petrochemicals are useful substances obtained from petroleum. They are used in the manufacture of detergents, fibres, polyethylene and other plastics.

7. In India, the Petroleum Conservation Research Association (PCRA) advises people how to save petrol/diesel while driving. Their tips are:
   (i) drive at a constant and moderate speed,
   (ii) switch off the engine at traffic lights,
   (iii) ensure correct tyre pressure, and
   (iv) ensure regular maintenance of the vehicle.

8. Use: (a) Charcoal: As a fuel.
   (b) Bone charcoal: Purification of brown coloured sugarcane juice in the manufacture of sugar.
   (c) Coke: Used as a reducing agent in the extraction of metals.

9. The advantages of judicious use of energy are:
   - it will delay the energy crisis, and
   - it will give the scientists more time to develop more efficient alternate sources of energy.

10. (a) Diesel is used as fuel for heavy motor vehicles and in electric generators.
    (b) Petrol can be used as a motor fuel, aviation fuel and as solvent for dry cleaning.

11. (a) Coal (b) Petroleum (c) Refinery (d) Kerosene (e) Sunlight.

Long Answer Type Questions

1. (a) Petroleum was formed by the sedimentation of dead remains of microscopic marine plants and animals, which were buried under the surface of the earth, millions of years ago.
   (b) All liquid fuels are formed from petroleum.
   (c) Butane is the main constituent of LPG.
   (d) CNG: Compressed Natural Gas. It is used as a fuel.
   (e) Green leaves do not catch fire easily because they contain moisture.

2. (a) They have been formed by decomposition of pre-historic plants and animals buried under the earth's crust millions of years ago.
   (b) Fossil fuels are non-renewable sources of energy as once exhausted, they cannot be re-created in a short period of time.
   (c) LPG is a better fuel than coal because:
   - it has a low ignition temperature,
   - it has a high calorific value, and
   - it does not cause pollution.
   (d) Smelling agent (ethyl mercaptan) is added to LPG to detect the leakage of gas as LPG is a colourless and odourless gas.
   (e) Advantages of the modern chullah over the traditional chullah.
   - High efficiency.
   - Consumes less fuel.
   - Causes less pollution.

3.

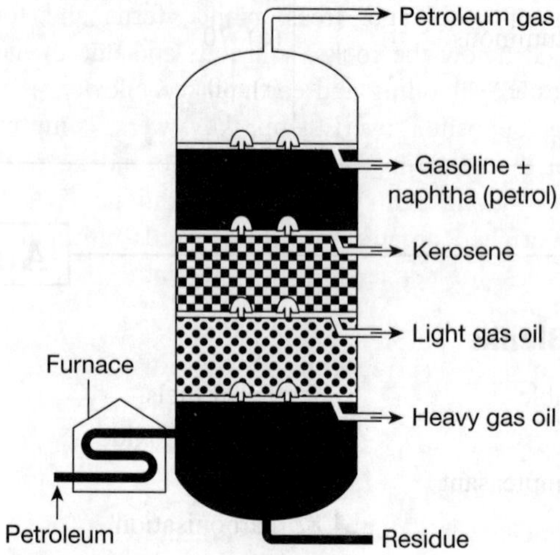

Uses

- Petrol is used as a motor fuel, aviation fuel and as solvent for dry cleaning.
- Kerosene is used as fuel at home and in jet aircrafts.
- Diesel is used as fuel for heavy motor vehicles and electric generators.

4. Burning of fuels is a major cause of global warming because they produce a lot of the greenhouse gas 'carbon dioxide' on burning.

   The earth's atmosphere allows most of the sunlight that falls on it to pass through and heat the surface. But greenhouse gases such as carbon dioxide, methane and water vapour trap the heat radiated from the earth. This leads to an increase in the earth's temperature. This is known as the *greenhouse effect*. When the concentration of greenhouse gases such as carbon dioxide increases, the greenhouse effect also increases. This is known as global warming. It may cause melting of ice caps and glaciers resulting in a rise in the sea level.

5. The natural resources which are present in a limited quantity in nature and can be exhausted (used up) by human activities are called exhaustible natural resources. For example, coal, petroleum, natural gas, minerals, forests and wildlife.

   It requires millions of years for the conversion of the dead organisms into petroleum from which petrol is separated. So, once the present stock gets exhausted, no new supplies will be available to us in the near future. However, sunlight is present in an unlimited quantity in nature, so it is an inexhaustible natural resource. It is not likely to be exhausted by human activities.

Exemplar Questions

Short Answer Type Questions

1. Fossil fuels take millions of years to be formed. They are limited in quantity and cannot be replenished easily, once consumed. Hence, they are considered as exhaustible natural resources.

2. Millions of years ago, trees, plants, ferns and forests got buried below the rocks, soil and sand due to natural processes like flooding and earthquakes. Slowly, as more soil was deposited over them, they were compressed. This led to conditions of high pressure and heat. These conditions along with anaerobic conditions turned the carbon-enriched organic matter of wood into coal. This slow process of conversion of wood into coal is called carbonisation.

3. The advantages of using CNG and LPG as fuels are:
   - they burn with a smokeless flame and so do not cause any pollution,
   - they leave no ash on burning, and
   - they are easy to handle and convenient to store.

4. **Characteristics of coke:** Coke is 98% pure carbon. It is a tough, porous and black substance. It produces a very little smoke.

   Uses of coke: Coke is very useful as a fuel. It is a good reducing agent. It is widely used in metallurgical processes to reduce metals from their oxides. It is used for producing water gas.

5. Petroleum is formed by the buried aquatic plants and animals below the sea bed. The marine animals and plants that died thousands of years ago settled down at the bottom of the sea. Under anaerobic conditions, microorganisms decomposed this organic matter. Due to high pressure and heat, the dead remains of these tiny plants and animals were slowly converted into petroleum.

6. Decantation can be used to separate a mixture of petroleum and water. The mixture is allowed to stand. Petroleum is lighter than water, so it floats over water. Carefully, the lighter layer of petroleum can be poured and separated without disturbing the heavier layer of water.

7. CNG stands for Compressed Natural Gas. It is considered to be a better fuel because it burns with a smokeless flame and causes no air pollution. It also does not produce any poisonous gases on burning.

8. Characteristics of coal are:
   - it is a hard, black, combustible fuel,
   - it is a fossil fuel,
   - it contains mainly carbon,
   - when heated in air, it burns and produces mainly carbon dioxide gas, and
   - it is a source of energy.

   Some important uses of coal are:
   - it is used as a fuel in homes and industry, and at thermal power plants for generating electricity,
   - it is used to make coal gas and coke.

9. Forests, floods, soil, soil, temperature, pressure, temperature.

Long Answer Type Questions

1. When coal is strongly heated in the absence of air in closed retorts, various useful products are obtained. These are coke, coal tar and coal gas.

   Uses of coke: It is used in the manufacture of steel and in the extraction of many metals. It acts as a fuel and is a better fuel than coal.

   Uses of coal tar: The various compounds obtained from coal tar are used as starting materials for manufacturing a large number of substances used in everyday life and in industry, such as, synthetic dyes, drugs, explosives, perfumes, plastics, paints, photographic materials and roofing materials. Naphthalene balls used to repel moths and other insects are also obtained from coal tar.

   Uses of coal gas: It is used as a fuel in many industries situated near the coal processing plants.

2. Fossil fuels will last only for a few hundred years because they are exhaustible natural resources. If consumed at a rate faster than the rate at which these are formed in nature, they will get exhausted very soon.

   Once the present stock of these gets exhausted, no new supplies of these fossil fuels will be available to us in the near future because it requires millions of years for the conversion of dead organisms into these fuels.

3. We must follow the following tips to save petrol/diesel/natural gas:
   - ensure correct air pressure in the tyres,
   - ensure regular maintenance of the vehicle,
   - drive at a constant and moderate speed as far as possible, and
   - switch off the engine at traffic lights or at a place where you have to wait.

4. If all the exhaustible natural resources are exhausted by human activities, then the survival of living beings would not be possible. Actually, fossil fuels are the most important source of energy for us today. They are a concentrated source of energy and give off heat and light on burning. The heat can be used to cook food or to run engines such as automobile engines. It can be used to generate electricity (in powerhouses, the most common fuel used is coal).

5. Some of the important uses are:
   - Petroleum gas in liquid form (LPG) is used as fuel in homes and industries.
   - Petrol is used as fuel for automobiles and aviation.
   - Kerosene is used as fuel for stoves, lamps and jet aircrafts.
   - Diesel is used as fuel for heavy motor vehicles, electric generators.
   - Lubricating oil is used for lubrication.
   - Paraffin wax is used in ointments, candles, vaseline.
   - Bitumen is used in paints and road surfacing.

6. Yes, the known resources of coal will last only for about 100 years. Once the present stock of coal present in the earth gets exhausted, no new supplies of these fossil fuels will be available to us in the near future because it takes millions of years to convert the dead organisms into fossil fuels in nature. So, coal should be used with care and caution and not wasted at all.

7. We should use fossil fuels only when absolutely necessary. We can also use natural gas as a substitute. The reserves of natural gas discovered by us have gone up ten times within 20 years.

   Alternative sources of energy such as solar, wind and biomass should be used in place of fossil fuels.

8.

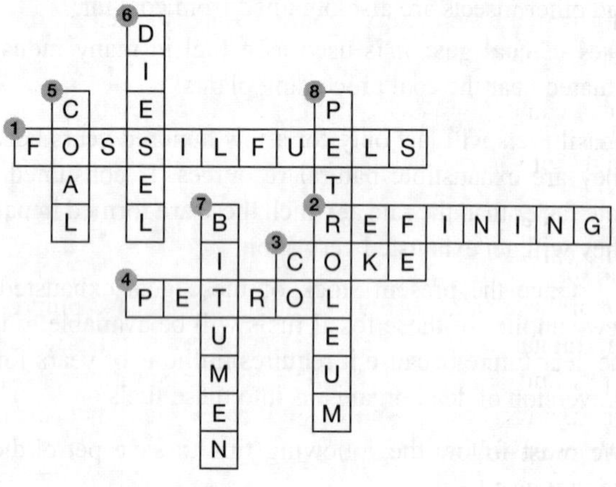

Across

1. Fossil fuels 2. Refining
3. Coke 4. Petrol

Down

5. Coal 7. Bitumen
6. Diesel 8. Petroleum

Competition Window (Objective Type)

Topic-wise MCQs

1. (c) **2.** (d) **3.** (c) **4.** (c) **5.** (d)
6. (c) **7.** (b) **8.** (d) **9.** (b) **10.** (d)
11. (c) **12.** (d) **13.** (d) **14.** (c) **15.** (b)
16. (a) **17.** (d) **18.** (c) **19.** (b) **20.** (d)
21. (a) **22.** (c) **23.** (c) **24.** (c) **25.** (c)
26. (c) **27.** (d) **28.** (a) **29.** (d) **30.** (c)
31. (a) **32.** (c) **33.** (d) **34.** (a) **35.** (d)
36. (a) **37.** (c) **38.** (c)

Miscellaneous

1. (d) **2.** (a) **3.** (d) **4.** (a) **5.** (d)
6. (d) **7.** (d) **8.** (d) **9.** (a) **10.** (a)
11. (c) **12.** (d) **13.** (c) **14.** (b) **15.** (d)
16. (a) **17.** (c) **18.** (c)

Advanced and Olympiads

Single Choice

1. (c) **2.** (b) **3.** (d) **4.** (c) **5.** (c)
6. (d)

Multiple Choice

7. (abc) **8.** (abc) **9.** (abc) **10.** (acd) **11.** (abc)
12. (abd) **13.** (bcd) **14.** (bcd) **15.** (abd)

Comprehension Type

16. (b) 17. (b) 18. (d)

Matrix Matching

19. (a)—(r), (b)—(s), (c)—(p), (d)—(q)
20. (a)—(t), (b)—(s), (c)—(q), (d)—(r)
21. (a)—(s), (b)—(r), (c)—(q), (d)—(p)

Integer Type

22. 5 **23.** 4 **24.** 5 **25.** 3 **26.** 6

Hints and Solutions

Competition Window (Objective Type)

Topic-wise MCQs

1. Exhaustible natural resources are present in a limited quantity in nature and can be completely used up by human activities. ***Note:*** Coal, petroleum, natural gas, minerals, forests and wildlife are exhaustible natural resources.

2. Fossil fuels are obtained from the remains of living organisms (plants and animals) buried under the earth millions of years ago. ***Note:*** Coal, petroleum and natural gas are fossil fuels.

3. Sunlight, air and water are inexhaustible natural resources. That is, they cannot be exhausted by human activities.

5. Wood can be replenished easily by growing more trees.

9. Coal tar is a black, thick liquid.

10. Nowadays, coal gas is used as a source of heat.

11. It is a hard coal and contains about 96% carbon. Peat, lignite and bituminous coal contain 60%, 67% and 88% of carbon respectively.

14. Methane is present in coal mines along with coal.

15. Coke is used in preparation of water gas, producer gas and reduction of iron oxide to iron.

16. When coal is heated in the absence of air, coke, coal gas and coal tar are the products.

17. Unburnt coal particles get deposited on lamps by condensation of burning oil or kerosene.

18. Coal is used in converting water into steam in a thermal power station.

20. Petroleum is a naturally occurring crude oil containing saturated, cyclic, unsaturated hydrocarbons with some other elements, such as sulfur, oxygen and nitrogen.

22. Since crude petroleum is a black viscous liquid and it is of utmost importance, it is called black gold.

23. Fractional distillation is used because the difference between the boiling points of the different components is less.

24. Kerosene is used as fuel in stoves, lamps and jet aircrafts.

28. Gasoline has the lowest boiling point.

29. Petroleum consists of petroleum gases, petrol, diesel, kerosene, lubricating oil, paraffin wax and bitumen. Coal tar and coal gas are obtained by the processing of coal.

30. Synthetic polymers, paints and plastics are obtained from petrochemicals.

33. Natural gas is a very important fossil fuel because it is easy to transport through pipes.

34. Natural gas consists mainly of methane.

35. Liquefied petroleum gas is a mixture of propane and butane.

36. Coal, petroleum and peat are fossil fuels, hence they cause pollution.

37. Fossil fuels produce greenhouse gases like carbon dioxide which are responsible for global warming.

Miscellaneous

1. Wildlife and minerals are exhaustible natural resources because they are present in a limited quantity in nature and can be exhausted by human activities.

2. The process of heating coal to a high temperature in the absence of air is called destructive distillation of coal. It yields products such as coke and coal gas.

3. Paraffin wax is used in making ointments and vaseline.

4. The constituent hydrocarbons of LPG do not have any smell. Therefore, ethyl mercaptan, a strong-smelling agent is added to it to detect leakages.

5. Concentrated H_2SO_4 dehydrates sugar to give charcoal.

6. Butane can be easily compressed and liquefied to be filled into cylinders.

7. Minerals, water and soil are natural resources but plastic is a man-made material. So, plastic is not a natural resource.

8. CNG (Compressed Natural Gas) is not obtained from coal. ***Note***: Natural gas is formed under the earth by the decomposition of vegetable matter lying under water.

9. Coal is formed by the decomposition of large land plants and trees buried under the earth about 300 million years ago.

10. Coal is formed by the carbonisation of dead vegetation.

11. Naphthalene balls are used to repel moths and other insects (in stored clothes).

12. Coke is not a constituent of petroleum. Coke is prepared by heating coal in the absence of air.

13. Petroleum was formed by the decomposition of the remains of tiny plants and animals buried under the sea millions of years ago.

15. Coke can be used as a reducing agent in metallurgy.

16. Different varieties of coal have different carbon content.

17. O, H, N and S are present in small amounts in coal along with carbon.

18. Natural gas is stored under high pressure as compressed natural gas. ***Note:*** It is easy to transport natural gas through pipes and it can be used directly for burning in homes where it can be supplied through pipes. Natural gas is used for power generation.

Advanced and Olympiads

Single Choice

3. All the statements are correct.

4. Coal tar has highest boiling point (> 350°C) and petrol has the lowest boiling point (40–75°C) while the boiling point of kerosene is 160–250°C and lubricating oil has 300–350°C.

6. A and B are carbon-rich materials so they could be coal or its derivatives. When coal is heated in the absence of air, coke is formed which is a tough, porous and black substance. So A is coal and B is coke.

$$\underset{(A)}{\text{Coal}} \xrightarrow{\text{Absence of air}} \underset{(B)}{\text{Coke}}$$

Multiple Choice

7. Except biogas, all the rest are obtained by refining of petroleum which is a fossil fuel.

8. We can save fossil fuels by using them sparingly so it is true.

9. On combustion, coal causes air pollution in the form of smoke.

12. Bees wax is obtained from beehives so it is not a petroleum product.

15. Petrol is a mixture of hydrocarbons while coke, charcoal and diamond consist of carbon only.

Integer Type

22. Petroleum, forests, coal, natural gas and minerals are examples of exhaustible natural resources.

23. Gasoline contains 4–10 carbon atoms. So, $x = 4$.

24. 1 g of LPG produces nearly 50 kJ of heat. So, $n \times 10 = 50$; $n = 5$.

CHAPTER 12 Combustion and Flame

LEARNING OBJECTIVES

After studying this unit, you will be able to understand:

- Combustion and its type
- Flames with special reference to the candle
- Fuels and their types, calorific value

Introduction

We use different types of fuels for various purposes at our houses, in industries and for running automobiles. For example, we use a candle or a lamp to provide light in the case of a power cut. We also use LPG (Liquified Petroleum Gas) in our houses to produce heat for our cooking needs. We use petrol and diesel in our cars and other vehicles. Coal is used at a very massive scale for the production of electricity in power stations. Natural gas is used for heating and lighting purposes through a network of pipelines. Now it is clear to us that burning of fuel provides us heat energy which can be used for several purposes. The most commonly used fuels are coal, charcoal, wood, petrol, diesel, compressed natural gas (CNG), LPG and cow dung cakes. Here are some common questions that may arise in your mind while reading this chapter.

- Can you name a few fuels used in our homes?
- Can you name a few fuels used in trade and industry?
- What fuels are used for running automobiles?
- What is the difference between the burning of a candle and the burning of a fuel like coal?
- Why does a candle burn with a flame whereas coal does not?

Let us study the chemical process of burning of different fuels and all the types of flames produced during this process (Fig. 12.1).

Burning of LPG   Wood   Candle   Power plant

Fig. 12.1 Burning of various fuels

Combustion

Let us recall the activity of the burning of a magnesium ribbon performed in Class VII. We learnt that magnesium burns to form magnesium oxide and produces heat and white dazzling light (Fig. 12.2). You can perform the same type of activity with a piece of charcoal. You will find that charcoal burns in air producing carbon dioxide, heat and light.

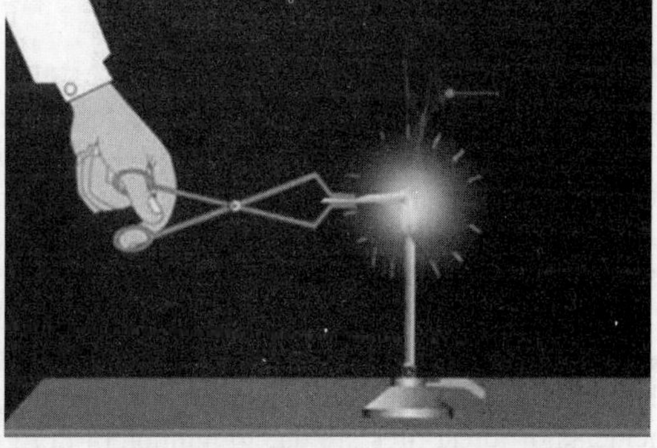

Fig. 12.2 Burning of magnesium

A chemical process in which a substance reacts with oxygen to give off heat is known as *combustion*. A substance which undergoes combustion is said to be combustible. It is also called a fuel. The fuel may be solid, liquid or gas. Sometimes, light is also given off during combustion, either as a flame or as a glow. Sometimes combustion may proceed with a hissing sound, a crackling sound or an explosion.

For example, $CH_4 + O_2 \rightarrow CO_2 + H_2O + \text{Heat}$

Combustible and Non-combustible Substances

Combustibility is the measure of how easily a substance can catch fire. On the basis of combustibility, substances are of two types—combustible and non-combustible.

Combustible

A material that undergoes combustion or burns is known as a *combustible material*. For example, in the reactions mentioned above, magnesium and charcoal are combustible substances. Some other examples are coal, wood, kerosene, diesel, petrol, CNG and LPG.

Non-combustible Material

Any material that does not undergo combustion or burn readily is known as a *non-combustible material*. For example, glass, metals, ceramics.

Conditions for Combustion

The following conditions are needed for combustion.

Combustible Substance

The presence of a combustible material is the primary condition for combustion and such substances are known as fuels. For example, petrol, CNG.

Supporter of Combustion

A substance that helps in the combustion of a fuel is a supporter of combustion and its presence is also needed. For example, oxygen and air are supporters of combustion. As air contains 78% N_2 (non-combustible and non-supporter) and 21% O_2, combustion occurs in air with a moderate rate while it is quite fast in the presence of oxygen.

Ignition or Kindling Temperature

Ignition temperature or ignition point is the temperature at which a combustible substance catches fire. Below the ignition temperature, even if a combustible substance is present along with oxygen, it will not undergo combustion. For example, a matchstick does not start burning on its own in air. However, it can be lighted by rubbing it on the side of the match box as on striking, friction is created that generates enough heat to light the matchstick or make the chemicals coated on it, catch fire. Incidents of forest fires are due to ignition of dry grass in extreme hot conditions.

- Every substance has a specific ignition temperature. For example, a paper can burn at once by bringing it in contact with a burning match stick or fire source. However, this does not occur in the case of coal or wood. Petrol or diesel ignites with a little spark but not coal or wood.

KNOWLEDGE BOOSTER

The sun produces light and heat but not because of combustion since oxygen is absent there. However, it is due to nuclear reactions.

A paper cup filled with or wetted with water cannot catch fire by bringing it near a burning match stick as the heat provided by match stick gets transferred to the water by condensation which does not allow the paper to attain its ignition temperature.

Inflammable Substance

A substance which has a very low ignition temperature and can easily catch fire with a flame is known as an *inflammable substance*. Both flammable and inflammable substances undergo combustion. However, inflammable substances have low ignition temperature. Examples of highly inflammable substances are petrol, kerosene, diesel, liquefied petroleum gas.

- A fuel with low ignition point is highly dangerous to use. For example, petrol has a very low ignition temperature compared to kerosene. That is, petrol can catch fire easily and cause explosions while kerosene due to its high ignition temperature burns smoothly. That is why petrol is not used in stoves and lamps while kerosene can be used.

KNOWLEDGE BOOSTER

Matchsticks

They are made of a suitable wood like white pine and aspen with a solidified mixture of flammable chemicals deposited on one of their ends. It is a mixture of antimony trisulfide and potassium chlorate with some glue. The rubbing surface has powdered glass and a little red phosphorus to make it less dangerous. When the head of a match stick is rubbed against the rubbing surface, the red phosphorus is converted into white phosphorus. This immediately reacts with potassium chlorate in the matchstick head to produce enough heat to ignite the antimony trisulfide and start the combustion.

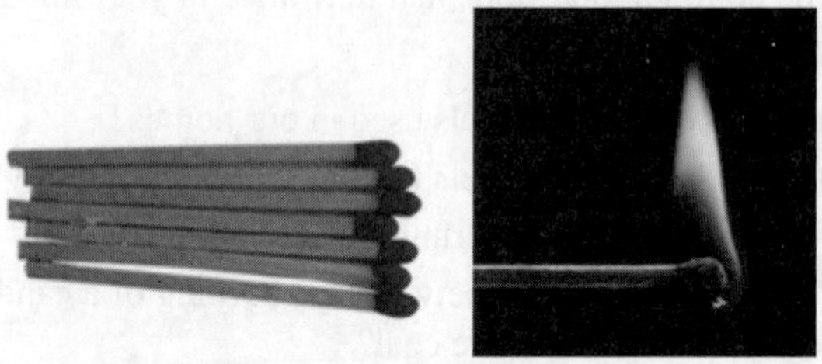

Control of Fire and Fire Extinguishers

Although fire is of great importance in our lives for many domestic purposes, it is quite dangerous. You must have seen or heard of fire breaking out in homes, shops, factories and forests. it is very essential that we must take necessary steps to control fire and to put off fire during any fire accident. The phenomenon of putting off fire is known as fire-fighting. To control fire, one or more of the conditions required for combustion should be removed. Any fire can be extinguished by

- removing the combustible substance,
- cutting off the supply of oxygen to the combustible substance, and
- reducing the temperature of the combustible substance below its ignition temperature.

Fire Extinguishers

Fire extinguishers are used for putting off fire. A few common fire extinguishers are discussed here.

(i) Water: Water is a very good substance in fire fighting as it cools down the fuel below its ignition temperature. Hence, it is sprayed over the burning material. It helps in lowering the ignition temperature and also putting off the flames by stopping the supply of oxygen to the flame. It is especially suitable for fires due to paper and wood. However, in the case of fire due to electrical faults, it cannot be used as it conducts electricity and can harm the fire fighters. It is also not suitable for fire due to oils since water is heavier than oil, the drops of oil are above the water surface and so the fire continues.

(ii) Sand and soil: Sand and soil may also be used as good fire extinguishers, particularly for fire due to burning oils as their spraying will cut off oxygen or air supply to stop fire.

(iii) Blanket: In case the clothes of a person catch fire, it may be put off with the help of a blanket as it will also cut off power supply of air.

(iv) Carbon dioxide: For fire involving electrical equipment and inflammable materials like petrol, carbon dioxide (CO_2) is the best extinguisher. Compressed CO_2 stored in a cylinder when released on a fire, causes the formation of a blanket around the fire and cools down the temperature of the fuel. Carbon dioxide (CO_2) being heavier than oxygen, covers the fire like blanket. As the contact between the fuel and oxygen is cut off, the fire is controlled. CO_2 does not harm the fire fighters and the electrical equipment.

Different types of Fire Extinguishers Providing CO_2

(a) Dry powder fire extinguisher: A dry powder fire extinguisher contains a mixture of sand and sodium bicarbonate (baking soda). When it is thrown over the burning substance, sodium bicarbonate gets decomposed to give carbon dioxide which acts like a fire extinguisher and cuts off air supply forming a blanket between air and the fire.

$$2NaHCO_3 \xrightarrow{\Delta} Na_2CO_3 + H_2O + CO_2$$

(b) Soda–acid fire extinguisher: Soda–acid type extinguisher contains dry sodium bicarbonate in a metallic cylinder. At the bottom of the cylinder there is a thin sealed glass tube containing concentrated sulfuric acid. This tube is surrounded by a fixed wire gauge. The cylinder has a plunger with a sharp end at the bottom of the cylinder and a nozzle on the top side of the cylinder. When the plunger is hit against the floor, its sharp tip breaks the glass tube. As a result, sodium bicarbonate reacts with sulfuric acid to release carbon dioxide with a lot of pressure which forms a layer over the burning material, cutting off oxygen supply.

$$2NaHCO_3 + H_2SO_4 \rightarrow Na_2SO_4 + 2H_2O + 2CO_2$$

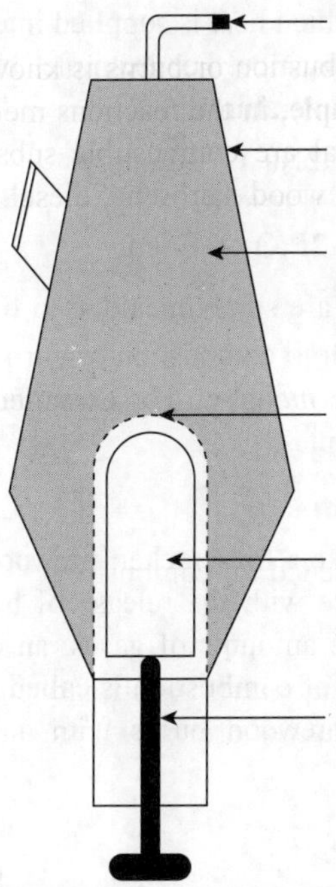

(c) Foam type fire extinguisher: It is just like the soda–acid fire extinguisher but containing sodium bicarbonate solution along with Turkey red oil (Saporin). It produces a foam with the gas and the solution issuing from the nozzle. Foam being light, will easily float on the surface of oil fires, hence cutting off the air supply easily.

Types of Combustion

Combustion can be further classified according to the rate or speed at which the combustible substance burns or undergoes combustion.

Slow Combustion

Such a combustion takes place at moderate speed. For example, burning of coal, wood, paper and candle. Here, the fuel is not completely burnt up. This results in the production of smoke with unburnt carbon particles. In a few cases, the rate of combustion is very slow also.

Food is fuel for our body. During respiration, the digested food, for example glucose, is broken down by reaction with oxygen in the body cells to give carbon dioxide, water and heat energy. This heat energy is utilised by our body for various functions. Hence, respiration is a type of slow combustion of food which takes place in the body to produce heat energy.

Rapid Combustion

When a substance or fuel burns in a short span of time, the combustion of the substance is almost complete and a large amount of heat and light are produced, then the combustion is known as rapid combustion. Examples are burning of LPG gas, sprit, petrol and camphor (highly inflammable fuels).

Spontaneous Combustion

The combustion in which a material suddenly bursts into flames, without any apparent cause is called *spontaneous combustion*. It occurs in the case of substances having low ignition temperature and in which heat is supplied internally by the slow oxidation of the substance. For example, burning of white phosphorus at room temperature.

$$\underset{\text{Sodium}}{4Na} + O_2 \rightarrow \underset{\text{Sodium oxide}}{2Na_2O} + \text{Energy}$$

$$4P + 5O_2 \rightarrow 2P_2O_5 + \text{Energy}$$

- Spontaneous forest fires are sometimes due to the heat of the sun or lighting strikes.
- Spontaneous combustion of coal dust has resulted in many disastrous fires in coal mines.

Explosive Combustion

On ignition of a substance like a fire-cracker, firewood or gun, a sudden reaction takes place with the release of heat, light, sound and evolution of large amounts of gases, an explosion is said to occur and this type of combustion is called *explosive combustion*. For example, firewood burns with a crackling sound, firing of a gun.

Depending on the amount of oxygen or air available for combustion, it can be further classified as:

(a) Complete combustion: When a carbon-containing substance burns in a sufficient supply of oxygen, it burns completely producing carbon dioxide, water and heat.

$$\underset{\text{Methane (from fuel)}}{CH_4} + \underset{\text{Oxygen (from air)}}{2O_2} \rightarrow \underset{\text{Carbon dioxide}}{CO_2} + \underset{\text{Water vapour}}{2H_2O} + \text{Heat}$$

$$\underset{\text{Glucose}}{C_6H_{12}O_6} + 6O_2 \rightarrow 6CO_2 + 6H_2O + \text{Heat}$$

(b) Incomplete combustion: When a carbon-containing substance burns in limited supply of oxygen, it burns incompletely producing carbon monoxide, water and heat. It also produces soot which contains unburnt carbon particles.

$$2CH_4 + \underset{\text{(Limited supply)}}{3O_2} \rightarrow 2CO + 4H_2O + \text{Heat}$$

$$CH_4 + \underset{\text{(Very limited supply)}}{O_2} \rightarrow \underset{\text{Carbon black}}{C} + 2H_2O$$

Illustrations

1. Select combustible and non-combustible material from these substances—paper, iron nail, glass, cloth, plastic, stone pieces, diesel, matchsticks, straw.

   Solution: Combustible substances are: paper, cloth, plastic, diesel, matchsticks, straw.

   Non-combustible substances are: iron nail, glass, stone pieces.

2. Explain the following:

   (i) Can the process of rusting be called combustion?

   (ii) It is difficult to burn a heap of green leaves but dry leaves catch fire easily. Explain.

   Solution:

   (i) As the process of rusting emits heat during the formation of its oxide, we can call the process of rusting as slow combustion.

   (ii) As green leaves hold some amount of water, their ignition temperature gets increased and they do not burn easily. On the other hand, dry leaves are waterless and having low ignition temperature, they catch fire easily.

3. Explain why fire-crackers on ignition produce sound?

   Solution: When a fire-cracker is ignited, a sudden (very rapid) reaction occurs with the evolution of heat, light and sound and a large amount of gas is liberated. Such a reaction is known as an explosion. The gases produced are heated by the heat evolved in the reaction. The hot gases expand rapidly and cause a loud sound.

Flame

The flame is a zone of combustion of gaseous substances accompanied by the evolution of heat and light. You already know that everything that burns cannot produce a flame. Only those substances which burn and vaporise produce a flame. The combustible substance and the supporter of combustion both have to be in the gaseous state to produce a flame. For example, kerosene and molten wax rise through the wick and are vaporised during burning and form the flame. Charcoal, on the other hand, does not vaporise and so does not produce a flame. A wick is a loosely twisted thread made from an absorbent fibre (cotton or a spongy material). One end of the wick is in a lamp and kept dipped in the liquid fuel while the other end is kept free in the air. The free end of the wick is also soaked in the oil or fuel and is lit. The wick catches fire and bears the flame (Fig. 12.3).

Fig. 12.3 Flames from various sources

Structure of a Flame

A candle flame is taken to study the various zones of a flame. A candle zone can be easily divided into three colours—yellow, black and blue. The three zones of a flame are discussed below.

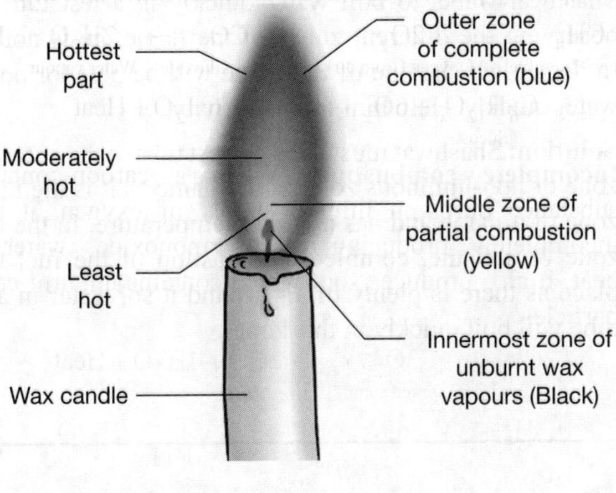

Outermost or Non-luminous Zone

This is the zone of complete combustion as adequate supply of oxygen is available. In this zone, carbon particles of the wax vapours are completely burnt to give carbon dioxide and water vapour. Since carbon dioxide is a colourless gas, this zone is the non-luminous zone and the flame appears to be pale blue. It is the hottest part of the flame also.

Middle Zone or Luminous Zone

This part of the flame is yellow in colour and in this zone, partial combustion of carbon vapour takes place as the supply of oxygen is non-adequate. It is moderately hot and black soot is produced by this zone of flame. Here the carbon particles glow by emitting a bright yellow light.

Innermost Zone or Dark Zone

This zone is close to the wick. It is the coldest zone of the flame due to unburnt wax vapour as no oxygen is available for burning in this zone. Due to the presence of carbon particles, it looks black.

KNOWLEDGE BOOSTER

- More the number of carbon atoms in a molecule of a fuel, the greater is the capacity of fuel to produce heat.
- A block pipe is used by a goldsmith to blow air into the outermost zone of the flame to make it hotter. Now this hot flame is used for melting silver or gold.

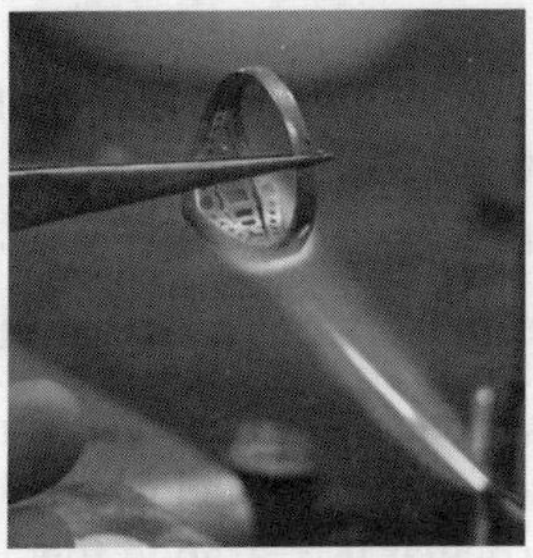

Chemical Description of Burning of a Candle

When a candle burns, nearly 25% of the energy created by combustion is given off as heat. A small fraction of the radiant heat from the candle flame feeds back into the candle and melts the wax to fuel the flame. The candle flame surface itself is the place where fuel (wax vapour) and oxygen mix and burn at high temperature, radiating heat and light. Heat from the flame is conducted down the wick and melts the wax at the wick base. The liquid wax rises up the wick due to capillary action. As the liquid wax nears the flame, the flame's heat causes it to vaporise and combine with oxygen. The vapours are drawn into the flame where they ignite. The heat produced melts more wax, and so on. Carbon dioxide and water are formed, and carbon rich particles (soot) are formed and re-formed. The soot is drawn up into the top part of the flame where increased temperature causes the carbon to luminesce and burn. Fresh oxygen from the surrounding air is drawn into the flame primarily because of convection currents that are created by the released heat. Hot gases produced during burning are less dense than the cooler surrounding air. They rise upward and in doing so, draw the surrounding air, which contains fresh oxygen, into the flame.

Solid particles of soot that form in the region between the wick and flame are also carried upward by the convection currents. They ignite and form the bright yellow tip of the flame. The upward flow of hot gases causes the flame to stretch out in a teardrop shape.

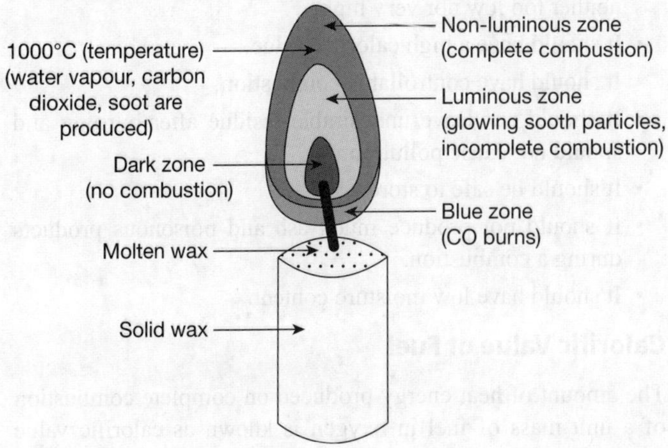

Candle flame

Illustrations

1. Khushi and Shanvi were doing an experiment in which water had to be heated in a beaker. Khushi kept the beaker near the wick in the yellow part of the candle flame. Shanvi kept the beaker in the outermost part of the flame. Whose water will get heated in a shorter time?

   Solution: The water which was put by Khushi will get heated in a shorter time as she had put it nearer to the hottest zone of the flame.

2. Two glass jars X and Y are filled with carbon dioxide and oxygen gases respectively. In each jar, a lighted candle is placed simultaneously. In which jar will the candle remain lighted for a longer time and why?

   Solution: In jar Y, the candle remains lighted for a longer time because oxygen is a supporter of combustion. So, we can say that oxygen (air) is necessary for combustion to take place.

3. Shashwat wants to boil water quickly in a test tube. On observing the different zones of the flame, he is not able to decide which zone of the flame will be best for boiling water quickly. Help him in this activity.

   Solution: Shashwat must keep his test tube in the outermost zone or non-luminous zone of the flame as it is the hottest zone of a flame and has a higher temperature. In the outer zone of a flame, complete combustion of the fuel takes place as there is plenty of air around it so, water in a test tube will boil quickly in this zone.

Fuel

A fuel is a substance that is used to produce heat or energy by burning or combustion. Many different types of fuels are used for providing heat and energy. Examples are petrol, diesel, natural gas, kerosene, coal, wood, CNG and LPG.

Characteristics of a Good Fuel

There is probably no fuel that could be considered an ideal fuel. We should look for a fuel which fulfils most of the requirements for a particular use. A good fuel is one which has the following characteristics.

- It should be easily available.
- It should be cheap and easy to transport.
- It should burn easily in air at a moderate rate.
- It should have a moderate ignition temperature that is neither too low nor very high.
- It should have a high calorific value.
- It should have controllable combustion.
- It should not leave undesirable residue after burning and should not cause pollution.
- It should be safe to store.
- It should not produce much ash and poisonous products during a combustion.
- It should have low moisture content.

Calorific Value of Fuel

The amount of heat energy produced on complete combustion of a unit mass of fuel in oxygen is known as calorific value of that fuel. It is measured in terms of kilojoules per kilogram that is kJ/kg and in kilojoules per gram that is kJ/g. It is used to compare the efficiencies of fuel. Hydrogen has the highest calorific value of 150 kJ/g. Any fuel having oxygen in it has a low calorific value as O_2 supports combustion but does not burn. A fuel having more mass percentage of hydrogen has higher calorific value (Table 12.1). For example, the calorific value of CH_4 (methane) is more than that of C_2H_6 (ethane).

- Usually, fuels are compounds of carbon and hydrogen, thus, they combine with oxygen on burning and liberate carbon dioxide and water vapour.
- Hydrogen is expected to be the future fuel due to its high calorific value and minimum pollution.

Table 12.1 Some common fuels and their calorific values

Fuel	Calorific value (kJ/kg)
Hydrogen	150000
Biogas	35000–40000
Coal	25000–33000
Petrol	45000
Kerosene	45000
Diesel	45000
Methane	50000
CNG	50000
LPG	55000
Wood	17000–22000
Cow dung cake	6000–8000

Classification of Fuels

Fuels may exist in all three forms—solid, liquid and gas.

Solid Fuels

Such fuels exist in the solid state at room temperature and they leave a residue after burning. Examples are firewood, coal, wood, cowdung cake, straw and other agricultural waste.

Liquid Fuels

Such fuels are liquids at room temperature and are mostly obtained from petroleum as a result of fractional distillation. They leave relatively very little residue or smoke on burning. For example, petrol, diesel, kerosene.

Gaseous Fuels

Such fuels occur in the gaseous state at room temperature and they do not leave any residue on burning. They do not emit any smoke even on complete combustion. For example, LPG, CNG, water gas, coal gas, biogas and natural gas (methane).

- Flame from acetylene (C_2H_2) is quite hot and is also used for welding and cutting metals.

Natural and Processed Fuels

Natural fuels

Natural fuels are also known as primary fuels as they are used in the same form as they exist in nature. For example, coal, wood, cowdung cakes.

Processed Fuels

Processed fuels are also known as secondary fuels as they are produced from natural fuels after processing by chemical methods. For example, petrol (fossil fuel), diesel (crude oil), biogas.

Composition of Some Processed Gaseous Fuels

(i) **Natural gas:** It is found in oil wells, present with the crude petroleum. It has methane and traces of ethane and propane.

(ii) **Biogas or gobar gas:** It has mainly methane and is produced by processing biomass.

(iii) **LPG:** It is liquefied petroleum gas and is a mixture of butane and a small amount of propane. As LPG does not have any smell, a strong-smelling compound, ethyl mercaptan (C_2H_5SH), is added to it to detect any leakage of LPG.

(iv) **CNG:** It is compressed natural gas and it is made by compressing methane extracted from natural gas. It can also be stored in cylinders like LPG.

(v) **Water gas:** It is a mixture of carbon monoxide and hydrogen ($CO + H_2$).

(vi) **Producer gas:** It is a mixture of carbon monoxide and nitrogen ($CO + N_2$).

(vii) **Coal gas:** It is produced during the destructive distillation of coal. It has mainly methane, hydrogen and carbon monoxide.

Impact of Burning of Carbon Fuel on the Environment

The increasing fuel consumption has following harmful effects on our environment:

- Carbon fuels like coal, wood and petroleum release unburnt carbon particles. These fine particles are suspended in the atmosphere and they are dangerous pollutants causing respiratory diseases such as asthma.
- Incomplete combustion of these fuels gives carbon monoxide gas which is a very poisonous gas and may be fatal. It is dangerous to burn coal in a closed room as the carbon monoxide gas produced can kill persons sleeping in that room.
- Combustion of most fuels releases carbon dioxide into the environment. The increased concentration of carbon dioxide in the air is believed to cause global warming.
- Burning of coal and diesel releases sulfur dioxide gas which is an extremely suffocating and corrosive gas. Moreover, petrol engines give off gaseous oxides of nitrogen. Oxides of sulfur and nitrogen dissolve in rain water and form acids. Such rain is known as acid rain and it is very harmful for crops, buildings and soil.

Controlling Pollution

Pollution caused due to burning of fuels can be controlled by applying the following methods.

- **Managing energy resources:** Presently most of the energy demand is fulfilled by using fossil fuels like coal and petroleum. Now, the use of solar energy is promoted for home needs (solar cooker for cooking and solar panels for generating electricity).
- Solid fuels must be replaced by gaseous fuel as gaseous fuels have not only more calorific value but also do not leave any residue. LPG and biogas are very good domestic fuels and good substitutes of solid fuels such as wood and coal.
- Use of CNG in vehicles in the place of petrol and diesel is a good method of controlling air pollution by vehicles.
- Less use of coal in power plants is the demand of today and it must be replaced by solar plants.
- Replacement of old traditional chulhas with improved chulhas (annat or smokeless chulha) will control smoke (air pollution).

KNOWLEDGE BOOSTER

Greenhouse effect

The phenomenon of abnormal heating up of the earth's surface due to the presence of excess of greenhouse gases which trap solar radiation is called *greenhouse effect*. The name greenhouse originates from glass structures known as greenhouses, that are used to grow green plants in the colder regions of the earth. The glass structure allows sunlight to enter it but does not allow the radiated heat to escape, hence heating up the green house.

(*Continued*)

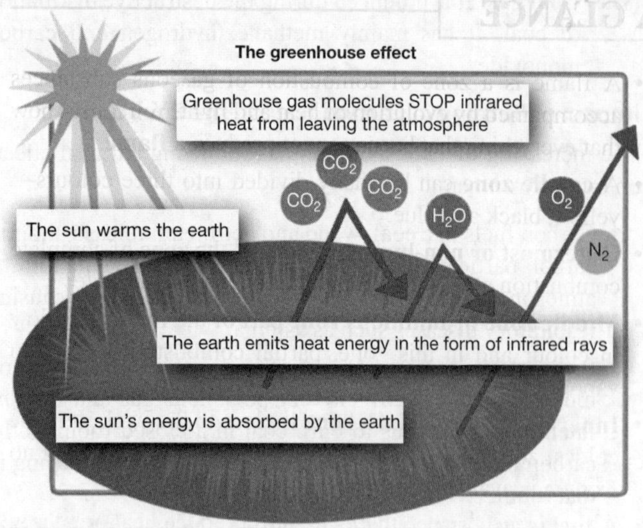

Greenhouse gases

Greenhouse gases are those that absorb and emit infrared radiation in the wavelength range emitted by earth. Carbon dioxide is the main greenhouse gas. Some other greenhouse gases are methane, CFCs (chlorofluorocarbons), O_3, NO and water vapour.

Global warming

Nearly three-fourths of the sun's energy that reaches the earth is absorbed by the earth's surface due to which there is an increase in the temperature of the earth's surface. Some part of this heat is also trapped by greenhouse gases like CO_2, CH_4 and water vapour that are present in the atmosphere. This causes global warming. A slight increase of 1–3°C in temperature can disturb the earth's environment.

Effect of global warming

Some of the effects of global warming are as follows:

- **Rise in sea level:** Due to the melting of glaciers and polar ice caps, there is an increase in sea level. It causes floods in coastal areas like the Maldives and India.
- **Decrease in water on earth:** Due to global warming, more water evaporates from water bodies so there is a decrease in amount of water and increase in water vapour.
- **Unseasonal rains:** Due to global warming, a change in the rain pattern is observed which affects trees and plants in forests which in turn disturbs wildlife.
- **Effects on agriculture:** Due to rapid depletion of surface water by global warming, the agriculture sector is also affected.

Methods of reducing global warming

A few important steps can be taken to reduce global warming.

- Minimise the use of automobiles based on petroleum and replace them with bicycles, electric vehicles and larger use of the public transport system.
- More tree planting must be done.
- Check or avoid burning of wood and dry leaves.
- Avoid release of carbon dioxide during industrial processes.

Illustrations

1. Write some major drawbacks of global warming.

Solution: Due to global warming, the average global temperature will increase to a level which may lead to melting of polar ice caps and flooding of low-lying areas. It may also increase the occurrence of diseases.

2. Explain how the use of CNG in automobiles has reduced pollution in our cities.

Solution: The use of CNG in automobiles has reduced pollution in our cities as it is a good quality fuel and has some benefits.

- It gives out less carbon dioxide, carbon monoxide, sulfur dioxide and nitrogen dioxide, which is beneficial as they play a crucial role in global warming and acid rain.
- It leaves behind no residue after its combustion.

3. People usually keep angithi/burning coal in their closed rooms during the winter. Why is it advised to keep the door open?

Solution: When coal burns in air, it produces carbon dioxide which is non-poisonous but when angithi or coal burns in a closed room, the air or oxygen cannot enter the room so, in the incomplete supply of oxygen, coal burns to produce carbon monoxide which is a highly poisonous gas. It causes suffocation in a closed room. Therefore, the door and windows should be open to avoid any mishaps.

CHAPTER AT A GLANCE

- A chemical process in which a substance reacts with oxygen to give off heat is known as **combustion**.
- The substance which undergoes combustion is said to be combustible. It is also called a **fuel**. The fuel may be solid, liquid or gas.
- Sometimes, light is also given off **during combustion**, either as a flame or as a **glow**. Sometimes combustion may proceed with a **hissing sound**, a **crackling sound** or an **explosion**.
- The material that undergoes combustion or burns is known as a **combustible** material. The material which does not undergo combustion or burn readily is known as a **non-combustible** material.
- **Combustion** is possible under these conditions—(i) using a combustible substance, (ii) a supporter of combustion, (iii) using ignition.
- A substance which has a very low ignition temperature and can easily catch fire with a flame is known as an **inflammable substance**.
- **Matchsticks:** They are made of a suitable wood such as white pine and aspen, with a solidified mixture of flammable chemicals deposited on one of their ends. It is a mixture of antimony trisulfide and potassium chlorate with some glue.
- When the head of a matchstick is rubbed against the rubbing surface, the red phosphorus is converted into white phosphorus. This immediately reacts with potassium chlorate in the matchstick head to produce enough heat to ignite antimony trisulfide and start the combustion or burning.
- **Fire extinguishers** are used for putting off fire.
- Fire involving electrical equipment and inflammable materials like petrol, **carbon dioxide (CO_2)** is the best extinguisher.
- A **dry powder fire extinguisher** has a mixture of sand and sodium bicarbonate (baking soda).
- **Soda–acid type extinguisher** contains dry sodium bicarbonate in a metallic cylinder.
- **Foam type fire extinguisher** is just like the soda–acid fire extinguisher and contains sodium bicarbonate solution along with Turkey red oil (Saporin).
- A flame is a zone of combustion of gaseous substances accompanied by evolution of heat and light. You must know that everything that burns cannot produce a **flame**.
- A **candle zone** can be easily divided into three colours—yellow, black and blue.
- **Outermost or non-luminous zone** is the zone of complete combustion as adequate supply of oxygen is possible.
- **Middle zone or luminous zone** part of the flame is yellow in colour and in this zone, partial combustion of carbon vapour takes place as supply of oxygen is non-adequate.
- **Innermost zone or dark zone** is close to the wick. It is the coldest zone of the flame due to unburnt wax vapour as no oxygen is available for burning in this zone.
- **A fuel** is a substance that is used to produce heat or energy by burning or combustion. Many different types of fuels are used for providing heat and energy.
- The amount of heat energy produced on complete combustion of a unit mass of fuel in oxygen is known as **calorific value** of that fuel. It is measured in terms of kilojoules per kilogram that is kJ/kg and in kilojoules per gram that is kJ/g.
- Fuels may exist in all three forms—solid (firewood, coal), liquid (petrol, diesel) and gas (LPG, CNG).
- **Natural fuels** are also known as primary fuels as they are used in the same form as they exist in nature.
- **Processed fuels** are also known as secondary fuels as they are produced from natural fuels after processing by chemical methods.
- The phenomenon of abnormal heating up of the earth's surface due to the presence of excess of greenhouse gases which trap solar radiation is called **greenhouse effect**.
- **Global warming:** Nearly three-fourths of the sun's energy reaching the earth is absorbed by the earth's surface due to which there is an increase in the temperature of the earth's surface. Some part of this heat is also trapped by greenhouse gases such as CO_2, CH_4 and water vapour that are present in the atmosphere. This causes global warming.

PRACTICE QUESTIONS

Analyse Your Concepts (School Exam Based)

Fill in the Blanks

Instructions: Complete the following statements with an appropriate word/term to be filled in the blank spaces.

1. A chemical process in which a substance reacts with oxygen to give off heat is called ________.
2. Wood, paper, CNG are ________ substances.
3. When the clothes of a person catch ________ the person is covered with a ________ to extinguish fire.
4. The ________ temperature at which a substance catches fire is called its ________ temperature.
5. The ignition temperature of ________ is lower than that of wood.
6. The substances which vaporise during ________ give a flame.
7. The substances which have a very low ________ temperature and can easily catch fire with a flame are called ________ substances.
8. The amount of heat energy produced on complete combustion of 1 kg of a fuel is called its ________.
9. The rubbing surface on a match box contains ________ phosphorus.
10. ________ zone of the candle flame is yellow in colour.
11. The calorific value of CNG is ________ kJ/kg.
12. The melting of polar ice will ________ the level of sea water.

True or False

Instructions: Read the following statements and write your answer as true or false.

1. A physical process in which a substance reacts with oxygen to give off heat is called combustion.
2. The sun produces heat and light; it is a case of combustion.
3. Water is the best extinguisher for fires involving electrical equipment.
4. Ceramics are non-combustible substances.
5. Alcohol, CNG and LPG are inflammable substances.
6. SO_2 is a fire extinguisher.
7. Increased concentration of nitrogen in air is believed to cause global warming.
8. Combustion of glucose is a case of complete combustion.
9. Greater the calorific value, the better is the fuel.
10. Global warming will also lead to a change in climate of almost every part on the earth.
11. The middle zone is the hottest zone of a flame.
12. The substances which vaporise during burning, give a flame.

Very Short Answer Type Questions

1. Write any three conditions under which combustion can take place.
2. Name the unit in which the calorific value of a fuel is expressed.
3. What are the main constituents of biogas and kitchen gas (LPG)?
4. What is biogas?
5. Give two examples of combustible substances.
6. How is heat and light produced in the sun?
7. What is meant by rapid combustion?
8. Why is a smelling agent added to LPG?
9. Why does yellow phosphorus catch fire on its own?
10. Is the food a fuel for our body?
11. Which substances burn with a flame?
12. Why is water poured to extinguish burning wood?
13. What is the function of a wick in the candle?

Short Answer Type Questions

1. What is a fuel? Write the name of any two liquid fuels and two gaseous fuels.
2. Explain how CO_2 is able to control fires.
3. What is acid rain and how is acid rain harmful?
4. Can you put off a fire due to electric short circuit by using water? Why?
5. Explain the following.
   (a) Does magnesium burn with a flame?
   (b) Camphor burns with a flame, but charcoal glows.
6. (a) Name the type of nuclear reaction taking place in the atmosphere of the sun.
   (b) Why should water not be used to put out a fire caused by burning of petrol.

7. Write any four characteristics of an ideal fuel. What is CNG? Mention one use.
8. (a) Which is the hottest zone and the coolest zone of a flame?
   (b) Why does the middle zone of the candle flame glow with yellow colour?
9. How does the foam type fire extinguisher work?
10. Fill in the blanks.
    (a) Burning of wood and coal causes _________ of air.
    (b) A liquid fuel used in homes is __________.
    (c) Fuel must be heated to its __________ before it starts burning.
    (d) Fire produced by oil cannot be controlled by __________.
11. Which zone of a flame does a goldsmith use for melting gold and silver and why?
12. In an experiment, 4.5 kg of a fuel was completely burnt. The heat produced was measured to be 180,000 kJ. Calculate the calorific value of the fuel.
13. Hydrogen has the highest calorific value, but it is not used as a fuel. Why?
14. Can the process of rusting be called combustion? Discuss.

Long Answer Type Questions

1. Explain the different zones of a flame with the help of a neat and well labelled diagram.
2. Give reasons for the following
   (a) LPG is a better domestic fuel than wood.
   (b) Paper by itself catches fire easily whereas a piece of paper wrapped around an aluminium pipe does not.
   (c) It is difficult to burn a heap of green leaves but dry leaves catch fire easily. Explain.
3. Draw a labelled diagram to show the construction of a soda–acid fire extinguisher and explain its working.
4. Compare LPG and wood as fuels.

NCERT Corner

Textbook Exercises

1. When fuels burn, what do they produce?
2. What name is given to the substances which can burn easily?
3. What name is given to the substances which do not burn?
4. When does a substance start burning?
5. When does a fire brigade arrive?
6. How does the fire brigade put off fire?
7. When water is poured on a fire, which condition of combustion is not fulfilled?
8. What name is given to combustion that takes place on its own?
9. What is meant by global warming?
10. Why is respiration referred to as slow combustion?
11. Give one reason why LPG is a better fuel than kerosene.
12. A person sleeping in a closed room with burning coal, feels suffocated after sometime. Why?
13. Distinguish between combustible and non-combustible substances.
14. Can you burn paper when a burning matchstick is brought near it? Why?
15. What is forest fire?
16. Why do different substances catch fire at different temperatures?
17. Kerosene catches fire faster than wood. Why?
18. Supposing the kerosene catches fire, how will you put it off?
19. How is fire extinguished at airports and petrol pumps?
20. What is meant by spontaneous combustion? Give one example of a substance which shows spontaneous combustion.
21. Which substances burn with a flame?
22. Why cannot petrol be used as a fuel in stoves at homes?
23. Amit and Sunil were doing an experiment in which water was to be heated in a beaker. Amit kept the beaker near the wick in the yellow part of the candle flame. Sunil kept the beaker in the outermost part of the flame. Whose water will get heated in a shorter time? Why?
24. Why do we use paper or kerosene to start a fire in wood or coal?
25. Explain how water extinguishes the fire.
26. You are given three substances A, B and C. How will you find out which of them is a combustible material?
27. When the clothes of a person catch fire, we cover him with a blanket. Why?

28. Red buckets containing sand are kept in offices and cinema halls. Why?

29. Explain how the use of CNG in automobiles has reduced pollution in our cities.

30. Give some effects of fuels combustion.

31. Why is wood used as a fuel in villages? What are the disadvantages of using wood as a fuel?

Exemplar Problems

Short Answer Type Questions

1. Some words (underlined) in the following sentences are jumbled up. Write them in their correct form.
   (a) Seldie is a combustible substance.
   (b) Slaas is a non-combustible material.
   (c) Chittsmack does not burn by itself.
   (d) Some substances on combustion produce thea and mafel.
   (e) The amount of heat energy produced on complete combustion of 1 kg of a fuel is called its ficalroic value.

2. Why is the use of diesel and petrol as fuels in automobiles being replaced by Compressed Natural Gas (CNG) in big cities?

3. Boojho wants to separate the following materials as combustible and non-combustible. Can you help him? Charcoal, chalk, stone, iron rod, copper coin, straw, cardboard, glass, paper, candle, wood.

4. State whether the following statements are True/False.
   (a) Air is necessary for combustion.
   (b) Magnesium is a non-combustible metal.
   (c) Carbon dioxide is an excellent fire extinguisher.
   (d) Calorific value of wood is higher than that of coal.

5. You are provided with three watch glasses containing milk, petrol and mustard oil respectively. Suppose you bring a burning candle near these materials one by one, which material(s) will catch fire instantly and why?

6. If you hold a piece of iron wire with a pair of tongs inside a candle flame or a Bunsen burner flame, what will you observe? Will it produce a flame?

7. What do you understand by fuel efficiency?

8. Match the items of Column I with the items of Column II.

Column I	Column II
(a) Oxides of sulphur and nitrogen	(i) Fire extinguisher
(b) CNG	(ii) Incomplete combustion of coal
(c) Oxygen	(iii) Very low ignition temperature
(d) Inflammable substance	(iv) Acid rain
(e) Carbon dioxide	(v) Necessary for combustion
(f) Carbon monoxide	(vi) Fuel for automobiles

9. Match the following for the flame of a candle.

Column I	Column II (zone)	Column III (colour)
(a) Hottest part	(i) Innermost zone of unburnt wax vapours	(x) Blue
(b) Moderately hot	(ii) Middle zone of partial combustion	(y) Black
(c) Least hot	(iii) Outer zone of complete combustion	(z) Yellow

Long Answer Type Questions

1. Manu was heating oil to fry potato chips. The cooking oil all of a sudden caught fire, he poured water to extinguish the fire. Do you think, this action was suitable? If yes, why? If not, why not? In such a condition what should Manu have done?

2. What are the three essential requirements to produce fire? How is a fire extinguisher useful for controlling the fire?

3. Give two examples each for a solid, liquid and gaseous fuel along with some important uses.

4. The calorific values of petrol and CNG are 45000 kJ/kg and 50000 kJ/kg respectively. If you have a vehicle which can run on petrol as well as CNG, which fuel will you prefer and why?

5. Although wood has a very high calorific value, we still discourage its use as a fuel. Explain.

6. Forest fire produces a lot of air pollution. Write in brief about the reasons of forest fires.

7. Complete the crossword with the help of the clues.

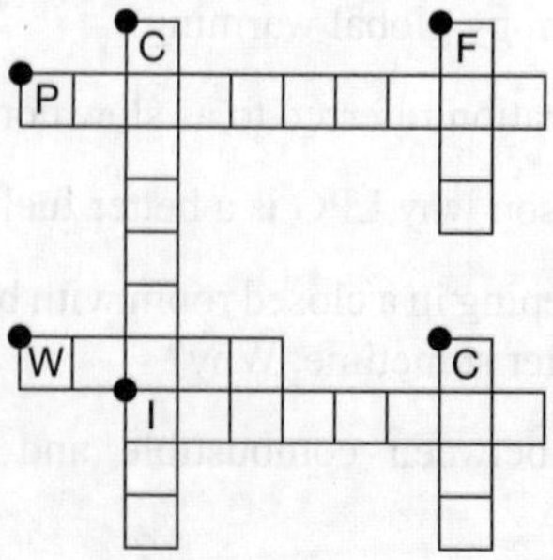

Across

1. Non-metal which catches fire if exposed to air. (10)
3. The lowest temperature at which a substance catches fire is called its __________ temperature. (8)
5. The most common fire extinguisher. (5)

Down

2. A chemical process in which a substance reacts with oxygen to give off heat. (10)
4. Petrol is used as a ________ in automobiles. (4)
6. It is as hard as stone and black in colour. (4)

Activity 1 (NCERT Textbook, Page 65)

Collect some materials like straw, matchsticks, kerosene, paper, iron nails, stone pieces and glass. Under the supervision of your teacher try to burn each of these materials one by one. If combustion takes place mark the material combustible, otherwise mark it non-combustible (Table).

Solution: Combustible and non-combustible substances

Material	Combustible	Non-combustible
Wood	✓	
Paper	✓	
Iron nails		✓
Kerosene	✓	
Stone piece		✓
Straw	✓	
Charcoal	✓	
Matchsticks	✓	
Glass		✓

Activity 2 (NCERT Textbook, Page 65) (Caution: Be careful while handling a burning candle).

Fix a lighted candle on a table. Put a glass chimney over the candle and rest it on a few wooden blocks in such a way that air can enter the chimney. Observe what happens to the flame. Now remove the blocks and let the chimney rest on the table. Again, observe the flame. Finally, put a glass plate over the chimney. Watch the flame again. What happens in the three cases? Does the flame flicker off? Does it flicker and give smoke? Does it burn unaffected? Can you infer anything at all about the role played by air in the process of burning?

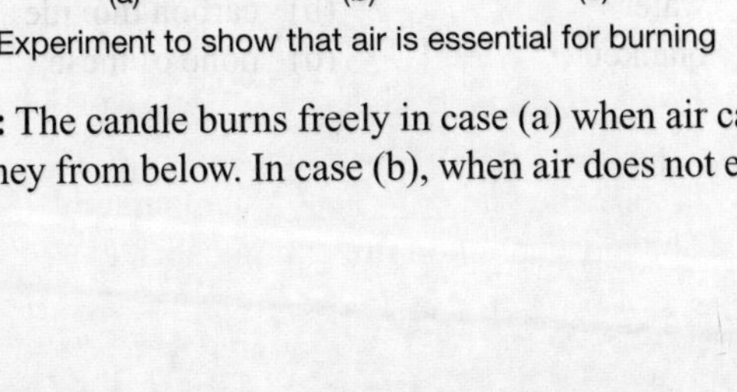

Experiment to show that air is essential for burning

Solution: The candle burns freely in case (a) when air can enter the chimney from below. In case (b), when air does not enter the chimney from below, the flame flickers and produces smoke. In case (c), the flame finally goes off because the air is not available.

Activity 3 (NCERT Textbook, Page 66)

Place a piece of burning wood or charcoal on an iron plate (tawa). Cover it with a glass jar or a tumbler, or a transparent plastic jar. Observe what happens. Does charcoal stop burning after sometime? Can you think of a reason why it stops burning?

Solution: After burning under the jar for some time, the wood/charcoal stops burning due to the discontinuous supply of air (or oxygen present in the air).

Activity 4 (NCERT Textbook, Page 67) (Caution: Be careful while handling a burning candle)

Make two paper cups by folding a sheet of paper. Pour about 50 mL of water in one of the cups. Heat both the cups separately with a candle. What do you observe?

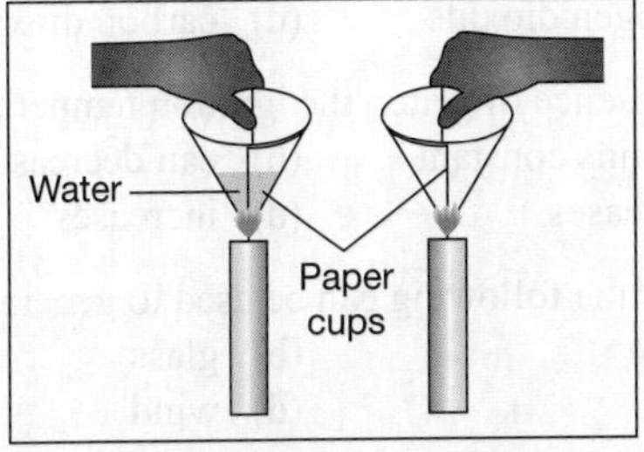

Heating water in a paper cup

Solution: It is observed that the empty cup catches fire immediately and begins to burn, whereas the cup containing water does not burn; instead, the water present in it becomes hot.

Activity 5 (NCERT Textbook, Page 77)

Light a candle (Caution: be careful). Hold a glass tube with a pair of tongs and introduce one end in the dark zone of a non-flickering candle flame. Bring a lighted matchstick near the other end of the glass tube. Do you see a flame? If so, what is it that produces a flame?

Solution: When we bring a lighted matchstick near the other end of the glass tube then we observe a flame. The vapourised wax from the candle produces this flame.

COMPETITION WINDOW (OBJECTIVE TYPE)

[For NEET, JEE (Main and Advanced), NTSE, KVPY and Olympiads]

Topic-wise MCQs

Combustion

1. A substance which reacts with oxygen giving heat is called a combustible substance. Which one of the following is a combustible substance?
(a) iron nail (b) glass
(c) stone piece (d) wood

2. The gas required for combustion is
(a) oxygen (b) nitrogen
(c) carbon dioxide (d) hydrogen

3. Burning of charcoal in a closed room will produce
(a) sulfur dioxide (b) carbon monoxide
(c) nitrogen dioxide (d) carbon dioxide

4. In the presence of water, the ignition temperature of paper
(a) remains constant (b) can decrease or increase
(c) decreases (d) increases

5. Which of the following can be used to generate electricity?
(a) stone (b) glass
(c) soil (d) wind

6. Place of piece of burning charcoal on an iron plate and cover it with a plastic jar. The charcoal stops burning because
(a) it becomes cold after some time
(b) its ignition temperature is lowered
(c) supply of oxygen is cut off
(d) none of the above

7. Respiration is a reaction which can be named as
(a) substitution (b) combustion
(c) carbonisation (d) polymerisation

8. Digestion of food is an example of
(a) slow combustion
(b) rapid combustion
(c) incomplete combustion
(d) complete combustion

9. Which one of the following has the highest calorific value?
(a) kerosene (b) biogas
(c) LPG (d) petrol

10. The substance expected to have the highest ignition temperature out of the following is
(a) kerosene (b) petrol
(c) coal (d) alcohol

11. Burning of hydrogen is an example of
(a) slow combustion (b) rapid combustion
(c) explosion (d) spontaneous combustion

Fire extinguishers

12. Fire extinguishers extinguish the fire by
(a) bringing down the temperature of the fuel
(b) cutting off the supply of air
(c) supplying air to the fuel
(d) both (a) and (b)

13. The salt used in the soda–acid type extinguisher is
(a) sodium bicarbonate
(b) potassium nitrate
(c) potassium carbonate
(d) sodium nitrate

14. Which one is the best fire extinguisher?
(a) NO_2 (b) O_2
(c) SO_2 (d) CO_2

15. Which of the following is best explanation of the statement—water should never be used to extinguish oil fires.
(a) water reacts with oil and forms a new flammable compound
(b) water is lighter than oil; on spraying water it sinks, and oil floats and the fire continues
(c) water is heavier than oil; on spraying water it sinks and oil floats and the fire continues
(d) oil fire is very intense, it converts water into steam which evaporates off

16. Why do fire brigades pour water on the fire?
(a) water vapour helps in cutting off the supply of air
(b) this prevents the fire from spreading
(c) water cools the combustible material so that its temperature is brought below its ignition temperature
(d) all of the above

17. If a person's clothes catch fire, the best way to extinguish the fire is to
(a) throw water on the clothes
(b) use a fire extinguisher
(c) cover the person with a woollen blanket
(d) cover the person with a polythene sheet

18. The gas produced in the working of a soda–acid type fire extinguisher is
(a) carbon dioxide (b) oxygen
(c) sulfur dioxide (d) hydrogen

19. Which of the following can be used to extinguish fire at the petrol pump?
(a) water (b) carbon dioxide
(c) blanket (d) none of these

Flame

20. LPG burns with a
(a) yellow flame (b) luminous flame
(c) non-luminous flame (d) both (a) and (b)

21. The hottest zone in the candle flame is the
(a) invisible zone
(b) luminous zone
(c) non-luminous zone
(d) all zones are equally hot

22. The zone that produces yellow light in the candle flame is
(a) luminous zone (b) non-luminous zone
(c) invisible zone (d) all zones give light

23. Coal burns with
(a) both flame and glow (b) does not burn
(c) flame (d) only glow

24. The central zone of the candle flame is luminous due to
(a) no combustion of wax vapour
(b) formation of carbon dioxide
(c) incomplete combustion of wax vapour
(d) complete combustion of wax vapour

25. On placing an inverted tumbler over a burning candle, the flame gets extinguished after some time. This is because of non-availability of
(a) oxygen (b) water vapours
(c) carbon dioxide (d) wax

26. While shaping gold into ornaments, which part of the flame is directly used by the goldsmith?
(a) non-luminous (b) luminous
(c) innermost zone (d) whole flame

Fuel

27. Which fuel is the ideal fuel to be used at home?
(a) LPG (b) CNG
(c) wood (d) coal

28. While cooking, if the bottom of the vessel is getting black on the outside, it means that
(a) the fuel has high calorific value
(b) the fuel is burning completely
(c) the food is not cooked completely
(d) the fuel is not burning completely

29. The calorific value of a fuel is expressed in a unit called
(a) kilojoule per litre (b) kilogram per millilitre
(c) kilojoule per gram (d) kilojoule per kilogram

30. In villages, people use wood as fuel because
(a) it is considered to be an ideal fuel
(b) of its easy availability and low cost
(c) it is environment friendly
(d) it catches fire easily

31. Which one among the following is considered as the cleanest fuel?
(a) cow dung cake (b) petrol
(c) kerosene (d) hydrogen gas

32. The fuel used in the human body to produce energy, is
(a) coal (b) food
(c) juices (d) paper

Miscellaneous

1. Burning of LPG at home is an example of
(a) slow combustion (b) rapid combustion
(c) spontaneous combustion (d) explosion

2. Which one of the following will show spontaneous combustion?
(a) sodium (b) calcium
(c) sulfur (d) carbon

3. Vaporised wax gets oxidised in the
(a) luminous zone (b) non-luminous zone
(c) blue zone (d) dark zone

4. Molten wax on heating releases
(a) carbon and hydrogen
(b) water and carbon dioxide
(c) water and carbon
(d) carbon dioxide and hydrogen

5. Place a piece of burning charcoal and iron plate and cover it with a plastic jar. The charcoal stops burning because
(a) it becomes cold after some time
(b) supply of oxygen is cut off
(c) its ignition temperature is lowered
(d) none of the above

6. When the matchstick strikes against the rubbing surface, what happens to the red phosphorus?
(a) it ignites antimony trisulfide
(b) it reacts with potassium chlorate
(c) it converts into white phosphorus
(d) no reaction takes place

7. The upper most part of matchstick is made up of
(a) antimony trisulfite and potassium perchlorate
(b) red phosphorus and white phosphorus
(c) antimony trisulfide and potassium perchlorate
(d) antimony trisulfide and potassium chlorate

8. Choose the correct statement about inflammable substances from the following.
They have
(a) low ignition temperature and cannot catch fire easily
(b) high ignition temperature and can catch fire easily
(c) low ignition temperature and can catch fire easily
(d) high ignition temperature and cannot catch fire easily

9. Which of the following oxides are present in acid rain?
(a) nitrogen oxides (b) sulfur oxides
(c) phosphorous oxides (d) both (a) and (b)

10. Which of the following phenomena is responsible for global warming?
(a) ozone formation
(b) acid rain
(c) greenhouse effect
(d) none of these

11. The use of which fuel in vehicles has reduced pollution?
(a) CNG (b) kerosene
(c) petrol (d) diesel

12. Which of the following are cases of spontaneous combustion?
(a) burning of phosphorus at room temperature
(b) burning of a cracker
(c) burning of LPG
(d) both (a) and (b)

13. The calorific value of petrol is nearly
(a) 6000 kJ/kg
(b) 150000 kJ/kg
(c) 45000 kJ/kg
(d) 17000 kJ/kg

14. Which of the following is coolest zone of the flame of a candle?
(a) non-luminous zone
(b) luminous zone
(c) dark zone
(d) blue zone

15. In incomplete combustion of methane, the products formed may be
(a) CO_2
(b) CO
(c) carbon black
(d) both (b) and (c)

16. Which of the following is the most corrosive and suffocating gas?
(a) CO_2 (b) SO_2
(c) CO (d) NO_2

Advanced and Olympiads

Single Choice

1. Magnesium ribbon on burning in air produces
(a) magnesium oxide, water and light
(b) magnesium oxide and heat
(c) magnesium oxide, heat and light
(d) magnesium oxide, water and heat

2. Hydrogen satisfies the conditions of an ideal fuel but cannot be used freely because
(i) it has high calorific value
(ii) its ignition temperature is very low
(iii) it is soluble in water
(iv) it is highly inflammable

Select the correct statements.
(a) (ii) and (iv)
(b) (iii) and (iv)
(c) (i) and (iii)
(d) (i) and (ii)

3. Read the following statements.
(i) Water is the best extinguisher for fires involving inflammable materials.
(ii) Carbon dioxide is used to extinguish fires involving oil and petrol.
(iii) Water is not suitable for fires involving electrical equipment.
(iv) Fires caused by short-circuit should be immediately put off by sprinkling water on it.

Which option represents the set of correct statements?
(a) (i) and (ii) (b) (ii) and (iv)
(c) (ii) and (iii) (d) (iii) and (iv)

4. Shanvi was cooking potato curry on a chulha. To her surprise, she observed that the copper vessel was getting blackened from outside. It may be due to
(a) proper combustion of fuel
(b) improper cooking of potato curry
(c) improper combustion of fuel
(d) burning of copper vessel

5. Identify the set having correct statements.
(i) Carbon dioxide is used to extinguish fires involving oil and petrol.
(ii) Sand is used for extinguishing fires caused by oil.
(iii) Fires caused by short-circuit should not be put off sprinkling water on it.
(iv) Water is not the best extinguisher for fires involving petrol and kerosene.

(a) (i), (ii) and (iii)
(b) (ii) and (iii)
(c) (ii), (iii) and (iv)
(d) (i), (ii), (iii) and (iv)

6. Calorific values of some fuels are given:

Fuel	Calorific value (kJ/kg)
Biogas	35000–40000
Diesel	45000
Hydrogen	150000
CNG	50000

On the basis of this data, find out the correct order of efficiency of fuels.
(a) Biogas > Diesel > CNG > Hydrogen
(b) CNG > Hydrogen > Biogas > Diesel
(c) Hydrogen > CNG > Diesel > Biogas
(d) Diesel > CNG > Biogas > Hydrogen

Multiple Choice

7. Which of the following is/are combustible substance(s)?
(a) camphor (b) glass
(c) straw (d) alcohol

8. Forest fires are usually due to:
(a) cutting of trees
(b) carelessness of humans
(c) heat of sun
(d) lightning strike

9. Select which of the following statements are true.
(a) Biogas is produced by the decay of biomass.
(b) The matter contained in the body of plants and animals is known as biomass.
(c) Petroleum is formed from the remains of dead animals in the sea.
(d) Fossil fuels should be conserved as they provide lot of energy.

10. The substances that burn with a flame are
(a) LPG
(b) camphor
(c) dry grass
(d) charcoal

11. Select the correct statements from the following:
A good fuel is one which
(a) is readily available
(b) produces a large amount of heat
(c) leaves behind many undesirable substances
(d) burns easily in air at a moderate rate

12. Which of the following are due to global warming?
(a) melting of ice on the earth's poles
(b) increase in sea level
(c) change in climate
(d) various type of diseases

Comprehension Type

Comprehension I: Carbon fuels such as wood, coal, petroleum release unburnt carbon particles into the environment. These particles are highly dangerous pollutants and cause respiratory diseases like asthma. When fuels are incompletely burnt, they release carbon monoxide gas into the atmosphere. This gas is very dangerous due to its poisonous nature. The combustion of fossil fuels also releases a large amount of carbon dioxide into the atmosphere. Carbon dioxide is a greenhouse gas which is responsible for global warming. Burning of coal and diesel releases sulfur dioxide gas. SO_2 is extremely corrosive and suffocating in nature. Petrol gives off oxides of nitrogen (NO, NO_2). The oxides of sulfur and nitrogen get dissolved in rain water and form acids to cause acid rain.

13. The release of unburnt carbon particles may cause
(a) respiratory problem
(b) skin problems
(c) stomach infection
(d) all of these

14. Which of the following is the most fatal gas?
(a) CO_2 (b) SO_2
(c) CO (d) NO_2

15. Which of the following can irritate the nose and throat?
(a) CO_2 (b) NO_2
(c) SO_2 (d) none of these

Comprehension II: On the basis of rate of combustion, it can be divided into rapid, slow, explosive and spontaneous types. Firewood burns with a crackling sound is a case of explosive combustion. Forest fires are a case of spontaneous combustion. Food is a fuel for our body. During respiration, the digested food (glucose) is broken down by reaction with oxygen in the body cells to give carbon dioxide, water and heat energy. This heat energy is utilised by our body for various functions. Hence, respiration is a type of slow combustion of food which takes place in the body to produce heat energy.

16. Which of the following is correct regarding combustion?
(a) it involves a chemical change
(b) it is exothermic by nature
(c) it may produce heat and sound
(d) all of these

17. Which of the following is an example of rapid combustion?
(a) burning of LPG
(b) burning of camphor
(c) rusting of iron
(d) both (a) and (b)

18. Which of the following involves slow combustion?
(a) respiration
(b) burning of camphor
(c) rusting of iron
(d) both (a) and (c)

Matrix Matching

19. Match the following

Column I (Fuel)	Column II (Calorific value in kJ/kg)
(a) Methane	(p) 150000
(b) Kerosene	(q) 45000
(c) Hydrogen	(r) 25000–33000
(d) Coal	(s) 50000

20. Match the following

Column I	Column II
(a) Slow combustion	(p) burning of white phosphorus
(b) Rapid combustion	(q) combustion of glucose
(c) Spontaneous combustion	(r) rusting of iron
(d) Complete combustion	(s) burning of LPG
	(t) respiration

Integer Type

21. How many of the following are processed fuels?
Petrol, wood, diesel, biogas, coal, LPG, CNG, cow dung cakes

22. Out of the following how many species are combustible?
Iron nail, diesel, paper, glass, straw, matchstick, stone, cloth, plastic

23. Which of the following are greenhouse gases?
Nitrogen, carbon dioxide, water vapour, methane, helium, nitrous oxide, aerosol, ammonia

24. Out of the following how many fuels can have a calorific value of 50000 kJ/kg or more?
Methane, petrol, CNG, LPG, diesel, hydrogen, wood, coal, biogas

25. In an experiment, W kg of a fuel was completely burnt. The heat produced was measured to be 2,00000 kJ and the calorific value of fuel was 40000 kJ/kg. Find the value of W.

Answer Keys

Fill in the Blanks

1. combustion	**2.** combustible
3. fire, blanket	**4.** lowest, ignition
5. petrol	**6.** burning
7. ignition, inflammable	**8.** calorific value
9. red	**10.** middle or luminous
11. 50000	**12.** raise

True or False

1. F	**2.** F	**3.** F	**4.** T	**5.** T
6. F	**7.** F	**8.** T	**9.** T	**10.** T
11. F	**12.** T			

Very Short Answer Type Questions

1. Combustion can take place in the presence of:
(a) a combustible substance
(b) oxygen, that is, the supporter of combustion
(c) attainment of ignition temperature of the substance

2. The unit in which the calorific value of a fuel is expressed is kilojoules per kilogram (kJ/kg).

3. The main constituent of biogas is methane and of kitchen gas (LPG) is butane.

4. Biogas is formed by the decomposition of plant and animal wastes.

5. Kerosene and wood.

6. In the sun, heat and light are produced by nuclear fusion reactions.

7. A combustion in which a substance burns rapidly and produces heat and light is known as rapid combustion.

8. LPG does not have any smell, so to detect the leakage of the gas, a smelling agent is added to it.

9. The ignition temperature of yellow phosphorus is very low, so it catches fire on its own when exposed to air.

10. Yes, in our body food is broken down by reaction with oxygen and heat is produced.

11. The substances which vaporise during burning, form a flame.

12. When water is poured on burning wood, the temperature of wood falls below its ignition temperature. This prevents the fire from spreading.

13. The wick draws up the molten wax, which then burns to form a flame.

Short Answer Type Questions

1. A substance which on burning produces heat energy is called a fuel. Kerosene and petrol are liquid fuels. LPG and CNG are two gaseous fuels.

2. As CO_2 is heavier than oxygen, it forms a blanket around the fire, because of which the supply of air is stopped. Moreover, it brings down the temperature of the burning substance. In these ways, it plays a significant role in controlling fire.

3. Sulfur dioxide and nitrogen oxides dissolve in rain water and form acids (H_2SO_4, HNO_3). Such rain is called acid rain. Acid rain corrodes buildings and reduces the fertility of soil by making it acidic.

4. We cannot put off a fire due to electric short circuit by using water because water is a conductor of electricity and may result in electric shock and can harm fire fighters.

5. (a) Magnesium does not burn with a flame because it burns as a solid.
(b) Camphor has low ignition temperature and vaporises to burn with a flame but charcoal does not vaporise.

6. (a) Nuclear fusion
(b) Petrol is lighter than water, so it floats on top of water and continues burning. So, water cannot be used for extinguishing fire due to burning of petrol.

7. The following are the features of an ideal fuel:
- it is cheap and easily available,
- it does not produce any harmful gas during burning,
- it has high calorific value, and
- it is safe and easy to store.

CNG is Compressed Natural Gas and it is used as a fuel for automobiles.

8. (a) The innermost or dark zone has minimum temperature while the outermost or non-luminous zone has maximum temperature.
   (b) In the middle zone, the unburnt carbon particles give the flame its yellow colour.

9. At airports and petrol pumps, the foam type fire extinguisher is used which is based on the principle of smothering the fire. In such an extinguisher, sodium bicarbonate contains Turkey red oil. When it is operated, carbon dioxide liberated in the reaction of dilute sulfuric acid with sodium bicarbonate comes out under pressure in the form of a foam and settles on the fire.

10. (a) pollution (b) kerosene (c) ignition temperature (d) water

11. A goldsmith uses the outermost zone of a flame, which is non-luminous, to melt gold and silver as it is the hottest zone of the flame, with a higher temperature.

12. Calorific value of a fuel = Heat produced / Amount of fuel = 1800004.5 kJ/kg = 40,000 kJ/kg

13. Hydrogen forms an explosive mixture with oxygen and it is difficult to handle. Therefore, it is not used as a fuel.

14. Rusting is an oxidation process but it cannot be called combustion because heat and light are not produced.

Long Answer Type Questions

1. The innermost zone is the dark zone. It contains unburnt vapours of wax. The second zone is the yellow zone where incomplete combustion takes place. It is known as the luminous zone. The thin outermost zone of the flame is blue in colour and complete combustion takes place. This is the non-luminous zone.

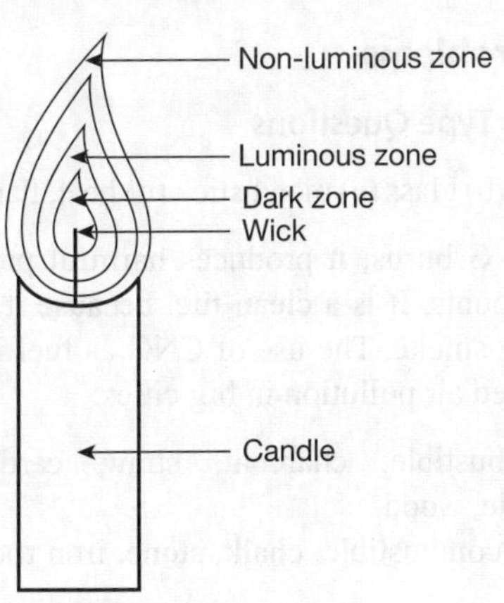

2. (a) LPG is a better domestic fuel than wood because:
   - it does not give out smoke,
   - it has a high heat content,
   - it does not leave ash behind, and
   - it is easily available.

   (b) Paper by itself catches fire easily because it is a highly combustible substance. But when it is wrapped on aluminium, it does not burn because aluminium is a good conductor of heat and it takes the heat from the paper.

   (c) Green leaves contain moisture, so they do not catch fire. Dry leaves do not contain moisture, so they catch fire easily. Moisture increases the ignition temperature of leaves.

3.

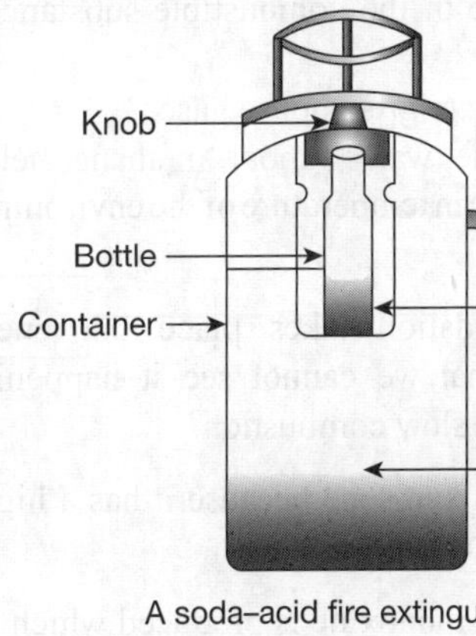

A soda–acid fire extinguisher

In the soda–acid fire extinguisher, carbon dioxide is produced by the reaction of sulfuric acid and sodium bicarbonate. The bottle containing sulfuric acid is struck with a knob and the reaction starts.

$$\underset{\text{Sodium bicarbonate}}{2NaHCO_3} + \underset{\text{Sulfuric acid}}{H_2SO_4} \longrightarrow \underset{\text{Sodium sulfate}}{Na_2SO_4} + \underset{\text{Water}}{H_2O} + \underset{\text{Carbon dioxide}}{2CO_2}$$

The carbon dioxide produced is neither combustible nor a supporter of combustion. Being heavier than air, it forms a blanket over the fire and cuts off the supply of air. The water formed cools the substance below its ignition temperature.

2.

LPG	Wood
(i) It does not cause pollution on combustion.	(i) It pollutes air on its combustion.
(ii) No smoke is produced.	(ii) It produces smoke.
(iii) It is a liquid fuel.	(iii) It is a solid fuel.
(iv) It has more calorific value (55000 kJ/kg).	(iv) It has less calorific value (17000 kJ/kg).
(v) It can be easily transported, as it is stored in cylinders.	(v) It can't be transported easily like LPG.

NCERT Corner

Textbook Exercises

1. They produce heat and light.
2. Combustible substances.
3. Non-combustible substances.

4. A substance starts burning when its ignition temperature is reached.
5. When a building catches fire, the fire brigade is called to put off the fire.
6. It puts off the fire by using water under pressure.
7. The ignition temperature of the combustible substance is lowered.
8. Spontaneous combustion.
9. Global warming is the rise in temperature of the environment of the earth.
10. During respiration, oxidation takes place but energy is released so slowly that we cannot see it happening. Therefore, it is known as slow combustion.
11. LPG is a better fuel than kerosene because it has a higher calorific value.
12. In a closed room, carbon monoxide is produced which is a poisonous gas.
13. Combustible substances: The substances which burn easily are called combustible substances; examples are paper, cloth.
    Non-combustible substances: The substances which do not burn easily are called non-combustible substances; examples are iron, glass.
14. A paper can be burnt with a burning matchstick because of its low ignition temperature.
15. During extreme heat of summer, at some places dry grasses catch fire. This fire then spreads to trees and the whole forest is on fire.
16. Different substances catch fire at different temperatures because they have different ignition temperatures.
17. The ignition temperature of kerosene is lower than that of wood, so it burns quickly.
18. A fire due to kerosene can be put out by throwing sand or soil over it.
19. At airports and petrol pumps, fire is extinguished by using the foam type fire extinguisher.
20. The combustion in which a material suddenly bursts into flame without the application of any apparent cause is called spontaneous combustion. Example, phosphorus.
21. Those solids or liquids which vaporise on being heated, burn with a flame.
22. Petrol vapourises easily and catches fire easily as its ignition temperature is low. Therefore, it is dangerous to use in stoves.
23. Sunil's water will get heated faster than Amit's water because the outermost part of candle flame is the hottest.
24. The ignition temperature of wood or coal is very high. So, we burn paper or kerosene to provide a large amount of heat, as they have a low ignition temperature.
25. Water cools the combustible material so that its temperature is brought below its ignition temperature. Water vapour also surrounds the combustible material, helping in cutting the supply of air. So, the fire is extinguished.
26. He will burn each substance with the help of a match stick. If it burns it is a combustible substance and if it does not burn it is a non-combustible substance.
27. When the person is covered with a blanket, the supply of oxygen is cut off. So, the fire is put off.
28. These are kept to extinguish the fires. When there is a fire, sand is thrown on it, so as to cut off the supply of oxygen.
29. CNG produces no pollutants like carbon dioxide, carbon monoxide, unburnt carbon particles or sulfur dioxide. Therefore, it has reduced pollution in our cities.
30. Carbon fuels release unburnt carbon particles and cause asthma. Incomplete combustion forms carbon monoxide which is a poisonous gas. Combustion of fuels releases carbon dioxide which causes global warming. Increased use of fuel causes acid rain which harm crops, buildings and soil.
31. In villages, wood is used because it is easily available and cheap.
    Its disadvantages are: It produces a lot of smoke which causes respiratory problems.
    Wood contains several important substances which are lost when it is burnt.
    Cutting of trees leads to deforestation.

Exemplar Problems

Short Answer Type Questions

1. (a) diesel (b) glass (c) matchstick (d) heat, flame (e) calorific
2. When CNG burns, it produces harmful products in very small amounts. It is a clean fuel because it burns without producing smoke. The use of CNG as fuel in automobiles has reduced air pollution in big cities.
3. (i) Combustible: charcoal, straw, cardboard, paper, candle, wood.
   (ii) Non-combustible: chalk, stone, iron rod, copper coin, glass.
4. (a) True, air contains oxygen and oxygen is a supporter of combustion.
   (b) False, magnesium is a combustible metal.
   (c) True
   (d) False, the calorific value of coal is higher than that of wood.
5. The watch glass containing petrol will catch fire instantly because its ignition temperature is very low. Also, petrol is

an inflammable substance, that is, it can easily catch fire with a flame.

6. Iron wire will become red hot and glow. It will not produce a flame.

7. Fuel efficiency is determined by its calorific value which is the amount of heat energy produced on complete combustion of 1 kg of a fuel. The calorific value of a fuel is expressed in kJ/kg.
A good fuel should have high calorific value. That means it should produce a large amount of heat per unit mass.

8. The correct matching is as given:
(a)—(iv), (b)—(vi), (c)—(v), (d)—(iii), (e)—(i), (f)—(ii)

9. The correct matching is as given:
(a)—(iii)—(x), (b)—(ii)—(z), (c)—(i)—(y)

Long Answer Type Questions

1. Pouring water to extinguish a fire due to oil is not a suitable action. This is because oil is lighter than water. So, water will settle down below the oil. The oil floats on water and continues to burn.

Under such conditions, sand or soil should be used. They cut off the supply of air to the fire. Thus, fire can be controlled. But sand or soil will not be available at that place. So, Manu should have switched off the flame of the burner and put a lid on the frying pan. By doing this, the contact between fuel and oxygen is cut off and the flame will go off.

2. Three essential requirements to produce fire are as follows:
(i) fuel (combustible substance),
(ii) air (or oxygen), and
(iii) heat to acquire the ignition temperature.
The job of a fire extinguisher is to cut off the supply of air or to bring down the temperature of fuel or both. The most common fire extinguisher is water. But water works only when things like wood and paper are on fire. Water extinguishes the fire by cooling the burning substance.

For fires involving electrical equipment and inflammable materials like petrol, carbon dioxide (CO_2) is the best extinguisher. CO_2 being heavier than oxygen, covers the fire like a blanket. Since the contact between the fuel and oxygen is cut off, the fire can be controlled. The added advantage of CO_2 is that in most cases, it does not harm the electrical equipment.

3. Examples of solid fuels are wood and coal. These are used to cook food in homes. Coal is also used in industries.
Examples of liquid fuels are kerosene and petrol. Kerosene is used in stoves to cook food and in lamps and petrol is used as a fuel in automobiles.
Examples of gaseous fuels are natural gas and petroleum gas. These are used in industries, CNG is used to run automobiles.

4. CNG (Compressed Natural Gas) is preferred because the calorific value of CNG is higher than that of petrol. CNG will produce a larger amount of heat energy than petrol. At the same lime, it produces the least air pollutants. CNG will be more economical.

5. Burning of wood has several disadvantages. These are as follows.
- Burning of wood produces a lot of smoke which causes respiratory diseases.
- The cutting down of trees to obtain wood as a fuel, leads to deforestation which is very harmful to the environment.
- Trees provide many useful substances. When trees are cut down to obtain fuel wood, all other useful substances which can be obtained from trees are lost.

6. Reasons of forest fires are:
- At high temperatures, sometimes dry grass catches fire which spreads throughout the forest.
- Camp fire may also be a reason.
- Due to a spark from lightning.
- The use of fires by villagers to ward off wild animals.
- Fires lit intentionally by people living around forests for recreation.
- Fires started accidentally by careless visitors to forests. When they throw away lighted cigarettes in the forest.
- The friction of bamboos due to high wind velocity and rolling stones.

7. **Across:** 1. Phosphorus 3. Ignition 5. Water
Down: 2. Combustion 4. Fuel 6. Coal

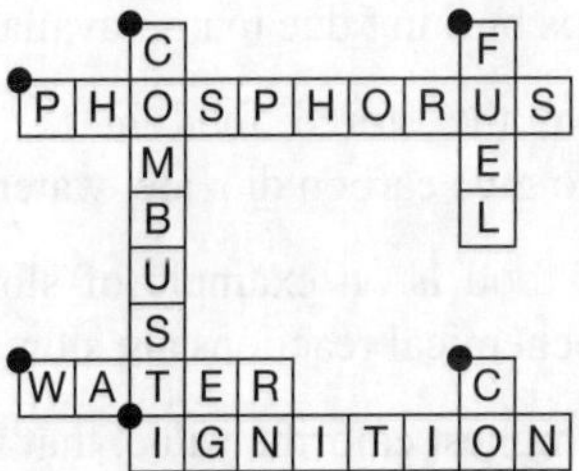

Competition Window (Objective Type)

Topic-wise MCQs

1. (d)	**2.** (a)	**3.** (b)	**4.** (d)	**5.** (d)
6. (c)	**7.** (b)	**8.** (a)	**9.** (c)	**10.** (c)
11. (c)	**12.** (d)	**13.** (a)	**14.** (d)	**15.** (c)
16. (d)	**17.** (c)	**18.** (a)	**19.** (b)	**20.** (c)
21. (c)	**22.** (a)	**23.** (d)	**24.** (c)	**25.** (a)
26. (a)	**27.** (a)	**28.** (d)	**29.** (d)	**30.** (b)
31. (d)	**32.** (b)			

Miscellaneous

1. (b)	**2.** (a)	**3.** (c)	**4.** (b)	**5.** (b)
6. (c)	**7.** (d)	**8.** (c)	**9.** (d)	**10.** (c)
11. (a)	**12.** (d)	**13.** (c)	**14.** (c)	**15.** (d)
16. (b)				

Advanced and Olympiads

Single Choice

1. (c) **2.** (a) **3.** (c) **4.** (c) **5.** (d) **6.** (c)

Multiple Choice

7. (acd) **8.** (bcd) **9.** (abc) **10.** (abc) **11.** (abd) **12.** (abc)

Comprehension Type

13. (a) **14.** (c) **15.** (c) **16.** (d) **17.** (d) **18.** (d)

Matrix Matching

19. (a)—(s), (b)—(q), (c)—(p), (d)—(r)

20. (a)—(rt), (b)—(s), (c)—(p), (d)—(q)

Integer Type

21. 5 **22.** 6 **23.** 5 **24.** 4 **25.** 5

Hints and Solutions

Competition Window (Objective Type)

Topic-wise MCQs

1. Wood is a combustible substance. It is a solid fuel and produces smoke during combustion.

3. Burning of charcoal in a closed room will produce carbon monoxide.

4. Water increases the ignition temperature of an object.

5. Wind energy can be used to generate electricity.

6. Charcoal stops burning due to non-availability of oxygen.

7. In respiration, the stored food in the body undergoes combustion to give carbon dioxide, water and energy.

8. Digestion of food is an example of slow combustion as nearly all biochemical reactions are slow reactions.

9. LPG has the highest calorific value, that is, 55000 kJ/kg.

10. Coal has the highest ignition temperature. Kerosene, petrol and alcohol have low ignition temperature.

11. Burning of hydrogen is an example of explosion.

13. The salt used in the soda–acid type extinguisher is sodium bicarbonate. It reacts with sulfuric acid to give carbon dioxide.

14. Carbon dioxide (CO_2) is the best fire extinguisher.

17. Cover the person with a woollen blanket so that the supply of air to the burning clothes is cut off and hence, the burning (or fire) stops.

20. Due to complete combustion, LPG burns with a non-luminous flame.

21. The non-luminous zone is the hottest zone of the candle flame.

22. The luminous zone produces yellow light in the candle flame.

24. The central zone does not get sufficient supply of oxygen.

25. The flame gets extinguished due to non-availability of oxygen.

26. While shaping gold into ornaments, the non-luminous part of the flame is directly used by the goldsmith.

28. Due to incomplete combustion of fuel, the unburnt carbon particles are deposited at the bottom of the vessel.

29. The units for the calorific value of a fuel is kilojoule per kilogram (kJ/kg).

30. In villages, people use wood as fuel because of its easy availability and low cost.

31. Hydrogen gas is considered as the cleanest fuel. Unlike carbon-based fuels, hydrogen produces no harmful byproducts on combustion. Only energy and clean water are produced.

Miscellaneous

3. Vaporised wax gets oxidised to carbon monoxide and carbon monoxide burns completely with a blue flame in this zone.

5. Due to non-availability of oxygen, charcoal stops burning.

6. Some of the red phosphorus is converted into white phosphorus which burns the matchstick.

7. The upper most part of the matchstick is made up of antimony trisulfide and potassium chlorate.

8. Inflammable substances have a low ignition temperature and can catch fire easily.

9. Oxides of sulfur and nitrogen form acids when dissolved in water and cause acid rain.

11. CNG does not produce any harmful gases.

13. The calorific value of petrol is 45000 kJ/kg.

Advanced and Olympiads

Single Choice

1. Magnesium is a very reactive metal and on burning in air it produces magnesium oxide, heat and light.

$$\underset{}{\text{Magnesium}} + \underset{\text{(From air)}}{\text{Oxygen}} \xrightarrow{\text{Combustion}} \text{Magnesium oxide} + \text{Heat} + \text{Light}$$

2. Hydrogen has a very low ignition temperature and it is highly inflammable so it cannot be used freely as a fuel.

4. Due to incomplete combustion of the fuel, the copper vessel gets blackened on the outside.

Multiple Choice

7. Glass is non-combustible while the rest are combustible.

8. Forest fires are due to the carelessness of humans, heat of the sun and lightning strikes.

9. Fossil fuels should be saved as they are in limited supply.

10. LPG, camphor and dry grass can burn with a flame. Charcoal is a solid fuel which does not vaporise on heating. So, charcoal does not burn by producing a flame. It only glows on combustion.

11. A good fuel does not produce any harmful gases or leave any residue after burning. So, statement (c) is incorrect while the rest are correct.

Integer Type

21. Petrol, diesel, biogas, LPG, CNG are processed fuels.

22. Diesel, paper, straw, matchstick, cloth and plastic are combustible.

23. Carbon dioxide, water vapour, methane, nitrous oxide, aerosols are greenhouse gases.

24. Methane, CNG, LPG, hydrogen have a calorific value of 50000 kJ/kg or more (see text part).

25. Calorific value of a fuel = Heat produced /Amount of fuel
Weight of fuel (W) = 200000/40000 = 5 kg
Hence, $W = 5$

CHAPTER 13 Pollution of Air and Water

LEARNING OBJECTIVES

After studying this unit, you will be able to understand:

- The environment and its segments
- Air pollution, air pollutants and their effects
- Acid rain and its effects
- Global warming and greenhouse effect
- Ozone layer and its depletion
- Water pollution
- Purification and conservation of water

Introduction

The term 'environment' simply means the surroundings and the conditions in which we live. Our environment constitutes the air, water, soil, atmosphere, and the plants and animals around us. Environmental chemistry deals with the changes in chemical dynamics taking place around us and the damage it causes to the environment. The environment consists of four segments (Fig. 13.1):

(i) atmosphere,
(ii) hydrosphere,
(iii) lithosphere, and
(iv) biosphere.

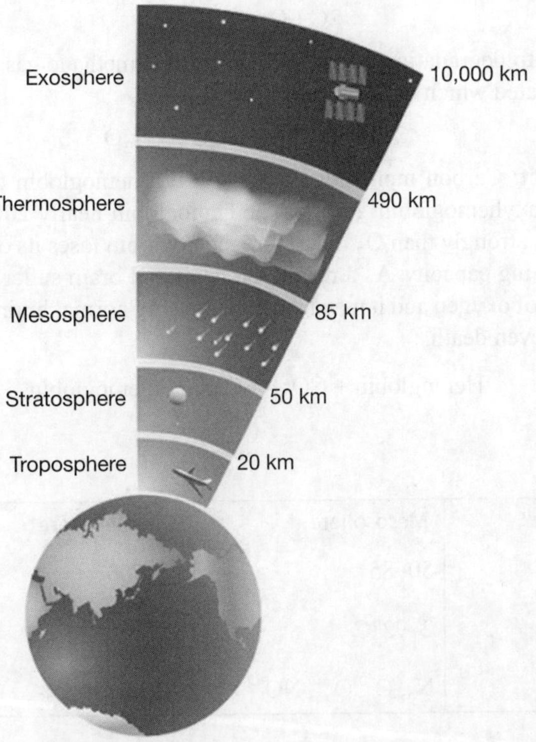

Fig. 13.1 Layers of the atmosphere

Environmental Pollution and Pollutants

The word pollution is derived from a Latin word 'pollutes' which means 'made dirty'. Environmental pollution is the effect of undesirable changes in our surroundings that have a harmful effect on human beings, animals and plants. Environmental pollution is caused by the addition of any undesirable substance to air, water or soil, naturally or by human activity to such an extent that its adverse effects are observed on human beings, animals and plants. A pollutant is a substance whose presence in undesirably higher concentrations causes pollution, in turn adversely affecting the environment. Hg, Pb, CO and SO_2 are some examples of pollutants. A contaminant is a substance which is not present in nature but is introduced into the environment by human activity and has adverse effects on the environment. For example, methyl isocyanate (MIC). A *receptor* is the medium affected by the pollutant. Human eyes are the receptors for the smoke released by automobiles which cause irritation to eyes.

Sink is the medium which interacts with the long-lived pollutant and removes pollution. Sea water acts as a sink for carbon dioxide. The main causes of pollution are the increase in population and depletion of natural resources, industrialisation, urbanisation and deforestation.

Types of Pollution

Some of the major types of pollutions are—air, water, soil and noise. On the basis of their origin, pollutants have two sources—natural and man-made.

Types of Pollutants

(i) Primary pollutants: These pollutants, after their formation, enter the environment and remain there unchanged. For example, NO, NO_2, CO, SO_2.

(ii) Secondary pollutants: They are formed as a result of the chemical reactions between primary pollutants present in the atmosphere and those in the hydrosphere. For example, peroxy acylnitrate (PAN).

(iii) **Biodegradable pollutants:** These pollutants can be easily decomposed with the help of microorganisms and are not harmful. For example, domestic sewage, cow dung.

(iv) **Non-biodegradable pollutants:** These pollutants cannot be decomposed and their presence is very harmful for animals and human beings. For example, DDT, mercury.

Air Pollution

The atmosphere is the layer of air enveloping the earth (Table 13.1). Air pollution means degradation of air quality. It is caused by the addition of undesirable substances into the atmosphere either naturally or by human activity. It is of two types.

(i) **Tropospheric pollution:** It is caused by gaseous air pollutants such as SO_2, NO_2, CO_2 or H_2S.

(ii) **Stratospheric pollution:** The damage done to the ozone layer by the detrimental action of compounds like nitric acid and chlorofuorocarbons constitute this particular type of pollution.

Some common air pollutants are:

- oxides of carbon such as CO and CO_2,
- oxides of nitrogen such as N_2O, NO,
- compounds of sulfur such as SO_2, SO_3, H_2S,
- chlorofluorocarbons (freons),
- hydrocarbons such as methane and butane,
- metals such as lead and mercury,
- organic pollutants such as benzopyrene, biocides.

Natural Sources of Air Pollutants

The following two are the natural sources of air pollutants.

(a) **Volcanoes:** Volcanoes release large amounts of air pollutants such as carbon monoxide, sulfur dioxide, hydrogen sulfide, chlorine, hydrocarbons and particulates.

(b) **Decaying vegetation:** Microbial action on organic matter in the soil releases pollutants such as nitrous oxide (N_2O). Forest fires release poisonous gases such as carbon monoxide. Winds and dust storms carry particulate matter such as sand and dust.

Man-made Sources of Air Pollutants

- Automobiles use petrol and diesel as fuel. The incomplete combustion of these fuels releases carbon monoxide, sulfur dioxide, nitrogen oxides and hydrocarbons and particulates like lead (now, petrol is lead free).
- Factories release carbon dioxide, sulfur dioxide, nitrogen monoxide and particulates.
- Industrial processes release different types of air pollutants, depending upon the type of process involved. For example, coal power plants release carbon monoxide, sulfur dioxide, ash and smoke. Fertiliser industries mainly release nitrogen oxides and ammonia.
- Burning of plastics releases carbon monoxide and other harmful gases.
- Decay of crop residue in rural areas is the main source of carbon monoxide and methane.

Carbon Monoxide (CO) as Air Pollutant

It is one of the most poisonous air pollutants. Most of the air pollution is because of automobiles in which the carbon fuels undergo incomplete combustion and liberate carbon monoxide.

$$2C + O_2 \rightarrow 2CO$$

Due to degradation of organic substances, methane gas will be liberated which oxidises into CO.

$$2CH_4 + 3O_2 \rightarrow 2CO + 4H_2O$$

Effect: Carbon monoxide combines with hemoglobin to form carboxyhemoglobin. It binds the hemoglobin nearly 200 times more strongly than O_2. Therefore, hemoglobin loses its oxygen-carrying capacity. As a result, the heart and brain suffer from a lack of oxygen and it may lead to paralysing normal brain action and even death.

$$\text{Hemoglobin} + CO \rightarrow \text{Carboxyhemoglobin}$$

Table 13.1 Regions of atmosphere

	Troposphere[a]	Stratosphere[b]	Mesosphere	Thermosphere
Altitude (km)	0–10	10–50	50–85	85–500
Temperature range	15–56°C	56–2°C	2–92°C	92–1200°C
Gases/species	N_2, O_2, CO_2, water vapour	N_2, O_2, O_2	N_2, O_2, O_2^+, NO^+	O_2^+, O^+, NO^+

Control of CO Pollution

CO pollution can be controlled by applying the following methods.

- Using electrically powered automobiles.
- Using pollution control devices in automobiles.
- Using gasoline as fuel as burning of gasoline does not give CO.

$$\underset{\text{Octane}}{2C_8H_{18}} + 25O_2 \rightarrow 16CO_2 + 18H_2O$$

- Using fuels like compressed natural gas (CNG) and liquified natural gas (LNG) and alcohols.
- Using catalytic converters like finely divided platinum which oxidise CO into CO_2.

$$CO \xrightarrow[\text{[O]}]{\text{Pt}} CO_2 + H_2O$$

Oxides of Nitrogen as Air Pollutants

Oxides of nitrogen such as N_2O, NO and NO_2 are liberated into the air during the combustion of fossil fuels, burning of fuel in internal combustion engines and also when N_2 and O_2 in the atmosphere react due to electric discharge due to lightning during storms.

$$N_2 + O_2 \rightarrow 2NO$$

$$2NO + O_2 \rightarrow 2NO_2$$

$$NO + O_3 \rightarrow NO_2 + O_2$$

Effect:

- Higher concentration of NO_2 damages the leaves of plants, retards the rate of photosynthesis, causes acute respiratory disease in children, and so on.
- It causes irritation in the mucous membrane.
- In sunlight, NO_2 oxidises hydrocarbons into photochemical smog which causes eye irritation, asthma attacks and nasal and throat infections.
- The irritant red haze in the high traffic places is due to oxides of nitrogen.
- Nitrogen oxide coming out of the supersonic jets directly enters the stratosphere and decomposes the ozone present there.

$$NO + O_3 \rightarrow NO_2 + O_2$$

$$NO_2 + O_3 \rightarrow NO + 2O_2$$

Control: It is possible by using electrical vehicles and using catalytic converters like Pt or Pd.

$$2NO \xrightarrow{\text{Pt}} N_2 + O_2$$

$$2NO_2 \xrightarrow{\text{Pt}} N_2 + 2NO_2$$

Sulfur Dioxide (SO_2) as Air Pollutant

SO_2 gas is released into the atmosphere directly by burning sulfur or by roasting sulfide ores or by burning fuels containing the sulfur.

$$S + O_2 \rightarrow SO_2$$

Effect:

- SO_2 causes headache, irritation of eyes, vomiting, respiratory tract diseases and may even cause death due to respiratory failure.
- SO_2 bleaches the chlorophyll and thus prevents photosynthesis.
- Higher concentration of SO_2 leads to stiffness of flower buds.
- It destroys vegetation and weakens building materials.
- It mixes with smoke and fog to form harmful smog.
- It also causes acid rain by forming H_2SO_4 as follows:

$$2SO_2(g) + O_2(g) \rightarrow 2SO_3(g)$$

$$SO_3(g) + H_2O(l) \rightarrow H_2SO_4(aq)$$

Control: It can be controlled by minimising combustion of sulfur-containing fuels such as coal and oil.

Hydrogen Sulfide (H_2S) as Air Pollutant

H_2S is produced by the decay of organic matter such as vegetables, sewage and industrial effluents.

Effect: It causes nausea, irritation of the eyes and throat and also destroys vegetable matter.

Smoke and Smog

A pollutant that is a combination of the oxides of nitrogen and sulfur and of partially oxidised hydrocarbons and their derivatives produced by industries and automobiles forms a dark, thick, dust and soot laden fog and is called *smog*. Smog is noxious and irritating. It reduces visibility, induces respiratory troubles and can cause death by suffocation also. Smog particles can cause asthma and other lung diseases. Cotton dust can cause lung fibrosis.

Illustration

1. Between CO and CO_2, which is a more dangerous air pollutant and why?

Solution: CO is more dangerous than CO_2 as it combines with hemoglobin in the blood, forming a stable complex—carboxyhemoglobin. It destroys the oxygen carrying capacity of blood to tissues.

Effects of Air Pollutants

Acid Rain

The term *acid rain* is used to describe all precipitation, that is, rain, snow, fog and dew, which are more acidic than normal water. According to Robert August, the pH of rain water is normally 5.6 due to the dissolution of CO_2 in it.

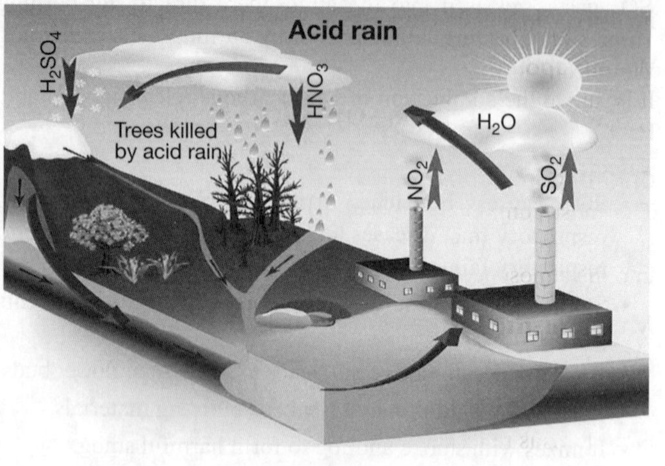

$$CO_2 + H_2O \rightleftharpoons H_2CO_3 \rightleftharpoons H^+ + HCO_3^-$$

Composition and Cause of Acid Rain

Acid rain is a result of the presence of acids such as HNO_3 and H_2SO_4 in polluted air. Acid rain is due to the large-scale emission of acidic gaseous oxides (SO_2, NO_2) into the atmosphere by thermal power plants, automobiles and industries. The oxides of nitrogen and sulfur dissolve in rain water forming nitric acid and sulfuric acid, resulting in acid rain. The reactions involved are:

HNO_3 formation

$$N_2 + O_2 \xrightarrow{\Delta} 2NO$$

$$2NO + O_2 \rightarrow 2NO_2$$

$$4NO_2 + 2H_2O + O_2 \rightarrow 4HNO_3$$

H_2SO_4 formation

$$S + O_2 \xrightarrow{\Delta} SO_2$$

$$SO_2 + \frac{1}{2}O_2 \rightarrow SO_3$$

$$SO_3 + H_2O \rightarrow H_2SO_4$$

The pH of the acid rain is a about 4–5.

Effects of Acid Rain

Acid rain changes the pH of the soil which removes basic minerals in soil (calcium and potassium), it affects the fertility of the soil. Loss of fertility of the soil damages forests and causes loss of nutrients from plants as a result of which their leaves also get damaged.

- It is toxic to vegetation and plant life.
- The life of buildings is considerably reduced by acid rain. For example, the Taj Mahal is affected due to the reaction between the rain and marble ($CaCO_3$).

$$CaCO_3 + H_2SO_4 \rightarrow CaSO_4 + CO_2 + H_2O$$

- It corrodes water pipes due to which heavy metals (Fe, Cu, Pb) mix in the water and cause toxic effects.
- It has serious ecological impacts as it affects water bodies also by making the water of rivers and lakes acidic, thus, disturbing aquatic life.
- It affects human life also as in high concentrations, it causes breathing problems, other respiratory disorders and injuries to the lung tissues.

Illustrations

1. When do we call rainwater as acid rain?

   Solution: When the presence of acids such as H_2CO_3, H_2SO_4, HNO_3 decreases the pH value of rain below 5.6, we call it acid rain.

2. Why is the pH of acid rain sometimes found to be as low as 2?

   Solution: The pH of rainwater is neutral but when the rain combines with the carbon dioxide in the atmosphere to form carbonic acid, it reduces the pH of rainwater to 3.5–5.6 which sometimes may reach as low as 2.

Greenhouse Effect or Global Warming

Greenhouse Effect

The phenomenon of abnormal heating up of the earth's surface due to the presence of excess of greenhouse gases which trap solar radiation is called *greenhouse effect* (Fig. 13.2). The name *greenhouse* originates from the glass structures known as greenhouses that are used to grow green plants in the colder regions of the earth. The glass structure allows sunlight to enter it but does not allow the radiated heat to escape, hence heating up the greenhouse.

Greenhouse Gases

Greenhouse gases are those that absorb and emit infrared radiation in the wavelength range emitted by the earth. Carbon

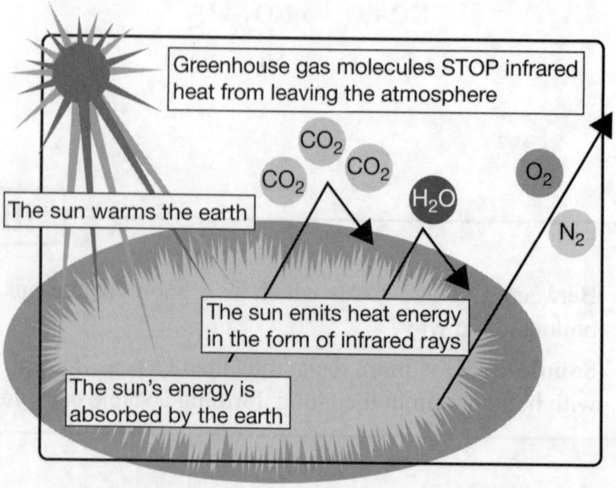

Fig. 13.2 The greenhouse effect

dioxide is the main greenhouse gas. Some other greenhouse gases are methane, CFCs (chlorofluorocarbons), O_3, NO and water vapour (Fig. 13.3).

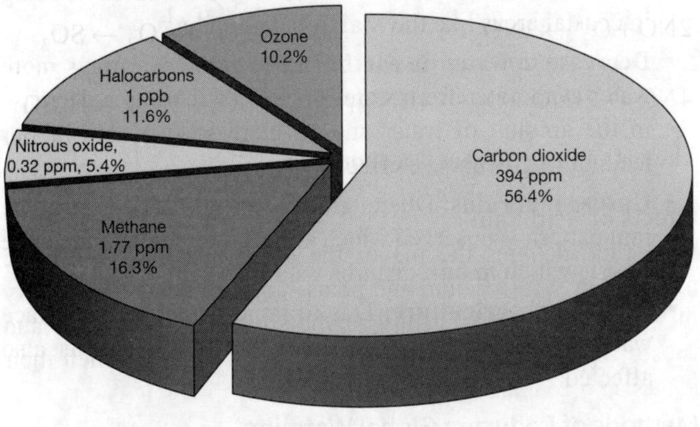

Fig. 13.3 Composition of greenhouse gases

- The composition of these gases differs slightly in different countries and in different time intervals.

Sources of Greenhouse Gases

The major sources of greenhouse gases are the following.

Carbon Dioxide (CO_2)

It is due to the burning of fossil fuels like coal, petroleum, and industrial processes involving the manufacture of lime, biological decay of plants and respiration by human beings and animals.

Methane (CH_4)

It is due to the anaerobic decomposition of organic matter present in soil, water and sediments and due to the incomplete combustion of fossil fuels.

Oxides of Nitrogen

It is due to the combustion of fossil fuels like petroleum in automobiles and power plants and due to some chemical industries producing nitrogen oxides as byproducts.

Water Vapour

It is due to the combustion of hydrocarbons releasing CO_2 and H_2O.

Chlorofluorocarbons (CFCs)

The most common source of CFCs are refrigerants, but fire suppression systems for aircrafts and aerosols also emit CFCs into the atmosphere.

Mechanism of Greenhouse Effect

Greenhouse effect occurs in these six steps (Fig. 13.4):

Step 1: Solar radiation reaches the earth's atmosphere and some part of this is reflected back into space.

Step 2: The rest of the sun's energy is absorbed by the land and the oceans, heating the earth.

Step 3: Heat radiates from the earth towards space.

Step 4: Some part of this heat is trapped by greenhouse gases in the atmosphere, keeping the earth warm enough to sustain life.

Step 5: Human activities such as burning of fossil fuels, agriculture and land clearing are increasing the amount of greenhouse gases released into the atmosphere.

Step 6: This involves trapping of extra heat and causing the earth's temperature to rise.

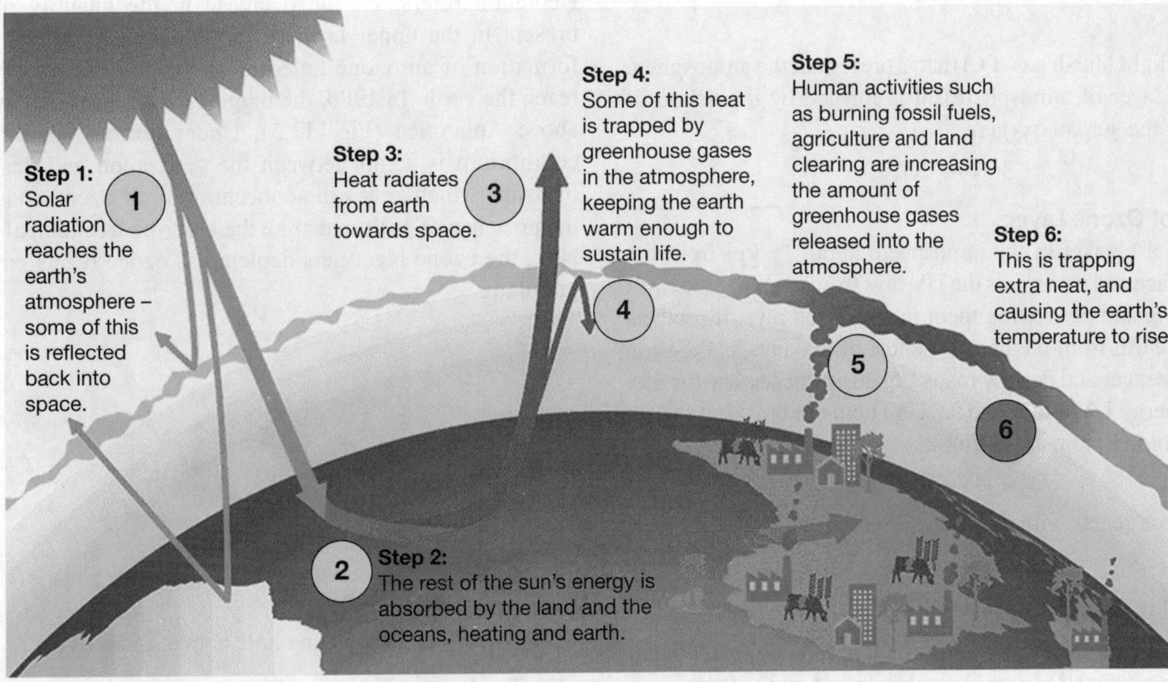

Fig. 13.4 Steps involved in greenhouse effect

Global Warming

Nearly three-fourths of the sun's energy reaching the earth is absorbed by the earth's surface due to which there is an increase in the temperature of the surface. Some part of this heat is also trapped by greenhouse gases like CO_2, CH_4 and water vapour present in the atmosphere. This causes global warming. A slight increase of 1–3°C in temperature can disturb the earth's environment.

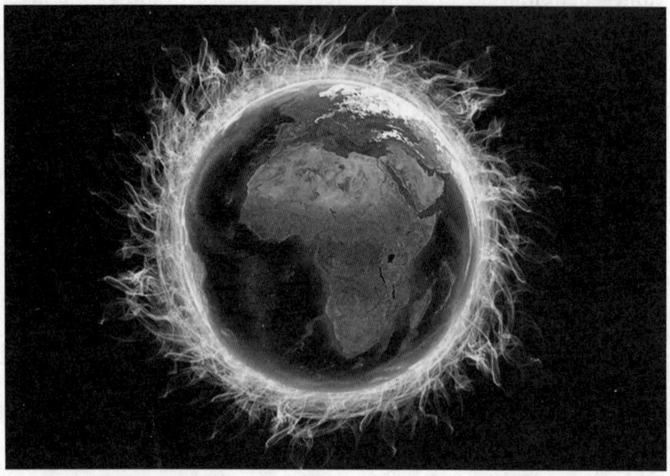

Effect of Global Warming

Some of the effects of global warming are as follows:

- **Rise in sea level:** Due to melting of glaciers and polar ice caps, there is an increase in the sea level. It causes flooding in coastal areas like the Maldives and India.
- **Decrease in water on earth:** Due to global warming, more water evaporates from water bodies; so there is a decrease in the amount of water and an increase in water vapour leading to greenhouse effect.
- **Unseasonal rains:** Due to global warming, a change in the rain pattern is observed which affects trees and plants in the forests which in turn disturbs wildlife.
- **Effects on agriculture:** Due to rapid depletion of surface water by global warming, the agriculture sector is also affected.

Methods of Reducing Global Warming

A few important steps can be taken to reduce global warming.

- Minimise the use of automobiles based on petroleum and replace them with bicycles, electric vehicles and promote the use of the public transport system.
- More planting of trees must be done.
- Check or avoid burning of wood and dry leaves.
- Avoid release of carbon dioxide during industrial processes.

Illustration

1. Write some major drawbacks of global warming.

   Solution: Due to global warming, the average global temperature will increase to a level which may lead to melting of polar ice caps and flooding of low-lying areas. It may also increase cases of dengue, malaria and yellow fever.

Depletion of Ozone Layer

Ozone

Ozone is a light bluish gas (O_3) that is present in the stratosphere (the upper layer of atmosphere). It is formed by the action of UV-rays of the sun on oxygen.

$$3O_2 \rightarrow 2O_3$$

Function of Ozone Layer

It acts like a blanket in the atmosphere about 16 km from the earth's surface and it prevents the UV-rays from reaching the earth by absorbing and converting them into infrared rays. It protects the life on earth from the harmful effect of UV-rays as they can cause skin cancer and destroy many organisms necessary for life.

High energy UV-radiation (far-UV) helps in breaking the O_2 molecule into O-atoms as follows:

$$O_2 + UV \rightarrow O + O$$

Oxygen atom reacts with O_2 molecules to give ozone.

$$O + O_2 \rightarrow O_3$$

The overall reaction is $3O_2 + UV \rightarrow 2O_3$

This ozone produced here absorbs UV radiations of longer wavelength giving rise to an O_2 molecule and an O-atom.

$$O_3 \rightarrow O_2 + O$$

Depletion of Ozone Layer

Depletion refers to the decrease in the quantity of ozone present in the upper layer of the atmosphere. It leads to the formation of an ozone hole due to which UV-rays of the sun reach the earth. In 1980, the biggest ozone layer hole was seen above Antarctica (Fig. 13.5). Under normal conditions, an equilibrium is setup between the generation and destruction of ozone which keeps the concentration of ozone constant. In the presence of pollutants like the chlorine free radical ($\dot{Cl}$) or NO_2, the ozone layer gets depleted as ozone reacts with these pollutants.

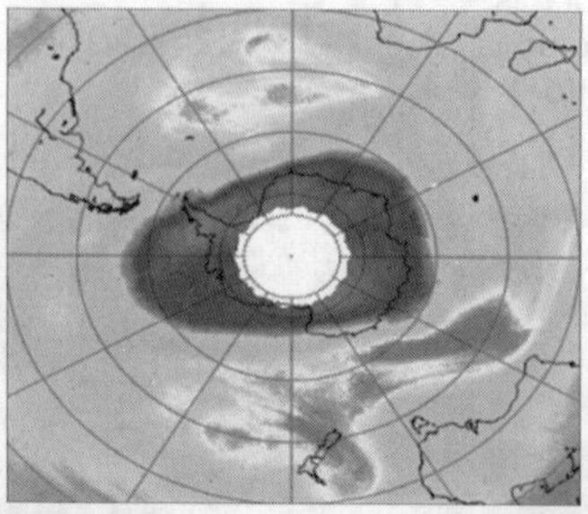

Fig. 13.5 Ozone hole above Antarctica

(i) **Depletion by chlorofluorocarbons (CF_2Cl_2):** Chlorofluorocarbons (freons), absorb the ultraviolet radiation and get photolysed to liberate chlorine atoms. The chlorine atoms catalyse the decomposition of ozone resulting in the depletion of the ozone layer (Fig. 13.6).

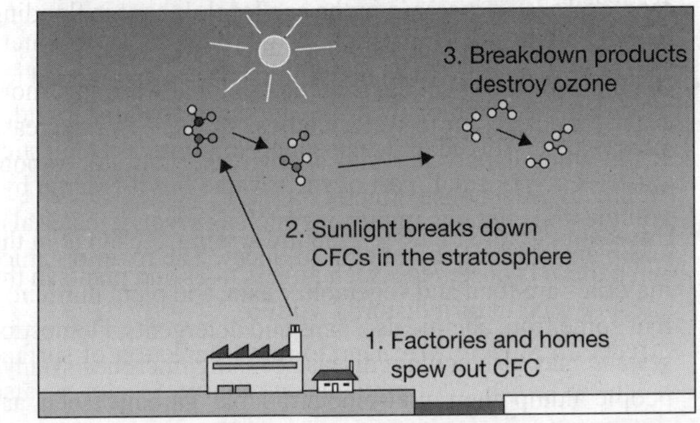

Fig. 13.6 Process of destruction of ozone

$$CF_2Cl_2 \xrightarrow{hv} CF_2Cl + \overset{o}{Cl}$$

$$CFCl_3 \xrightarrow[uv]{hv} CFCl_2 + \overset{o}{Cl}$$

$$\overset{o}{Cl} + O_3 \rightarrow ClO^\circ + O_2$$

$$ClO^\circ + [O] \rightarrow Cl^\circ + O_2$$

(ii) Aviation fuel: Burning of aviation fuel emits a large quantity of NO and other gases into the atmosphere. The presence of oxides of nitrogen in the atmosphere increases the decomposition of ozone as follows:

$$NO + O_3 \rightarrow NO_2 + O_2$$

$$O_2 \xrightarrow{hv} O + O$$

$$NO_2 + O \rightarrow NO + O_2$$

It has been noticed that one molecule of CFC can destroy more than one thousand O_3 molecules in the stratosphere. Due to this reaction, a huge ozone hole was created in the ozone layer over Antarctica. However, in other parts of the stratosphere, the ozone hole is not observed due to the fact that both ClO and Cl° can be consumed as follows:

$$ClO^\circ + NO_2 \rightarrow ClONO_2$$

$$Cl^\circ + CH_4 \rightarrow \overset{o}{C}H_3 + HCl$$

This locking of ClO° by NO_2 and Cl° by CH_4 is known as *scavenging*.

Some chemicals used in industries, such as CH_3Cl (methyl chloride), CH_3Br (methyl bromide), CCl_4 (carbon tetrachloride) and gases causing global warming also deplete the ozone layer.

Effect of Depletion of Ozone Layer

Due to the damage done to the ozone layer, the ultraviolet light from the sun directly falls on the earth, causing skin cancer, irritation to the eyes and is harmful to vegetation.

Illustration

1. Why is an ozone hole formed over Antarctica?

Solution: It was observed that a unique set of conditions was responsible for the formation of the ozone hole. In summer, NO_2 and CH_4 react with ClO° and Cl° forming chlorine that sinks preventing much O_3 depletion. In winter, polar stratospheric clouds are formed over Antarctica which provide a surface for reactions to occur. These are responsible for O_3 depletion.

Controlling Air Pollution

- Microorganisms and enzymes that can naturally degrade pollutants must be developed. Such control measures are called bio-remedies.
- Clean and green, environment friendly (eco-friendly) technology must be developed.
- By dissolving or absorbing harmful gases and chemicals.
- Large-scale plantation of trees must be carried out.
- Industries must be situated far away from urban areas.
- Use of vehicles should be limited.
- Petrol without lead and diesel with less sulfur must be used.
- Automobiles must be fitted with tune ups (for high air–fuel ratio) and catalytic convertors (to change CO into CO_2 and NO_2 into N_2).
- Bagasse and rice husk should not be used as fuel.
- Fly ash (coal-thermal plants) should be removed by the wet method and should be used in building materials.
- Industrial wastes can be checked by using tall chimneys, wet scrubbers, cyclone collectors, bag filters, electrostatic precipitators, and so on.

Water Pollution

Water is the most important natural resource required for sustenance of life. Water provides a habitat for many living organisms. *Water pollution* is the contamination for water by undesirable substances like sewage, silt, oil, grease, toxic chemicals and so on, which make water unsuitable for drinking and other domestic uses. Contamination of water bodies such as lakes, rivers, oceans and groundwater can be harmful to organisms and plants that live in them. Pollution occurs when contaminants are discharged directly into water bodies without first treating them.

Sources of Water Pollution

There are two significant sources of pollution of water.

Point Sources

The sources such as industries, municipal sewage treatment plants and septic tanks which are readily identified as single sources are referred to as point sources of pollution.

Non-point Sources

The sources whose location cannot be identified are referred to as non-point sources. They are agriculture, forests, mines, construction areas, domestic wastes, and so on.

Causes of Water Pollution

The main causes of water pollution are given below.

(i) **Use of chemicals in farming:** The quality of groundwater is greatly affected due to the use agricultural practices followed during cultivation. Modern methods of cultivation of crops makes the chemicals from fertilisers and pesticides to seep into the groundwater. This process is called *leaching*. Significant sources of water pollution have been recognised as regular application of fertilisers and pesticides for agriculture and the indiscriminate disposal of industrial and domestic waste into water. The high nitrate content in groundwater is mainly due to irrigation overflow from agricultural fields where chemical fertilisers have been used randomly.

(ii) **Industrial wastes:** Industries located alongside water bodies discharge their wastes into them. The wastewater discharged from the industries is called industrial effluent (Fig. 13.7). Industrial effluents discharged into water bodies contain many harmful chemicals such as lead and arsenic. These chemicals make the water unsuitable for human consumption and for the survival of animals and plants living there. The soil is also affected by the polluted water, as it changes the acidity of the soil. Waste water from industries is a significant water pollutant. Industrial waste water usually contains many harmful chemical compounds. During the last few years, there has been a rapid growth in the number of industries in India. These industries discharge their effluents directly into the water bodies. This is the reason water pollution is mainly in the form of toxic wastes and organic pollutants. Out of this, a large portion can be traced to the processing of industrial chemicals and to the food products industry. Many large and medium sized industries in the region covered by the Ganga Action Plan do not have proper effluent treatment facilities. Most of these polluting industries have been identified as sugar mills, distilleries, leather processing industries, fertiliser industries and thermal power stations. Although major industries have treatment facilities for industrial effluents, this is not the case with small-scale industries, which cannot afford enormous investments in pollution control equipment as their profit margin is very small. Some of the industries contain high levels of nitrates in their effluents thereby increasing the percentage of nitrate in water. Water is often contaminated by fluoride (due to fluoride bearing rock minerals such as phosphate rock), which is released during the manufacture of fertilisers.

Fig. 13.7 Industrial waste discharged into a river

(iii) **Sewage disposal:** Domestic sewage refers to waste water that is thrown away from households. It is also referred to as sanitary sewage or waste. Such water contains a wide variety of dissolved and suspended impurities. Domestic waste is a very small part of the sewage but it is large by volume and contains impurities such as organic materials and plant nutrients that begin to decay. The main organic materials are food and vegetable waste, and plant nutrients that come from chemicals, soaps and detergents. Domestic sewage may also contain disease causing microbes. Many people dump their non-biodegradable garbage such as cans, bottles, plastics and other household products into nearby water bodies. The various household cleansing products that we use add to water pollution as they contain harmful chemicals.

These wastes contain many disease-causing microorganisms. The bacteria present in the feces of mammals are indicators of the quality of water. If water has these bacteria, it means that it has been contaminated by fecal matter. People, after consuming water from these sources, often suffer from diarrhea, dysentery, cholera, typhoid and jaundice. The sewage must be treated with chlorine before discharge in order to kill the microorganisms.

(iv) **Natural disasters:** Water pollution can be the consequence of natural disasters also. For example, hurricanes often involve water contamination from sewage and petrochemical spills from ruptured boats or ships. Leakage of oil from ships and tankers also pollute water bodies.

(v) **Unawareness of public:** Ignorance among the general public about the cleanliness of water is one of the major causes of water pollution. Activities like using water bodies for bathing animals, washing clothes and utensils, and throwing garbage, degrade the water quality.

There are two sources of water contamination—direct and indirect contaminants. Effluents discharged from factories, refineries and waste treatment plants are the direct sources that release chemicals directly into the water bodies. Indirect sources include contaminants that enter the water bodies through soil or groundwater systems and from the atmosphere during rainfall. Soil and groundwater contain the leftovers of agricultural activities such as fertilisers and pesticides, and improperly disposed of industrial wastes.

Hence, the major water pollutants are:

- industrial pollutants: toxic chemicals (lead and arsenic),
- agricultural pollutants: pesticides, insecticides, and so on,
- domestic pollutants: detergents and sewage that contains food wastes, detergents and microbes, and
- leakage of oil from tankers and ships.

KNOWLEDGE BOOSTER

Biochemical Oxygen Demand (BOD)

The amount of oxygen required by bacteria to break down the organic matter present in a certain volume of a sample of water, is known as *Biochemical Oxygen Demand* (*BOD*). The amount of BOD in the water is a measure of the amount of organic material in the water, in terms of how much oxygen will be required to break it down biologically. Clean water would have a BOD value of less than 5 ppm whereas highly polluted water could have a BOD value of 17 ppm or more.

Chemical Oxygen Demand (COD)

Chemical oxygen demand is the amount of oxygen needed to oxidise the organic matter present in water. Chemical oxygen demand testing is used to determine the amount of oxidation that will occur and the amount of organic matter in a water sample. Chemical oxygen demand testing is also used to determine the amount of inorganic chemicals in a sample.

CASE STUDY

How does Water Get Polluted?

Ganga is one of the most famous rivers of India (Fig. 13.8). It sustains most of the northern, central and eastern Indian population. Millions of people depend on it for their daily needs and livelihood. However, a recent study by the World-Wide Fund for Nature (WWF) found that the Ganga is one of the ten most endangered rivers in the world. The pollution levels have been rising for many years. We have reached this stage because the towns and cities, through which the river flows, throw large quantities of garbage, untreated sewage, dead bodies and many other harmful things, directly into the river. In fact, the river is 'dead' at many places where the pollution levels are so high that aquatic life cannot survive.

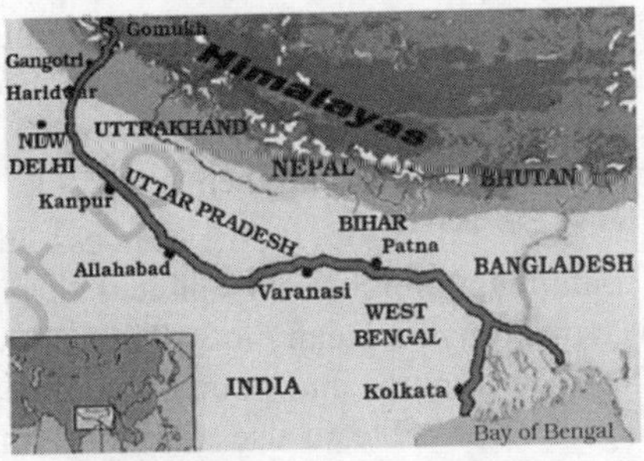

Fig. 13.8 Course of the river Ganga

An ambitious plan to save the river, known as the *Ganga Action Plan* was launched in 1985. It aimed at reducing the pollution levels in the river. However, the increasing population and industrialisation have already damaged this mighty river beyond repair. The Government of India has launched a new initiative called the National Mission for Clean Ganga (NMCG) in 2016.

Let us take a specific example to understand the situation. The Ganga at Kanpur in Uttar Pradesh (UP), has one of the most polluted stretches of the river (Fig. 13.9). Kanpur is one of the most populated towns in UP. People bathe, wash clothes and defecate in the river. They also throw garbage, flowers, idols of gods and goddesses and non-biodegradable polythene bags into the river.

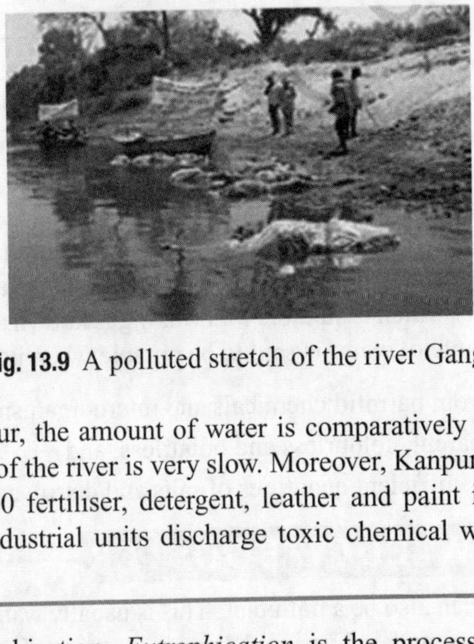

Fig. 13.9 A polluted stretch of the river Ganga

At Kanpur, the amount of water is comparatively small and the flow of the river is very slow. Moreover, Kanpur has more than 5000 fertiliser, detergent, leather and paint industries. These industrial units discharge toxic chemical wastes into the river.

Effect of Water Pollution on Life and Environment

Water pollution has a great impact on the life of living organisms when untreated sewage with many harmful microorganisms is released into water bodies causing many water-borne diseases when this water is consumed. Leakage of oil from ships and tankers affect the aquatic animals (such as fish) and plants. Presence of chemicals like arsenic, lead and fluorine may cause permanent deformities in children. Water contaminated with mercury may also lead to blurred vision and mental disorders. Excess of unwanted nutrients in water bodies results in a thick growth of algae and is known as algal bloom. Soon these algae cover the entire surface of water and this is known as eutrophication. *Eutrophication* is the process of nutrient enrichment of water bodies and subsequent growth of algae (type of plants) on the surface of water. Such algae use oxygen in such huge amounts that the animals and plants living in water bodies die due to the lack of it.

There has been a sharp increase in agricultural production in India. This has been possible due to the enormous use of fertilisers applied in the fields. As a result, the chemical content in groundwater is continuously increasing through soil. Groundwater in most areas is badly affected by nitrates and fluorides. Leaching of chemicals from the soil contaminates the groundwater.

Control Measures

As water is absolutely essential for life, it is our moral duty to save water from pollution. The situation is becoming alarming. We can handle such problems by the following methods.

- Laws for industrial units should be strictly implemented so that polluted water is not disposed off directly into rivers and lakes.
- We must avoid the overuse and misuse of pesticides and fertilisers and use only eco-friendly products.
- The effluents of factories should be treated before being discharged into water bodies.
- Installation of waste water treatment plants should be compulsory. Water treatment plants should be installed in all industrial areas (Fig. 13.10). At the individual level, we should consciously save water and not waste it. Reduce, reuse and recycle should be our mantra!

Fig. 13.10 Water treatment plant

- Domestic waste should be treated chemically before releasing them into water bodies.
- There should be a check on washing of clothes and bathing of animals in water bodies.

Illustrations

1. What is a point source of water pollution?

   Solution: See text part.

2. State at least two major pollutants of water.

   Solution: See text part.

3. How does water get contaminated by fluoride?

   Solution: See text part.

4. Mention a cause of water pollution due to the ignorance of the masses.

   Solution: See text part.

What is Potable Water and How is Water Purified?

Boojho is very upset. He tells Paheli that he drank water which looked clear and was without any smell, but he still fell sick! Paheli explains that water that looks clean may still have disease-carrying microorganisms and dissolved impurities. So, it is essential to purify water before drinking, for example, by boiling. Water which is suitable for drinking is known as potable water. A sample of water is said to be potable when it

- is free from harmful chemicals and microorganisms,
- is transparent, colourless and odourless, and
- contains sufficient quantities of salts and dissolved oxygen.

KNOWLEDGE BOOSTER

Hot water can also be a pollutant! This is usually water from power plants and industries. When it is released into the rivers, it raises the temperature of the water body, adversely affecting the animals and plants living in it.

Purification of Water

The water that we get from the taps is not clean enough for drinking. We should only drink pure water. We may fall sick if we drink dirty water. There are many ways by which we can purify the water for drinking and cooking.

- **By boiling:** We should boil the water for at least 10 minutes and this is the simplest method of purification of water. Boiling of water kills most types of disease-causing organisms and is the most recommended purification technique. Boil the water and then let it cool.
- **By filtration:** Filtration is the process in which the water containing impurities is passed through a cloth. Filter water through a clean cloth. Then allow it to settle for at least 30 minutes and pour off the clear water for purifying. It is possible to use a cotton cloth such as a clean handkerchief or white muslin cloth. A popular household filter is a candle type filter.
- **By portable water purifier:** Portable water filters have ultraviolet light to kill germs so it is quite necessary to use them.
- **By chlorination:** Chlorination is a commonly used chemical method for purifying water. It is done by adding chlorine tablets or bleaching powder to the water. We must be cautious. We should not use more chlorine tablets than specified.
- **Reverse osmosis:** Reverse or RO is a popular water purifier in almost every household. Here, the water is passed through a semi-permeable membrane which has very fine pores. Water can pass through it but it does not allow impurities to pass. These purifiers can remove solid impurities as well as germs.

Water Conservation

The most important step in the conservation of water is to change our attitude and habits. We can follow some simple rules and contribute to water conservation. We can make a difference by saving even a single drop of water. Here are a few simple things that can be practiced for water conservation.

- Use water judiciously. That is, use only the amount you actually need. Make sure that all taps are leak-free at your home. Do not leave any leaking pipe unnoticed.
- Do not leave the tap running while you are brushing our teeth or washing your face.
- Check that there are no leaks in the toilet tank and do not flush the toilet unnecessarily.
- Use water from a bucket and not a hosepipe while washing vehicles.
- Do not throw away water that has been used for washing vegetables, rice or dal; use it to water plants or to clean the floor.
- Water can be stored in a variety of ways for future use.
- Keep looking for new ways to conserve water in and around your home. Promote water conservation in community newsletters and on bulletin boards. Encourage your friends, neighbours and co-workers also, to contribute.

Illustrations

1. What should be done with domestic wastes before releasing into water bodies?

   Solution: See text part.

2. What is potable water?

   Solution: See text part.

3. How can water be disinfected?

   Solution: See text part.

4. Water is the reason for lowering of oxygen in water?

   Solution: See text part.

CHAPTER AT A GLANCE

- The term '**environment**' means, the surroundings and conditions in which we live. Air, water, soil, atmosphere, plants and animals around us constitute the environment.
- The environment consists of **four segments**—atmosphere, hydrosphere, lithosphere and biosphere.
- The word **pollution** is derived from a Latin word 'pollutes' which means, made dirty. Environmental pollution is the effect of undesirable changes in our surroundings that have a harmful effect on human beings, animals and plants.
- A **pollutant** is a substance whose presence in undesirably higher concentrations causes pollution in turn, adversely affecting the environment.
- A **contaminant** is a substance which is not present in nature but is introduced into the environment by human activity and has adverse effects on the environment.
- **Pollutants** are of the following types—primary, secondary, biodegradable and non-biodegradable.
- **Air pollution** denotes the degradation of air quality. It is caused by the addition of undesirable substances into the atmosphere either naturally or by human activity.
- **Carbon monoxide (CO)** is one of the most poisonous air pollutants. Most of the air pollution is because of automobiles in which the carbon fuels undergo incomplete combustion and liberate carbon monoxide. Carbon monoxide combines with hemoglobin to form carboxyhemoglobin, thus reducing the oxygen carrying capacity of hemoglobin.
- **Oxides of nitrogen** (N_2O, NO and NO_2) are liberated into air during the combustion of fossil fuels, burning of fuel in internal combustion engines and also when N_2 and O_2 in the atmosphere react due to electric discharge due to lightning. Higher concentration of NO_2 damages leaves of plants, retards rate of photosynthesis and causes acute respiratory disease in children.
- **Sulfur dioxide (SO_2)** is released into the atmosphere directly by burning sulfur, by roasting sulfide ores or by burning fuels containing sulfur. SO_2 causes headache, irritation of eyes, vomiting, respiratory tract diseases and may even cause death due to respiratory failure.
- **Hydrogen sulfide (H_2S)** is produced by the decay of organic matter such as vegetables, sewage and industrial effluents. It causes nausea, irritation of eyes and throat and also destroys vegetation.
- **Smoke and smog:** A pollutant that is a combination of the oxides of nitrogen and sulfur and of partially oxidised hydrocarbons and their derivatives produced by industries and automobiles forms a dark, thick, dust and soot-laden fog and is called smog. Smog is noxious and irritating. It reduces visibility, induces respiratory troubles and can cause death by suffocation.
- The term **acid rain** is used to describe all precipitation – rain, snow, fog and dew – which is more acidic than normal water.
- **Acid rain** results due to the presence of acids like HNO_3, H_2SO_4 in polluted air. Acid rain is due to the large-scale emission of acidic gaseous oxides (SO_2, NO_2) into the atmosphere by thermal power plants, automobiles and industries.

- The phenomenon of abnormal heating up of the earth's surface due to the presence of excess of greenhouse gases which trap solar radiation is called **greenhouse effect**.
- **Greenhouse gases** are those that absorb and emit infrared radiation in the wavelength range emitted by the earth. Carbon dioxide is the main greenhouse gas. Some other greenhouse gases are methane, CFCs (chlorofluorocarbons), O_3, NO and water vapour.
- **Global warming:** Nearly three-fourths of the sun's energy reaching the earth is absorbed by the earth's surface due to which there is an increase in the temperature. Some part of this heat is also trapped by greenhouse gases like CO_2, CH_4 and water vapour present in the atmosphere. This causes global warming. A slight increase of 1–3°C in temperature can disturb the earth's environment.
- **Depletion of ozone layer:** Depletion of ozone present in the upper layer of the atmosphere leads to the formation of an ozone hole due to which the UV rays of the sun reach the earth and cause skin cancer.
- **Controlling air pollution**
  (i) Microorganisms and enzymes (bio-remedies) which can naturally degrade pollutants must be developed.
  (ii) Clean and green, environment friendly (eco-friendly) technology must be developed.
  (iii) By dissolving or absorbing harmful gases and chemicals.
- **Water pollution:** Water pollution is the contamination of water by undesirable substances such as sewage, silt, oil, grease and toxic chemicals which make water unsuitable for drinking and other domestic uses.
- **Point sources:** The sources such as industries, municipal sewage treatment plants and septic tanks which are readily identified as a single source are referred as point sources of pollution.
- **Non-point source:** The sources whose location cannot be identified are referred to as non-point sources. They are agriculture, forests, mines, construction areas and domestic waste.
- **Causes of water pollution:** Use of chemicals in farming, industrial wastes, sewage disposal, natural disasters and ignorance of the public.
- **Potable water:** Water which is suitable for drinking is known as potable water.
- **Purification of water:** By boiling, filtration, portable water purifier, chlorination and reverse osmosis.

PRACTICE QUESTIONS

Analyse Your Concepts (School Exam Based)

Fill in the Blanks

Instructions: Complete the following statements with an appropriate word/term to be filled in the blank spaces.

1. In Latin, the word 'pollutes' means ______.
2. The pollutants such as NO_2, SO_2 and SO_3 dissolved in the moisture of air are the cause of ______.
3. Locking of chlorine monoxide and free radical of chlorine is known as ______.
4. Excessive release of carbon dioxide into the atmosphere is the cause of the ______ effect which causes global warming.
5. Smoke particles can cause asthma and other ______ diseases.
6. The ozone layer prevents the harmful ______ radiation of the sun from reaching the earth.
7. Carbon monoxide reacts with hemoglobin to form ______.
8. Decrease of the concentration of ozone in the stratosphere is the cause of the formation of ______ holes.
9. The pH of rain water is in between 3.5 and ______.
10. Ozone depletion is mainly caused by the active ______ atoms generated from CFCs in the presence of UV radiation.
11. ______ sewage causes water pollution.
12. Industrial waste is an example of ______ sources.
13. Potable water is free from harmful ______ and

True or False

Instructions: Read the following statements and write your answer as true or false.

1. Pollutants are toxic and harmful substances.
2. Volcanoes release large amounts of chlorofluorocarbons.
3. Fluorides can cause destruction of vegetation.
4. Cotton dust produces lung fibrosis.
5. Impact of acid rain can be reduced by using scrubbers.
6. Oxygen is a greenhouse gas.

7. Ozone is a light yellowish gas.
8. Acetic acid is a component of acid rain.
9. Acid rain can cause severe fever.
10. Existence of coastal countries like the Maldives are in danger due to global warming.
11. Unawareness among the general population about the cleanliness of water is one of the major causes of water pollution.
12. We must avoid the overuse and misuse of pesticides and fertilises.
13. You can make a difference by saving even a single drop of water.

Very Short Answer Type Questions

1. Name any four gaseous pollutants.
2. Name the natural sources of air pollution.
3. Name the man-made sources of air pollution.
4. Why does rainwater have a pH of less than 7?
5. Write the significance of growing plants in greenhouses.
6. Name some particulate pollutants.
7. Why is cigarette smoking injurious to health?
8. Define the term ppm.
9. Name any two gases which are responsible for the formation of acid rain.
10. Write any two effects of ozone layer depletion.
11. Write an advantage of CNG (Compressed Natural Gas) as a fuel.
12. Name a substance used for disinfecting water.
13. Name the light that portable water filters use, to kill germs.
14. Name the chemical added to purify water in the form of tablets.

Short Answer Type Questions

1. Define the following terms.
   (a) Pollution
   (b) Pollutant
   (c) Air pollution
2. Write any three compounds of sulfur that cause air pollution and also write the harmful effects of these compounds.
3. Write the main sources and effects of carbon monoxide.
4. Explain the formation of acid rain due to:
   (i) oxides of sulfur
   (ii) oxides of nitrogen
5. Write few ways of reducing the presence of greenhouse gases.
6. Write only one effect each of the following pollutants on living beings.
   (a) Fluorides
   (b) Nitrogen oxides
   (c) Mercury compounds
   (d) Lead
   (e) Smog
   (f) Smoke particles
7. Define air pollution. Why does it occur?
8. Write the origin and health impacts of smog.
9. What are the major air pollutants? Explain how carbon monoxide pollutes our environment.
10. Explain how CFCs deplete the ozone layer?
11. Differentiate between point and non-point sources with at least two examples for each.
12. What is potable water? Write its important features.
13. How water can be purified by boiling?
14. Name any two water-borne diseases. Name any two chemicals used to kill microorganisms in water.

Long Answer Type Questions

1. How do the oxides of nitrogen enter the atmosphere? Also write their harmful effects.
2. (a) Give the mechanism of action of carbon monoxide.
   (b) How can you control carbon monoxide poisoning?
3. (a) What are the causes of acid rain?
   (b) Give the impact of acid rain:
   (i) on plants, (ii) on soil and (iii) on water bodies.
   (c) How does a scrubber help in reducing the formation of acid rain?
4. (a) What do you understand by the greenhouse effect?
   (b) What are greenhouse gases? How are they responsible for global warming?
   (c) State the sources and effects of the following gases:
   (i) carbon dioxide, (ii) methane and (iii) water vapour.
5. (a) Explain the formation of ozone in the atmosphere.
   (b) What is the function of ozone in the atmosphere?
   (c) State the chemicals responsible for ozone layer destruction.
6. How does water gets dirty due to chemical contamination?
7. What is the effect of water pollution on the life of all living organisms?
8. Explain the various ways of purification of water.
9. What is conservation of water and how it is possible?

COMPETITION WINDOW (OBJECTIVE TYPE)

[For NEET, JEE (Main and Advanced), NTSE, KVPY and Olympiads]

Topic-wise MCQs

Environment

1. The gaseous envelope around the earth is known as the atmosphere. The lowest layer of this extends up to 10 km above the sea level. This layer is the ______.
(a) stratosphere (b) troposphere
(c) mesosphere (d) hydrosphere

2. Biosphere includes
(a) plants and animals
(b) rocks and minerals
(c) atmosphere and lithosphere
(d) water sources

3. The major components of the atmosphere are
(a) CO_2 and N_2
(b) O_3 and SO_2
(c) CO and CO_2
(d) N_2 and O_2

4. The uppermost region of the atmosphere is
(a) troposphere (b) exosphere
(c) stratosphere (d) ionosphere

5. The point of temperature inversion between the troposphere and the ionosphere is called
(a) mesopause (b) stratopause
(c) ionopause (d) tropopause

6. Hydrosphere includes various forms of water as
(a) polar ice caps and ground water
(b) oceans and lakes
(c) sea and rivers
(d) all of these

7. Which of the following is a contaminant?
(a) CO (b) CO_2
(c) SO_2 (d) MIC

8. Ozone layer is present in the
(a) stratosphere (b) troposphere
(c) mesosphere (d) exosphere

Air pollution

9. The medium which interacts with the long-lived pollutant is
(a) source (b) receptor
(c) sink (d) all

10. Major sources of NO_2 pollutants are
(a) natural gas
(b) gasoline
(c) combustion of coal and oil
(d) all of these

11. Which of the following is a primary pollutant?
(a) PAN (b) aldehydes
(c) CO (d) H_2SO_4

12. Which of the following is not regarded as a pollutant?
(a) CO_2 (b) O_3
(c) NO_2 (d) hydrocarbons

13. Which of the following is not an air pollutant?
(a) N_2O (b) N_2
(c) CO (d) NO

14. The industrial usage of fluorocarbons is very high because
(a) they are unstable
(b) they are gases
(c) they can be manufactured cheaply
(d) their reactivity is high

15. SO_2 is a dangerous air pollutant and harms plant life. What changes in the plant indicate its toxic effects?
(a) falling of leaves (b) bleaching of leaves
(c) darkening of leaves (d) withering of leaves

16. Photochemical smog occurs in warm, dry and sunny climate. One of the following is not amongst the components of photochemical smog, identify it.
(a) NO_2 (b) O_3
(c) SO_2 (d) unsaturated hydrocarbon

17. Dinitrogen and dioxygen are the main constituents of air but these do not react with each other to form oxides of nitrogen because ______.
(a) the reaction is endothermic and requires very high temperature
(b) the reaction can be initiated only in the presence of a catalyst
(c) oxides of nitrogen are unstable
(d) N_2 and O_2 are unreactive

18. The pollutants which come directly into the air from sources are called primary pollutants. Primary pollutants are sometimes converted into secondary pollutants. Which of the following are secondary air pollutants?
(a) CO
(b) hydrocarbons
(c) peroxyacetyl nitrate
(d) NO

Depletion of ozone layer, acid rain, greenhouse effect

19. Ozone hole refers to
(a) hole in the ozone layer
(b) reduction in the thickness of the ozone layer in the stratosphere

(c) reduction in the thickness of the ozone layer in the troposphere
(d) increased concentration of ozone

20. Ozone hole is maximum over
(a) Africa (b) Europe
(c) Antarctica (d) India

21. Excessive release of carbon dioxide in the atmosphere is the cause of
(a) depletion of ozone
(b) formation of polar vertex
(c) global warming
(d) formation of smog

22. Which of the following statements is wrong?
(a) Ozone is not responsible for the greenhouse effect.
(b) Ozone can oxidise the sulfur dioxide present in the atmosphere to sulfur trioxide.
(c) Ozone hole is the thinning of the ozone layer present in the stratosphere.
(d) Ozone is produced in the upper stratosphere by the action of UV rays on oxygen.

23. Ozone is an important constituent of the stratosphere because it
(a) removes poisonous gases of the atmosphere by reacting with them
(b) destroys bacteria which are harmful to human life
(c) prevents the formation of smog over large cities
(d) absorbs ultraviolet radiation which is harmful to human life

24. Acid rain
(a) retards the growth of trees
(b) affects big marble constructions
(c) results in the loss of flora and fauna
(d) all of these

25. Inhalation of air polluted with carbon monoxide is dangerous because
(a) CO combines with O_2 dissolved in the blood
(b) CO combines with hemoglobin of blood
(c) CO removes water from the body and causes dehydration
(d) CO causes coagulation of proteins in the body

26. A decrease in the amount of ozone in the stratosphere is called depletion of ozone and it is caused by
(a) UV radiations of the sun
(b) use of CFCs
(c) excessive use of detergents
(d) use of polychlorinated biphenyls

27. Which of the following gases is not a greenhouse gas?
(a) CO (b) O_3
(c) CH_4 (d) H_2O vapour

28. Which of the following is not a greenhouse gas?
(a) water vapour (b) CO_2
(c) O_2 (d) CH_4

29. The greenhouse effect is
(a) rise in temperature of the earth
(b) rise in pressure on the earth
(c) decrease in oxygen content of the earth's atmosphere
(d) decrease in CO_2 content of the earth's atmosphere

30. Which of the following is responsible for depletion of the ozone layer in the upper strata of the atmosphere?
(a) freons (b) ferrocene
(c) fullerenes (d) polyhalogens

Water pollution

31. The rock mineral that causes contamination of water is
(a) phosphate (b) carbonate
(c) coal (d) quartz

32. Which of the following is a point source of water pollution?
(a) agriculture (b) forests
(c) mines (d) industries

33. Which of the following are examples of non-point sources?
(a) agriculture (b) forests
(c) domestic waste (d) all of these

34. Industrial waste increases the amount of which of the following in water?
(a) oxygen
(b) nitrate
(c) fluorides
(d) both (b) and (c)

35. Which of the following is not a characteristic of water?
(a) it is slightly coloured with a characteristic smell
(b) it is transparent, colourless and odourless
(c) it contains sufficient quantities of salts and dissolved oxygen
(d) it is free from harmful chemicals and microorganisms

36. Which of these are methods of purification of water?
(a) boiling
(b) filtration
(c) reverse osmosis
(d) all of these

37. Which of the following can be used to purify water?
(a) fluorine
(b) chlorine
(c) bleaching powder
(d) both (b) and (c)

Miscellaneous

1. Ozone in the stratosphere is depleted by
(a) C_7F_{16} (b) C_6F_6
(c) CF_2Cl_2 (d) $C_6H_6Cl_6$

2. In Antarctica, ozone depletion is due to the formation of the compound,
(a) SO_2 and SO_3 (b) chlorine, nitrate
(c) acrolein (d) formaldehyde

3. Which of the following statements is not true about classical smog?
(a) Its main components are produced by the action of sunlight on emissions of automobiles and factories.
(b) Produced in cold and humid climate.
(c) It contains compounds of reducing nature.
(d) It contains smoke, fog and sulfur dioxide.

4. Which of the following statements about photochemical smog is wrong?
(a) It has a high concentration of oxidising agents.
(b) It has a low concentration of oxidising agents.
(c) It can be controlled by controlling the release of NO_2, hydrocarbons, ozone, and so on.
(d) Plantation of some plants such as pines helps in controlling photochemical smog.

5. Which of the following statements is correct?
(a) Ozone hole is a hole formed in the stratosphere from which ozone oozes out.
(b) Ozone hole is a hole formed in the troposphere from which ozone oozes out.
(c) Ozone hole is the thinning of the ozone layer of the stratosphere at some places.
(d) Ozone hole means vanishing of ozone layer around the earth completely.

6. A contaminant is
(a) a pollutant released from industries
(b) a pollutant
(c) a component originally not present in environment but released into the environment by human activity
(d) a pollutant released into the environment during natural calamites

7. Ozone layer of the stratosphere requires protection from the indiscriminate use of
(a) aerosols and high-flying jets
(b) balloons
(c) pesticides
(d) atomic explosions

8. There is a possibility of melting of polar ice caps and increase in the level of sea water due to
(a) greenhouse effect
(b) acid rain
(c) depletion of ozone layer
(d) any one of these

9. Incomplete combustion of petrol or diesel in automobile engines can be best detected by testing the fuel gases for the presence of
(a) sulfur dioxide
(b) nitrogen dioxide
(c) carbon monoxide
(d) carbon monoxide and water vapour

10. Among the following, all except which causes pollution?
(a) nuclear power plant
(b) thermal power plant
(c) hydroelectric plant
(d) automobiles

11. $CFCl_3$ is responsible for the decomposition of ozone into oxygen. Which of the following reacts with ozone to form oxygen?
(a) Cl_2 (b) Cl^-
(c) F^- (d) Cl°

Answer Keys

Fill in the Blanks

1. made dirty
2. acid rain
3. scavenging
4. greenhouse effect
5. lung
6. ultraviolet
7. carboxyhemoglobin
8. ozone
9. 5.6
10. chlorine
11. domestic
12. point
13. chemicals, microorganisms

True or False

2. F	3. T	4. T	5. T
7. F	8. F	9. F	10. T
11. T	12. T	13. T	

Very Short Answer Type Questions

1. Carbon monoxide, nitrogen oxide, sulfur dioxide and hydrogen sulfide.
2. Natural sources of air pollution are volcanoes, decaying vegetation, forest fires and dust storms.
3. Man-made sources of air pollution are automobiles, factories, industrial processes and decay of crop residue in rural areas.
4. Carbon dioxide in the atmosphere combines with rainwater to give carbonic acid (H_2CO_3) which is acidic in nature. So, the pH of the rainwater is less than 7.
5. Greenhouse gases increase the temperature of the earth to the optimum level. Temperature moderation provides a controlled environment for plants to grow and thrive.
6. Dust, smoke, mist, spray and fumes are examples of some particulate pollutants.
7. Cigarette smoking is injurious to health because it increases the environmental pollution and smoking causes asthma, respiratory diseases and cancer in humans.

8. ppm is parts per million and it describes the concentration of a substance.

9. SO_2 and NO_2 are gases responsible for acid rain.

10. Effects of ozone layer depletion are:
- it causes respiratory problems, and
- it damages plants and trees.

11. CNG is a complete fuel, and it burns without soot and emission of greenhouse gases.

12. Chlorine.

13. Ultra violet rays/light.

14. Chlorine.

Short Answer Type Questions

1. (a) Presence or introduction of a harmful or poisonous substance into the environment is known as pollution.
(b) Toxic or harmful substances that have an adverse effect on the environment and living beings are known as pollutants.
(c) Degradation of air quality due to an increase in the concentration of harmful contaminants is known as air pollution.

2. Hydrogen sulfide, sulfur dioxide and sulfur trioxide are the compounds of sulfur that cause air pollution.
Harmful effects of sulfur compounds are:
- Hydrogen sulfide causes nausea and irritates the eyes and throat.
- Sulfur dioxide affects crop yield, causes damage to the lungs and causes irritation to human eyes.
- Sulfur trioxide combines with water to form H_2SO_4 which causes acid rain.

3. The main sources of carbon monoxide are incomplete combustion of fuels in homes, factories and automobiles.
The harmful effects caused by carbon monoxide are:
- it decreases the oxygen-carrying capacity of the lungs,
- a high amount of carbon monoxide paralyses normal brain function.

4. (a) Sulfur dioxide and sulfur trioxide react with water to form H_2SO_4 which causes acid rain.
$S + O_2 \rightarrow SO_2$
$2SO_2 + O_2 \rightarrow 2SO_3$
$SO_3 + H_2O \rightarrow H_2SO_4$
(b) Nitrogen dioxide combines with water to form a mixture of nitrous acid and nitric acid which causes acid rain.
$2NO_2 + H_2O \rightarrow HNO_2 + HNO_3$

5. These are a few steps to reduce the presence of greenhouse gases.
- Use of automobiles should be reduced by using public transport, bicycles and electric vehicles.
- Trees should be planted and grown.
- Burning of dry leaves and wood should be avoided.
- Smoking should be avoided.
- Educate people about the harmful effects caused by the greenhouse effect.

6. (i) Fluorides adversely affect teeth and bones.
(ii) Nitrogen oxides cause cancer in humans.
(iii) Mercury compounds cause Minamata disease.
(iv) Lead impairs our metabolic process, thereby affecting our health.
(v) Smog can cause respiratory diseases and also leads to suffocation.
(vi) Smoke particles cause asthma and other respiratory diseases.

7. See text part.

8. Pollutants which are a combination of oxides of nitrogen and sulfur, and partially of hydrocarbons produced by industries and automobiles. This forms a dark, thick soot laden fog known as smog.
Impacts of smog are as follows:
- it is noxious and irritating,
- it reduces visibility,
- it causes respiratory problems,
- it can cause suffocation and death.

9. Major air pollutants are:
sulfur dioxide, hydrogen sulfide, fluorides, nitrogen oxides, carbon monoxide, lead, cotton dust particles, smog, nitric oxide and sulfur trioxide.
Carbon monoxide combines with atmospheric oxygen to increase carbon dioxide concentration in the atmosphere. This will adversely affect the environment by leading to global warming and ozone depletion.

10. CFCs are decomposed by UV rays to highly reactive chlorine, which is produced in its atomic form.
$CF_2Cl_2(g) \xrightarrow{\text{UV rays}} CF_2\,Cl(g) + Cl(g)_{(\text{free radical})}$
These free radicals of chlorine react with ozone to form chlorine monoxide.
$Cl(g) + O_3(g) \rightarrow ClO(g) + O_2(g)$
This causes ozone depletion.

11. See text part.

12. See text part.

13. See text part.

14. Two water-borne diseases are dysentery and cholera.
Two chemicals used to kill microorganisms are chlorine and bleaching powder.

Long Answer Type Questions

1. Oxides of nitrogen enter the atmosphere in the following ways:
(i) On the burning of fuels in furnaces, the temperature increases. At high temperature, nitrogen and oxygen present in the air combine to form oxides of nitrogen.

(ii) Oxides of nitrogen are produced during the burning of fuel in an internal combustion engine. They enter the atmosphere as exhaust gases.

(iii) During thunderstorms, nitric oxide is formed by the reaction between atmospheric nitrogen and oxygen in the presence of electric discharge.

(iv) Nitric oxide further reacts with atmospheric oxygen and ozone to form nitrogen dioxide.

The harmful effects caused by the oxides of nitrogen are:

(i) Nitrogen oxides cause irritation of the mucous membrane.

(ii) Large concentrations of nitrogen oxide cause lung problems in humans.

(iii) Nitrogen oxide oxidises hydrocarbons in the presence of sunlight, which causes eye irritation, asthma attacks, nasal and throat infections.

(iv) It causes injuries to vegetation by damaging leaves.

2. (a) When we inhale carbon monoxide, it combines with hemoglobin (which carries oxygen to tissues) forming carboxyhemoglobin. Hemoglobin binds carbon monoxide 200 times more strongly than it does oxygen. This reduces the oxygen carrying capacity of the blood (see text part).

(b) We can control carbon monoxide poisoning by taking the following measures.

- By switching over from internal combustion engines to electrically powered cars or vehicles.
- Many pollution control devices are now installed in cars. Most of these devices help in reducing pollution by burning gasoline completely. Complete combustion of gasoline produces only carbon dioxide and water vapour.

  $2C_8H_{18} + 5O_2 \rightarrow 16CO_2 + 18H_2O$
- By using substitute fuels for gasoline—natural gas in both compressed (CNG) and liquefied (LNG) forms is now increasingly being used as fuel. Alcohols are other feasible substitutes.
- By using catalytic converters.

  (i) Nitrogen oxide is reduced to nitrogen and oxygen in the presence of finely divided platinum or palladium as a catalyst.

  $$2NO \xrightarrow{Pt} N_2 + O_2$$

  $$2NO_2 \xrightarrow{Pt} N_2 + 2O_2$$

  (ii) Carbon monoxide changes to carbon dioxide in the presence of finely divided platinum as a catalyst.

  $$CO \xrightarrow[{[O]}]{Pt} CO_2 + H_2O$$

3. (a) Formation of mineral acids like carbonic acid (H_2CO_3), nitric acid (HNO_3), nitrous acid (HNO_2) and sulfuric acid (H_2SO_4) and their mixture with rain is the main cause of acid rain.

(b) Impact on acid rain are as follows:

(i) Acid rain causes nutrient loss and causes damage to leaves.

(ii) Acid rain results in calcium and potassium loss from the soil which affects the soil fertility.

(iii) Acid rains make the water acidic which will affect the aquatic life adversely.

(c) The impact of acid rain can be reduced by checking the root cause of acid rain, that is, by reducing the emission of the oxides of sulfur and nitrogen. This can be done by using coal or oil with low sulfur content. We can also reduce such emissions by using a scrubber—a device that absorbs gaseous pollutants. A scrubber is used for removing sulfur dioxide from a smokestack. It usually consists of a fine spray of water and gas rising from the stack is passed through the scrubber and the water absorbs the sulfur dioxide. Thus, the formation of this constituent of acid rain is reduced. (See figure in text part.)

4. See text part.

5. See text part.

6. See text part.

7. See text part.

8. See text part.

9. See text part.

Competition Window (Objective Type)

Topic-wise MCQs

1. (b)	2. (a)	3. (d)	4. (b)	5. (d)
6. (d)	7. (d)	8. (a)	9. (c)	10. (d)
11. (c)	12. (a)	13. (b)	14. (c)	15. (b)
16. (c)	17. (a)	18. (c)	19. (b)	20. (c)
21. (c)	22. (a)	23. (d)	24. (d)	25. (b)
26. (b)	27. (a)	28. (c)	29. (a)	30. (a)
31. (a)	32. (d)	33. (d)	34. (d)	35. (a)
36. (d)	37. (d)			

Miscellaneous

1. (c)	2. (b)	3. (a)	4. (b)	5. (c)
6. (c)	7. (a)	8. (a)	9. (c)	10. (c)
11. (d)				

Hints and Solutions

Competition Window (Objective Type)

Topic-wise MCQs

4. The uppermost region of the atmosphere is the exosphere.
5. Tropopause is the point of temperature inversion between troposphere and ionosphere.
8. Ozone layer is present in the stratosphere (at an altitude of 25–0 km).
11. CO is a primary pollutant.
12. CO_2 is generally not regarded as a pollutant.
13. N_2 is not an air pollutant.
20. The ozone hole is maximum over Antarctica.
27. Greenhouse gases absorb solar energy near the earth's surface and radiate it back to the earth.
28. O_2 is not a greenhouse gas.

Miscellaneous

7. Aerosols use CFCs and high-flying jets release NO which are responsible for the depletion of the ozone layer.
9. Incomplete combustion of petrol or diesel in automobile engines produce CO.

Appendices

Some Useful Conversion Factors

Common Units of Mass and Weight

1 pound = 453.59 grams
1 pound = 453.59 grams = 0.45359 kilogram
1 kilogram = 1000 grams = 2.205 pounds
1 gram = 10 decigrams = 100 centigrams
= 1000 milligrams
1 gram = 6.022×10^{23} atomic mass units or u
1 atomic mass unit = 1.6606×10^{-24} gram
1 metric tonne = 1000 kilograms = 2205 pounds

Common Units of Volume

1 quart = 0.9463 litre
1 litre = 1.056 quarts
1 litre = 1 cubic decimetre = 1000 cubic centimetres = 0.001 cubic metre
1 millilitre = 1 cubic centimetre = 0.001 litre
= $1.056 \times 10\ 3$ quart
1 cubic foot = 28.316 litres = 29.902 quarts
= 7.475 gallons

Common Units of Energy

1 joule = 1 × 107 ergs
1 thermochemical calorie
= 4.184 joules
= 4.184×10^{7} ergs
= 4.129×10^{-2} litre-atmospheres
= 2.612×10^{19} electron volts
1 ergs = 1×10^{-7} joule = 2.3901×10^{-8} calorie
1 electron volt = 1.6022×10^{-19} joule
= 1.6022×10^{-12} erg

Common Units of Length

1 inch = 2.54 centimetres (exactly)
1 mile = 5280 feet = 1.609 kilometres
1 yard = 36 inches = 0.9144 metre
1 metre =100 centimetres = 39.37 inches
= 3.281 feet
= 1.094 yards
1 kilometre = 1000 metres = 1094 yards
= 0.6215 mile
1 Angstrom = 1.0×10^{-8} centimetre
= 0.10 nanometre
= 1.0×10^{-10} metre
= 3.937×10^{9} inch

Common Units of Force and Pressure

1 atmosphere = 760 millimetres of mercury
= 1.013×10^{5} pascals
= 14.70 pounds per square inch
1 bar = 10^{5} pascals
1 torr = 1 millimetre of mercury
1 pascal = 1 kg/ms^{2} = 1 N/m^{2}

Temperature

SI Base Units: kelvin (K)
K = –273.15 C
K = C + 273.15
F = 1.8(C) +32

$$°C = \frac{°F - 32}{1.8}$$

Elements, their Atomic Numbers and Molar Masses

Element	Symbol	Atomic Number	Molar mass/ (g mol^{-1})
Actinium	Ac	89	227.03
Aluminium	Al	13	26.98
Americium	Am	95	(243)
Antimony	Sb	51	121.75
Argon	Ar	18	39.95
Arsenic	As	33	74.92
Astatine	Al	85	210

Element	Symbol	Atomic Number	Molar mass/ (g mol^{-1})
Barium	Ba	56	137.34
Berkelium	Bk	97	(247)
Beryllium	Be	4	9.01
Bismuth	Bi	83	208.98
Bohrium	Bh	107	(264)
Boron	B	5	10.81
Bromine	Br	35	79.91

(Continued)

Element	Symbol	Atomic Number	Molar mass/ ($g\ mol^{-1}$)
Cadmium	Cd	48	112.40
Caesium	Cs	55	132.91
Calcium	Ca	20	40.08
Californium	Cf	98	251.08
Carbon	C	6	12.01
Cerium	Ce	58	140.12
Chlorine	Cl	17	35.45
Chromium	Cr	24	52.00
Cobalt	Co	27	58.93
Copper	Cu	29	63.54
Curium	Cm	96	247.07
Dubnium	Db	105	(263)
Dysprosium	Dy	66	162.50
Einsteinium	Es	99	(252)
Erbium	Er	68	167.26
Europium	Eu	63	151.96
Fermium	Fm	100	(257.10)
Fluorine	F	9	19.00
Francium	Fr	87	(223)
Gadolinium	Gd	64	157.25
Gallium	Ga	31	69.72
Germanium	Ge	32	72.61
Gold	Au	79	196.97
Hafnium	Hf	72	178.49
Hassium	Hs	108	(269)
Helium	He	2	4.00
Holmium	Ho	67	164.93
Hydrogen	H	1	1.0079
Indium	In	49	114.82
Iodine	I	53	126.90
Iridium	Ir	77	192.2
Iron	Fe	26	55.85
Krypton	Kr	36	83.80
Lanthanum	La	57	138.91
Lawrencium	Lr	103	(262.1)
Lead	Pb	82	207.19

Element	Symbol	Atomic Number	Molar mass/ ($g\ mol^{-1}$)
Lithium	Li	3	6.94
Lutetium	Lu	71	174.96
Magnesium	Mg	12	24.31
Manganese	Mn	25	54.94
Meitneium	Mt	109	(268)
Mendelevium	Md	101	258.10
Mercury	Hg	80	200.59
Molybdenum	Mo	42	95.94
Neodymium	Nd	60	144.24
Neon	Ne	10	20.18
Neptunium	Np	93	(237.05)
Nickel	Ni	28	58.71
Niobium	Nb	41	92.91
Nitrogen	N	7	14.0067
Nobelium	No	102	(259)
Osmium	Os	76	190.2
Oxygen	O	8	16.00
Palladium	Pd	46	106.4
Phosphorus	P	15	30.97
Platinum	Pt	78	195.09
Plutonium	Pu	94	(244)
Polonium	Po	84	210
Potassium	K	19	39.10
Praseodymium	Pr	59	140.91
Promethium	Pm	61	(145)
Protactinium	Pa	91	231.04
Radium	Ra	88	(226)
Radon	Rn	86	(222)
Rhenium	Re	75	186.2
Rhodium	Rh	45	102.91
Rubidium	Rb	37	85.47
Ruthenium	Ru	44	101.07
Rutherfordium	Rf	104	(261)
Samarium	Sm	62	150.35
Scandium	Sc	21	44.96
Seaborgium	Sg	106	(266)

(Continued)

Element	Symbol	Atomic Number	Molar mass/ (g mol^{-1})
Selenium	Sc	34	78.96
Silicon	Si	14	28.08
Silver	Ag	47	107.87
Sodium	Na	11	22.99
Strontium	Sr	38	87.62
Sulphur	S	16	32.06
Tantalum	Ta	73	180.95
Technetium	Tc	43	(98.91)
Tellurium	Te	52	127.60
Terbium	Tb	65	158.92
Thallium	T1	81	204.37
Thorium	Th	90	232.04
Thulium	Tm	69	168.93
Tin	Sn	50	118.69
Titanium	Ti	22	47.88
Tungsten	W	74	183.85
Ununbium	Uub	112	(277)
Ununnilium	Uun	110	(269)
Unununium	Uuu	111	(272)
Uranium	U	92	238.03
Vanadium	V	23	50.94
Xenon	Xe	54	131.30
Ytterbium	Yb	70	173.04
Yttrium	Y	39	88.91
Zinc	Zn	30	65.37
Zirconium	Zr	40	91.22

(Continued)

Some Important Common Monoatomic Ions

+1 Charge Name of ion	Formula	+2 Charge Name of ion	Formula	+3 Charge Name of ion	Formula
Lithium ion	Li^+	*Lead (II) ion	Pb^{2+}	Arsenic ion	As^{3+}
		Cadmium ion	Cd^{2+}	Bismuth ion	Bi^{3+}
		Magnesium ion	Mg^{2+}	Antimony ion	Sb^{3+}
Sodium ion	Na^+	*Copper (II) ion	Cu^{2+}	Scandium ion	Sc^{3+}
Potassium ion	K^+	Strontium ion	Sr^{2+}	Chromium (III) ion	Cr^{3+}
Silver ion	Ag^+	Iron (II) ion (Ferrous ion)	Fe^{2+}	Iron (III) ion (Ferric ion)	Fe^{3+}
Copper ion (Cuprous ion)	Cu^+	Barium ion Cobalt ion	Ba^{2+} Co^{2+}	Aluminium ion Auric ion	Al^{3+} Au^{3+}
Aurous	Au^+	Manganese (II) ion	Mn^{2+}		
		*Mercury (I) ion	Hg_2^{2+}		
		Zinc ion	Zn^{2+}		

–1 Charge Name of ion	Formula	–2 Charge Name of ion	Formula	–3 Charge Name of ion	Formula
Chloride ion	Cl^-	Sulfide ion	S^{2-}	Phosphide ion	P^{3-}
Fluoride ion	F^-			Boride ion	B^{3-}
Bromide ion	Br^-	Oxide ion	O^{2-}	Nitride ion	N^{3-}
Iodide ion	I^-				

Some Common Important Polyatomic Ions

–1 Charge Name of ion	Formula	–2 Charge Name of ion	Formula	–3 Charge Name of ion	Formula
Nitrate ion	NO_3^-	Sulfite ion	SO_3^{2-}		
Hydrogen carbonate or bicarbonate ion	HCO_3^-	Carbonate ion Manganate ion	CO_3^{2-} MnO_4^{2-}	Phosphate ion Arsenate ion	PO_4^{3-} AsO_4^{3-}
Hydrogen sulfate or (bisulfate ion)	HSO_4^-	Thiosulfate ion Silicate ion	$S_2O_3^{2-}$ SiO_3^{2-}	Arsenite ion	AsO_3^{3-}
Hydroxide ion	OH^-	Sulfate ion	SO_4^{2-}	Phosphite ion	PO_3^{3-}
Acetate ion	CH_3COO^-	Oxalate ion	$C_2O_4^{2-}$		
Chlorate ion	ClO_3^-	Chromate ion	CrO_4^{2-}	Borate ion	BO_3^{3-}
Nitrite ion	NO_2^-	Dichromate ion	$Cr_2O_7^{2-}$	Ferricyanide ion	$[Fe(CN)_6]^{3-}$
Permanganate ion	MnO_4^-	Hydrogen phosphate ion	HPO_4^{2-}		
Ammonium ion	NH_4^+				
Cyanide ion	CN^-				
Hypophosphite ion	$H_2PO_2^-$				– 4 Charge
Metaaluminate ion	AlO_2^-			Carbide ion	C^{4-}
				Ferrocyanide ion	$[Fe(CN)_6]^{4-}$